Foundations of Mathematics

LICENSE, DISCLAIMER OF LIABILITY, AND LIMITED WARRANTY

Foundations of Mathematics

Algebra, Geometry, Trigonometry, Calculus

Philip Brown

Texas A&M University at Galveston

Mercury Learning and Information

Dulles, Virginia
Boston, Massachusetts
New Delhi

Publisher: David Pallai
Mercury Learning and Information
22841 Quicksilver Drive
Dulles, VA 20166
info@merclearning.com
www.merclearning.com
(800) 232-0223

Philip Brown. *Foundations of Mathematics: Algebra, Geometry, Trigonometry, Calculus*
ISBN: 978-1-942270-75-1

Library of Congress Control Number: 2015957711
16171832 This book is printed on acid-free paper.

My mother, Ria Brown, who passed away on February 3, 2014, was the source of much of the inspiration for this book. Indeed, there are many exercises and examples included in this book that she collected over a period of forty years as a high school mathematics teacher in South Africa. Her greatest passion was for Euclidean Geometry. Some of her insights and approach to teaching geometry are included in Chapter 9 of this book. This book is dedicated to her, in loving memory.

CONTENTS

PREFACE

This book is intended primarily for university students. In particular, this book can be used as a textbook or an additional reference book by university students attending a course in algebra, trigonometry, geometry, or calculus. As a calculus textbook, this book is unique in that it contains all the mathematics (and more) that students will need to know in order to be successful in a calculus course. For this reason, the book will be invaluable for students who may need to fill in some "gaps" in their mathematical background, or review certain topics, while attending a university level calculus course. (Calculus professors will normally not have the time to do this!)

Mathematics can be appreciated and enjoyed more when it is presented as a development of ideas. Hence, there is a strong emphasis in this book on the unfolding and development of the concepts, and there are comments throughout the book to help the reader trace the historical development of mathematics. Of course, students also need to develop their skills. For this reason, there are many exercises included at the ends of the chapters, and students are encouraged to do all of them as an essential part of working through the book. The range of topics in this book, and examples that demonstrate their interplay and interconnectedness is another unique aspect of the book. This is a rewarding aspect of learning mathematics, which university students are typically not exposed to because of the way current-day university courses are organized.

The four chapters that make up the introduction to algebra are Chapters 1 (The Laws of Algebra), 2 (The Cartesian Plane), 3 (Solving Equations and Factorizing Polynomials), and 6 (Techniques of Algebra). The chapters dealing with trigonometry are Chapters 4 (Trigonometry) and 10 (Spherical Trigonometry). They can be read independently (after the first two chapters have been read, if needed). Chapter 5 (Functions), which is an introduction to functions, can be regarded as the beginning of calculus because the operations of calculus are applied to functions. The concepts and methods relating to the calculation of limits, which underlie the operations of calculus, are introduced in Chapter 7 (Limits). In Chapter 8 (Differential Calculus), the definition of the derivative, the rules for computing derivatives, and the formulas for derivatives of all the standard types of function (i.e., polynomial, rational, trigonometric, root, absolute value, exponential, and logarithmic functions) are introduced. Chapter 10 (Euclidean Geometry) presents many of the theorems that are important in Euclidean Geometry, along with some guides to solving problems in geometry.

Because many students have difficulties with algebra when they enroll in a university calculus course, a first semester course in calculus may well consist of Chapters 5–8. It should also be mentioned that Chapter 6 includes an introduction to partial fractions, a topic that is usually not taught to students until they reach their second semester of calculus.

In the following two paragraphs, some comments are made regarding the presentation of the material in Chapters 7 and 8.

It is usual, in most calculus textbooks, for the *rules for limits* to be taken as the starting point for the evaluation of limits. In Chapter 7, the slightly different approach to evaluating limits is to take as a fact the continuity of all of the standard functions and any algebraic combinations and compositions of the standard functions. (This fact can be proved using methods of real analysis, which is too advanced for this book.) This means that a limit to any point in the domain of any of these functions can be evaluated simply by making a substitution into the function (i.e., by an application of the *equation of continuity*). The *rules for limits* are introduced, instead, at the end of the chapter, where they are used to evaluate certain limits involving trigonometric functions.

In Chapter 8, the derivative of a function is first defined in the special case that the graph of the function passes through the origin, and the tangent line (through the origin) is defined in a precise way as the *best approximating line* to the graph near the origin. Then, in the general case, the formula for the derivative at any point in the domain of a function is obtained by means of an appropriate horizontal and vertical shift of the function. With this approach, it is possible for students to focus on the special case where the formula for the derivative is the simplest possible. Furthermore, the concept of a tangent line as a "best approximating line" is a much more flexible concept than the concept of the tangent line as a limit of secant lines, as it is sometimes defined. Consequently, students should be able to identify a tangent line more easily in certain situations. For example, they shouldn't have too much difficulty identifying the x-axis as the tangent line to the basic cubic graph at the origin. The *chain rule* is also presented first in the special case (at the origin), where it can be demonstrated clearly that a tangent line to the graph of a composition of functions is the composition of the tangent lines to the graphs of the functions.

One of the innovative aspects of this book is the introduction of vectors early on, in Chapter 2. The concept of a vector is very natural, and vectors have practical applications; however, the notation and terminology relating to vectors may be confusing at first. The hope is that a gradual introduction to vectors will give students more time to become comfortable with vectors. Vectors are typically introduced for the first time in a course in vector calculus, causing great alarm to the students!

Another surprise in this book is the inclusion of Spherical Trigonometry (Chapter 10). Much of the material in this chapter was obtained from the book "Plane and Spherical Trigonometry," by K. Nielsen and J. Vanlonkhuyzen, first published in 1944, which is one of a few texts available on this topic. The purpose of including a chapter on spherical trigonometry is to give students the opportunity to acquire a deeper understanding of trigonometry and also to give students some exposure to non-Euclidean geometry. Chapter 10 also introduces vectors in three dimensions, including the definition of the cross product of vectors, which students normally do not see until they take a course in vector calculus. Spherical trigonometry is interesting from a historical point of view because many of the formulas relating to spherical trigonometry were discovered by Arabic and Iranian mathematicians from the ninth to the thirteenth centuries.

A few comments regarding format, notation, and mathematical language are in order: new terminology is presented for the first time in italics, the most important definitions are presented as numbered definitions, statements that make important clarifications are presented as numbered remarks, and examples, theorems, corollaries, and lemmas are also numbered. A theorem is a general mathematical statement which is proved on the basis of known mathematical truths. A corollary is a consequence or a special case of a theorem that is important enough to be stated separately from the theorem. Lemmas are statements that can be used as stepping stones toward the proofs of more general statements (theorems). Throughout this book, the biconditional phrase "if and only if" is used when two statements are implied by each other.

This book can serve as a textbook, a reference book, or a book that can be read for fun! Please tell your friends about it, if you like it.

The author should be contacted regarding any errors in the book so that corrections can be made for future editions. Suggestions for future editions will also be welcome.

Philip Brown
Galveston, Texas
February 2016

ACKNOWLEDGMENTS

This book could not have been published without a considerable amount of assistance from certain individuals and institutions. First, I would like to thank my employers, Texas A&M University, for granting me leave in 2009 to begin writing this book and I would like to thank Professor Johann Engelbrecht for arranging that my leave time be spent at the Department of Mathematics at the University of Pretoria in South Africa. Second, I would like to express a great deal of appreciation to my editors, Ria Brown, Ronelda Jordaan, Laura Robichaux and most of all, Ian White, for correcting errors, improving the writing, and making many helpful suggestions. Third, the enthusiasm and encouragement of many of my students, in particular, Sean Hennigan, while I was working on this book made it a much lighter and enjoyable task. Last, I am grateful to Mercury Learning and Information for the copyediting, formatting, cover design, production, and marketing of the book.

CHAPTER 1

THE LAWS OF ALGEBRA

1.1 INTRODUCTION

This chapter is an overview of mathematics that schoolchildren typically learn up to about grade nine. The presentation, however, is more axiomatic and set theoretic than is taught in schools because this book is intended for students preparing for a university course in mathematics.

This chapter begins (section 1.2) with a brief discussion of different historical concepts of *number* and the relationship of these concepts to mathematics and philosophy in general. It is convenient to identify the set of real numbers with the infinite number line (sections 1.3–1.5) because this leads easily to the notion of negative numbers, the ordering of real numbers, and the identification of fractions (rational numbers) with fractional distances on the number line. The disadvantage of this approach is that it does not explain how the irrational numbers fill in the "gaps" in the number line. A satisfactory explanation of the existence of irrational numbers requires a formal approach that involves definitions of sequences of rational numbers, which is more suited to an advanced study of real numbers that is not within the scope of this chapter.

The decimal number system is introduced in section 1.2 but not much is said about it, because the ability to understand, read, and write large decimal numbers is a skill that a reader of this book is expected to have (read aloud the number 3,678,501).

Absolute value notation is introduced in section 1.6; the basic operations of addition, subtraction, multiplication, and division are introduced in section 1.7; and the properties of real numbers with respect to these operations are explained in section 1.8. The multiplication principle known to high-school students as "FOIL" is introduced early in the chapter (section 1.8.1) because of its importance as a skill that students would best learn at the beginning of this book. This is followed by the definition of an exponent (section 1.9) and a discussion of the order of operations (section 1.10).

The relationship of fractional notation to the division operation is explained in section 1.11. It will be worthwhile to read this section very carefully so that the proofs are well understood. Anybody who has struggled with math at school knows that adding fractions can be very confusing! Decimal and scientific notations and rounding of decimals are reviewed in section 1.12.

Mathematicians have, for thousands of years, been fascinated by the properties of prime numbers. It is a difficult problem, in general, to determine whether a large number is a prime number or to find the prime factors of a large number. This has made it possible for cryptographers to design cryptosystems for the coding of Internet transactions that banks and other institutions depend upon every day. We only give the definition of a prime number and the explanation of the prime decomposition of a natural number in section 1.13, followed by a paragraph on tests for divisibility of natural numbers by small prime numbers.

This chapter on algebra ends with the laws for integer exponents (section 1.14) and an explanation of radicals (section 1.15).

1.2 NUMBERS

Numbers are expressed in terms of numerals or symbols. Many kinds of numeral have existed through the ages. We will be using the *Arabic* numerals (or *digits*) 0, 1, 2, 3, 4, 5, 6, 7, 8, and 9. These are known as Arabic because they were used by Arabic mathematicians and merchants for many centuries, before being introduced to western Europe during the time of the Italian Renaissance. Originally, these numerals were used (perhaps invented) in India, in particular, the first known instance of the use of the symbol 0 is an inscription that was made about fifteen centuries ago on a rock in a cave in northern India.

Here are some useful and interesting ways we can think about numbers, depending on our objectives.

1. *Numbers as points on a number line*. Imagine an infinitely long straight line, as indicated in figure 1.1, with an arbitrary point on the line labeled as "0" (zero). At some distance to the right of zero, we can label a point "1." To the left of zero, the same distance away, we label a point as "–1." This is called a *negative number*. Continuing in this way we label at equal distances apart, to the right of zero, the points "2" up to "9," and, to the left of zero, the successive values "–2," "–3" down to "–9." We normally work in a base 10, that is, the *decimal* number system. This means that the successor of "9" on the number line is labeled as "10." After that we label points "11," "12," and so on and do the same for negative values.

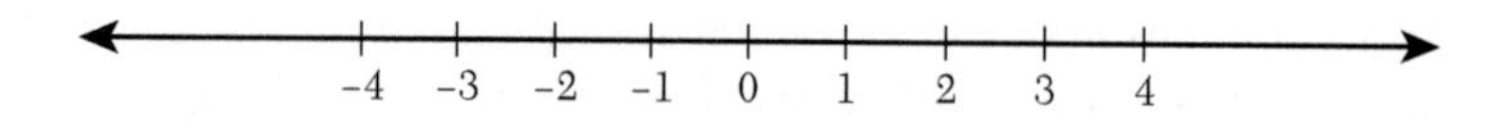

FIGURE 1.1. The number line.

2. *Numbers as quantifiers*. At the dawn of our civilization people managed to keep track of quantities by making scratches on a bone, by tying knots in a rope, or by keeping sacks of pebbles. For instance, a shepherd who let sheep out of an encampment in the morning could have placed a pebble in a sack for each sheep that had left and, in the evening, removed a pebble from the sack for each sheep that had returned. If any pebbles were left in the sack at the end of the day, then the shepherd would know that some sheep had not returned.

3. *Numbers as counters*. In figure 1.2, we have three urns with a different number of marbles in each. One can see at a glance that there are three marbles in the first urn and five marbles in the second urn, but how many marbles are in the third urn? (Why can we not see at a glance how many marbles are in this urn?) We have to *count* the number of marbles in the third urn!

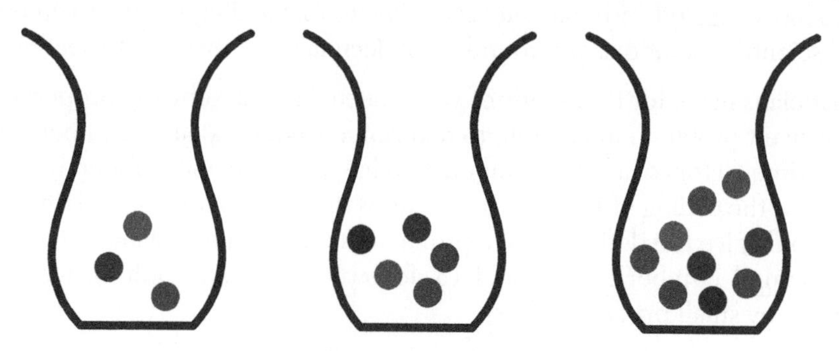

FIGURE 1.2. Marbles in urns.

4. *Numbers as symbols.* The Pythagoreans (about 550 BC) believed that the meaning of all things was related to numbers, and they attached great significance to certain numbers. For example, the number 7, or heptad, was regarded as *the number of religion*, because it was believed that seven spirits, or archangels, were controlling the planets, to whom mankind had to make offerings.

5. *Numbers as entities.* Philosophers after Pythagoras thought about numbers more abstractly. For instance, Socrates and his followers, including Plato, believed in a realm of perfect forms or entities. In this realm, every number may be considered as its own entity. For example, try to imagine the number "7" as an entity that exists by itself (see figure 1.3).

FIGURE 1.3. Number seven.

6. *Numbers as elements of sets.* This is a modern view of numbers. In this book, we will define different kinds of sets of numbers, for example, the set of *natural, whole, rational, real, irrational numbers and integers*.

7. *Numbers as elements of an algebraic system.* For the purposes of doing algebra, we will regard numbers as the targets of operations such as addition, subtraction, multiplication, and division.

1.3 FRACTIONS AND INEQUALITIES

There were gaps between the points labeled on the number line in figure 1.1 above. How do we label the points in these gaps? It is not possible to label all the points, but we can label some fractional distances as shown in figure 1.4, which demonstrates fractional distances of one-half, one-third, one-quarter, one-fifth, and multiples of these fractions inside the interval from zero to one. Note that $\frac{1}{3}$ is to the *left* of $\frac{2}{5}$ (or that $\frac{2}{5}$ is to the *right* of $\frac{1}{3}$). This can be indicated using the inequality symbol "<," by writing $\frac{1}{3} < \frac{2}{5}$, meaning one-third is *less than* two-fifths. Alternatively, we can write $\frac{2}{5} > \frac{1}{3}$, meaning two-fifths is *greater than* one-third.

FIGURE 1.4. Fractions on the number line.

1.4 NOTATION FOR SETS

Table 1.1 gives a complete description of the sets of numbers mentioned above. All of these sets are infinite sets, so we represent missing elements using an ellipsis (three dots). The natural numbers ($\mathbb{N}$) are also known as *counting numbers*. The set of whole numbers ($\mathbb{N}^0$) is the set of natural numbers including zero. The set of integers ($\mathbb{Z}$) is the extension of the set of natural numbers that includes the corresponding negative values and zero. The set of rational numbers ($\mathbb{Q}$), which includes the set of integers, is the set of all possible fractions. All points on the number line that cannot be expressed as fractions (there is an infinite number of these) are called irrational numbers. We include all of these numbers with the set of rational numbers to form the set of real numbers ($\mathbb{R}$). This is an *uncountable set*, so we do not give a partial listing of the elements in the table.

The symbol "$\cup$" is used to form the *union* of two given sets, that is, the smallest set that contains both of the given sets.

An example of an irrational number is the ratio of the circumference of any circle to the length of its diameter. This number is represented by the Greek symbol π.

TABLE 1.1. Sets of numbers

Symbol	Description of set	Notation
$\mathbb{N}$	Natural numbers	$\{1, 2, 3, ...\}$
$\mathbb{N}^0$	Whole numbers	$\{0, 1, 2, 3, ...\}$
$\mathbb{Z}$	Integers	$\{..., -3, -2, -1, 0, 1, 2, 3, ...\}$
$\mathbb{Q}$	Rational numbers	$\{0, \pm\frac{1}{2}, \pm\frac{1}{3}, \pm\frac{2}{3}, \pm\frac{1}{4}, \pm\frac{3}{4}, \pm\frac{3}{4}, \pm\frac{3}{4}, \pm\frac{1}{5}, \pm\frac{2}{5}, ...\}$
$\mathbb{R}$	Real numbers	$\mathbb{Q} \cup \{\text{irrational numbers}\}$

1.5 INTERVALS OF THE REAL LINE AND SET INTERSECTIONS

An *interval* is a segment of the real line. Possible notations for an interval are:

- $[a, b]$, the set of all real numbers between a and b including a and b.
- (a, b), the set of all real numbers between a and b but not including a and b.
- $[a, b)$, the set of all real numbers between a and b including a, but not including b.
- $(a, b]$, the set of all real numbers between a and b including b, but not including a.
- $(-\infty, b]$, the set of all real numbers to the left of b and including b.
- $[-\infty, b)$, the set of all real numbers to the left of b but not including b.
- $[a, \infty)$ the set of all real numbers to the right of a and including a.
- (a, ∞), the set of all real numbers to the right of a but not including a.

The symbol "∞" that has been used above is the symbol that mathematicians use for infinity. The meaning of infinity varies with the context. Here, $-\infty$ means there is no lower bound to the left of the point b on the number line, and ∞ means there is no upper bound to the right of the point a on the number line.

The symbol "$\cap$" is used to form the *intersection* of two given sets, that is, the largest set that both of the given sets contain. For example, if $a < c < b < d$, then $[a, b) \cap [c, d]$ is the interval $[c, b)$ because the interval $[a, b)$ includes all numbers between c and b (including c but not including b) and $[c, d)$ also includes all numbers between c and b (including c). If $a < c < d < b$, then $[a, b) \cap [c, d] = [c, d]$. Another possibility is $a < b < c < d$, then $[a, b) \cap [c, d]$ is an *empty set*. The symbol for this is $\varnothing$, so we would write $[a, b) \cap [c, d] = \varnothing$.

1.6 ABSOLUTE VALUE OF A REAL NUMBER

If a is a real number (i.e., a point on the number line), then the distance on the number line between zero and the number is denoted by $|a|$. For example, $|3| = 3$ and $|-3| = 3$. In general, if x is any real number, then:

$$|x| = \max\{x, -x\}.$$

meaning that we take the maximum of x and its negative because one of these numbers will be non-negative. For example, because the value of π is a little more than 3,

$$|3 - \pi| = \max\{3 - \pi, \pi - 3\} = \pi - 3.$$

If a and b are two different points on the number line, then the distance between them is $|a - b|$ because it is same as the distance between $a - b$ and zero.

1.7 ALGEBRAIC OPERATIONS

A *binary operation* is a calculation involving two elements of a set that produces another element of the set. The familiar binary operations in algebra are addition, subtraction, multiplication, and division. In the following paragraphs, we explain briefly how these operations are applied to natural numbers, integers, and rational numbers expressed as fractions or decimals. There is a comment at the end of this section regarding the addition, subtraction, multiplication, and division of irrational numbers. We will be jumping the gun a little because we will be using terminology and properties introduced later in this chapter but which should already be familiar from mathematics taught in schools.

1.7.1 Addition and Subtraction

Interpreted on the number line, addition can be described as moving to the right from a given number by the length equivalent to the number being added ("$2 + 3 = 5$" means moving three units to the right from 2, for example) and subtraction as moving to the left from a given number by the length equivalent to the number being subtracted ("$2 - 3 = -1$" means moving three units to the left from 2, for example). What's more, we can think of subtraction as the "undoing" of addition ($2 + 3 - 3 = 2$, for example). It follows from this and properties (III) and (IV) from table 1.3 in section 1.8 (try to verify this!) that subtracting a number is the same as adding a number with the opposite sign, for example, $2 - 3 = 2 - 3 + 3 + (-3) = 2 + (-3)$.

The addition (or subtraction) of two rational numbers as fractions is property (6) in section 1.11. If two positive rational numbers can be expressed as decimals without recurring digits, that is, if they have finite decimal expansions (see section 1.12), then in practice they can be added by the familiar method of stacking the numbers above one another by matching the decimal positions of their digits and then adding the digits in each column (starting from the right and moving to the left). If the sum of digits in a column is larger than nine, then the unit in the 10's position is carried to the next column. Similarly, if two positive rational numbers can be expressed as decimals without recurring digits, then the smaller of the two numbers can be subtracted from the larger by stacking it below the larger and then subtracting the lower digit from the upper digit in each column (starting from the right and moving to the left). If the lower digit is larger than the upper digit in any column, then a unit in the 10's position can be taken from the next column and added to the upper digit so that the lower digit can be subtracted. (We omit examples, because these techniques are drilled in schools.)

1.7.2 Multiplication

Multiplication of natural numbers is repeated addition. Many symbols are used for multiplication, for example, we can write $2 \cdot 3 = 6$, $2 \times 3 = 6$, or $(2)(3) = 6$. We get the answer 6 by calculating either $2 + 2 + 2 = 6$ or $3 + 3 = 6$. When two numbers, or several numbers, are multiplied together, each number is called a *factor*. It is helpful to memorize the factors of some numbers (multiplication tables!). Table 1.2, for example, contains the products of some pairs of small *prime numbers*. Prime numbers are natural numbers that have no factors smaller than themselves, other than 1. (There is more information about prime numbers in section 1.13.)

TABLE 1.2. Pairwise products of small prime numbers

$2 \times 7 = 14$	$2 \times 11 = 22$	$2 \times 13 = 26$	$2 \times 17 = 34$	$2 \times 19 = 38$	$2 \times 23 = 46$
$3 \times 7 = 21$	$3 \times 11 = 33$	$3 \times 13 = 39$	$3 \times 17 = 51$	$3 \times 19 = 57$	$3 \times 23 = 69$
$5 \times 7 = 35$	$5 \times 11 = 55$	$5 \times 13 = 65$	$5 \times 17 = 75$	$5 \times 19 = 95$	$5 \times 23 = 115$
$7 \times 7 = 49$	$7 \times 11 = 77$	$7 \times 13 = 91$	$7 \times 17 = 119$	$7 \times 19 = 133$	$7 \times 23 = 161$
	$11 \times 11 = 121$	$11 \times 13 = 143$	$11 \times 17 = 187$	$11 \times 19 = 209$	$11 \times 23 = 253$
		$13 \times 13 = 169$	$13 \times 17 = 221$	$13 \times 19 = 247$	$13 \times 23 = 299$
			$17 \times 17 = 289$	$17 \times 19 = 323$	$17 \times 23 = 391$
				$19 \times 19 = 361$	$19 \times 23 = 437$
					$23 \times 23 = 529$

It is also true that $(-1) \cdot a = -a$ for any real number a (exercise 1.7 at the end of this chapter). This fact along with properties (IV) and (XI) from table 1.3 in section 1.8 can be used to prove that the sum of two negative numbers is the negative of the sum of the corresponding positive numbers, for example,

$$-2-3=(-1)2+(-1)3=(-1)(2+3)=-(2+3).$$

The multiplication of rational numbers as fractions is property (7) in section 1.11. If two rational numbers can be expressed as decimals without recurring digits, then they can be multiplied by ignoring their decimals points, that is, by multiplying them as integers and then moving the decimal point to the left (starting to the right of the last digit) by the number of decimal positions equal to the sum of the decimal positions in each of the factors. This might be easier to understand by means of an example: the product of the rational numbers 2.5 and 1.32 is 3.8 because the product of 25 and 132 is 3,800 and the decimal point is moved (starting to the right of the last digit) three decimal positions to the left resulting in 3.800. (We do not need to write the final two zeros to the right of the decimal point.)

1.7.3 Division

Division can be described as "un-multiplication" because it reverses what multiplication does. Again, there are different ways to write division; usually, the symbols "÷" or "/" are used. For example, we write "28 ÷ 7 = 4" or "28/7 = 4," because $4 \times 7/7 = 4$. We can express a division operation a/b as a fraction $\frac{a}{b}$ (where a is called the *numerator* or *dividend* and b is called the *denominator* or *divisor*). If a and b are two integers, then, in general, the division of a by b can be carried out by the method of *long division*, resulting in a rational number expressed as a decimal (which might have recurring digits). This can be a lengthy process with many divisions. (School children typically spend a year learning to do it.) If an integer a is divisible by an integer b (this is explained in section 1.13), then the division of a by b (by the method of long division) results in an integer.

The division of two rational numbers expressed as fractions is property (8) in section 1.11. If two rational numbers can be expressed as decimals without recurring digits, then, as with multiplication, they can be divided by ignoring their decimals points, that is, by dividing them as integers (using the method of long division) and then moving the decimal point to the left by the number of decimal positions equal to the number of decimal positions in the dividend minus the number of decimal positions in the divisor. For example, 2.17 divided by 1.5 is $1.44\dot{6}$ because 217 divided by 15 is $14.4\dot{6}$, and the decimal point is moved one decimal position to the left. (The dot above the 6 means it is recurring. This is explained in section 1.12.1.)

1.7.4 Recurring Decimals and Irrational Numbers

Rational numbers with recurring decimal digits and irrational numbers, which have infinitely many nonrecurring decimal digits (see section 1.12.3), can be added (or subtracted) and multiplied (or divided) as decimal numbers after truncating or rounding their decimal expansions (see section 1.12.5) according to the degree of accuracy required. (We omit the details of this. Calculators do it automatically!)

1.8 AN ALGEBRAIC SYSTEM ON THE SET OF REAL NUMBERS

We list in table 1.3, the fundamental properties of the binary operations mentioned in section 1.7 applied to elements of the set of real numbers R, thus defining an algebraic system on the set of real numbers. These are fundamental properties because all other properties of the algebraic system can be deduced from them. We also introduce the concept of a variable. Because we are listing properties of operations applied to *any* real numbers, we represent these numbers in the table using letters (variables) a, b and c. Wherever operations occur in parentheses, they are to be performed first. We use the terms *commutative* and *associative* to mean "interchangeable" in a precise mathematical sense, as indicated in the third column of the table.

TABLE 1.3. The algebra of real numbers

(I)	$a+b = b+a$	Addition is commutative
(II)	$a+(b+c) = (a+b)+c$	Addition is associative
(III)	$a+0 = a$	Zero is a *unique* additive identity
(IV)	$a+(-a) = 0$	Every element has a *unique* additive inverse
(V)	$a\cdot b = b\cdot a$	Multiplication is commutative
(VI)	$a(b\cdot c) = (a\cdot b)c$	Multiplication is associative
(VII)	$a\cdot 1 = a$	1 is a *unique* multiplicative identity
(VIII)	$a\cdot 0 = 0$	Multiplication by zero always results in zero
(IX)	$a\cdot\frac{1}{a} = 1, \text{ if } a\neq 0$	Every non-zero element has a *unique* multiplicative inverse
(X)	$(a+b)c = a\cdot c+b\cdot c$	Multiplication distributes through addition (from the right)
(XI)	$a(b+c) = a\cdot b+a\cdot c$	Multiplication distributes through addition (from the left)

The reason we make a list of these seemingly obvious statements is that mathematicians know about algebraic systems where all of these properties do not hold. We will also refer to these properties in order to logically prove new properties. For instance, most of us have probably been taught that "a negative times a negative is a positive," but why should this be true? In particular, why should it be true that $(-1)(-1) = 1$? Below is a short proof in the form of a sequence of equations. Each equation is true according to the property from table 1.3 stated next to the equation.

$$\begin{aligned}(-1)(-1)+(-1) &= (-1)(-1)+(-1)(1) \quad \texttt{property (VII)}\\ &= (-1)(-1+1) \quad \texttt{property (XI)}\\ &= (-1)\cdot 0 \quad \texttt{property (IV)}\\ &= 0 \quad \texttt{property (VIII)}.\end{aligned}$$

So we have proved logically that $(-1)(-1) + (-1) = 0$. This statement says that "$(-1)(-1)$" is an additive inverse for "-1." It now follows from property (IV) that $(-1)(-1) = 1$. As an application, we have the following example.

Example 1.8.1. One number can be subtracted from another by reversing the subtraction and taking the negative of the result, for example,

$$2-3=(-1)(-1)2+(-1)3=(-1)(-2)+(-1)3=(-1)(-2+3)=-(3-2).$$ ◈

1.8.1 Foil

The distributive properties (X) and (XI) in section 1.8 can be generalized to a more general multiplicative property known as FOIL, which can be expressed as

$$(p+q)(r+s)=p\cdot r+p\cdot s+q\cdot r+q\cdot s.$$

This is a formula for multiplication of a pair of *binomial factors*. (Any expression that is a sum of two *terms* is called a binomial.) FOIL is an acronym for "First, Outer, Inner, Last," where "First" means that the first terms of each binomial factor should be multiplied together, "Outer" means that the product of the two outer terms should be added to this, and so on. The sequence of equations below is a proof of FOIL. (When it is understood that an operation on variables is multiplication, the multiplication operator may be omitted.)

Proof of FOIL.

$$\begin{aligned}&\text{Let } m=p+q,\\ &\text{then } (p+q)(r+s)=m(r+s)\\ &\qquad = mr+ms \quad \texttt{property (XI)}\\ &\qquad = (p+q)r+(p+q)s\\ &\qquad = pr+qr+ps+qs \quad \texttt{property (X)}\\ &\qquad = pr+ps+qr+qs \quad \texttt{property (I)}.\end{aligned}$$ □

Example 1.8.2. If p, q, r, and s are replaced with x, -4, y, and 3, respectively, in the formula for FOIL, then we get $(x-4)(y+3) = xy + 3x - 4y - 12$. ◈

1.9 NATURAL NUMBERS AS EXPONENTS

Repeated multiplication by a number is usually expressed by means of a natural number as a superscript, called an *exponent* or a *power*, above and to the right of the number. For example, $2^5 = 32$ (2 to the power 5 equals 32) is an abbreviation for $2\cdot 2\cdot 2\cdot 2\cdot 2 = 32$. There are more examples in table 1.4. It is worthwhile to memorize them.

TABLE 1.4. Powers of two and three

$2^1 = 2$	$2^2 = 4$	$2^3 = 8$	$2^4 = 16$	$2^5 = 32$
$2^6 = 64$	$2^7 = 128$	$2^8 = 256$	$2^9 = 512$	$2^{10} = 1{,}024$
$3^1 = 3$	$3^2 = 9$	$3^3 = 27$	$3^4 = 81$	$3^5 = 243$

A power of two is very often called a *square* because the area of a geometric square is the product of the side length (of the square) with itself. For the same reason, a power of three is called a *cube* because the volume of a geometric cube is the side length times itself, times itself again.

EXAMPLE 1.9.1. A special case of FOIL results in a difference of two squares:

$$(a-b)(a+b) = a^2 + ab - ab - b^2 = a^2 - b^2.$$

A natural number is called a perfect square if it can be expressed as the square of a smaller natural number (or the square of the same number in the case of the number 1). Some natural numbers that are perfect squares are 1, 4, 9, 16, 25, 36, 49, 64, and 81 because they are the squares of 1, 2, 3, 4, 5, 6, 7, 8, and 9, respectively. ◈

EXAMPLE 1.9.2. The memorization of some perfect squares can be an aid to performing mental calculations; for example, the calculation 24×34 can be done mentally as follows:

$$24 \times 34 = (29-5)\times(29+5) = 29^2 - 5^2 = 841 - 25 = 816.$$

◈

1.10 ORDER OF OPERATIONS

In order to evaluate a sequence of operations like

$$3 \cdot 2^3 / 4 \cdot 5 - (6+7) + 221/13, \qquad (1.1)$$

there is a convention for deciding which operations to perform first. We use a mnemonic device called PEMDAS, an acronym in which each letter has a specific meaning:

- P = parentheses,
- E = exponents,
- M = multiplication,
- D = division,
- A = addition,
- S = subtraction.

This means that, in a sequence of operations such as in formula (1.1), all calculations inside parentheses take precedence. Thereafter, any exponent is computed. Thus formula (1.1) reduces to

$$3 \cdot 8 / 4 \cdot 5 - 13 + 221/13. \qquad (1.2)$$

We continue by doing all multiplication and division operations. These are interchangeable, however, a number should never be separated from the division symbol preceding it. For example, $18/2/3$ is equivalent to $18/3/2 = 3$ but *not* equivalent to $18/(2/3) = 27$ (see section 1.11 below). So formula (1.2) now evaluates to

$$3 \cdot 2 \cdot 5 - 13 + 17 = 30 - 13 + 17 = 34.$$

The addition and subtraction operations are performed last (in any order). We can use PEMDAS to verify the next example.

EXAMPLE 1.10.1. $22-2^2\cdot(6^{38/19}\cdot(69/23-4))=166$. ◈

The interchangeability of multiplication and division is a useful tool when doing mental arithmetic.

EXAMPLE 1.10.2. The calculation $17\cdot 12\cdot 25$ is more easily done by rewriting it as $(17\cdot 3)\cdot(4\cdot 25) = 51\cdot 100 = 5{,}100$. ◈

1.11 LAWS OF DIVISION

The properties (1) to (9) listed below include formulas for adding fractions. A careful proof is given for each statement. In these proofs, we employ the basic algebraic principle that equality in an equation is preserved if the same algebraic operation is applied to each side of the equation. By convention, expressions such as $\frac{(x+y)}{w}$ and $\frac{x}{(y+w)}$ are expressed without the parentheses as $\frac{x+y}{w}$ and $\frac{x}{y+w}$, respectively.

(1) $\frac{a}{b}=c$ is equivalent to $a = bc$

PROOF. $\frac{a}{b}=c$ can also be expressed as $a/b=c$. If both sides are multiplied by b, then $a/b\cdot b=c\cdot b$, and so $a=b\cdot c$. □

(2) $\frac{a}{b}=a\cdot\frac{1}{b}$

PROOF. $\frac{a}{b}=a/b=a\cdot 1/b=a\cdot\frac{1}{b}$. □

(3) $\frac{a}{bc}=a/b/c$

PROOF. Let $\frac{a}{b\cdot c}=x$, then

$$a=x(b\cdot c),$$
$$a/c=x\cdot b\cdot c/c,$$
$$a/c/b=x\cdot b/b,$$
$$a/b/c=x.$$
□

(4) $\frac{a}{-b} = \frac{-a}{b} = -\frac{a}{b}$ (moving a negative sign inside a fraction)

PROOF. $\frac{a}{-b} = \frac{a}{(-1)b} = a \cdot 1/(-1)/b = a(-1)/b = (-a)/b = (-1)a/b = \frac{-a}{b}$. □

(5) $\frac{a}{b} + \frac{c}{b} = \frac{a+c}{b}$ (adding fractions with the same denominator)

PROOF. $\frac{a}{b} + \frac{c}{b} = a \cdot \left(\frac{1}{b}\right) + c \cdot \left(\frac{1}{b}\right) = (a+c) \cdot \left(\frac{1}{b}\right) = (a+c)/b = \frac{a+c}{b}$. □

(6) $\frac{a}{b} + \frac{c}{d} = \frac{ad+bc}{bd}$ (adding fractions with different denominators)

PROOF. $\frac{a}{b} + \frac{c}{d} = \frac{ad}{bd} + \frac{bc}{bd} = \frac{ad+bc}{bd}$. □

(7) $\frac{a}{b} \cdot \frac{c}{d} = \frac{ac}{bd}$ (multiplying fractions)

PROOF. $\frac{a}{b} \cdot \frac{c}{d} = (a/b) \cdot (c/d) = a \cdot c/b/d = \frac{ac}{bd}$. □

(8) $\frac{a}{b} \div \frac{c}{d} = \left(\frac{a}{b}\right) \Big/ \left(\frac{c}{d}\right) = \frac{a}{b} \cdot \frac{d}{c} = \frac{ad}{bc}$ (dividing fractions)

PROOF. Let $\left(\frac{a}{b}\right) \Big/ \left(\frac{c}{d}\right) = x$, then $\frac{a}{b} = \left(\frac{c}{d}\right)x$, and so $x = \frac{ad}{bc}$. □

(9) $\frac{ad}{bd} = \frac{a}{b}$ (simplifying a fraction)

PROOF. $\frac{a \cdot d}{b \cdot d} = a \cdot d/b/d = a \cdot d/d/b = \frac{a}{b}$. □

EXAMPLE 1.11.1.

(i) $\frac{323}{19} = 17$ is equivalent to $323 = 19 \cdot 17$

(ii) $\frac{114}{2 \cdot 19} = 114/2/19 = 3$

(iii) $\frac{377}{483} = \frac{13 \cdot 29}{17 \cdot 29} = \frac{13}{17}$

(iv) $\frac{11}{-x} = \frac{-11}{x} = -\frac{11}{x}$

(v) $\frac{2}{119} + \frac{5}{119} = \frac{2+5}{119} = \frac{7}{119} = \frac{7}{7 \cdot 17} = \frac{1}{17}$

(vi) $\frac{7}{13} + \frac{12}{11} = \frac{7 \cdot 11}{13 \cdot 11} + \frac{12 \cdot 13}{11 \cdot 13} = \frac{77}{143} + \frac{156}{143} = \frac{77+156}{143} = \frac{233}{143}$

(vii) $\frac{7}{11} \cdot \frac{19}{29} = \frac{133}{483}$

(viii) $\left(\frac{4x}{23}\right) / \left(\frac{x^2}{16}\right) = \frac{4x}{23} \cdot \frac{16}{x^2} = \frac{64}{23x}$.

Note that the second and third steps of (vi) above could have been omitted because the fourth step could have been obtained immediately using property (6). above. However, we encourage the writing of these steps because it's more important to produce the correct answer than to save a little bit of time by not writing them! ◈

1.12 DECIMAL NOTATION

1.12.1 Decimal Representation of Fractions

The *decimal* expressions for $\frac{1}{10}$, $\frac{1}{100}$, and $\frac{1}{1{,}000}$ are 0.1, 0.01, and 0.001, respectively. The "." in each of these expressions is called a *decimal point*. In more generality, fractions such as $\frac{3}{10}$, $\frac{11}{100}$, and $\frac{799}{1{,}000}$ can be expressed in decimal form as 0.3, 0.11, and 0.799, respectively. Note that $\frac{799}{1{,}000}$ can be expanded in the form

$$\frac{799}{1{,}000} = \frac{7}{10} + \frac{9}{100} + \frac{9}{1{,}000} = \frac{7}{10} + \frac{9}{10^2} + \frac{9}{10^3}.$$

In other words, a *decimal expression* or *expansion* is an abbreviation for a sum of fractions of increasing powers of 10. We frequently refer to a number expressed in decimal form as a "decimal," for convenience.

We can write a decimal equivalent for $\frac{1}{2}$ by writing it as $\frac{1 \times 5}{2 \times 5} = \frac{5}{10} = 0.5$. Converting $\frac{1}{3}$ into decimal form is more difficult. We begin by finding decimal numbers that approximate $\frac{1}{3}$. For instance,

$$0.333 = \frac{3}{10} + \frac{3}{100} + \frac{3}{1{,}000} = \frac{333}{1{,}000} = \frac{999}{3{,}000},$$

which is very close to $\frac{1,000}{3,000} = \frac{1}{3}$. We can get decimal values even closer to $\frac{1}{3}$ by increasing the number of repetitions of the digit 3, for example 0.3333 or 0.33333, and so on. In fact, any approximation of this type can be improved by writing even more 3's. We indicate this phenomenon by writing the decimal equivalent for $\frac{1}{3}$ as $0.\dot{3}$, where the dot above the 3 means that the digit 3 is *recurring*.

The decimal expansions for $\frac{1}{4}$ and $\frac{1}{5}$ are finite because $4 \times 25 = 100$ and $5 \times 20 = 100$, that is, $\frac{1}{4} = 0.25$ and $\frac{1}{5} = 0.2$. The decimal expansion for $\frac{1}{6}$ is $\frac{1}{6} = 0.1\dot{6}$. This means that only the 6 is recurring, that is, $\frac{1}{6} = 0.1666666\ldots$ (the three dots at the end are another way to indicate infinite repetition).

Some fractions have a recurring finite sequence of digits. In this case, we draw a horizontal line over the recurring digits, as we see in table 1.5 for decimal expansions of one-seventh, two-sevenths, up to six-sevenths.

TABLE 1.5. Sevenths as decimals

$\frac{1}{7} = 0.\overline{142857} = 0.142857142857\ldots$
$\frac{2}{7} = 0.\overline{285714} = 0.285714285714\ldots$
$\frac{3}{7} = 0.\overline{428571} = 0.428571428571\ldots$
$\frac{4}{7} = 0.\overline{571428} = 0.571428571428\ldots$
$\frac{5}{7} = 0.\overline{714285} = 0.714285714285\ldots$
$\frac{6}{7} = 0.\overline{857142} = 0.857142857142\ldots$

Any fraction (such as the fractions of 7 in table 1.5) can be converted into decimal form by the method of long division.

We can also express a number as an integer plus a fractional part. For example, we can say "two and two-thirds" or write $2\,{}^{2}\!/_{3}$ to mean $2 + \frac{2}{3}$. This can also be expressed as the decimal $2.\dot{6}$.

An *improper* fraction (as opposed to a *proper* fraction) is a fraction with a larger numerator than denominator. Any such fraction can easily be converted to an integer part plus a proper fractional part, which can be expressed as a decimal.

EXAMPLE 1.12.1. $\frac{1,012}{437} = \frac{2 \times 437 + 138}{437} = \frac{2 \times 437}{437} + \frac{138}{437} = 2 + \frac{138}{437} = 2.\overline{315789473684210526}$.

(Check this on a computer!) ◈

1.12.2 Scientific Notation and Precision

We might prefer to express very big and very small numbers using a variation of decimal notation called *scientific notation*. Here are some examples:

$$513 = 5.13 \times 10^2,\ 93{,}000{,}000 = 9.3 \times 10^7,\ \text{and}\ 0.00043 = 4.3 \times 10^{-4}.$$

Note that one digit is placed before the decimal point, the remaining digits after the point, and then the actual position of the decimal point is indicated by multiplication by a power of 10 equal to the number of digits the point should move to the right if the power is positive or the number of digits the point should move to the left (with zeros filling in after the decimal point) if the power is negative.

It is the case in experimental science that measurements are made with limited precision. The number of decimal digits recorded for a measurement should be an indication of the precision obtained in the measurement.

EXAMPLE 1.12.2. If the measurement of the speed of light in a vacuum (usually represented using the letter c) is recorded in a certain laboratory as $c = 2.998 \times 10^8$ m/s/ and, in another laboratory, $c = 2.997925 \times 10^8$ m/s, then we know that the scientists in the second laboratory are much more confident of the precision of their measurement than the scientists in the first laboratory. For most of us, it would suffice to know that the speed of light in a vacuum is approximately 3×10^8 m/s. An approximation of any number can be indicated using the symbol "$\approx$" (a wavy equal sign). Hence, we can write $c \approx 3 \times 10^8$ (300 million) m/s. ◈

EXAMPLE 1.12.3. Scientists frequently need to measure very large and very small quantities, and these quantities would most conveniently be expressed in scientific notation. A very large number used by chemists is Avogadro's number, which equals 6.02×10^{23}. This is the number of atoms contained in a mole of a substance. For example, if some kind of measurement informs us that one mole of hydrogen gas has a mass of 1.01 g, then we can determine the mass of one hydrogen atom by dividing 1.01 by Avogadro's number, that is, $\frac{1.01}{6.02 \times 10^{23}} = \frac{1.10}{6.02} \times \frac{1}{10^{23}} \approx 1.67774 \times 10^{-23}$ g. ◈

1.12.3 Decimal Representation of Irrational Numbers

We explained in section 1.12.1 how to convert a fraction to a decimal. Real numbers that are not fractions or integers (that is, irrational numbers) cannot be expressed with a finite number of decimal digits, nor with a recurring pattern of decimal digits. The reason for this is that the decimal digits of any irrational number are infinite and unpredictable! Any finite decimal expression for the number π, for instance, is an approximation. Here are thirty digits:

$$\pi \approx 3.14159265358979323846264338327.$$

It might seem like a fuss to work out so many digits, but mathematicians have taken this very seriously for thousands of years, because the number of digits that can be computed is an indication of the state of the art of mathematics. The Egyptians of the Old Kingdom knew the approximation $3\,{}^1\!/_7$, and the Greek Mathematician Archimedes (287–212 BC) proved that π was between $\frac{223}{71}$ and $\frac{22}{7}$. Currently, more than a trillion (10^{12}) decimal digits for π are known.

Another irrational number of particular interest is *Euler's number e* (after the Swiss Mathematician Leonhard Euler who lived during the eighteenth century). The first 30 digits of e are shown below. An ordinary pocket calculator would typically show nine decimal digits leading one to suspect that there is a recurring pattern (which is not the case!).

$$e \approx 2.71828182845904523536028771828.$$

1.12.4 Conversion of Decimals into Fractions

There is a precise technique for converting a decimal with recurring digits (or a finite number of digits) into a fraction. The idea is to represent the decimal as an unknown value x, then to multiply x by a power of 10 until the repeating digits start immediately after the decimal point, thereby creating a new number that can be represented as another unknown value, y. Then, y can be multiplied by another power of 10 so that exactly one repetition of the recurring string of digits occurs *before* the decimal point. If y is subtracted from this new number, an integer value remains. It is then possible to obtain x as a fraction. Here is an example:

EXAMPLE 1.12.4.

$$
\begin{aligned}
x &= 0.2\overline{35} \\
y = 10x &= 2.\overline{35} \\
100y &= 235.\overline{35} \\
100y - y = 99y &= 233 \\
10x = y &= \frac{233}{99} \\
x &= 233/990
\end{aligned}
$$

◈

EXAMPLE 1.12.5. It is useful to know a fraction that is a good approximation for e. We can approximate e by the slightly smaller recurring decimal value $2.7\overline{1828}$, which agrees with the decimal value for e up to the ninth decimal position. Using the method above, we can determine that this recurring decimal is an expression of the fraction $\frac{271,801}{99,990}$. This fraction cannot be simplified; however, the fraction $\frac{271,800}{99,990}$ is a tiny bit smaller, and its numerator and denominator have a common factor of 90, which means it can be simplified to $\frac{3,020}{1,111}$. The decimal expression of $\frac{3,020}{1,111}$ is $2.7\overline{1827}$ (check this on a computer), which matches the decimal expression for e up to the fifth decimal digit. In summary, we have

$$2.718271 < \frac{3,020}{1,111} < \frac{271,801}{99,990} < e < 2.718282.$$

From this, it is clear that the difference $2.718282 - 2.718271 = 0.000011$ is bigger than the difference between e and $\frac{3,020}{1,111}$. We write this formally as

$$\left| e - \frac{3,020}{1,111} \right| < 0.000011.$$

The number $0.000011 = 1.1 \times 10^{-5}$ can be called the *error of the approximation* ◈

1.12.5 Rounding Off Decimals

Whenever we give an approximate value of a number by selecting a finite number of digits from a possibly infinite (decimal) representation, we use a method called *rounding off*. Suppose, we decide for some purpose that six decimal digits of a particular decimal number would be an accurate enough approximation. We then look at the seventh decimal digit. If this digit is 5 or bigger, we *round up* by increasing the sixth decimal digit by one unit. If the seventh digit is a 4 or less, we *round down* by leaving the sixth digit unchanged. The only exception is when the sixth digit is followed by $4\dot{9}$. In this case, we regard the seventh digit as a 5 and round up. If the digit we need to round up is a 9, then we round up the digit before the 9 and replace the 9 with a 0. This can be done repeatedly and we keep the 0's to indicate the accuracy of rounding. The number of accurate digits is called the number of *significant digits*. The counting of significant digits begins from the first nonzero digit from the left and can include digits before the decimal point. If there is no decimal point, then the final significant digit is underlined.

EXAMPLE 1.12.6.

(i) π rounded to seven significant digits: $\pi \approx 3.141593$

(ii) π rounded to nine significant digits: $\pi \approx 3.14159265$

(iii) $1{,}095{,}487$ rounded to four significant digits: $1{,}095{,}487 \approx 1{,}09\underline{5}{,}000$

(iv) $1{,}095{,}487$ rounded to three significant digits: $1{,}095{,}487 \approx 1{,}1\underline{0}0{,}000$

(v) $\frac{5}{6} = 0.8\dot{3}$ rounded to three significant digits: $\frac{5}{6} \approx 0.833$

(vi) $\frac{1}{60} = 0.01\dot{6}$ rounded to three significant digits: $\frac{1}{60} \approx 0.0167$

(vii) 76.34999981 rounded to seven significant digits: 76.35000

(viii) $0.44\dot{9}$ rounded to one significant digit: 0.5 ◈

1.13 DIVISIBILITY OF NATURAL NUMBERS

1.13.1 Prime Decomposition of a Natural Number

You might have wondered why the fraction $\frac{271{,}801}{99{,}990}$ in example 1.12.5 could not be simplified. This is a question concerning the divisibility of natural numbers. If a natural number can be expressed as a product of two other natural numbers (called factors), then we say that the number is *divisible* by each of these factors. Any or all of these factors could in turn be divisible by smaller factors, and then the original number is also divisible by these smaller factors. For example, the natural number 506 is equal to 2×253. This means that 506 is divisible by the natural numbers 2 and 253. The diligent student who has memorized the products in table 1.2 will know that $253 = 11 \times 23$. This means that 506 is also divisible by 11 and 23. In fact, we can form the product $506 = 2 \times 11 \times 23$. A number which is only divisible by itself and one is called a *prime number*. Every natural number can be expressed as a product of prime numbers (called *prime factors*) after fully testing divisibility of the number, and this product is called the *prime decomposition* of the number. A prime number is its own prime decomposition.

EXAMPLE 1.13.1. Here are the prime decompositions of some numbers:

(i) $506 = 2 \times 11 \times 23$

(ii) $612 = 2 \times 2 \times 3 \times 3 \times 17 = 2^2 \times 3^2 \times 17$

(iii) $1{,}111 = 11 \times 101$

(iv) $3{,}020 = 2^2 \times 5 \times 151$

(v) $271{,}801 = 47 \times 5{,}783$

(vi) $99{,}990 = 90 \times 1{,}111 = 2 \times 3^2 \times 5 \times 11 \times 101$

(vii) $271{,}800 = 90 \times 3{,}020 = 2^3 \times 3^2 \times 5^2 \times 151$

(viii) $31 = 31$ (because 31 is a prime number) ◈

The number of times a factor appears in the prime decomposition of a number is called the *multiplicity* of the factor.

EXAMPLE 1.13.2. The prime factors of 612 (in (ii) above) are 2, 3, and 17, where the factor 2 occurs with multiplicity 2, 3 occurs with multiplicity 3, and 17 occurs with multiplicity 1, and of 3,020 (in (iv) above) are 2, 5, and 151 and their multiplicities are 2, 1, and 1, respectively. Furthermore, we observe that the prime decompositions of 271,801 and 99,990 have no prime factors in common. This is the reason the fraction $\frac{271,801}{99,990}$ cannot be simplified. (The fraction $\frac{3,020}{1,111}$ also cannot be simplified.) ◈

1.13.2 Finding the Prime Factors of a Natural Number

It is, in general, not an easy task to find the prime factors of large (nonprime) numbers when all the prime factors are also large numbers. The straightforward but laborious method to find a prime factor for a given large number n is to test every prime number p with the property $p^2 < n$ for divisibility.

EXAMPLE 1.13.3. To find the prime factors of 271,801, we generate the following list of prime numbers that satisfy the property stated above: 2, 3, 5, 7, 11, 13, 17, 19, 23, 29, 31, 37, 41, 43, 47, 53, 59, 61, 67, 71, 73, 79, 83, 89, 91, 97, 101, 103, 107, 109, 113, 127, 131, 137, 139, 149, 151, 157, 163, 167, 173, 179, 181, 191, 193, 197, 199, 211, 223, 227, 229, 233, 239, 241, 251, 257, 263, 269, 271, 277, 281, 283, 293, 307, 311, 313, 317, 331, 337, 347, 349, 353, 359, 367, 373, 379, 383, 389, 397, 401, 409, 419, 421, 431, 433, 439, 443, 449, 457, 461, 463, 467, 479, 487, 491, 499, 503, 509, and 521. (These are the first 99 prime numbers.) Starting from 2, 15 prime numbers have to be checked (using a calculator) before the prime factor 47 is found (a calculator will then yield $271,801/47 = 5,783$). Next, we can verify that $5,783$ is a prime number. Because $73^2 = 5,328$ and $79^2 = 6,421$, this involves checking that $5,783$ is not divisible by any prime number up to 73. Consequently, the prime decomposition of 271,801 is $27801 = 47 \times 5,783$. ◈

1.13.3 Testing for Divisibility by Small Prime Numbers

There are some quick methods for testing a number for divisibility by small prime numbers. Here are some examples:

- A number is divisible by 3, if its digits add up to a multiple of 3 (e.g., 8,175 is divisible by 3 because $8+1+7+5=21$).
- A number is divisible by 5, if its last digit is 5 or 0.
- A test for divisibility by 7 is to *subtract two times the last digit* from the remaining digits. If the result is divisible by 7, then the original number is also divisible by 7. For example, 1,001 is divisible by 7 because $100-2=98$ and $98=7\times14$.
- To test a number for divisibility by 11, we can add every other digit starting from the first digit to obtain one total, add the remaining digits to obtain a second total, and then subtract the second total from the first. If the result is divisible by 11, then the original number is also divisible by 11. (We consider 0 to be divisible by all natural numbers.) For example, 6,773,294 is divisible by 11 because $(6+7+2+4)-(7+3+9)=19-19=0$.
- Divisibility by 13 can be tested by *adding four times the last digit* to the remaining digits and testing the resulting number again for divisibility by 13.
- Divisibility by 17 can be tested by *subtracting five times the last digit* from the remaining digits and testing the resulting number again for divisibility by 17.
- Divisibility by 19 can be tested by *adding two times the last digit* to the remaining digits and testing the resulting number again for divisibility by 19.

1.13.4 Adding Fractions Using Their Lowest Common Denominator

Property (6) of the laws of division in section 1.11 is a formula for adding any pair of fractions. However, if the prime decompositions of the denominators of two fractions are known, then the fractions can be added more easily by finding the *lowest common denominator* (LCD) of the fractions. This is the *smallest* number that is divisible by each of the denominators, and it can be expressed as a product of all the prime numbers occurring in each of the denominators, with the multiplicity of each prime number being the larger of the multiplicities that occurs in each denominator.

EXAMPLE 1.13.4. If the denominators of two fractions are $45=3^2\times5$ and $150=2\times3\times5^2$, respectively, then the LCD is $450=2\times3^2\times5^2$. Now, if the fractions we want to add are $\frac{13}{45}$ and $\frac{7}{150}$, then we will perform the addition this way:

$$\begin{aligned}\frac{13}{45}+\frac{7}{150}&=\frac{13}{45}\times\frac{10}{10}+\frac{7}{150}\times\frac{3}{3}\\&=\frac{130}{450}+\frac{21}{450}\\&=\frac{130+21}{450}=\frac{151}{450}.\end{aligned}$$

That is, we multiply and divide each fraction by the appropriate number so that its new denominator is equal to the LCD, and then we add the fractions. Note that 151 is not divisible by 2, 3, or 5 (why?), so the answer cannot be simplified. ◈

1.14 LAWS FOR EXPONENTS

Exponents were introduced briefly in section 1.9. Table 1.6 is a list of *laws for exponents* that can be derived from the multiplicative property of exponents. (We assume that a and b are nonzero real numbers and that m and n are integers.)

TABLE 1.6. Laws for exponents

(I)	$a^m\cdot a^n=a^{m+n}$	Exponents with the same base can be added
(II)	$\left(a^m\right)^n=a^{m\cdot n}$	A power of an exponent becomes a product of exponents
(III)	$(a\cdot b)^n=a^n\cdot b^n$	An exponent distributes through a product
(IV)	$\left(\frac{a}{b}\right)^n=\frac{a^n}{b^n}$	An exponent distributes through a quotient
(V)	$\frac{a^m}{a^n}=a^{m-n}$	An exponent in the denominator changes sign in the numerator
(VI)	$\frac{a^m}{a^n}=\frac{1}{a^{n-m}}$	An exponent in the numerator changes sign in the denominator

If we set $m=0$, then (I) becomes $a^0\cdot a^n=a^{0+n}=a^n$. For this reason, we set $a^0=1$ for any nonzero real number a. The first statement of (V) (with $m=0$) can thus be expressed as $\frac{1}{a^n}=a^{-n}$.

This means that a negative exponent is interpreted as the reciprocal of the corresponding positive exponent.

EXAMPLE 1.14.1.

(i) $2^{-6}=\frac{1}{2^6}=\frac{1}{64}$.

(ii) $\left(\frac{3}{2}\right)^{-5}=\left(\frac{2}{3}\right)^{5}=\frac{2^5}{3^5}=\frac{32}{243}$.

(iii) $\frac{2^{-1}a^3b^{-2}}{a^{-2}b^4}=\frac{a^5}{2b^6}$. ◈

1.15 RADICALS

It is unusual in nature for a process to be reversible. As a familiar saying goes, it's no use crying over spilled milk (because the milk cannot be un-spilled). In mathematics, however, it is frequently the case that processes are reversible. In fact, when mathematicians learn a new process they also strive to learn how to reverse the process. In doing so, they might invent notation for this purpose. This has already become a theme of this book: subtraction and division can be thought of as the reversals of the processes of addition and multiplication, respectively. The notation we introduce now is called *radical* notation.

A specific instance of this notation is a *root*. The most common root is the *square root*, with symbol $\sqrt{\ }$, which is the notation for the operation that reverses the squaring of a number. For example, we write "$\sqrt{16}=4$" because $4^2=16$. In the same way, we can compute the *cube root* of a number using the notation $\sqrt[3]{729}=9$ (because $9^3=729$). In general, for any natural number n, we use the notation $\sqrt[n]{\ }$ for the reversal of an nth power. It is always possible to take the nth root of any real number so long as n is an *odd* number, for example, $\sqrt[3]{-27}=-3$ (because $(-3)^3=-27$); however, if n is any even number, then the nth root of a negative real number cannot be expressed as another real number. For example, $\sqrt{-4}$ is not a real number. Furthermore, any even root—in particular the square root—of a positive number is, by convention, also a positive number. This last statement has to be considered very carefully because it means, for example, that $\sqrt{(-11)^2}=\sqrt{121}=11$.

The most general statement of this kind is $\sqrt{x^2}=|x|$, whenever x is a real number (which means x could be a negative number). An expression such as $\sqrt{144x^2y^3}$, where x is a real number and y is any positive real number or zero (i.e., any *nonnegative* real number), can be simplified to $12|x|y\sqrt{y}$.

Table 1.7 is a list of laws for radicals. These are variations of the laws for exponents. We are supposing that m and n are natural numbers.

TABLE 1.7. Laws for radicals

(I)	$\sqrt[n]{a\cdot b}=\sqrt[n]{a}\cdot\sqrt[n]{b}$	An nth root distributes through a product
(II)	$\sqrt[n]{\frac{a}{b}}=\frac{\sqrt[n]{a}}{\sqrt[n]{b}}$	An nth root distributes through a quotient
(III)	$\sqrt[m]{\sqrt[n]{a}}=\sqrt[mn]{a}$	An mth root of an nth root is an mnth root

EXAMPLE 1.15.1.

(i) $\sqrt[5]{x^7} = \sqrt[5]{x^5} \cdot \sqrt[5]{x^2} = x\left(\sqrt[5]{x^2}\right)$

(ii) $\sqrt[3]{x^7} = \sqrt[3]{(x^2)^3} \cdot \sqrt[3]{x} = x^2\left(\sqrt[3]{x}\right)$

(iii) $\sqrt{x^6} = \sqrt{(x^3)^2} = |x^3| = |x|^3$

(iv) $\sqrt[4]{x^6 y^3} = \sqrt[4]{x^4 x^2 y^3} = |x|\left(\sqrt[4]{x^2 y^3}\right)$

(v) $\sqrt[3]{320} = \sqrt[3]{64 \cdot 5} = 4(\sqrt[3]{5})$

(vi) $\sqrt{3a^2b^3} \cdot \sqrt{6a^5b} = \sqrt{2 \cdot 9(a^3)^2 a(b^2)^2} = 3|a|^3 b^2 \sqrt{2a}$

(vii) $\sqrt{\sqrt[3]{512}} = \sqrt[6]{(2\sqrt{2})^6} = 2\sqrt{2}.$ ◈

EXERCISES

1.1. On the same number line, label all fractions that are integer multiples of $\frac{1}{7}$ and $\frac{1}{8}$ between 0 and 1. Do this as accurately as you can.

1.2. Order the elements in each of the following sets from smallest to largest.

(a) $\left\{-\frac{4}{7},\ \pi,\ \frac{22}{7},\ -\frac{7}{4},\ 3.1\right\}$

(b) $\left\{\frac{5}{6},\ \frac{\sqrt{2}}{2},\ \frac{16}{27},\ \frac{2}{3},\ 0.59\right\}$

1.3. Decide whether each of the following statements is *true* or *false*.

(a) Every natural number is an integer.

(b) A recurring decimal is an irrational number.

(c) The ratio of the radius to the circumference of the circle is an integer.

(d) There are infinitely many rational numbers.

(e) Every integer is a rational number.

(f) $\frac{\pi}{2}$ is a rational number.

1.4. Decide whether each of the following numbers is irrational or rational.

(a) $\sqrt{7}-2$

(b) $-\frac{7}{2}$

(c) $0.\overline{12345}$

(d) $\sqrt[3]{9}$

(e) $\sqrt{289}$

(f) $\frac{3}{\sqrt{3}}$

(g) $\frac{15{,}979}{19}$

(h) $\frac{1}{\pi}$

1.5. Evaluate each of the following interval intersections. Write your answer as an interval, a set with one element or an empty set. You can suppose that a, b, c, d are real numbers and that $a < b < c < d$.

(a) $(a,c] \cap (b,d]$

(b) $(a,d) \cap (b,c]$

(c) $[a,d) \cap [b,d)$

(d) $[a,d) \cap (a,b)$

(e) $(a,b] \cap [b,d)$

(f) $(a,c] \cap (c,d)$

1.6. Rewrite each of the following expressions without an absolute value sign. If possible, evaluate the expression.

(a) $|5-3\pi|$

(b) $3-|7-11|$

(c) $|(113-221)-|113-221||$

(d) $\pi-|3-|\pi-3||$

(e) $|2-|4+|2-|4-|2-4|||||$

1.7. Use properties (VII), (X), (IV), and (VIII) in section 1.8 to prove that $(-1)\cdot a=-a$ for any real number a. (Hint: $a+(-1)a=1\cdot a+(-1)a$.)

1.8. Compute the following products using FOIL.

(a) $(11-4)(6+13)$

(b) $(x-13)(x+7)$

(c) $(2x-1)(6+x)$

(d) $(4-y)(y+12)$

(e) $(y-10)(y+10)$

(f) $(x+3y)(x+3y)$

(g) $(7x-3y)(19x+5y)$

1.9. Evaluate the following exponents.

(a) 2^{12}

(b) 6^4

(c) 5^3

(d) 25^2

(e) 6^5

(f) 11^3

1.10. Use the PEMDAS rule to simplify each of the following expressions. Write your answers as a single integer or fraction.

(a) $22-4^{-3}\cdot(14^{4/2}\cdot 6/3/2)$

(b) $1\cdot 2^3/4\cdot 5-(6+7)$

(c) $6/2(1+2)$

(d) $6(5(4(3(2(1-2)-3)-4)-5)-6)$

(e) $(1-(1+(1-(1+\sqrt{2}))))$

(f) $120/2\cdot 3/4\cdot 5/6/7\cdot 8/9$

(g) $1+\cfrac{24}{1+\cfrac{1}{1+\cfrac{1}{2}}}$

1.11. Simplify the following fractions.

(a) $\dfrac{121}{132}$

(b) $\dfrac{7a}{6a}$

(c) $\dfrac{7(a+3)}{119(a+4)}$

(d) $\dfrac{118(a+3)}{119(a+3)}$

(e) $\dfrac{6\cdot 8+2\cdot 7}{7\cdot 14}$

(f) $\dfrac{341+132}{22+99}$

1.12. Multiply the following fractions, as indicated. Write each answer as a single, simplified fraction.

(a) $\left(\dfrac{15}{17}\right)\left(\dfrac{4}{3}\right)$

(b) $\left(\dfrac{15}{17}\right)\left(\dfrac{3}{4}\right)$

(c) $\left(\dfrac{5x}{21}\right)\left(\dfrac{3}{15x}\right)$

(d) $\left(\dfrac{b+2}{16(b+2)}\right)\left(\dfrac{3(b+2)}{2(b+2)}\right)$

(e) $\left(\dfrac{a^2b}{c}\right)\left(\dfrac{bc^2}{a}\right)$

(f) $\left(\dfrac{31xy^3}{33x^2y}\right)\left(\dfrac{21x}{9y}\right)$

1.13. If Curtis cuts his birthday cake into seven slices of equal size and gives one slice to his sister, Celeste, and Celeste gives two-fifths of her slice to her puppy, how much of Curtis' birthday cake does his sister's puppy have? Write your answer as a decimal, rounded to five significant digits. (Hint: multiply the fractions!)

1.14. Add the following fractions, as indicated. Write each answer as a single, simplified fraction.

(a) $\frac{1}{3}+\frac{2}{7}$

(b) $\frac{2}{9}+\frac{1}{3}$

(c) $\frac{16}{11}+\frac{5}{13}$

(d) $\frac{5}{12}+2$

(e) $\frac{2}{7}+\frac{7}{2}$

(f) $\frac{x}{4}+\frac{4}{x}$

(g) $\frac{2}{9y}+\frac{9y}{2}$

(h) $\frac{19}{20}+\frac{20}{21}$

(i) $\frac{19x}{3}+\frac{1}{2y}$

1.15. If a blue Creepy Crawly can clean a swimming pool in 3 h and a red Creepy Crawly can clean the same swimming pool in 4 h, how long will it take both Creepy Crawlies working together, and starting simultaneously, to clean the swimming pool? (Assume the Creepy Crawlies start at opposite ends of the pool and do not get entangled with each other!) Leave your answer as a fractional number of hours. (Hint: How much of the job can each Creepy Crawly get done in 1 h? What fraction of the job can both get done in 1 h if they work simultaneously?)

1.16. Write all of the following fractions in decimal form. (A calculator will be required for some of these.)

(a) $\frac{7}{100}$

(b) $\frac{77}{100}$

(c) $\frac{77}{1{,}000}$

(d) $\frac{24}{25}$

(e) $\frac{1}{8}$

(f) $\frac{3}{8}$

(g) $\frac{5}{8}$

(h) $\frac{7}{8}$

(i) $\frac{1}{9}$

(j) $\frac{2}{9}$

(k) $\frac{7}{9}$

(l) $\frac{77}{99}$

(m) $\frac{777}{999}$

(n) $\frac{11}{7}$

(o) $\frac{22}{7}$

(p) $\frac{302}{111}$

(q) $\frac{30{,}201}{11{,}111}$

1.17. Try to find a (simplified) fractional approximation for π that has four digits in the denominator.

1.18. Without using a calculator, rewrite the following numbers using scientific notation.

(a) 1011.001

(b) -0.000345

(c) $0.100\overline{305}$

(d) 602×10^{21}

(e) $2^{11}+1$

(f) $4.5\times10^{-3}\cdot9\times10^{-7}$

1.19. Write 10 significant digits for each of the following irrational numbers. (A calculator will be required for all of these.)

(a) $\sqrt{2}$

(b) $\sqrt{3}$

(c) $\sqrt{5}$

(d) $\sqrt{7}$

(e) 2π

(f) e^{π}

1.20. Convert the following decimals into fractions.

(a) $0.\dot{1}$

(b) $0.\overline{111}$

(c) 1.111

(d) $6.30\overline{44}$

(e) $2.7\overline{183}$

(f) $3.1\overline{416}$

(g) $0.\dot{9}$

1.21. Round off the following decimal expressions to four decimal digits.

(a) 7.667788

(b) 0.001019

(c) 0.79999

(d) $0.78884\dot{9}$

(e) $0.78884\overline{95}$

1.22. Determine whether each natural number below is prime or *composite* (i.e., not prime). If the number is composite, work out its prime decomposition.

(a) 523

(b) 527

(c) 529

(d) 531

(e) 533

(f) 537

(g) 123456789

1.23. Which integers have the following property: If the final digit is deleted, the integer is divisible by the new number?

1.24. Find the prime decompositions of the following natural numbers.

(a) 507

(b) 613

(c) 112

(d) 3,021

(e) 271,802

(f) 99,991

(g) 99,997

1.25. Add the fractions, as indicated, by finding their lowest common denominator.

(a) $\frac{2}{15}+\frac{8}{21}$

(b) $\frac{2}{33}+\frac{8}{143}$

(c) $\frac{1}{30}+\frac{1}{100}$

(d) $\frac{1}{31}+\frac{1}{101}$

(e) $\frac{5}{34}+\frac{5}{202}$

(f) $\frac{11}{15}+\frac{12}{33}$

(g) $\frac{615}{612}+\frac{612}{615}$

1.26. Simplify the following expressions so that each variable appears only once, all exponents are positive, no powers of exponents appear and fractions appear in front of the expression.

(a) $(3x^3y^4)(4xy^5)^2$

(b) $(2a^3b^3c)^4$

(c) $\frac{8x^3y^{-5}}{4x^{-1}y^2}$

(d) $\left(\frac{u^2}{2v}\right)^{-3}$

(e) $\frac{4xy}{\left(\frac{x}{2}\right)}$

(f) $\frac{\left(\frac{84xz^2}{57z^{-1}}\right)}{3z}$

1.27. Evaluate the following expressions using the laws for radicals. Leave your answer in the simplest radical form.

(a) $\sqrt{361}$

(b) $\sqrt[3]{64}$

(c) $\sqrt[3]{6^{11}}$

(d) $\sqrt[11]{6^{12}}$

(e) $\frac{\sqrt[3]{8}}{\sqrt[3]{54}}$

(f) $\frac{\sqrt[3]{8}}{\sqrt[3]{56}}$

(g) $3\sqrt{2^3}$

1.28. Simplify each expression below by reducing the magnitude of the exponent inside the radical.

(a) $\sqrt[3]{x^5}$

(b) $\sqrt[5]{x^3}$

(c) $\sqrt{x^2y^3}$

(d) $\sqrt[3]{x^3y^2}$

(e) $\sqrt[3]{-8(xy)^9}$

(f) $\frac{\sqrt{56x^5}}{\sqrt{72x}}$

(g) $\sqrt[7]{\sqrt{77x^{13}y^{15}}}$

(h) $\sqrt{\sqrt[5]{128}}$

1.29. How many three-digit perfect squares are there? Which is the smallest and largest?

1.30. A jug contains three cups of mango juice and another jug contains three cups of guava juice. In order to make a mixed fruit juice, one cup of mango juice is taken from the first jug and poured into the second jug. What is the proportion of mango juice in the second jug? Suppose now that the juice in the second jug is thoroughly mixed and one cup of this mixture is removed and poured into the first jug. What is the proportion of guava juice in the first jug? If this process is repeated so that one cup of the mixture in the first jug is removed to the second jug, what is the new proportion of mango juice in the second jug? If one cup of the mixture in the second jug is now poured into the first jug, what is the new proportion of guava juice in the first jug?

1.31. Prove that $\sqrt{8}+\sqrt{18}=\sqrt{50}$. (This was proved by the Arabic Mathematician Alkarki in about 1000 AD)

1.32. Explain carefully why $0.6\left(\frac{a+b+c}{3}\right)=0.2a+0.2b+0.2c$. Mention PEMDAS, the appropriate law(s) of division and the relevant properties of the algebra of real numbers.

CHAPTER 2

THE CARTESIAN PLANE

2.1 INTRODUCTION

In the seventeenth century, the Philosopher and Mathematician René Descartes had the brain wave of joining two (infinite) number lines at right angles so that they intersected at their respective zero points, as shown in figure 2.1. This creates a *rectangular coordinate system* that, nowadays, is called the *Cartesian plane*. The horizontal number line is usually labeled the x-axis, and the vertical number line is usually labeled the y-axis. Each axis has an increasing direction indicated by means of an arrow. The point where the axes intersect is called the *origin* and labeled O. Descartes and mathematicians following him, including Isaac Newton, discovered a rich interplay of algebra and geometry in the Cartesian plane, one outgrowth of which was the discovery of calculus.

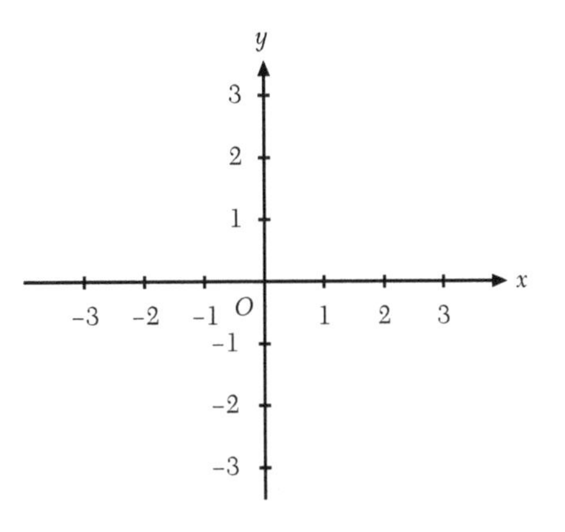

FIGURE 2.1. The Cartesian plane.

After a brief explanation of how a rectangular coordinate system works (section 2.2), we will make a complete study of the relationship between graphs of lines in the Cartesian plane and linear equations in section 2.3. This is followed by an introduction to circles (section 2.4) and conic sections (section 2.5) in the Cartesian plane. Vectors are a useful tool for doing geometric analysis in the Cartesian plane, and this topic takes up the remainder of the chapter (section 2.6).

The mathematical methods in a two-dimensional coordinate system that we introduce in this chapter can be generalized to a three-dimensional coordinate system (although we will not do it in this book). A three-dimensional coordinate system can be used as a mathematical representation of the three-dimensional physical space in which we humans move around; and it is in this coordinate representation of three-dimensional space that mathematicians and scientists can carry out advanced mathematical simulations of dynamic processes like ocean currents, planetary weather patterns, motions of projectiles, and exploding stars.

2.2 WORKING IN A COORDINATE SYSTEM

Any given point in the Cartesian plane can be labeled by means of two numbers called coordinates. The first coordinate is called the x-coordinate, and the second is called the y-coordinate. The x- and y-coordinates together form a *coordinate pair*. The x-coordinate can be found by following a vertical line from the given point to a point on the x-axis and reading its position on the number line. Similarly, the y-coordinate can be found by following a horizontal line from the given point to a point on the y-axis and reading its position on the number line. Points in the Cartesian plane are usually labeled using uppercase letters with the coordinate values following in parentheses, as shown in figure 2.2. The axes separate the Cartesian plane into four *quadrants* (first, second, third, and fourth quadrants) that can be labeled as I, II, III, and IV, respectively, in a counterclockwise order, starting with the upper right quadrant.

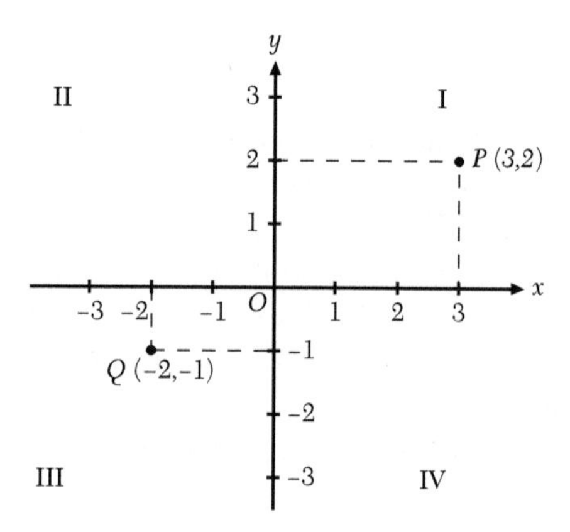

FIGURE 2.2. Points in the Cartesian plane.

2.3 LINEAR EQUATIONS AND STRAIGHT LINES

Much of mathematics involves the study of related quantities in the form of equations. For this reason, we begin with a study of linear equations and the graphs (infinite straight lines) of linear equations in the Cartesian plane because these quantify the simplest kind of relationship (linear) that related quantities can have. We will refer to an "infinite straight line" (in the Cartesian plane) as a "line."

When a line is drawn in the Cartesian plane, it can be described as a line leaning to the right if the y-coordinates of points on the line increase as the corresponding x-coordinates increase, as a line leaning to the left if the y-coordinates of points on the line decrease as the corresponding x-coordinates increase, or as a horizontal or vertical line (refer to figure 2.3.). The point where a line crosses the x-axis (if any) is called an x-intercept, and a point where the line crosses the y-axis (if any) is called a y-intercept.

There is a relationship between lines in the Cartesian plane and *linear equations*.

DEFINITION 2.3.1. *A linear equation is an equation of the form:*

$$y = mx + c,$$

(or one of the equivalent forms stated below), where m *and* c *are real numbers and* x *and* y *are variables. The variable* x *is called the independent variable and the variable* y *is called the dependent variable.*

In the case that $m = 0$, the linear equation reduces to $y = c$.

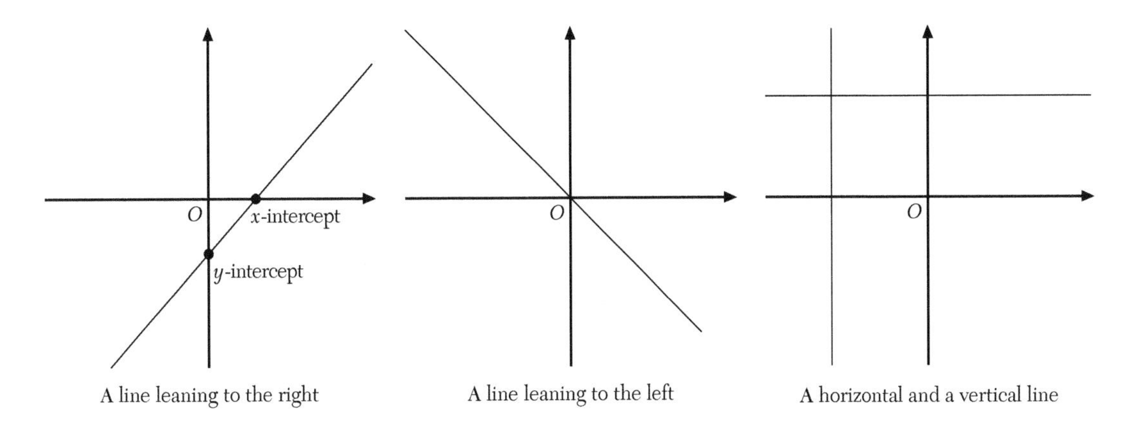

FIGURE 2.3. Lines in the Cartesian plane.

DEFINITION 2.3.2. *A solution of the equation* $y = mx + c$ *is a pair of coordinates* (a, b) *for which the equation is a true statement if the value* a *is substituted for* x *and the value* b *is substituted for* y*, that is, the equation* $b = ma + c$ *is a true statement.*

EXAMPLE 2.3.1. One solution of the equation $y = 5x - 1$ is the pair of values $(1, 4)$. A pair of values that is not a solution of the equation is $(1, 3)$. ◈

2.3.1 The Graph of a Linear Equation

If two solutions of a linear equation $y = mx + c$ are obtained, that is, two pairs of coordinates (a_1, b_1) and (a_2, b_2) so that $b_1 = ma_1 + c$ and $b_2 = ma_2 + c$ are true statements, then the *graph* of the linear equation can be obtained by plotting the coordinate pairs (a_1, b_1) and (a_2, b_2) in the Cartesian plane and drawing a line through them.

EXAMPLE 2.3.2. The graph of the equation $y = 3x - 5$ is drawn in figure 2.4 by plotting the points corresponding to the two solutions $P(3, 4)$ and $Q(-2, -11)$ of the equation. It is important to realize that every point on the line determines a solution of the linear equation $y = 3x - 5$ and, conversely, that every solution of this equation determines a point on the line. For example, observe that the point $(0, -5)$ is on the line; therefore, the coordinate pair $(0, -5)$ is a solution for the equation. ◈

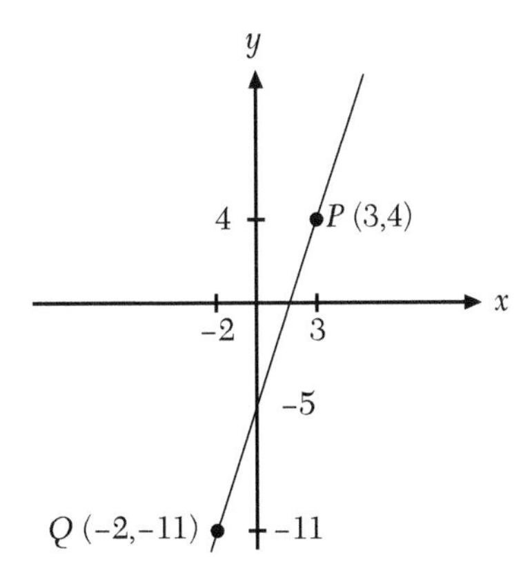

FIGURE 2.4. A line through two points.

Very often, the most convenient way to draw the graph of a line is to determine the x- and y-intercepts of the line and to draw the line through these two points. In the following example, we show how to determine the x- and y-intercepts. The basic observation is that the x-coordinate is zero for all points on the y-axis and the y-coordinate is zero for all points on the x-axis.

EXAMPLE 2.3.3. Draw the graph of the equation $y=\frac{3}{2}x-4$ by finding the x- and y-intercepts of the straight line in the Cartesian plane.

Answer: First, if we set $x=0$ in the equation $y=\frac{3}{2}x-4$, then $y=\frac{3}{2}(0)-4=-4$. We conclude that the y-intercept is the coordinate pair $(0,-4)$. Second, if we set $y=0$ in the equation $y=\frac{3}{2}x-4$, then $0=\frac{3}{2}x-4$. By inspection, this is a true statement if $x=\frac{8}{3}$, and we conclude that the x-intercept is the coordinate pair $\left(\frac{8}{3},0\right)$. The graph is shown in figure 2.5. ◈

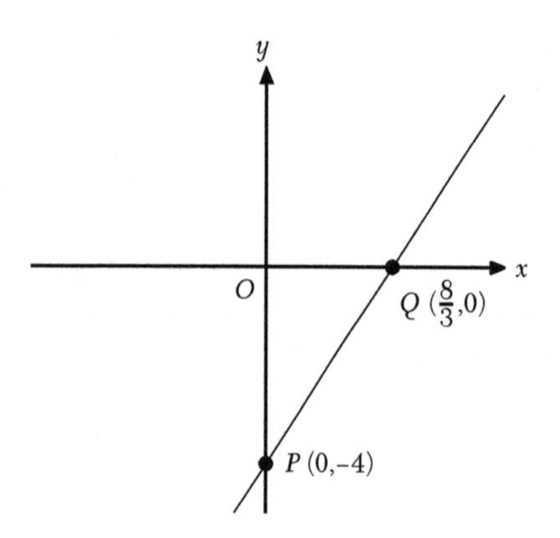

FIGURE 2.5. The x- and y-intercepts of a line.

2.3.2 The Linear Equation of a Line

We showed in section 2.3.1 how to draw a line corresponding to a given linear equation. On the other hand, we could be shown a line, or given some information about a line, and be required to find the corresponding linear equation. Typically, we are given the coordinates of two points $P(x_1,y_1)$ and $P(x_2,y_2)$ on the line, where one of the points, say $P(x_1,y_1)$, is to the left of the other in the Cartesian plane. We associate a number with the line called the *slope* of the line. We calculate the slope by noting the change (using the notation Δ) in the y value from $P(x_1,y_1)$ to $P(x_2,y_2)$, that is, $\Delta y = y_2 - y_1$, and dividing this number by the corresponding change in the x value, that is, $\Delta x = x_2 - x_1$. This ratio is normally denoted by the letter m, thus:

$$\boxed{\text{Slope of a line} = m = \frac{\Delta y}{\Delta x} = \frac{y_2 - y_1}{x_2 - x_1}}.$$

We make the following observations regarding the direction in which the line leans and the value of the slope of the line:

- If a line leans to the *left*, its slope is *negative*.
- If a line leans to the *right*, its slope is *positive*.
- If a line is *horizontal*, its slope is *zero*.
- If a line is *vertical*, its slope is *infinite*.

EXAMPLE 2.3.4. The slope of the line through the points $P(-3,-2)$ and $Q(3,4)$ is $m=\frac{\Delta y}{\Delta x}=\frac{4-(-2)}{3-(-3)}=\frac{6}{6}=1$. ◈

If the slope m of a line is known, then an expression relating the coordinates of any other point (x,y) on the line to the coordinates of any given point (x_1,y_1) on the line is $m=\frac{y-y_1}{x-x_1}$. Another way to write this is:

$$\boxed{\text{The point-slope form of a line: } y-y_1=m(x-x_1)}. \quad (2.1)$$

EXAMPLE 2.3.5. To find the equation of the line through the points $P(-3,-2)$ and $Q(3,4)$, we substitute $m=1$ from example 2.3.4 and the coordinates of Q, that is, $x_1=3$ and $y_1=4$ (we could also use the coordinates of P) in formula (2.1), resulting in $y-4=(1)(x-3)$. This can be simplified to $y=x+1$. The graph of this line is shown in figure 2.6. ◈

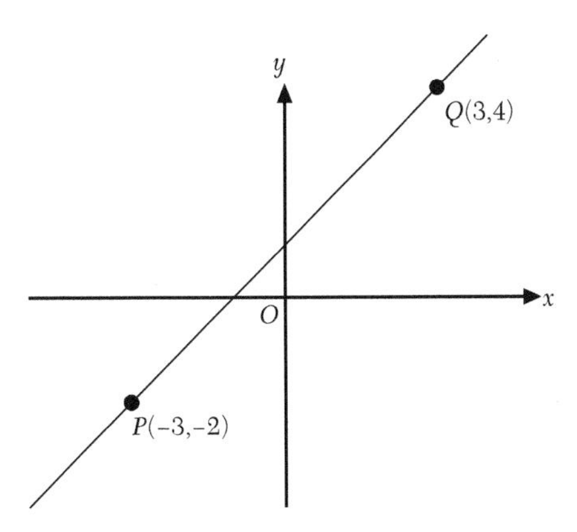

FIGURE 2.6. A line through two given points.

If we only know the slope of a line and the value of the y-intercept of the line, then we can substitute these values into another form of the equation of a line known as:

$$\boxed{\text{The slope-intercept form of a line: } y=mx+c}, \quad (2.2)$$

where c is the y-intercept of the line because this corresponds to the value of y when $x=0$.

EXAMPLE 2.3.6. If the slope of a line is $m=-2$ and it cuts the y-axis at $c=3$, then its equation is $y=-2x+3$. ◈

In the special case that the y-intercept is zero, the line passes through the origin. Figure 2.7 shows a number of lines passing through the origin, with the slope of each line labeled next to it.

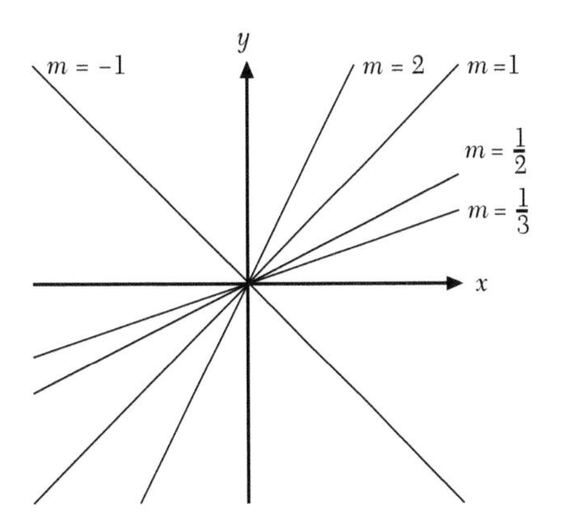

FIGURE 2.7. Lines through the origin.

Another form of the equation of a line that we will take note of here is:

$$\boxed{\text{The standard form of a line: } Ax+By=C}, \tag{2.3}$$

which can be obtained by algebraic manipulation of the point-slope form or slope-intercept form and renaming the constants.

EXAMPLE 2.3.7. The equation of the line in example 2.3.5 can be expressed in the standard form $-x+y=1$, and the equation of the line in example 2.3.6 can be expressed in the standard form $2x+y=3$. ◈

2.3.3 Linear Relationships in Statistical Analysis

Linear relationships are important in statistical analysis. For example, if data are obtained from some kind of experiment, it might be the case that a dependent variable (usually y) is observed to be linearly related to the independent variable (usually x). Coordinate pairs corresponding to measured data can be plotted in the Cartesian plane, but the points might not all lie exactly on the same line because of experimental errors or fluctuations. In this case, a straight line, called a *regression line*, can be fitted through the data points in order to determine an approximation of the linear relationship. The best method for fitting the line is to minimize the sum of squares of the vertical distances (i.e., the differences in the y values) of the data points from the fitted line. This is called the *principle of least squares*, and it was proposed by Carl Friedrich Gauss, a famous mathematician of the eighteenth century. The fitted line is called a regression line because the British scientist Francis Galton (1822–1911) used this method in his investigation of the (linear) relationship between the heights of men and their sons (as an example of an hereditary trait), and he noticed that tall men generally had tall sons, but these sons were on average not as tall as their fathers. Galton inferred that there was a regression of the height of tall men toward the average height of all men. He thus named his fitted line a "regression line" and this term is still in use.

EXAMPLE 2.3.8. Here are some data that could have been obtained from an actual population survey:

TABLE 2.1. Heights of fathers and sons

Father's height (inches) x	59	63	67	71	75
Average height of sons (inches) y	62	67	68	70	73

Based on the information given in table 2.1, we might want to predict the average height of sons whose fathers are 70 inches tall. The average of x values (denoted $\bar{x}$) and the average of y values (denoted $\bar{y}$) are:

$$\bar{x}=\frac{59+63+67+71+75}{5}=67, \quad \bar{y}=\frac{62+67+68+70+73}{5}=68. \tag{2.4}$$

According to the method of least squares (the details of which are not give here), the slope m of the regression line is:

$$m=\frac{(59-\bar{x})(62-\bar{y})+(63-\bar{x})(67-\bar{y})+(67-\bar{x})(68-y)+(71-\bar{x})(70-y)+(75-\bar{x})(73-\bar{y})}{(59-\bar{x})^2+(63-\bar{x})^2+(67-\bar{x})^2+(71-\bar{x})^2+(75-\bar{x})^2}$$

$$=\frac{(-8)(-6)+(-4)(-1)+(0)(0)+(4)(2)+(8)(5)}{(-8)^2+(-4)^2+(0)^2+(4)^2+(8)^2}=\frac{100}{68}=\frac{5}{8}=0.625, \tag{2.5}$$

and the equation of the regression line is:

$$y - \overline{y} = m(x - \overline{x}). \tag{2.6}$$

Substituting the value for m from formula (2.5) and the values for $\overline{x}$ and $\overline{y}$ from formula (2.4) into formula (2.6) yields $y-68=\frac{5}{8}(x-67)$. Now, if we set $x=70$ in this equation, we get $y=69.875$, which we take as the predicted value for the average height of sons whose fathers are 70 inches tall. ◈

2.3.4 Parallel and Perpendicular Lines

DEFINITION 2.3.3. *Lines are parallel if and only if they have the same slope.*

In the first diagram in figure 2.8 are parallel lines with slope $m=-\frac{1}{2}$ but with different y-intercepts, $y=-1$ and $y=1$.

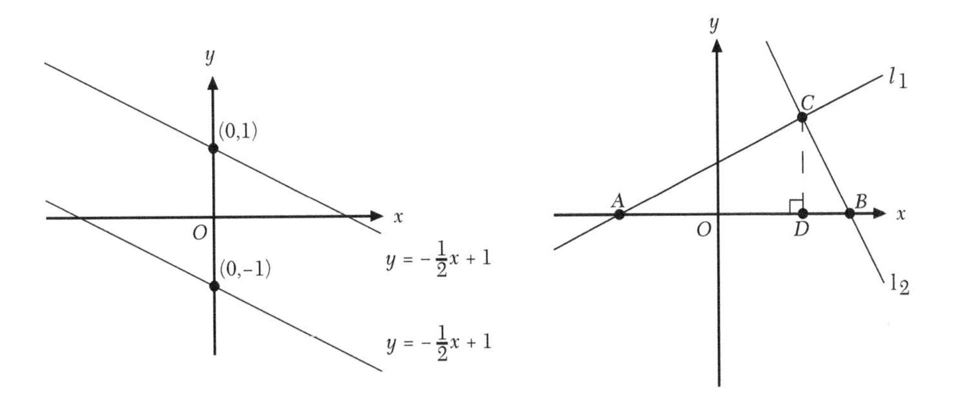

FIGURE 2.8. Parallel and perpendicular lines.

DEFINITION 2.3.4. *Two lines are perpendicular if and only if the slope of one line is the negative reciprocal of the other.*

The reason for this is explained with the help of the second diagram in figure 2.8, in which the two lines labeled l_1 and l_2 are perpendicular and meet at the point C. The x-intercept of l_1 is the point A and the x-intercept of l_2 is the point B. The dotted line drawn from C is perpendicular to the x-axis at the point D. We note that right triangles ACB and CDB have an angle at B in common and right triangles ACB and ADC have an angle at A in common, which means that they are all *similar* triangles, that is, $ACB \,|||\, CDB \,|||\, ADC$. If we denote the length of the line segment from C to D as $|CD|$ and the length of the line segment from A to D as $|AD|$ and so on, then from properties of similar triangles (see theorem 9.5.14(a) and the explanation following it), we can write $\frac{|CD|}{|AD|}=\frac{|BD|}{|CD|}$. In terms of the slopes m_1 and m_2 of lines l_1 and l_2, respectively, we can write this equation as $m_1=-\frac{1}{m_2}$ or, alternatively, as $m_1 m_2 = -1$ (note that m_2 is negative).

EXAMPLE 2.3.9. Find the equation of the line that is perpendicular to the line $y=2x+1$ and that meets this line at its y-intercept.

Answer: The slope of the line we need is $-\frac{1}{2}$, and it should pass through the point $(0,1)$. By substitution into the point-slope form of a line, formula (2.1), we find $y-1=-\frac{1}{2}(x-0)$, which we can also write as $y=-\frac{x}{2}+1$. ◈

EXAMPLE 2.3.10. Find the standard form of the line that passes through the point $(5,-7)$ and is perpendicular to the line $6x+3y=4$.

Answer: We first rewrite the equation for the given line in slope-intercept form. This is $y=-2x+\frac{4}{3}$, and so the slope of this line is -2. Thus, the slope of the line we want is $\frac{1}{2}$. By substituting $m=\frac{1}{2}$ and $(x_1,y_1)=(5,-7)$ into the point-slope form of a line, formula (2.1), we find $y-(-7)=\frac{1}{2}(x-5)$, which we can write in standard form as $x-2y=19$. ◈

As an exercise, carefully draw the graphs of the two lines in the previous example. Try to determine the coordinates of the point where the lines intersect.

2.3.5 The Distance between Points on a Line

If a line is horizontal, it has the equation $y=c$, where c is the y-intercept. The distance between points $P_1(x_1,c)$ and $P_2(x_2,c)$ (denoted $|P_1P_2|$) is the difference in the x-coordinates, that is, $|P_1P_2|=|x_2-x_1|$. (We write this as an absolute value because, in general, x_2 need not be larger than x_1.) If a line is vertical, it has the equation $x=c$, where c is the x-intercept of the line. The distance between points $Q_1(c,y_1)$ and $Q_2(c,y_2)$ is the difference in the y-coordinate, that is, $|Q_1Q_2|=|y_2-y_1|$ (see figure 2.9).

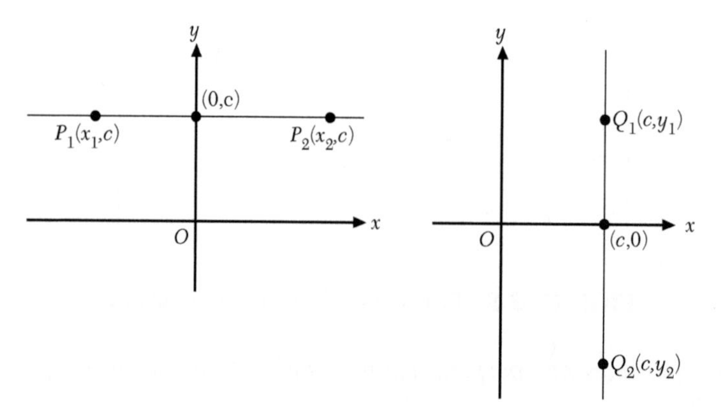

FIGURE 2.9. Horizontal and vertical lines.

If a line is neither vertical nor horizontal, it has the equation $y=mx+c$, where $m\neq 0$. In this case, the distance between points $P_1(x_1,y_1)$ and $P_2(x_2,y_2)$ is determined by means of the Pythagorean theorem (theorem 9.5.8(a)) applied to the right triangle shown in figure 2.10. The length of the horizontal side of the right triangle is $|P_1P|=|x_2-x_1|$, and the length of the vertical side of the right triangle is $|P_2P|=|y_2-y_1|$. Thus, we have:

$$\boxed{\text{The distance formula: } |P_1P_2|=\sqrt{(y_2-y_1)^2+(x_2-x_1)^2}}. \quad (2.7)$$

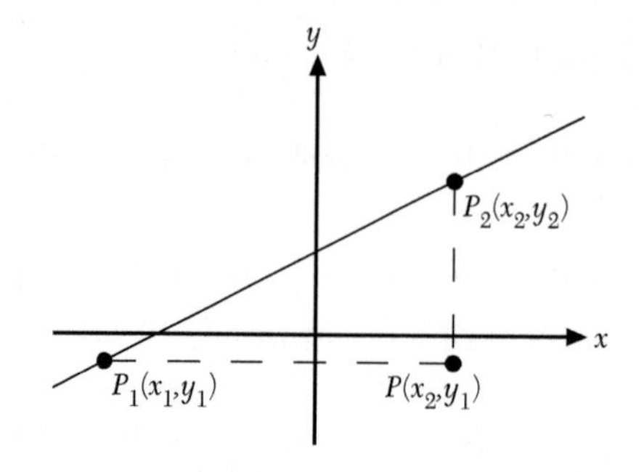

FIGURE 2.10. The distance between two points on a line.

Formula (2.7) is used to find the distance between any two points in the Cartesian plane.

EXAMPLE 2.3.11. The distance between the points $A(-3,6)$ and $B(5,1)$ is
$|AB| = \sqrt{(5-(-3))^2 + (1-6)^2} = \sqrt{64+25} = \sqrt{89} \approx 9.43.$ ◈

2.3.6 The Equation of a Perpendicular Bisector

We find the midpoint of the line segment joining two points $P_1(x_1,y_1)$ and $P_2(x_2,y_2)$ by computing the average of the x- and y-coordinates to produce the point $P\left(\frac{x_1+x_2}{2}, \frac{y_1+y_2}{2}\right)$.

DEFINITION 2.3.5. *The perpendicular bisector of a line segment is the line that passes through the midpoint of the segment and is perpendicular to it.*

EXAMPLE 2.3.12. Given points $A(-3,1)$ and $B(5,4)$, find the equation of the perpendicular bisector l of the line segment AB.

Answer: The coordinates of the midpoint M are $\left(\frac{-3+5}{2}, \frac{1+4}{2}\right) = \left(1, \frac{5}{2}\right)$. The slope of the line through A and B is $\frac{4-1}{5-(-3)} = \frac{3}{8}$. Therefore, the slope of l is $m = -\frac{8}{3}$. We now substitute this value for m and $x_1 = 1$ and $y_1 = \frac{5}{2}$ into the point slope form of a line to obtain $y - \frac{5}{2} = -\frac{8}{3}(x-1)$ or $16x + 6y = 31$ as the equation for l. ◈

2.4 CIRCLES IN THE CARTESIAN PLANE

DEFINITION 2.4.1. *If* C(a,b) *is a point in the Cartesian plane, then the set of all points in the plane that are a distance* r *from* C *determines the graph of a circle with center* C(a,b) *and radius* r *(as shown in figure* 2.11*).*

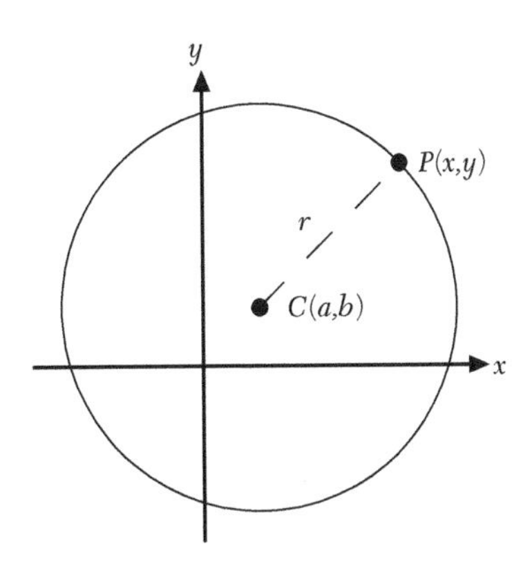

FIGURE 2.11. A circle in the Cartesian plane.

To find the equation of the circle, we use the fact that the distance between any point $P(x,y)$ on the circle and the center $C(a,b)$ of the circle is $|PC| = \sqrt{(x-a)^2 + (y-b)^2}$. Because this distance is the same as the radius of the circle, we obtain:

$$\boxed{\text{The equation of a circle: } (x-a)^2 + (y-b)^2 = r^2}.$$

This is the form in which the equation of a circle is usually expressed. Every point on the circle is a pair (x,y) for which the equation is a true statement and, conversely, any pair (x,y) for which the equation is a true statement is a point on the circle.

EXAMPLE 2.4.1. The circle with its center at the origin and unit radius (i.e., $r = 1$) is commonly referred to as the unit circle. Its corresponding equation is $x^2 + y^2 = 1$. Figure 2.12 shows the unit circle with x- and y-intercepts labeled. ◈

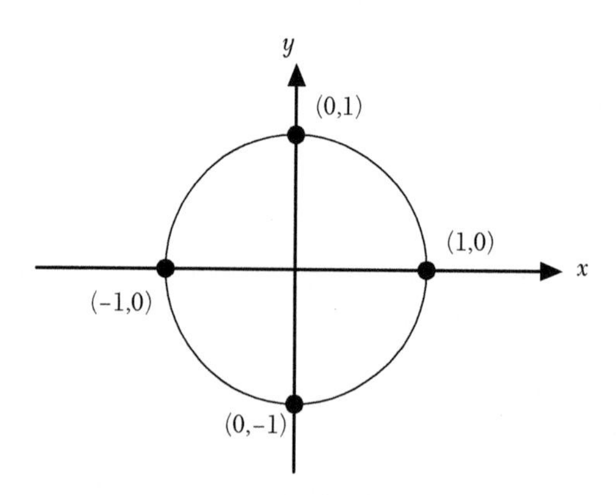

FIGURE 2.12. The unit circle.

EXAMPLE 2.4.2. The equation of the circle centered at $C(-1,1)$ with radius $r = 1$ is $(x-(-1))^2 + (y-1)^2 = 1^2$, which is the same as $(x+1)^2 + (y-1)^2 = 1$. We can expand the squares using FOIL to obtain the equation $x^2 + 2x + y^2 - 2y + 1 = 0$. ◈

EXAMPLE 2.4.3. The equation of the circle centered at $C(2,-2)$ and that passes through the origin is $(x-2)^2 + (y-(-2))^2 = (\sqrt{8})^2$, which is $(x-2)^2 + (y+2)^2 = 8$, because the distance from the center $C(2,-2)$ of the circle to the origin O (the radius of the circle) is $|CO| = \sqrt{2^2 + (-2)^2} = \sqrt{8}$. ◈

It can be verified, by means of a method called *completing the square*, that the graph of an equation of the form $cx^2 + dy^2 + ex + fy = g$, where c, d, e, f, and g are constants and c and d are not zero, is also a circle. This method will be explained fully in section 3.3, but in the meantime we will consider one example.

EXAMPLE 2.4.4. Consider the equation $x^2 + y^2 - 4x + 6y = 3$. This can be expressed in the form $(x^2 - 4x + 4) + (y^2 + 6y + 9) = 16$ by adding 13 to both sides. Each trinomial (an expression consisting of three terms added together) in parentheses on the left side of the equation is a perfect square trinomial, that is, $(x^2 - 4x + 4) = (x-2)^2$ and $(y^2 + 6y + 9) = (y+3)^2$. The equation is now $(x-2)^2 + (y+3)^2 = 4^2$, that is, the graph is a circle with center $C(2,-3)$ and radius $r = 4$. ◈

2.5 CONIC SECTIONS IN THE CARTESIAN PLANE

DEFINITION 2.5.1. *A conic section is the intersection of a plane with a solid (infinite) double cone. As long as the plane does not pass through the vertex of the cone, the boundary of the conic section is an ellipse, a parabola, or a hyperbola.*

Any ellipse, parabola, or hyperbola can be expressed as a Cartesian equation. In their most basic form, these equations are:

- $\boxed{\text{Ellipse: } \frac{x^2}{a^2} + \frac{y^2}{b^2} = 1}$ (a and b are nonzero constants),
- $\boxed{\text{Parabola: } 4py = x^2}$ (p is a nonzero constant),
- $\boxed{\text{Hyperbola: } \frac{x^2}{a^2} - \frac{y^2}{b^2} = 1}$ (a and b are nonzero constants).

The graphs of an ellipse, a parabola, and a hyperbola are shown in figures 2.13 and 2.14. The geometrical properties of each of these curves can be explained in terms of a certain point (or points) called the *focus* (or *foci*).

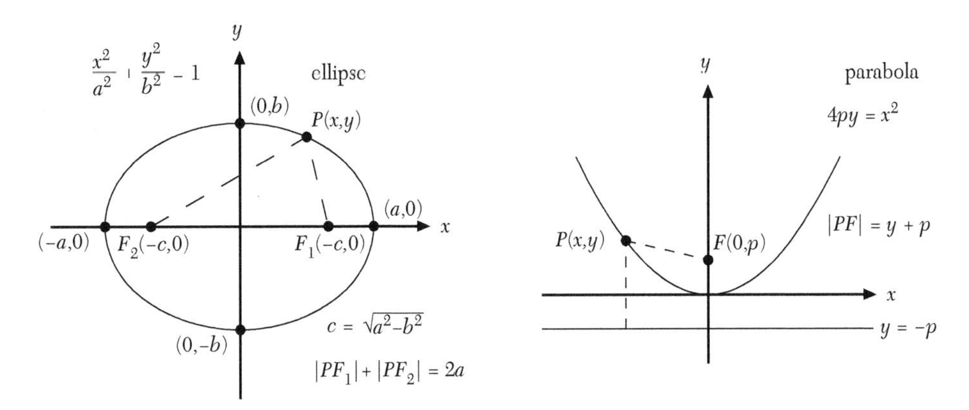

FIGURE 2.13. The ellipse and the parabola.

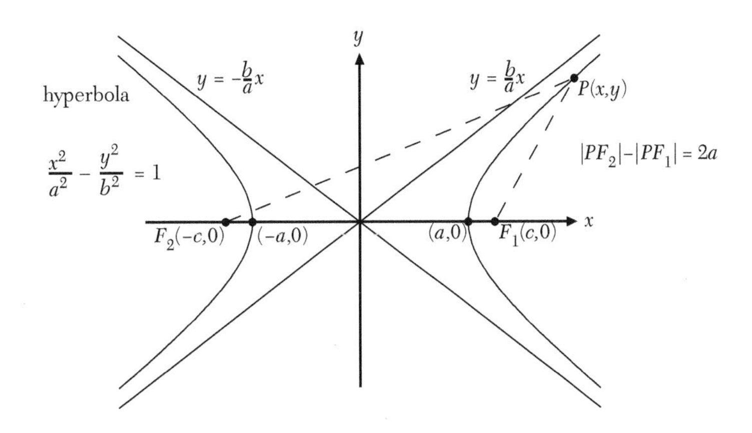

FIGURE 2.14. The hyperbola.

An ellipse, as defined above, has one focus at point $F_1(c,0)$ and another focus at point $F_2(-c,0)$, where $c=\sqrt{a^2-b^2}$. The graph of an ellipse is the set (or graph) of all points $P(x,y)$ in the plane with the property that the sum of the distances from P to each focus is the constant value $2a$, that is, all coordinate pairs $P(x,y)$, such that $|PF_1|+|PF_2|=2a$. The constants a and b, respectively, determine the width and height of the ellipse. Note that if $a=b=1$, then the ellipse is a circle.

The graph of a parabola is the set (or graph) of all points in the plane that are equidistant from the focus $F(0,p)$ and the horizontal line $y=-p$, that is, all coordinate pairs $P(x,y)$, such that $|PF|=y+p$. The constant p determines the width of the parabola.

A hyperbola has two *branches*. The focus of one branch is at the point $F_1(c,0)$ and the focus of the other branch is at the point $F_2(-c,0)$, where $c=\sqrt{a^2+b^2}$. The graph of a hyperbola is the set of all points in the plane with the property that the difference of the distances from the point to each focus is the constant value $\pm 2a$, that is, all coordinate pairs $P(x,y)$, such that $|PF_1|-|PF_2|=\pm 2a$. A point $P(x,y)$ is on the left branch or right branch of the hyperbola depending on whether the plus or the minus sign is taken on the right-hand side of this equation. The two straight lines $y=\pm\frac{b}{a}x$ shown in figure 2.14 are *asymptotes* for the hyperbola, that is, lines that the graph of the hyperbola gradually approaches as x gets larger and larger (in the positive or negative direction).

The conic sections have certain reflection properties. For example, if a light beam is emitted from one focus of an ellipse, it will reflect off the ellipse toward the other focus and telescopes have parabolic light-collecting dishes because any light beam from a distant object entering the dish is reflected toward the focus of the parabola.

EXAMPLE 2.5.1. An interesting historical problem relating to conic sections is the problem of doubling a cube. The Pythagorean mathematicians were perplexed by this problem because it was supposedly given by the Greek god Apollo. The story goes that, in 430 BC, a plague of typhoid fever caused suffering in the city of Athens. The Athenians consulted an oracle about a way to stop the plague. The oracle replied that they had to double the size of Apollo's altar, which was in the shape of a cube. The Athenian mathematicians did not know how to solve the problem so apparently the typhoid pestilence worsened. The Mathematician Hippocrates of Chios observed that the problem was equivalent to the problem of finding two means a and b between two given values p and $2p$ (where p would be the width of the cube), so that $p:a=a:b=b:2p$ (the Greek mathematicians stated many of their theorems in terms of proportions, because they did not have the algebraic symbolism that we use today, for example, these equations state that the proportion of p to a is the same as the proportion of a to b, which is the same as the proportion of b to $2p$.) In our algebraic notation, this statement can be represented as $\frac{p}{a}=\frac{a}{b}=\frac{b}{2p}$. The first equation is equivalent to $a^2=pb$, the second equation is equivalent to $b^2=2ap$; and, after elimination of b, these equations state that $a^3=2p^3$. Hippocrates and his contemporaries could not find a way to construct a and b (given the value of p). A few decades later, a solution to the problem was given by the Mathematician Archytas by means of a three-dimensional construction, involving a cylinder and a cone, but a more elegant two-dimensional solution was later given by Menaechmus (375 BC–325 BC) in terms of two intersecting parabolas in a plane. What he proved precisely, in terms of our notation, is that, if the parabolas $2px=y^2$ and $py=x^2$ are constructed, they will intersect in a coordinate pair $P(a,b)$, where $a^3=2p^3$, as shown in figure 2.15. (In other words, a cube constructed with width a would solve the problem.) Menaechmus was also the first person to classify the conic sections, although the names we use for the conic sections were given by Apollonius (260 BC–200 BC). ◈

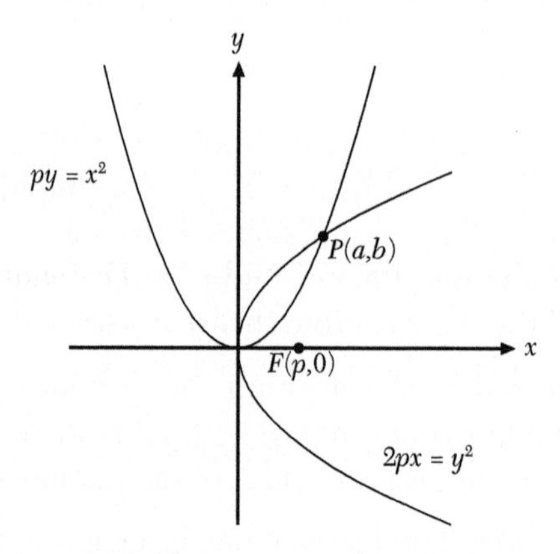

FIGURE 2.15. Doubling a cube.

2.6 VECTOR ALGEBRA

A *vector* is generally known as a quantity that determines a *magnitude* and a *direction*. We prefer the following definition:

DEFINITION 2.6.1. *A vector is one element of a set (a vector space) for which the properties of vector algebra are satisfied.*

The properties of vector algebra relate to the operations of *vector addition* and *scalar multiplication*, and the vector space is closed with respect to these operations, that is, the sum of any two vectors is another vector belonging to the vector space, and a scalar multiple of any vector is another vector belonging to the vector space. The properties of vector algebra are listed formally in table 2.2 in section 2.6.1.

A vector can be represented geometrically by means of an arrow. As an illustration, imagine a car on a racing circuit, with the speed and direction of the car at a position on the circuit represented by means of an arrow whose length is proportional to the speed of the car. Figure 2.16 shows how these arrows could be drawn. The arrows are shorter in the bends of the circuit, where the car slows down, and longer in the straight stretches, where the car can go faster.

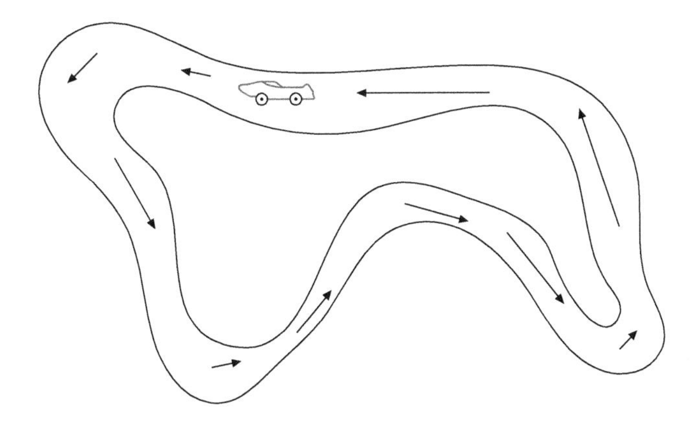

FIGURE 2.16. Vectors in a racetrack.

For the purposes of representing vectors in the Cartesian plane, we will describe a vector by means of a pair of values called *components* (an x *component* and a y *component*). We will also write an arrow above a letter for the variable representation of a vector in order to avoid confusion with a real variable. For example, we write $\vec{a} = \langle 3,2 \rangle$ to denote $\vec{a}$ as a vector with x component equal to 3 and y component equal to 2. We also define the *zero vector* $\vec{0} = \langle 0,0 \rangle$.

DEFINITION 2.6.2. *A representation of a vector* $\vec{a} = \langle a_1, a_2 \rangle$ *is a directed line segment* $\overrightarrow{AB}$ *in the Cartesian plane from any point* $A(x,y)$ *to a point* $B(x+a_1, y+a_2)$. *The point* A *is called the initial point (or tail) and the point* B *is called the terminal point (or tip).*

We talk interchangeably about a "vector" and a "representation of a vector." It is important to understand, however, that there are infinitely many representations possible for the same vector, depending on the point in the plane that is chosen as the initial point.

EXAMPLE 2.6.1. Three representations of the vector $\vec{a} = \langle 3,1 \rangle$ are shown in figure 2.17. The initial points chosen are $A(-3,-2)$, $C(-1,1)$, and $O(0,0)$, with corresponding directed line segments $\overrightarrow{AB}$, $\overrightarrow{CD}$, and $\overrightarrow{OP}$ terminating at points $B(0,-1)$, $D(2,2)$, and $P(3,1)$, respectively. ◈

DEFINITION 2.6.3. *The particular representation* $\overrightarrow{OP}$ *of a vector obtained by choosing the initial point to be the origin* O *and the terminal point to be any point* P *is called the position vector for the point* P.

REMARK 2.6.1. It is clear that there is a position vector associated with every point in the plane.

REMARK 2.6.2. Given any two points $A(x_1, y_1)$ and $B(x_2, y_2)$ in the plane, the directed line segment $\overrightarrow{AB}$ is a representation of the vector $\langle x_2 - x_1, y_2 - y_1 \rangle$ (note that the coordinates of A are subtracted from the coordinates of B).

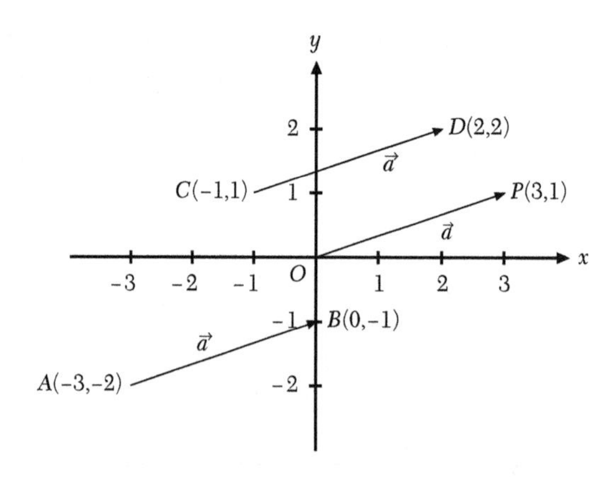

FIGURE 2.17. Representations of a vector.

EXAMPLE 2.6.2. Shown in figure 2.18 are points $A(2,-2)$ and $B(-2,1)$. The directed line segment $\overrightarrow{AB}$ is a representation for the vector $\vec{a} = \langle -2-2, 1-(-2) \rangle = \langle -4, 3 \rangle$. Note that a possible rectangular path from A to B follows a horizontal line four units to the left and then three units up. These directions correspond to the negative and positive signs of the x and y components of $\vec{a}$, respectively. ◈

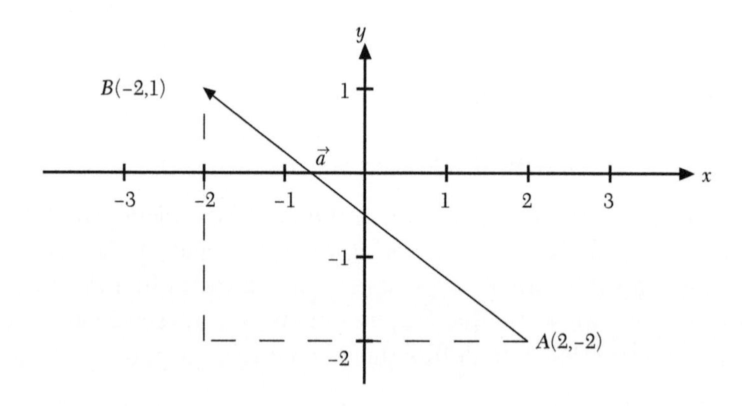

FIGURE 2.18. Representations of a vector.

DEFINITION 2.6.4. *The magnitude of a vector (denoted $|\vec{v}|$ for a vector $\vec{v}$) is the length of the directed line segment of any of its representations.*

The magnitude of a vector is computed using the distance formula, that is, if $\vec{v} = \langle v_1, v_2 \rangle$, then we have the following formula:

The magnitude of a vector: $|\langle v_1, v_2 \rangle| = \sqrt{v_1^2 + v_2^2}$.

EXAMPLE 2.6.3. The magnitude of the vector $\vec{a}$ in example 2.6.2 is $|\vec{a}| = \sqrt{(-4)^2 + (3)^2} = 5$. ◈

2.6.1 Addition and Scalar Multiplication of Vectors

In vector algebra, the sum of two vectors produces another vector. We first look at the geometric definition of vector addition with the help of figure 2.19, in which certain vectors $\vec{u}$ and $\vec{v}$ are drawn so that the tail of $\vec{v}$ is placed at the tip of $\vec{u}$. The sum $\vec{u} + \vec{v}$ is the vector that joins the tail of $\vec{u}$ to the tip of $\vec{v}$. From this, it is clear that vector addition can be understood as *completing a triangle*. We would expect that the commutative property $\vec{u} + \vec{v} = \vec{v} + \vec{u}$ holds. This can also be verified from the diagram. Because, if $\vec{u}$ is added to $\vec{v}$ in the fashion that has been explained, then

$\vec{v}+\vec{u}$ coincides with the vector $\vec{u}+\vec{v}$ as the diagonal of the parallelogram in the diagram. Thus, the commutative property of vector addition can be interpreted geometrically as *the completion of a parallelogram.*

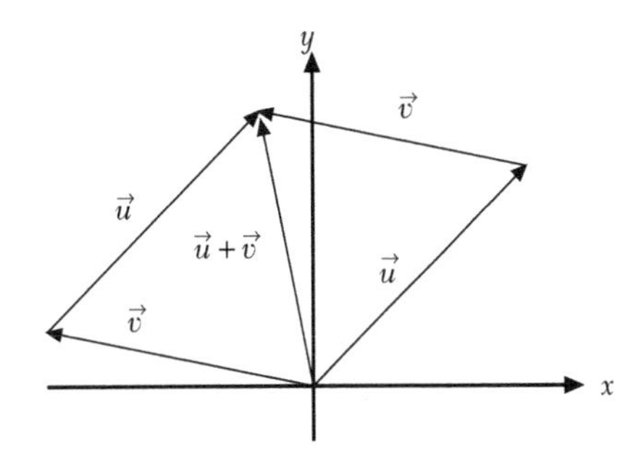

FIGURE 2.19. Addition of vectors.

EXAMPLE 2.6.4. *In physics, two vectors $\vec{u}$ and $\vec{v}$ could represent forces acting simultaneously on some object. The vector $\vec{u}+\vec{v}$ is then called the resultant force.* ◈

In algebraic form, vector addition is the addition of corresponding components, that is, if $\vec{u}=\langle u_1,u_2\rangle$ and $\vec{v}=\langle v_1,v_2\rangle$, then we have:

The addition of two vectors: $\vec{u}+\vec{v}=\langle u_1,u_2\rangle+\langle v_1,v_2\rangle=\langle u_1+v_1,u_2+v_2\rangle$

EXAMPLE 2.6.5. $\langle -1,11\rangle+\langle 7,2\rangle=\langle 6,13\rangle$. ◈

DEFINITION 2.6.5. *The term scalar is used in the context of vector algebra to mean a real number (as opposed to a vector).*

A vector can be multiplied by a scalar. Geometrically, this is interpreted as the lengthening or shortening of the vector or the switching of the direction of the vector, if the scalar is a negative number. For example, if $\vec{u}$ is a vector, then $2\vec{u}$ is the vector that is twice as long (that is., $|2\vec{u}|=2|\vec{u}|$), $\frac{1}{2}\vec{u}$ is half the length of $\vec{u}$, and $-\vec{u}$ points in the opposite direction to $\vec{u}$. These vectors are all shown in figure 2.20. Note that all scalar multiples of $\vec{u}$ have representations contained in the same line.

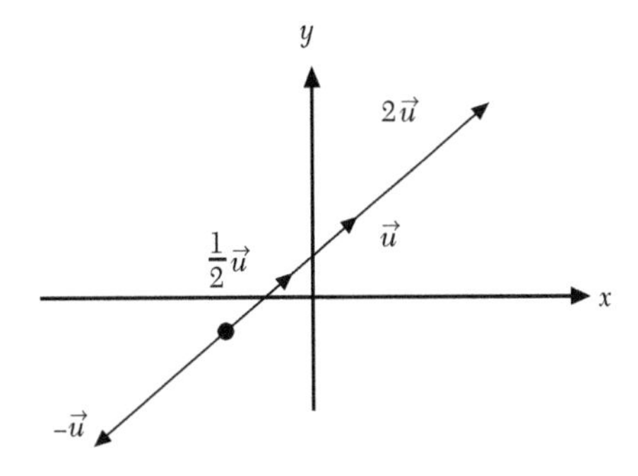

FIGURE 2.20. Scaling of vectors.

In algebraic form, scalar multiplication is multiplication of corresponding components, that is, if $\vec{u} = \langle u_1, u_2 \rangle$ and c is a scalar, then we have:

$$\boxed{\text{A vector multiplied by a scalar: } \quad c\vec{u} = c\langle u_1, u_2 \rangle = \langle cu_1, cu_2 \rangle}.$$

Example 2.6.6. $3\langle -4,6 \rangle = \langle -12,18 \rangle$. ◈

The properties of vector algebra are listed in table 2.2. They resemble the properties of the algebra of real numbers in table 1.3. Note that scalars are represented by the Greek letters α and β. It is left as an exercise to verify that the operations of addition and scalar multiplication do indeed satisfy all of these properties.

TABLE 2.2. The algebra of vectors

(I)	$\vec{u} + \vec{v} = \vec{v} + \vec{u}$	Vector addition is commutative
(II)	$\vec{u} + (\vec{v} + \vec{w}) = (\vec{u} + \vec{v}) + \vec{w}$	Vector addition is associative
(III)	$\vec{u} + 0 = \vec{u}$	The zero vector is a *unique* additive identity
(IV)	$\vec{u} + (-\vec{u}) = \vec{0}$	Every vector has a *unique* additive inverse
(V)	$\alpha(\vec{u} + \vec{v}) = \alpha\vec{u} + \alpha\vec{v}$	Scalar multiplication is distributive
(VI)	$(\alpha + \beta)\vec{u} = \alpha\vec{u} + \beta\vec{u}$	Scalar multiplication is distributive
(VII)	$(\alpha\beta)\vec{u} = \alpha(\beta\vec{u})$	Scalar multiplication is associative
(VIII)	$(1)\vec{u} = \vec{u}$	1 is a *unique* scalar multiplicative identity

2.6.2 Subtraction of Vectors

Because we can add vectors, it should also be possible to subtract vectors. Vector subtraction is done in a way consistent with vector addition. Hence, $\vec{u} - \vec{v}$ is the vector that should be added to $\vec{v}$, so that the result is $\vec{u}$, that is, $\vec{v} + (\vec{u} - \vec{v}) = \vec{u}$. Geometrically, this means that a representation of $\vec{u} - \vec{v}$ completes the triangle formed when representations of $\vec{u}$ and $\vec{v}$ are joined at their initial points, as shown in figure 2.21.

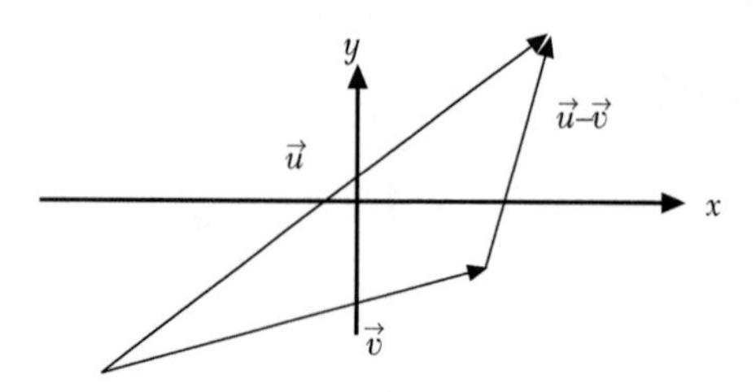

FIGURE 2.21. Subtracting vectors.

In algebraic form, vector subtraction is subtraction of corresponding components, that is, if $\vec{u} = \langle u_1, u_2 \rangle$ and $\vec{v} = \langle v_1, v_2 \rangle$, then:

$$\boxed{\vec{u} - \vec{v} = \langle u_1, u_2 \rangle - \langle v_1, v_2 \rangle = \langle u_1 - v_1, u_2 - v_2 \rangle}.$$

Example 2.6.7. $\langle -1,11 \rangle - \langle 7,2 \rangle = \langle -8,9 \rangle$. ◈

A geometric figure can be analyzed by introducing vectors into the figure and discovering relationships among the vectors according to the geometric properties of vector addition, subtraction, and scalar multiplication. This is called *vector geometry*. The next example is an illustration of the methods of vector geometry.

EXAMPLE 2.6.8. Let R be a point on a line segment PQ that is three times as far from Q as it is from P. If $\vec{u}=\overrightarrow{OP}$, $\vec{v}=\overrightarrow{OQ}$, and $\vec{w}=\overrightarrow{OR}$ prove that $\vec{w}=\frac{3}{4}\vec{u}+\frac{1}{4}\vec{v}$.

Answer: From figure 2.22, it can be deduced that $\vec{w}+\frac{1}{4}(\vec{u}-\vec{v})=\vec{u}$. This equation is equivalent to $\vec{w}=\vec{u}-\frac{1}{4}\vec{u}+\frac{1}{4}\vec{v}$ and simplification of the right-hand side of this equation gives the required formula. ◈

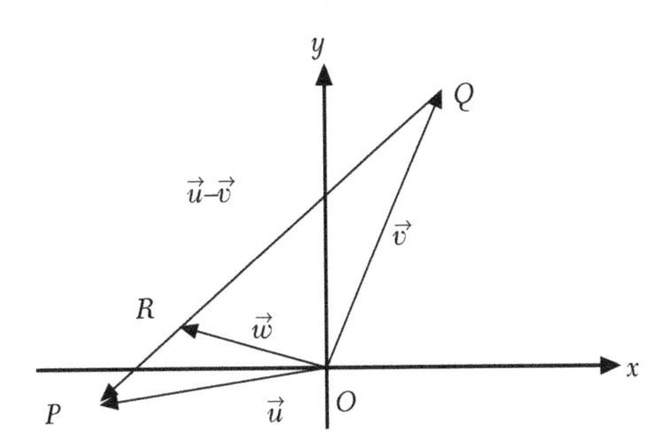

FIGURE 2.22. Vector geometry.

2.6.3 The Standard Basis Vectors

DEFINITION 2.6.6. *The standard basis vectors are the vectors* $\vec{i}=\langle 1,0\rangle$ *and* $\vec{j}=\langle 0,1\rangle$.

Thus, the vector $\vec{i}$ is the position vector for the point $(1,0)$ and the vector $\vec{j}$ is the position vector for the point $(0,1)$ in the Cartesian plane. If $\vec{a}=\langle a_1,a_2\rangle$ is any given vector then, by the properties of scalar multiplication and vector addition, we can write:

$$\begin{aligned}\vec{a}&=\langle a_1,a_2\rangle\\&=a_1\langle 1,0\rangle+a_2\langle 0,1\rangle\\&=a_1\vec{i}+a_2\vec{j}.\end{aligned}$$

Therefore, all vectors in the plane can be expressed in terms of the standard basis vectors $\vec{i}$ and $\vec{j}$.

EXAMPLE 2.6.9. Let $\vec{a}=\vec{i}+\vec{j}$ and $\vec{c}=2\vec{i}-\vec{j}$. Then:

(i) $|\vec{a}|=|\langle 1,1\rangle|=\sqrt{1^2+1^2}=\sqrt{2}$

(ii) $\vec{a}+\vec{c}=3\vec{i}=\langle 3,0\rangle$

(iii) $\vec{a}-\vec{c}=-\vec{i}+2\vec{j}=\langle -1,2\rangle$

(iv) $2\vec{a}=2\vec{i}+2\vec{j}=\langle 2,2\rangle$

(v) $3\vec{a}+4\vec{c}=11\vec{i}-\vec{j}=\langle 11,-1\rangle$ ◈

2.6.4 The Dot Product of Vectors

Vectors cannot be multiplied in the sense that real numbers can be multiplied. Instead, we define the *dot product* as a way to multiply vectors so that the result is a scalar. The usefulness of the dot product can be partly understood from its geometric interpretation, which is presented in section 4.14.2.

DEFINITION 2.6.7. *The dot product* $\vec{u}\cdot\vec{v}$ *of vectors* $\vec{u}$ *and* $\vec{v}$ *is defined by the formula:*

$$\boxed{\vec{u}\cdot\vec{v}=\langle u_1,u_2\rangle\cdot\langle v_1,v_2\rangle=u_1v_1+u_2v_2}.$$

It will suffice for now to give one example of an application in physics, in which the dot product is used.

EXAMPLE 2.6.10. A force measured in Newtons is given by a vector $\vec{F} = 2\vec{i} + 5\vec{j}$ and acts on a particle as it moves from a point $P(2,1)$ to a point $Q(3,6)$ measured in meters. Find the work done and express the answer in joules.

Answer: We first determine the displacement vector $\vec{d}$. This is the directed line segment $\overrightarrow{PQ}$. Thus, $\vec{d} = \langle 3-2, 6-1 \rangle = \langle 1,5 \rangle$. The work done is the dot product of the force and displacement vectors, which is $\vec{F} \cdot \vec{d} = \langle 2,5 \rangle \cdot \langle 1,5 \rangle = 2 \times 1 + 5 \times 5 = 27$ joules. ◈

Table 2.3 provides a list of the algebraic properties of the dot product. Each of these can be proved by writing components for the vectors and comparing the left- and right-hand sides of the equation. As a demonstration, proofs will be given below for properties (I), (III), and (VI).

Property (VI) is one form of the statement of an important and basic inequality known as the *Schwarz inequality*. Here, it states that the *absolute value* of the dot product of any two vectors is always less than or equal to the product of the *magnitudes* of the two vectors.

TABLE 2.3. Properties of the dot product

(I)	$\vec{u} \cdot \vec{u} = \lvert \vec{u} \rvert^2$	The dot product of a vector with itself
(II)	$\vec{u} \cdot \vec{v} = \vec{v} \cdot \vec{u}$	The dot product is commutative
(III)	$\vec{w} \cdot (\vec{u} + \vec{v}) = \vec{w} \cdot \vec{u} + \vec{w} \cdot \vec{v}$	The dot product distributes through addition
(IV)	$(c\vec{u}) \cdot \vec{v} = c(\vec{u} \cdot \vec{v})$	A scalar distributes through a dot product
(V)	$\vec{0} \cdot \vec{a} = 0$	A dot product with the zero vector gives zero
(VI)	$\lvert \vec{u} \cdot \vec{v} \rvert \le (\lvert \vec{u} \rvert)(\lvert \vec{v} \rvert)$	The Schwarz inequality

PROOF OF I. Let $\vec{u} = \langle u_1, u_2 \rangle$, then $\vec{u} \cdot \vec{u} = u_1^2 + u_2^2 = \lvert \vec{u} \rvert^2$. □

PROOF OF III. Let $\vec{u} = \langle u_1, u_2 \rangle$, $\vec{v} = \langle v_1, v_2 \rangle$, and $\vec{w} = \langle w_1, w_2 \rangle$, then:

$$\begin{aligned}
\vec{w} \cdot (\vec{u} + \vec{v}) &= \langle w_1, w_2 \rangle \cdot \langle u_1 + v_1, u_2 + v_2 \rangle \\
&= w_1(u_1 + v_1) + w_2(u_2 + v_2) \\
&= w_1u_1 + w_2u_2 + w_1v_1 + w_2v_2 \\
&= \vec{w} \cdot \vec{u} + \vec{w} \cdot \vec{v}.
\end{aligned}$$

□

PROOF OF VI. Let $\vec{u} = \langle u_1, u_2 \rangle$ and $\vec{v} = \langle v_1, v_2 \rangle$, then:

$$\begin{aligned}
\lvert \vec{u} \cdot \vec{v} \rvert^2 &= \lvert \langle u_1, u_2 \rangle \cdot \langle v_1, v_2 \rangle \rvert^2 \\
&= (u_1v_1 + u_2v_2)^2 \\
&= u_1^2v_1^2 + 2(u_1v_2)(u_2v_1) + u_2^2v_2^2 \\
&\le u_1^2v_1^2 + u_1^2v_2^2 + u_2^2v_1^2 + u_2^2v_2^2 \\
&= (u_1^2 + u_2^2)(v_1^2 + v_2^2) \\
&= \lvert \vec{u} \rvert^2 \lvert \vec{v} \rvert^2.
\end{aligned}$$

Note that the inequality in the fourth step above follows from the fact that $2(u_1v_2)(u_2v_1) \leq u_1^2v_2^2 + u_2^2v_1^2$, which is a consequence of the inequality $0 \leq (y-x)^2$ because, after expanding the square on the right-hand side, we get $2xy \leq x^2 + y^2$, for any two real numbers x and y. □

The next example is a demonstration of the Schwarz inequality.

EXAMPLE 2.6.11. Verify the Schwarz inequality for vectors $\vec{u} = \langle -1, 2 \rangle$ and $\vec{v} = \langle 3, -4 \rangle$.

Answer: We first calculate the value on the left side of the inequality:

$$|\vec{u} \cdot \vec{v}|^2 = |\langle -1, 2 \rangle \cdot \langle 3, -4 \rangle|^2 = (-1 \times 3 + 2 \times (-4))^2 = (-11)^2 = 121.$$

Now, we calculate the value on the right side of the inequality:

$$|\vec{u}|^2 |\vec{v}|^2 = |\langle -1, 2 \rangle|^2 \times |\langle 3, -4 \rangle|^2 = \left((-1)^2 + 2^2\right)\left(3^2 + (-4)^2\right) = 5 \times 25 = 125.$$

Because $121 < 125$, the inequality is verified. ◈

2.6.5 The Triangle Inequality and the Parallelogram Law

We prove two geometric statements by application of the properties of the dot product. The first of these is the *triangle inequality*, which is the statement that the length of one side of a triangle in the plane is less than or equal to the sum of the lengths of the other two sides. In vector form, this is:

$$\boxed{\text{The triangle inequality: } |\vec{u} + \vec{v}| \leq |\vec{u}| + |\vec{v}|}$$

for any vectors $\vec{u}$ and $\vec{v}$.

PROOF OF THE TRIANGLE INEQUALITY.

$$\begin{aligned} |\vec{u} + \vec{v}|^2 &= (\vec{u} + \vec{v}) \cdot (\vec{u} + \vec{v}) \\ &= \vec{u} \cdot \vec{u} + 2\vec{u} \cdot \vec{v} + \vec{v} \cdot \vec{v} \\ &\leq |\vec{u}|^2 + 2|\vec{u} \cdot \vec{v}| + |\vec{v}|^2 \\ &\leq |\vec{u}|^2 + 2|\vec{u}||\vec{v}| + |\vec{v}|^2 \\ &= (|\vec{u}| + |\vec{v}|)^2, \end{aligned}$$

where the first inequality is a consequence of the fact that $\vec{u} \cdot \vec{v} \leq |\vec{u} \cdot \vec{v}|$ (any number is smaller than or equal to its absolute value), and the second inequality is an application of the Schwarz inequality to the middle term of the expression. □

The *parallelogram law* states that the sum of the squares of the lengths of the diagonals of a parallelogram is equal to the sum of the squares of the lengths of the sides. In vector form, this is:

$$\boxed{\text{The parallelogram law: } |\vec{u} + \vec{v}|^2 + |\vec{u} - \vec{v}|^2 = 2|\vec{u}|^2 + 2|\vec{v}|^2}$$

for any vectors $\vec{u}$ and $\vec{v}$.

The proof of the parallelogram law is exercise 2.40 below.

EXERCISES

2.1. Plot the points $P_1(4,5)$, $Q(-4,5)$, $R(-4,-5)$, and $S(4,-5)$ in the Cartesian plane.

2.2. Which of the following are linear equations? (In (e) and (f) assume that m is an unknown constant.)

(a) $3-x=4y+1$

(b) $xy=7x+1$

(c) $x-y=7x+1$

(d) $y^2=x^2+49$

(e) $y=mx^2+11$

(f) $x=\frac{y}{m}+1$

2.3. Which of the following pairs of coordinates is not a solution of $y=13x-4$?

(a) $(4,48)$

(b) $(7,87)$

(c) $(13,165)$

(d) $(23,295)$

(e) $(29,375)$

2.4. Draw lines in the Cartesian plane passing through each of the pairs of points P and Q.

(a) $P(4,2)$ and $Q(-11,-2)$

(b) $P(-11,2)$ and $Q(4,-2)$

(c) $P(2,4)$ and $Q(-2,-11)$

(d) $P(-2,2)$ and $Q(11,-11)$

2.5. Draw the graphs of the following linear equations by finding the x- and y-intercepts.

(a) $y=2x+1$

(b) $y=3x-2$

(c) $y=-4x+3$

(d) $y=-\frac{1}{3}x+6$

2.6. Find the slope of each of the lines in Exercise 2.4 above.

2.7. Find the slope of the line passing through points (r,s) and $(r+s,2s)$. Simplify your answer.

2.8. Find the equation for a line with slope $\frac{2}{3}$ and passing through the point $(3,0)$.

2.9. Find the equation for a line with slope $-\frac{3}{2}$ and passing through the point $(2,-1)$.

2.10. Write each of the linear equations in Exercise 2.2 in slope-intercept form. What is the y-intercept of each of these lines?

2.11. Write each of the linear equations in Exercise 2.10 in standard form.

2.12. Find the standard equation of the line in each of the following cases.

(a) The line passes through the origin and the point $(1,1)$.

(b) The line is horizontal and the y-intercept is $(0,2)$.

(c) The line has slope $m=-7$ and passes through the y-axis at $(0,7)$.

(d) The line has slope $m=-7$ and passes through the x-axis at $(7,0)$.

2.13. Suppose that a population survey of mothers and daughters obtains the data:

TABLE 2.4. Heights of mothers and daughters

Mother's height (inches) x	57	61	65	69	73
Average height of daughters (inches) y	59	62	66	69	71

Under the assumption of a linear relationship in the heights of mothers and their daughters, use formulas (2.4)–(2.6) to find the equation for the regression line that fits the data in table 2.4. Predict the average height of daughters whose mothers are 67 inches tall.

2.14. Decide whether the following pairs of linear equations correspond to lines that are parallel, perpendicular, or neither.

(a) $\{y = 2x + 1,\ y - 2 = \frac{1}{2}x\}$

(b) $\{2y - x = 3,\ y + 2x = -4\}$

(c) $\{y - 2x = 3,\ y + 2x = -4\}$

(d) $\{y - x = 6,\ y = x + 4\}$

2.15. If l is a line determined by the linear equation $y = -\frac{1}{2}x + 3$, then

(a) Write the equation for a line parallel to l, with y-intercept $(0,-1)$.

(b) Write the equation for a line perpendicular to l and passing through the origin.

2.16. Find the distance between the following pairs of points A and B.

(a) $A(4,2)$ and $B(4,3)$

(b) $A(-7,2)$ and $B(-7,-2)$

(c) $A(2,4)$ and $B(3,5)$

(d) $A(-2,2)$ and $B(7,11)$

2.17. Relating to points $A(-3,4)$, $B(0,5.5)$, $C(5,1.5)$, and $D(4,-2)$, answer the following questions.

(a) Which point is closest to the origin?

(b) Which two points are closest to each other?

(c) Which two points are farthest from each other?

2.18. Determine the coordinates at which the following pairs of lines intersect.

(a) $\{y = 2,\ x = -4\}$

(b) $\{y - x = 3,\ y + x = 3\}$

(c) $\{y = x,\ y = 2x\}$

(d) $\{x = 6,\ y = 3x - 4\}$

(e) $\{x = y,\ y = 3x - 4\}$

2.19. Given the points A and B, find the equation of the perpendicular bisector l of the line segment AB.

(a) $A(-2,3)$ and $B(4,4)$

(b) $A(6,1)$ and $B(3,-1)$

2.20. Find (the coordinates of) two points on the unit circle with

(a) $x = \frac{\sqrt{2}}{2}$

(b) $x = \frac{\sqrt{3}}{4}$

2.21. Show that the equation of the circle with center $C(-2,3)$ and passing through the point $D(4,5)$ is $(x+2)^2+(y-3)^2=40$.

2.22. Find the equation of the circle in each of the following cases:

(a) center $C(2,0)$ and radius $r=\sqrt{5}$

(b) center $C(-1,1)$ and passing through the origin

(c) passing through points $(-1,0)$, $(1,0)$, and $(0,\sqrt{3})$ (Hint: find the circumcenter of the triangle with vertices at these three points; see section 9.5.4)

2.23. Find the center and radius of a circle with equation $x^2+2x+y^2-2y=2$.

2.24. Does the parabola $4py=x^2$ become wider or narrower as the value of p increases?

2.25. Verify that all points $P(x,y)$ on the graph of the ellipse $\frac{x^2}{a^2}+\frac{y^2}{b^2}=1$ satisfy the property $|PF_1|+|PF_2|=2a$, where F_1 and F_2 are the foci of the ellipse.

2.26. Verify that all points $P(x,y)$ on the graph of the parabola $4py=x^2$ satisfy the property $|PF|=|y+p|$, where F is the focus of the parabola.

2.27. Verify that all points $P(x,y)$ on the graph of the hyperbola $\frac{x^2}{a^2}-\frac{y^2}{b^2}=1$ satisfy the property $|PF_1|-|PF_2|=\pm 2a$, where F_1 and F_2 are the foci of the hyperbola.

2.28. Sketch the graphs of the following conic sections in the Cartesian plane. In each case, label the focus or foci and the x- and y-intercepts and draw the asymptotes, if appropriate.

(a) $\frac{x^2}{9}+\frac{y^2}{4}=1$

(b) $16y=x^2$

(c) $\frac{x^2}{9}-\frac{y^2}{4}=1$

(d) $25y^2=25-x^2$

(e) $x^2-y^2=1$

(f) $49y^2=196+4x^2$

2.29. Plot the following vectors all with initial point $A(1,-1)$.

(a) $\vec{a}=\langle 2,0\rangle$

(b) $\vec{b}=\langle 3,-1\rangle$

(c) $\vec{c}=\langle -1,1\rangle$

(d) $\vec{d}=\langle 1,-1\rangle$

(e) $\vec{e}=\langle 0,-4\rangle$

(f) $\vec{f}=\langle -3,0\rangle$

2.30. Find the components of the directed line segment $\overrightarrow{AB}$ in each of the following cases.

(a) $A(-2,2)$, $B(0,4)$

(b) $A(2,2)$, $B(-2,-2)$

(c) $A(0,1)$, $B(4,0)$

2.31. Examine the first vector diagram in figure 2.23 and answer the following questions.

(a) Write the components for the vectors $\vec{u}$, $\vec{v}$, $\vec{w}$, $\vec{u}+\vec{v}$, $\vec{u}+\vec{v}+\vec{w}$, and $\vec{v}+\vec{w}$.

(b) Compute the following values: $|\vec{x}|$, $|\vec{y}|$ and $\vec{x}\cdot\vec{y}$.

(c) If vectors $\vec{u}$, $\vec{v}$, $\vec{w}$, $\vec{x}$, $\vec{y}$, and $\vec{z}$ represent forces acting simultaneously, determine a vector $\vec{r}$ so that the resultant force is $\vec{0}$.

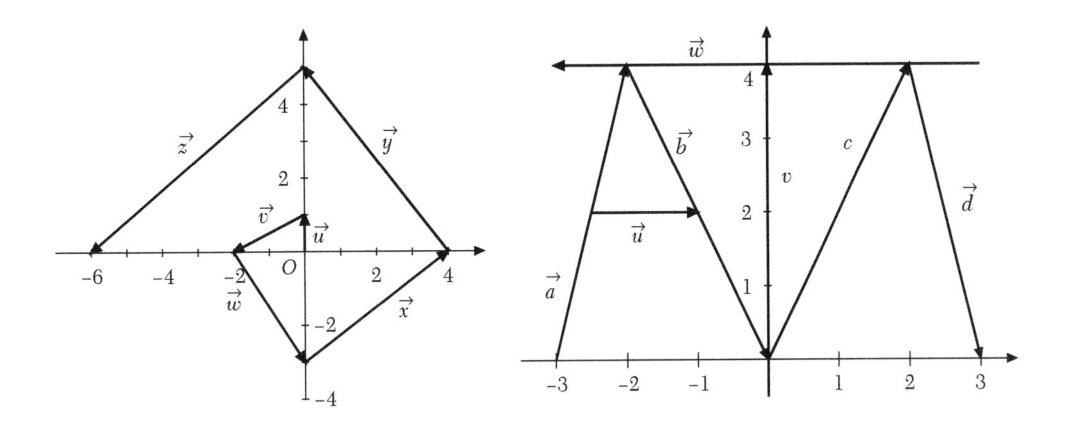

FIGURE 2.23. Vector diagrams.

2.32. Examine the second vector diagram in figure 2.23 and answer the following questions.

(a) Write the components for the vectors $\vec{a}$, $\vec{b}$, $\vec{v}$, $\vec{w}$, $\vec{a}+\vec{b}$, and $-\vec{c}-\vec{d}$.

(b) Compute the values $\vec{b}\cdot\vec{v}$, $\vec{u}\cdot\vec{v}$, $|\vec{a}|$, and $|\vec{w}|$.

(c) If vectors $\vec{a}$, $\vec{b}$, $\vec{c}$, $\vec{d}$, and $\vec{w}$ represent forces acting simultaneously on an object, draw a representation of vector $\vec{w}$ in the diagram to show that the resultant force is $\vec{0}$.

(d) Verify the parallelogram law using vectors $\vec{c}$ and $\vec{d}$.

2.33. Draw (accurately!) the three vectors $\vec{u}=\langle 5,2\rangle$, $\vec{v}=\langle 4,-1\rangle$, and $\vec{w}=\langle 8,2\rangle$. Show, by means of a sketch that there are scalars s and t so that $\vec{w}=s\vec{u}+t\vec{v}$ and find the exact values of s and t. (Hint: use scalars s and t to stretch or shrink the vectors $\vec{u}$ and $\vec{v}$ by precisely the right amount so that vectors $s\vec{u}$ and $t\vec{v}$ complete a parallelogram with $\vec{w}$ as a diagonal.)

2.34. In figure 2.24, add the vectors $\vec{a}=\langle -3,-3\rangle$, $\vec{b}=\langle -1,4\rangle$, $\vec{c}=\langle 7,0\rangle$, $\vec{d}=\langle 1,1\rangle$, $\vec{e}=\langle 2,-1\rangle$, $\vec{f}=\langle -3,-2\rangle$, $\vec{g}=\langle -1,-2\rangle$, and $\vec{h}=\langle -2,3\rangle$ by forming a chain starting at the origin (the vectors $\vec{a}$ and $\vec{b}$ have already been added).

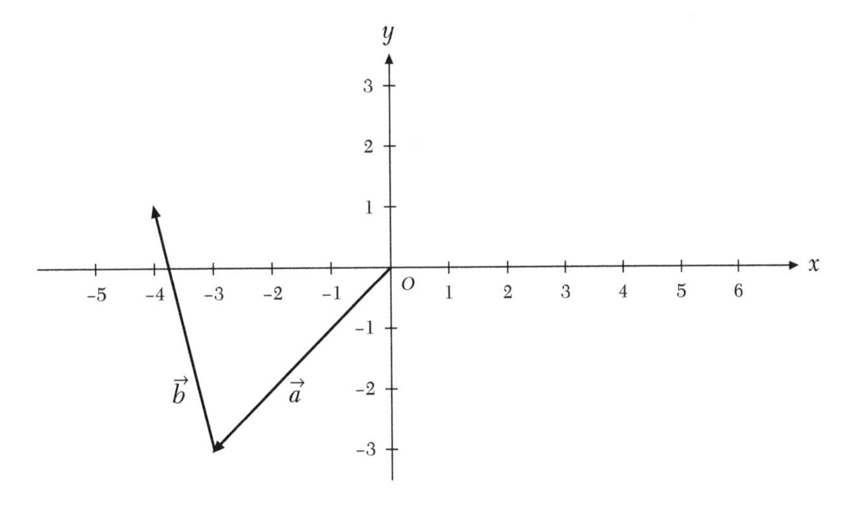

FIGURE 2.24. Plotting vectors.

2.35. Draw a vector diagram that verifies the associativity of vector addition (property (II) in the list of properties of vector algebra).

2.36. A constant force $\vec{F} = 3\vec{i} - 4\vec{j}$ measured in Newtons moves an object along a straight line from a point $P(2,3)$ to a point $Q(1,-2)$ measured in meters. Find the work done in joules by $\vec{F}$.

2.37. Verify the Schwarz inequality for vectors $\vec{a} = \langle 5,-3\rangle$ and $\vec{b} = \langle -2,3\rangle$.

2.38. Draw a vector diagram from which the vector statement of the triangle inequality can be deduced.

2.39. Draw a vector diagram from which the vector equation for the parallelogram law can be deduced.

2.40. Prove the parallelogram law using the properties of the dot product.

2.41. Suppose that $|\vec{u}| = |\vec{v}|$ for vectors $\vec{u}$ and $\vec{v}$ shown in figure 2.25.

(a) Prove, using methods of vector geometry, that the vector $\vec{w}$ that bisects the angle between $\vec{u}$ and $\vec{v}$ can be expressed as $\vec{w} = \frac{\vec{u}}{2} + \frac{\vec{v}}{2}$

(b) Find a formula for the area of the large triangle in terms of the magnitudes of the sum and difference of the vectors $\vec{u}$ and $\vec{v}$.

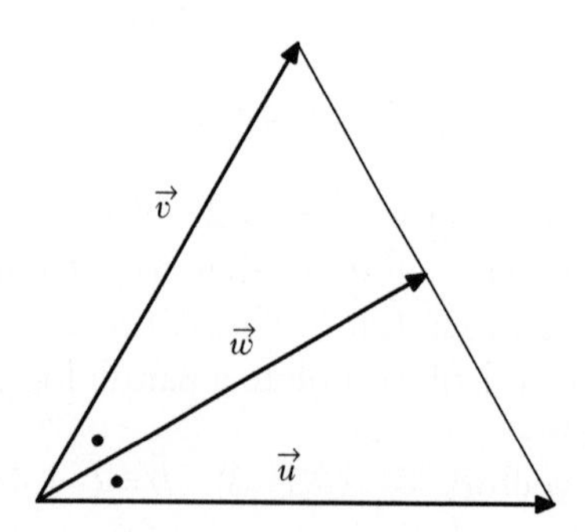

FIGURE 2.25. A vector geometry problem.

CHAPTER 3

SOLVING EQUATIONS AND FACTORIZING POLYNOMIALS

3.1 INTRODUCTION

Chapters 1 and 2 deal with the description of algebraic objects like numbers, sets, equations, and vectors as well as algebraic operations like addition, multiplication, the dot product for vectors, and so on. In this chapter, we are going to discover some of the useful techniques of algebra. The word *algebra* derives from the Arabic word *al-jabr*, which roughly translated means "transposition" or "reduction" and which was part of the title of a mathematical treatise written by the Arabic scholar Mohammed ibn Mûsâ al-Khowârizmî during the eighth century AD. Europeans who read this treatise in later centuries adopted this word as the name for the science of equations. What al-Khowarizmi meant by al-jabr is the method of adding or subtracting the same quantity to both sides of an equation. This is part of the process of *solving an equation*.

The notion of solving an equation by the application of formal techniques is usually inculcated into school children soon after the age of 10. It is, therefore, surprising that mathematicians did not fully employ these techniques until the sixteenth century. One reason for this is that solving an equation such as

$$x+3=0,$$

leads immediately to the problem of the existence of negative numbers. Indeed, the solution of this equation is obtained by subtracting 3 on each side of the equation

$$x+3-3=0-3,$$

with the result

$$x=-3.$$

Mathematicians were reluctant for a long time to acknowledge the existence of negative numbers. For example, René Descartes referred to the negative solutions of equations as "false roots," and he never extended his coordinate system (our Cartesian plane) to negative numbers.

An even more, difficult problem arises with the solutions of the equation

$$x^2+1=0,$$

which we will discuss later.

Another reason for the slow development of algebra, compared with that of geometry and trigonometry, was the cumbersome algebraic notation that mathematicians used for a long time. To give an example, a statement in our notation such as

$$3ba^2 - 2ba + a^3 = d,$$

would have been written by François Viète, the famous French mathematician of the sixteenth century, in the form

$$b\ 3 \text{ in } a \text{ quad} - b\ 2 \text{ in } a + a \text{ cub aequator } d \text{ solido.}$$

It is hard for us to imagine doing algebra with this laborious way of writing, yet the great Italian algebraists of the sixteenth century, such as Scipione del Ferro, Nicolo of Brescia (also known as Tartaglia), Girolamo Cardano, Lodovico Ferrari, and Rafael Bombelli, managed to find methods for solving cubic and quartic equations (see section 3.9 below) using this kind of notation.

The usual practice of using the initial letters of the alphabet $a,b,c,\ldots$ to denote constants and the last letters of the alphabet $\ldots,x,y,z$ to denote variables was a custom fixed by Descartes, and the use of the = sign became common practice after it was used by the English Mathematician Robert Recorde in 1557. (The symbol $\neq$ can also be used to state that two quantities are not equal to each other.)

This chapter deals with the general problem of solving a polynomial equation. We begin with the problem of solving a linear equation in section 3.2 and then move on to the problem of solving a quadratic equation in section 3.3. Following this, the basic algebraic properties of polynomials (including the Remainder Theorem and Factor Theorem) are introduced in section 3.4. A full treatment of the algebraic properties (including graphs) of quadratic polynomials is given in section 3.5.

Because the most general solution of a polynomial equation is a complex number (as explained in section 3.7), complex numbers are introduced in section 3.6. The approach taken in section 3.6 is to define complex numbers concretely as elements of a special class of 2×2 matrices. This requires an introduction to the algebra of matrices (which is very useful to know, in any case).

The important theorems of this chapter, namely the Fundamental Theorem of Algebra, the Complete Factorization Theorem for Polynomials, and the theorem that every polynomial with real coefficients and positive degree can be expressed as a product of linear and irreducible quadratic factors with real coefficients, are all stated in section 3.7. A few basic comments regarding the graphs of polynomials are made in section 3.8.

Some students might find section 3.9 rather challenging. However, it is natural to move on to the problem of solving cubic, quartic, and quintic equations. In fact, the study of algebra is not complete without it!

3.2 SOLVING LINEAR EQUATIONS

Our plan of action is to begin in the most basic situation, which is to solve a linear equation. Here is a simple problem that leads to a linear equation:

EXAMPLE 3.2.1. What number when multiplied by 3 is equal to that number plus 8?

Answer: If we denote the unknown value as x, then the problem can be restated in the form: for what value of x is the equation

$$3x = x + 8$$

a true statement?

The answer can be obtained by solving the equation. The way to do this is to isolate x on one side of the equation by applying the same algebraic operations to each side of the equation. First, subtract x from each side of the equation:

$$3x - x = (x + 8) - x.$$

Each side of the equation can be simplified, and the result is

$$2x = 8.$$

Next, we divide each side of the equation by 2:

$$\frac{2x}{2} = \frac{8}{2}$$

to obtain the answer

$$x = 4.$$

◈

Problems of this type have been solved since the times of the earliest known human civilizations. Here is a problem from the *Rhind papyrus*, an Egyptian mathematical text containing 85 problems copied by the scribe Ahmes in about 1650 BC.

Example 3.2.2. A quantity, its 2/3, its 1/2, and its 1/7, added together, becomes 33. What is the quantity?

Answer: We solve this once again, by denoting the unknown quantity as x and writing the statement of the problem as an equation:

$$x + \frac{2x}{3} + \frac{x}{2} + \frac{x}{7} = 33.$$

The way to proceed from here is a matter of choice, but the easiest way to isolate x is to multiply each side of the equation by the lowest common multiple of 2, 3, and 7, which is 42. Thus,

$$42\left(x + \frac{2x}{3} + \frac{x}{2} + \frac{x}{7}\right) = 42(33).$$

By expansion of the left side of the equation, this becomes

$$42x + 28x + 21x + 6x = 1{,}386$$

which simplifies to

$$97x = 1{,}386.$$

The final answer is

$$x = \frac{1{,}386}{97}$$

◈

Some equations might not be linear to begin with but after some simplifications or reductions have been carried out, the answer can be obtained by solving a linear equation. Here are two examples of this:

Example 3.2.3. Solve for x in the equation $(3x - 2)(8x + 4) = (6x + 1)(4x - 3)$.

Answer: Expand each side using the FOIL method and then subtract $24x^2$ from each side. This results in a linear equation that can be solved for x:

$$\begin{aligned}24x^2-4x-8&=24x^2-14x-3\\-4x-8&=-14x-3\\10x-8&=-3\\10x&=5\\x&=\frac{1}{2}.\end{aligned}$$

◈

EXAMPLE 3.2.4. Solve for x in the equation $\dfrac{1}{4x-2}-\dfrac{3}{x+5}=\dfrac{2}{2x-1}$.

Answer: To begin with, the fractions on the left side can be added in the same way that ordinary fractions can be added, that is, by finding their lowest common denominator (see section 1.13.4), as shown in the following steps:

$$\begin{aligned}\frac{1}{4x-2}\left(\frac{x+5}{x+5}\right)-\frac{3}{x+5}\left(\frac{4x-2}{4x-2}\right)&=\frac{2}{2x-1},\\\frac{x+5}{2(2x-1)(x+5)}-\frac{12x-6}{2(2x-1)(x+5)}&=\frac{2}{2x-1},\\\frac{x+5-(12x-6)}{2(2x-1)(x+5)}&=\frac{2}{2x-1}\left(\frac{x+5}{x+5}\right),\\\frac{11-11x}{2(2x-1)(x+5)}&=\frac{2(2x+10)}{2(2x-1)(x+5)}.\end{aligned}$$

Two fractions with the same denominator are equal if and only if their numerators are equal, and therefore, we can write down the linear equation containing only the numerators on each side. The value for x that solves this equation will be the solution for the original equation (provided it does not result in any zero denominators in the fractions).

$$\begin{aligned}11-11x&=4x+20,\\-11x-4x&=20-11,\\-15x&=9,\\x&=-\frac{9}{15}=-\frac{3}{5}.\end{aligned}$$

◈

3.3 SOLVING QUADRATIC EQUATIONS BY COMPLETING THE SQUARE

The next level of difficulty with solving equations arises from the problem of solving a general *quadratic equation*.

DEFINITION 3.3.1. *A quadratic equation is an equation of the type*

$$ax^2+bx+c=0,$$

where a solution is required for x *in terms of the constants* a, b, *and* c.

The special technique that makes it possible to solve for x is called *completing the square*. It is helpful to look at a geometric interpretation of this in the special case in which we want to solve for x in the simplified quadratic equation

$$x^2 + bx = 0.$$

If b and x are positive real numbers, then we can regard the value $x^2 + bx$ as the area of a rectangular sheet with the short side of length x and the long side of length $x + b$, as shown in figure 3.1. The long side can then be divided into one segment of length x and two segments of length $\frac{b}{2}$. This creates a square with side length x and two small rectangles with dimensions $x \times \left(\frac{b}{2}\right)$ inside the rectangular sheet. The lower of these small rectangles can be cut off and pasted alongside the sheet so that its side of length x matches the square (as shown in second diagram). In this way, a new square of side length $x \times \left(\frac{b}{2}\right)$ can be formed if a missing square (shaded in third diagram) of dimensions $\left(\frac{b}{2}\right) \times \left(\frac{b}{2}\right)$ is filled in. We now conclude that the area of the original sheet can be expressed as the area of the new (biggest) square minus the area of the small shaded square. Mathematically stated, this is the identity

$$\boxed{x^2 + bx = \left(x + \frac{b}{2}\right)^2 - \left(\frac{b}{2}\right)^2}. \tag{3.1}$$

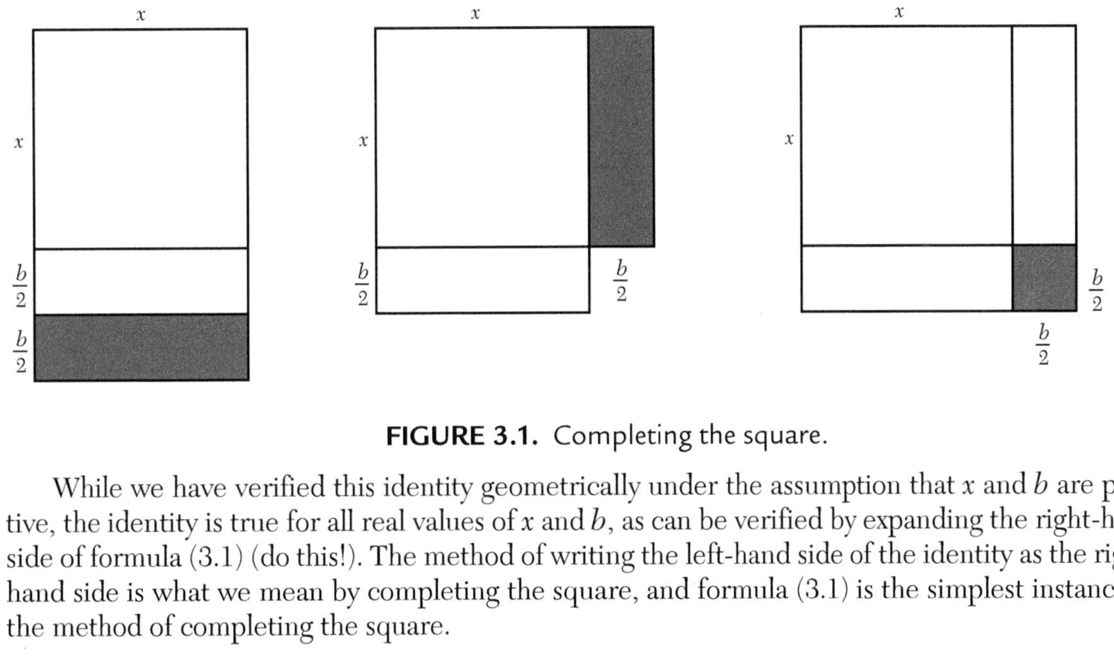

FIGURE 3.1. Completing the square.

While we have verified this identity geometrically under the assumption that x and b are positive, the identity is true for all real values of x and b, as can be verified by expanding the right-hand side of formula (3.1) (do this!). The method of writing the left-hand side of the identity as the right-hand side is what we mean by completing the square, and formula (3.1) is the simplest instance of the method of completing the square.

EXAMPLE 3.3.1. Complete the square for each of the following expressions:

(i) $x^2 + 3x$

(ii) $x^2 - 10x$.

Answers:

(i) $x^2 + 3x = \left(x + \frac{3}{2}\right)^2 - \left(\frac{3}{2}\right)^2 = \left(x + \frac{3}{2}\right)^2 - \frac{9}{4}$

(ii) $x^2 - 10x = (x - 5)^2 - 25.$ ◈

The general method of completing the square requires the preliminary step of factoring the coefficient of x^2 in order to isolate an expression of the form that appears on the left-hand side of the identity in formula (3.1), as in the following examples.

EXAMPLE 3.3.2. Complete the square for each of the following expressions:

(i) $3x^2 + 12x - 1$

(ii) $-4x^2 - 13x + 7$

Answers:

(i) $3x^2 + 12x - 1 = 3\left[x^2 + 4x\right] - 1 = 3\left[(x+2)^2 - 4\right] - 1 = 3(x+2)^2 - 13$

(ii) $$-4x^2 - 13x + 7 = -4\left[x^2 + \frac{13}{4}x\right] + 7 = -4\left[\left(x + \frac{13}{8}\right)^2 - \frac{169}{64}\right] + 7$$
$$= -4\left(x + \frac{13}{8}\right)^2 + \frac{169}{16} + 7$$
$$= -4\left(x + \frac{13}{8}\right)^2 + \frac{281}{16}.$$

◈

We now derive the general formula for completing the square of a quadratic expression:

$$ax^2 + bx + c = a\left[x^2 + \frac{b}{a}x\right] + c$$
$$= a\left[\left(x + \frac{b}{2a}\right)^2 - \frac{b^2}{4a^2}\right] + c$$
$$= a\left(x + \frac{b}{2a}\right)^2 - \frac{b^2}{4a} + \frac{4ac}{4a}$$
$$= a\left(x + \frac{b}{2a}\right)^2 - \frac{b^2 - 4ac}{4a}.$$

We record this result as the following identity:

$$\boxed{\text{Completing the square: } ax^2 + bx + c = a\left(x + \frac{b}{2a}\right)^2 - \frac{b^2 - 4ac}{4a}}. \qquad (3.2)$$

We are now prepared to solve quadratic equations by completing the square.

EXAMPLE 3.3.3. Solve the following equations by completing the square:

(i) $3x^2 + 12x - 1 = 0$

(ii) $-4x^2 - 13x + 7 = 0$

Answers: We rewrite each equation by completing the square of the quadratic expression on the left-hand side of each equation, as in example 3.3.2.

(i) $3(x+2)^2 - 13 = 0$

(ii) $-4\left(x + \frac{13}{8}\right)^2 + \frac{281}{16} = 0$

Now, in order to solve (i), we add 13 to each side of the equation and then divide both sides by 3, resulting in the new equation $(x+2)^2 = \frac{13}{3}$. This states that the square of the quantity $x+2$ is equal to $\frac{13}{3}$. What does this mean? It means that there are two possibilities: $x+2$ is equal to $\sqrt{\frac{13}{3}}$ or $x+2$ is equal to $-\sqrt{\frac{13}{3}}$. Therefore, the final solution is

$$x = -2 + \sqrt{\frac{13}{3}} \quad \text{or} \quad x = -2 - \sqrt{\frac{13}{3}}.$$

Similarly, the solution for (ii) is

$$x = -\frac{-13}{8} + \sqrt{\frac{281}{64}} \quad \text{or} \quad x = \frac{-13}{8} - \sqrt{\frac{281}{64}},$$

which can also be expressed as

$$x = \frac{-13 + \sqrt{281}}{8} \quad \text{or} \quad x = \frac{-13 - \sqrt{281}}{8}.$$

◈

In the next example, we will encounter a difficulty that did not arise in example 3.3.3.

Example 3.3.4. Solve for x in the equation $x^2 + 3x + 4 = 0$.

Answer: After completing the square, the equation can be expressed as

$$\left(x + \frac{3}{2}\right)^2 = -\frac{7}{4},$$

and here is the difficulty: solving this equation would require us to take the square root of a negative real number. Do not feel alone if you find this very troubling. Mathematicians were alarmed by this for thousands of years. Their solution was to ignore it and leave the equation unsolved. The resolution of this difficulty leads into new realms of algebra that mathematicians began to explore only about 300 years ago. Some of this will be explained in section 3.6. For now, we will proceed formally and take the square root of $-\frac{7}{4}$ in order to write the solution:

$$x = -\frac{3}{2} + \sqrt{-\frac{7}{4}} \quad \text{or} \quad x = -\frac{3}{2} - \sqrt{-\frac{7}{4}}.$$

If we allow the usual algebraic processes for taking the square roots of fractions, that is, $\sqrt{\frac{-7}{4}} = \frac{\sqrt{-7}}{\sqrt{4}} = \frac{\sqrt{-7}}{2}$, then the solution simplifies to

$$x = \frac{-3 + \sqrt{-7}}{2} \quad \text{or} \quad x = \frac{-3 - \sqrt{-7}}{2}.$$

These numbers involving $\sqrt{-7}$ are called *complex numbers*. A convenient way to write complex numbers is to use the notation $\boldsymbol{i} = \sqrt{-1}$. Then, we can write $\sqrt{-7} = \sqrt{-1}\sqrt{7} = \boldsymbol{i}\sqrt{7}$ and so the solution above can be expressed as

$$x = \frac{-3 + i\sqrt{7}}{2} \quad \text{or} \quad x = \frac{-3 - i\sqrt{7}}{2}.$$

◈

Complex numbers and the notation **i** will be explained further in section 3.6.

We now derive the general formula for the solution of a quadratic equation, known as the *quadratic formula*. In order to solve the equation $ax^2+bx+c=0$, we replace the left-hand side of the equation with the expression from formula (3.2). Thus, we need to solve the equation

$$a\left(x+\frac{b}{2a}\right)^2-\frac{b^2-4ac}{4a}=0.$$

As in the examples above, this means that

$$x=\frac{-b}{2a}+\sqrt{\frac{b^2-4ac}{4a^2}} \quad \text{or} \quad x=\frac{-b}{2a}-\sqrt{\frac{b^2-4ac}{4a^2}}.$$

The more usual way to write this is to use the $\pm$ notation, which means $+$ or $-$, that is,

$$\boxed{\text{The quadratic formula: } x=\frac{-b\pm\sqrt{b^2-4ac}}{2a}}. \tag{3.3}$$

The quadratic formula provides a shortcut for solving a quadratic equation. All one has to do is identify the values of a, b, and c with the coefficient of x^2, the coefficient of x and the constant term, respectively, and substitute them into the quadratic formula.

REMARK 3.3.1. The quantity b^2-4ac that appears inside the square root in formula (3.3) is called the *discriminant*. More about this later.

EXAMPLE 3.3.5. We can solve the equation $-4x^2-13x+7=0$ by substituting $a=-4$, $b=-13$, and $c=7$ in the quadratic formula, that is,

$$x=\frac{-(-13)\pm\sqrt{(-13)^2-4(-4)(7)}}{2(-4)}=-\frac{13\pm\sqrt{281}}{8}. \qquad \diamond$$

3.4 POLYNOMIALS

Polynomials and the algebraic operations that can be applied to them are the main concern of this chapter. We will begin with the formal definition of a polynomial (in one variable) and then explain how polynomials can be added, subtracted, multiplied, and divided.

DEFINITION 3.4.1. *A polynomial in* x *is a sum of the form*

$$a_nx^n+a_{n-1}x^{n-1}+\cdots+a_1x+a_0,$$

where n *is a nonnegative integer and each coefficient* a_j *for* j = 1, … n *is a real number. If* a_n *is not equal to zero, then the polynomial is said to be of degree* n *and* a_n *is called the leading coefficient of the polynomial.*

EXAMPLE 3.4.1. In table 3.1 are a few polynomials and their corresponding degrees.

TABLE 3.1. Polynomials

Polynomial	Degree
$3x^4+5x^3-7x+4$	4
$-3x^3+x$	3
$-x^2+\sqrt{2}x+1$	2
$11x+\pi$	1
2	0

◈

DEFINITION 3.4.2. *Polynomials of degree 2, 3, 4, and 5 are called quadratic, cubic, quartic, and quintic polynomials, respectively.*

DEFINITION 3.4.3. *A polynomial with two terms is called a binomial, and a polynomial with three terms is called a trinomial.*

EXAMPLE 3.4.2. $x-3x^3$ is a binomial, and $1+\sqrt{2}x-x^2$ is a trinomial. ◈

3.4.1 Addition and Subtraction of Polynomials

Polynomials are added or subtracted by adding like powers of x. Remember your school teacher saying "add apples to apples, bananas to bananas", and so on!

EXAMPLE 3.4.3.

(i) $(x^4+\sqrt{6}x^2-4x+5)+(2x^3+x^2-3x-1)=x^4+2x^3+(\sqrt{6}+1)x^2-7x+4$

(ii) $(x^4+\sqrt{6}x^2-4x+5)-(2x^3+x^2-3x-1)=x^4-2x^3+(\sqrt{6}-1)x^2-x+6$ ◈

3.4.2 Multiplication of Polynomials

We have already seen an example of polynomial multiplication using the FOIL rule in section 1.8.1. In general, multiplication of polynomials uses the same distributive property. Examine the following examples.

EXAMPLE 3.4.4.

(i) $(4x+5)(3x-1)=4x(3x-1)+5(3x-1)=12x^2-4x+15x-5=12x^2+11x-5$

(ii) $(2x^2+4x+5)(3x-1)=2x^2(3x-1)+4x(3x-1)+5(3x-1)=6x^3+10x^2+11x-5$

(iii) $(x^3+2x)^2=x^6+2(x^3)(2x)+4x^2=x^6+4x^4+4x^2$ ◈

REMARK 3.4.1. The word expand is used synonymously with multiply; so, in the previous example, if we expand $(4x+5)(3x-1)$, then the result will be $12x^2+11x-5$.

3.4.3 Polynomials in More Than One Variable

We have introduced polynomials in one variable (the variable x). Next, we give an example of a polynomial in *two* variables (x and y). The order of terms can be given in descending powers of x, descending powers of y, or in descending sums of the powers of x and y. The choice of the order of terms is a matter of preference.

EXAMPLE 3.4.5.

$$2x^4y-6x^3y^4+2x^2-3xy^3+6y-21=-6x^3y^4-3xy^3+2x^4y+6y+2x^2-21$$
$$=-6x^3y^4+2x^4y-3xy^3+2x^2+6y-21 \qquad \diamond$$

The expansion of the polynomials in table 3.2 should be memorized. The reason for memorizing them is that they are used for factorizing polynomials (the topic of section 3.5.3).

TABLE 3.2. Products of polynomials

Product	**Name**
$(x+y)(x-y)=x^2-y^2$	Difference of two squares
$(x+y)^2=x^2+2xy+y^2$	Perfect square trinomial
$(x-y)^2=x^2-2xy+y^2$	Perfect square trinomial
$(x+y)(x^2-xy+y^2)=x^3+y^3$	Sum of two cubes
$(x-y)(x^2+xy+y^2)=x^3-y^3$	Difference of two cubes
$(x+y)^3=x^3+3x^2y+3xy^2+y^3$	Perfect cube
$(x-y)^3=x^3-3x^2y+3xy^2-y^3$	Perfect cube

The expansion of $(x+y)^n$, for any natural number n, is called a *binomial expansion of degree n*, and the coefficients of such an expansion can be determined by writing $n+1$ rows of *Pascal's triangle* (named after the French Mathematician Blaise Pascal, who lived from 1623 to 1662).

EXAMPLE 3.4.6. Expand $(x+y)^5$.

Answer: We write six rows of Pascal's triangle as follows:

```
                1
             1     1
          1     2     1
       1     3     3     1
    1     4     6     4     1
 1     5    10    10     5     1
```

Note that the number 1 is the only number in the first row of Pascal's triangle. The numbers in the second row are two 1's placed diagonally either side of the number 1 in the first row. After that, each row begins and ends with a 1 and, in between them, each number in the row is the sum of the two numbers diagonally above it in the previous row. For example, in the sixth row the numbers are 1, 5, 10, 10, 5, and 1, for which 5 = 4 + 1, 10 = 4 + 6, and so on. Now, with reference to table 3.2, note that the numbers 1, 2, 1 in the third row and the numbers 1, 3, 3, 1 in the fourth row are the coefficients of the terms of the binomial expansions of degrees 2 and 3, respectively. Similarly, the numbers 1, 4, 6, 4, 1 are the coefficients of the binomial expansion of degree 4 (do the expansion!), and the numbers 1, 5, 10, 10, 5, 1 are the coefficients of the binomial expansion of degree 5. Furthermore, when we write the expansion of $(x+y)^5$, the powers of x start from 5 and decrease to 0 while the corresponding powers of y start from 0 and increase to 5. Thus, the expansion of $(x+y)^5$ is

$$(x+y)^5=\boxed{1}x^5y^0+\boxed{5}x^4y^1+\boxed{10}x^3y^2+\boxed{10}x^2y^3+\boxed{5}x^1y^4+\boxed{1}x^0y^5$$

or

$$(x+y)^5 = x^5 + 5x^4y + 10x^3y^2 + 10x^2y^3 + 5xy^4 + y^5. \qquad \diamond$$

The binomial expansions such as those given in table 3.2 can be used as templates for the expansions of binomials, in general, as we now demonstrate.

EXAMPLE 3.4.7. The expansion of $(2a-3)^4$ can be obtained from the expansion of $(x+y)^4$ by replacing x with $2a$ and replacing y with -3. Thus

$$\begin{aligned}(2a-3)^4 &= ((2a)+(-3))^4 \\ &= (2a)^4 + 4(2a)^3(-3) + 6(2a)^2(-3)^2 + 4(2a)(-3)^3 + (-3)^4 \\ &= 16a^4 - 96a^3 + 216a^2 - 216a + 81. \end{aligned} \qquad \diamond$$

3.4.4 Long Division of Polynomials

The division of a polynomial by another polynomial with the same or smaller degree can be done by means of the same method of long division that is used for dividing a natural number by a smaller natural number. This is demonstrated in the next example, in which the dividend is the polynomial x^4+4x^2-16, the divisor is the polynomial x^2+3x+1, and the quotient is the polynomial $x^2-3x+12$. The process of long division continues until a remainder, which is a polynomial with a smaller degree than the divisor, is reached. In the example, the remainder is $-33x-28$. At each stage of the process, a term is added to the quotient (the top row) which, when multiplied by the divisor, results in a polynomial with the same leading term as the polynomial in the previous row. When these polynomials are subtracted, the new polynomial has a smaller degree. The next term in the dividend is carried down and added to this polynomial, in preparation for the next stage. In the final answer, the dividend is expressed as the product of the divisor and the quotient, plus the remainder.

EXAMPLE 3.4.8. Divide x^4+4x^2-16 by x^2+3x+1.

Answer:

$$\begin{array}{r|l}
 & x^2 \quad -3x \quad +12 \\ \hline
x^2+3x+1 & x^4 + 0x^3 + 4x^2 + 0x - 16 \\
 & \underline{x^4 + 3x^3 + x^2} \ \downarrow \ \downarrow \\
 & \quad -3x^3 + 3x^2 + 0x \\
 & \quad \underline{-3x^3 - 9x^2 - 3x} \downarrow \ \downarrow \\
 & \qquad\qquad 12x^2 + 3x - 16 \\
 & \qquad\qquad \underline{12x^2 + 36x + 12} \\
 & \qquad\qquad\qquad -33x - 28
\end{array}$$

Therefore,

$$x^4 + 4x^2 - 16 = (x^2 - 3x + 12)(x^2 + 3x + 1) - 33x - 28$$

or

$$\frac{x^4 + 4x^2 - 16}{x^2 + 3x + 1} = (x^2 - 3x + 12) + \frac{-33x - 28}{x^2 + 3x + 1}. \qquad \diamond$$

In the special case of division of polynomials in which the divisor is a linear polynomial of the form $x-c$, the method of synthetic division is a quicker method to carry out the long division. The method of synthetic division uses a table with three rows. The first element in the first row is the number c. The remaining entries of the first row are the coefficients of the dividend, including zero coefficients, if there are any. The first coefficient is carried into the third row of the table. It is then multiplied by c, and the result is placed in the second row in the next column. The values in the first two rows of the third column are added, and the result is placed below them in the third row. The process is repeated until the last column is reached. The final value of the third row is the remainder, and the values in the third row (except the remainder) are the coefficients for the descending powers of x of the quotient. The next example will make this clear.

EXAMPLE 3.4.9. Divide x^3-5x^2+x-3 by $x-2$ using the method of long division and the method of synthetic division.

Answer 1 (the method of long division):

$$
\begin{array}{r|l}
 & x^2 - 3x - 5 \\
\cline{2-2}
x-2 & x^3 - 5x^2 + x - 3 \\
 & x^3 - 2x^2 \\
\cline{2-2}
 & -3x^2 + x \\
 & -3x^2 + 6x \\
\cline{2-2}
 & -5x - 3 \\
 & -5x + 10 \\
\cline{2-2}
 & -13.
\end{array}
$$

Therefore,

$$x^3-5x^2+x-3=(x-2)(x^2-3x-5)-13$$

or

$$\frac{x^3-5x^2+x-3}{x-2}=x^2-3x-5+\frac{-13}{x-2}.$$

Answer 2 (the method of synthetic division):

The divisor is $x-2$ and the coefficients of the dividend, corresponding to descending powers of x, are 1, –5, 1, and –3. Therefore, the table for synthetic division is

2	1	–5	1	–3
		2	–6	–10
	1	–3	–5	–13

The coefficients of the quotient, corresponding to descending powers of x, starting with the coefficient of x^2 are 1, –3, and –5. The remainder is –13. ◈

EXAMPLE 3.4.10. Divide $2x^5 - x^3 + 50x^2 + 8$ by $x+3$ using the method of synthetic division.

Answer: We write $x+3$ as $x-(-3)$; then the table is

–3	2	0	–1	50	0	8
		–6	18	–51	3	–9
	2	–6	17	–1	3	–1

Therefore,

$$2x^5 - x^3 + 50x^2 + 8 = (x+3)(2x^4 - 6x^3 + 17x^2 - x + 3) - 1$$

or

$$\frac{2x^5 - x^3 + 50x^2 + 8}{x+3} = 2x^4 - 6x^3 + 17x^2 - x + 3 - \frac{1}{x+3}.$$ ◈

3.4.5 The Remainder Theorem and the Factor Theorem

The method of division of polynomials, as described above, reveals, upon closer examination, something interesting about polynomials. To see this, it is helpful to write a general expression for division of polynomials as

$$\boxed{f(x) = p(x)q(x) + r(x)}, \tag{3.4}$$

where (as in the division of numbers) $f(x)$ is the dividend, $p(x)$ is the divisor, $q(x)$ is the quotient, and $r(x)$ is the remainder (these are all polynomials). We will assume that $p(x)$ is not a constant, then $\deg(p) > \deg(r)$, where $\deg(p)$ is the degree of p, and so on.

REMARK 3.4.2. We are using function notation here. This has not been introduced yet, but the meaning here is clear. All we need to understand here is that $f(2)$, for instance, is the evaluation of the polynomial $f(x)$ when x is replaced with the value 2.

In the case that the divisor $p(x)$ is a linear polynomial of the form $x-c$, then $r(x)$ is a constant (a polynomial of degree zero) and then, instead, we write

$$f(x) = (x-c)q(x) + r.$$

If we replace the value x with c in this equation, then

$$f(c) = (c-c)q(c) + r = 0q(c) + r = r,$$

that is, $f(c) = r$. This useful observation is as follows:

THEOREM 3.4.1. *The Remainder Theorem: If a polynomial $f(x)$ is divided by a linear polynomial $x-c$, then the remainder is $f(c)$.*

EXAMPLE 3.4.11. Use the Remainder Theorem to find the remainder, if $f(x) = 2x^5 - x^3 + 50x^2 + 8$ is divided by $x+3$.

Answer: The remainder was found in example 3.4.10; however, according to the Remainder Theorem, we can also get the remainder this way:

$$f(-3) = 2(-3)^5 - (-3)^3 + 50(-3)^2 + 8 = -486 + 27 + 450 + 8 = -1,$$

that is, the remainder is -1. ◈

It can happen that the division of one polynomial by another leaves zero remainder. In the case that division is by a linear polynomial $x-c$, we have (as above)

$$f(x)=(x-c)q(x)+0=(x-c)q(x).$$

The polynomial $x-c$ is now called a *linear factor* of the polynomial $f(x)$.

EXAMPLE 3.4.12. Show that $x+3$ is a linear factor of $2x^5-x^3+50x^2+9$.

Answer: This is just a variation of the previous example. The synthetic division table is

–3	2	0	–1	50	0	9
		–6	18	–51	3	–9
	2	–6	17	–1	3	0

Instead of doing synthetic division, we can verify this using the Remainder Theorem:

$$f(-3)=2(-3)^5-(-3)^3+50(-3)^2+9=-486+27+450+9=0.$$ ◈

We have just seen that, if division of a polynomial $f(x)$ by $x-c$ leaves zero remainder, then $x-c$ is a linear factor of $f(x)$. The converse statement is also true, that is, if we know that $x-c$ is a linear factor of a polynomial $f(x)$, then the division by $x-c$ will leave zero remainder (convince yourself!).

THEOREM 3.4.2. *Factor Theorem: A polynomial $f(x)$ has the factor $(x-c)$ if and only if $f(c)=0$.*

The Factor Theorem is a basic tool for *factorizing* polynomials. This is the topic of section 3.5.3, but here is a preview:

EXAMPLE 3.4.13. Find a linear factor of the polynomial $f(x)=3x^4+8x^3-2x^2-10x+4$.

Answer: According to the Factor Theorem, we need to find a value c such that $f(c)=0$. Here are the values $f(1)$, $f(-1)$, $f(2)$, and $f(-2)$:

$$\begin{aligned}
f(1)&=3+8-2-10+4=3\\
f(-1)&=3(-1)^4+8(-1)^3-2(-1)^2-10(-1)+4=3-8-2+10+4=7\\
f(2)&=3(2)^4+8(2)^3-2(2)^2-10(2)+4=48+64-8-20+4=88\\
f(-2)&=3(-2)^4+8(-2)^3-2(-2)^2-10(-2)+4=48-64-8+20+4=0.
\end{aligned}$$

Therefore, a linear factor of $f(x)$ is $x+2$. ◈

DEFINITION 3.4.4. *For a polynomial* f(x)*, a value* c *for which* f(c) = 0 *is called a root of the polynomial.*

REMARK 3.4.3. The quadratic formula (formula (3.3)) is an expression for the roots of a quadratic polynomial.

3.5 THE PROPERTIES OF QUADRATIC POLYNOMIALS

We will now examine quadratic polynomials in detail, because they are a special class of polynomials.

3.5.1 The Graphs of Quadratic Polynomials

In section 2.5, a parabola is defined as the graph (in the Cartesian plane) of the equation $4py = x^2$ $\left(\text{or } y = \frac{1}{4p}x^2\right)$ for any nonzero real number p. This is, in fact, the general equation for a parabola that passes through the origin and is symmetric with respect to the y-axis, that is, the *axis of symmetry* of the parabola is the y-axis. The most general equation for a parabola is the quadratic equation

$$y = ax^2 + bx + c,$$

where $a \neq 0$ and b and c are real numbers. It is not hard to see why this is true because the equation can be expressed in the completed square form

$$y = a\left(x + \frac{b}{2a}\right)^2 - \frac{b^2 - 4ac}{4a}.$$

Evidently, the graph of this equation is a parabola that, together with its axis of symmetry, is shifted $\frac{b}{2a}$ units according to the following formula.

$$\boxed{\text{The axis of symmetry of a parabola: } x = -\frac{b}{2a}}. \tag{3.5}$$

Furthermore, for this value of x, y attains the value $-\frac{b^2 - 4ac}{4a}$, which is the minimum possible value for y if a is positive, or the maximum possible value for y if a is negative. The coordinates at which y attains its maximum or minimum value is called

$$\boxed{\text{The turning point of a parabola: } \left(-\frac{b}{2a}, -\frac{b^2 - 4ac}{4a}\right)}. \tag{3.6}$$

EXAMPLE 3.5.1. The graphs of six parabolas and their corresponding quadratic equations are shown in figure 3.2. In table 3.3, each polynomial is expressed in completed square form (column 3). Also given in the table are the axis of symmetry (column 4), the roots (column 5), the turning point (column 6), and the value of the discriminant (column 7). Recall that the discriminant is the value $b^2 - 4ac$ that occurs inside the square-root term in the quadratic formula. The symbol Δ is usually used to denote the value of the discriminant. ◈

TABLE 3.3. The properties of quadratic polynomials

	Quadratic polynomials	Square completed	Axis	Roots	T.p.	Δ
(i)	$2x^2 - 4x - 6$	$2(x-1)^2 - 8$	$x = 1$	$\{-1, 3\}$	$(1, -8)$	64
(ii)	$-x^2 - 4x - 4$	$-(x+2)^2$	$x = -2$	$\{-2, -2\}$	$(-2, 0)$	0
(iii)	$6x^2 - x - 15$	$6\left(x - \frac{1}{12}\right)^2 - \frac{361}{24}$	$x = \frac{1}{12}$	$\left\{-\frac{3}{2}, \frac{5}{3}\right\}$	$\left(\frac{1}{12}, -\frac{361}{24}\right)$	361
(iv)	$x^2 - x - 11$	$\left(x - \frac{1}{2}\right)^2 - \frac{45}{4}$	$x = \frac{1}{2}$	$\left\{\frac{1 \pm \sqrt{45}}{2}\right\}$	$\left(\frac{1}{2}, -\frac{45}{4}\right)$	45
(v)	$-x^2 - 2$	$-x^2 - 2$	$x = 0$	$\{\pm i\sqrt{2}\}$	$(0, -2)$	−8
(vi)	$x^2 + x + \frac{5}{2}$	$\left(x + \frac{1}{2}\right)^2 + \frac{9}{4}$	$x = -\frac{1}{2}$	$\left\{\frac{-1 \pm 3i}{2}\right\}$	$\left(-\frac{1}{2}, \frac{9}{4}\right)$	−9

We will make some remarks about table 3.3.

REMARK 3.5.1. Note that the root "−2" in row (ii) in table 3.3 is listed twice. The reason for this is that a quadratic polynomial must have two roots. (In this case, we say that the polynomial $-x^2-4x-4$ has a double root.) We see that the corresponding parabola turns on the x-axis.

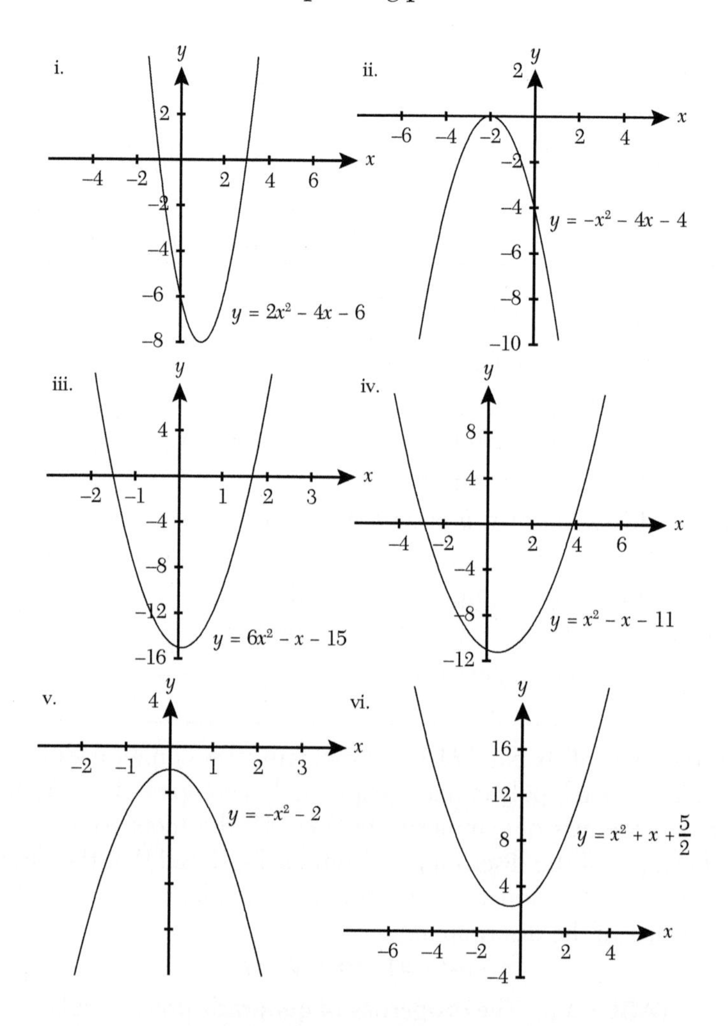

FIGURE 3.2. Six parabolas.

REMARK 3.5.2. If the leading coefficient (the coefficient of x^2) is positive (as we have in (i), (iii), (iv), and (vi) in table 3.3), then the parabola opens upward, whereas if the leading coefficient is negative, then the parabola opens downward (as in (ii) and (v) in the table).

REMARK 3.5.3. If the discriminant is negative (as in (v) and (vi) in the table), then the quadratic polynomial has complex roots and the corresponding parabola does not cut the x-axis—it lies entirely above or below the x-axis, depending on whether the sign of the leading coefficient is positive or negative, respectively.

3.5.2 The Nature of the Roots

As we observe in table 3.3, there is an interesting correspondence between the nature of the roots of a quadratic polynomial ax^2+bx+c and the value of the discriminant Δ.

Remark 3.5.3. If $\Delta < 0$ (as in (v) and (vi)), then the roots are complex numbers and, if $\Delta \geq 0$ (as in (i), (ii), (iii), and (iv)), then the roots are real numbers. Only in the particular case that $\Delta = 0$ (as in (ii)) are the roots equal.

The statement of remark 3.5.3 is summarized in table 3.4.

TABLE 3.4. The discriminant and the nature of the roots

Sign of Δ	Nature of the roots
$\Delta < 0$	The roots are complex numbers
$\Delta = 0$	The roots are real (and equal)
$\Delta > 0$	The roots are real (and unequal)

Furthermore, we have the following additional fact about the nature of the roots of a quadratic polynomial $ax^2 + bx + c$.

Remark 3.5.4. If a, b, and c are rational numbers and Δ is the square of a rational number (as in example (i) above, where $\Delta = 8^2$, and example (iii) above, where $\Delta = 19^2$), then the roots are integers or rational numbers (otherwise, the roots are irrational numbers or complex numbers).

We now demonstrate that certain problems relating to quadratic polynomials can be solved using the information given in table 3.4.

Example 3.5.2. Prove that the solutions of the equation $x^2 - cx - dx + cd = p^2$ are real (not complex numbers), if c, d, and p are real numbers.

Answer: We first write the equation in the standard form

$$x^2 + (-c - d)x + (cd - p^2) = 0$$

and then find the value of Δ for the quadratic polynomial on the left-hand side of the equation above:

$$\begin{aligned}\Delta &= (-c-d)^2 - 4(1)(cd - p^2)\\ &= c^2 + 2cd + d^2 - 4cd + 4p^2\\ &= c^2 - 2cd + d^2 + 4p^2\\ &= (c-d)^2 + 4p^2.\end{aligned}$$

Because $(c-d)^2 + 4p^2$ is a sum of two nonnegative terms, we have shown that the discriminant is nonnegative and, therefore, the solutions of the equation are real. ◈

3.5.3 Factorizing Quadratic Polynomials

According to the Factor Theorem (theorem 3.4.2), if $x = c$ is a root of a quadratic polynomial, then $x - c$ is a linear factor of the polynomial. Because any quadratic polynomial has two roots, it can be expressed as a product of two linear factors multiplied by a constant. As a demonstration in table 3.5, we give the factorized form of each of the quadratic polynomials from table 3.3.

TABLE 3.5. Quadratic polynomials in factorized form

	Quadratic polynomials	Factorized form
(i)	$2x^2 - 4x - 6$	$2(x+1)(x-3)$
(ii)	$-x^2 - 4x - 4$	$-(x+2)^2$
(iii)	$6x^2 - x - 15$	$6\left(x+\frac{3}{2}\right)\left(x-\frac{5}{3}\right)$
(iv)	$x^2 - x - 11$	$\left(x-\frac{1+\sqrt{45}}{2}\right)\left(x-\frac{1-\sqrt{45}}{2}\right)$
(v)	$-x^2 - 2$	$-(x-\sqrt{2}\mathrm{i})(x+\sqrt{2}\mathrm{i})$
(vi)	$x^2 + x + \frac{5}{2}$	$\left(x-\frac{1+3\mathrm{i}}{2}\right)\left(x-\frac{1-3\mathrm{i}}{2}\right)$

EXAMPLE 3.5.3. The factorized form for $6x^2 - x - 15$ (row (iii)) can also be expressed as

$$6x^2 - x - 15 = 2\left(x+\frac{3}{2}\right)3\left(x-\frac{5}{3}\right) = (2x+3)(3x-5).$$ ◈

Finding the factors of a polynomial is called *factorizing* or *factoring* (we will use the former). It is possible to factorize quadratic polynomials with integer coefficients *by inspection* if they factorize into a pair of linear factors that also have integer coefficients (e.g., the polynomials in rows (i) and (iii) of the table above). This is a very useful skill, and it is also fun! The trick is to find the correct pairs of factors for the leading coefficient and the constant term. For example, for the polynomial $6x^2 - x - 15$ (row (iii) in the table), the correct pair of factors of the leading coefficient is 3 and 2 and the correct pair of factors of the constant term is 3 and 5. Finding the correct pairs of factors can be done by trial and error, but it does help to be methodical. We demonstrate this with four sets of examples.

EXAMPLE 3.5.4. Factorize, by inspection, each of the following quadratic polynomials. Each one has a positive leading coefficient and a positive constant term:

(i) $12x^2 - 29x + 15$

(ii) $16x^2 - 56x + 49$

(iii) $45x^2 + 38x + 8$

Answers:

(i) Write down any pair of factors of 15, for example 3 and 5 or 1 and 15, then form all possible pairwise products of this pair with the pairs of factors of 12 and *add* the results, like this:

$3(3) + 4(5) = 9 + 20 = 29$	$3(1) + 4(15) = 3 + 60 = 63$
$4(3) + 3(5) = 12 + 15 = 27$	$4(1) + 3(15) = 4 + 45 = 49$
$1(3) + 12(5) = 3 + 60 = 63$	$1(1) + 12(15) = 1 + 180 = 181$
$12(3) + 1(5) = 36 + 5 = 41$	$12(1) + 1(15) = 12 + 15 = 27$
$1(3) + 6(5) = 3 + 30 = 33$	$1(1) + 6(15) = 1 + 90 = 91$
$6(3) + 1(5) = 18 + 5 = 23$	$6(1) + 1(15) = 6 + 15 = 21$

We look for the combination in which the sum of pairwise products is, without consideration of the sign, the coefficient of the middle term of the quadratic polynomial, that is, 29. This is the first combination in the table above. Now, we factorize the quadratic polynomial, essentially by undoing a FOIL operation:

$$\begin{aligned} 12x^2 - 29x + 15 &= 12x^2 - 9x - 20x + 15 \\ &= (12x^2 - 9x) - (20x - 15) \\ &= 3x(4x - 3) - 5(4x - 3) \\ &= (3x - 5)(4x - 3) \end{aligned}$$

Note that, in the first step, the value 29 breaks up according to the sum of the products of the pairs of factors that we selected. In the second step, the four terms are grouped in pairs, as shown, and in the third and fourth steps, the distributive laws are used (in reverse).

(ii) We compile a table of sums of pairwise products of all possible pairs of factors of 16 and 49:

$1(7) + 16(7) = 7 + 112 = 119$	$1(1) + 16(49) = 1 + 784 = 785$
$4(7) + 4(7) = 28 + 28 = 56$	$4(1) + 4(49) = 4 + 196 = 200$
$2(7) + 8(7) = 14 + 56 = 70$	$2(1) + 8(49) = 2 + 392 = 394$

The sum 56, in the second row, is, without consideration of the sign, the coefficient of x, so, again we factorize the quadratic polynomial by undoing a FOIL operation:

$$\begin{aligned} 16x^2 - 56x + 49 &= 16x^2 - 28x - 28x + 49 \\ &= (16x^2 - 28x) - (28x - 49) \\ &= 4x(4x - 7) - 7(4x - 7) \\ &= (4x - 7)(4x - 7) \\ &= (4x - 7)^2. \end{aligned}$$

A shorter way to factorize this quadratic polynomial is to identify it as a *perfect square trinomial* by writing it as $(4x)^2 - 2(4x)(7) + (7)^2$ and then comparing it with the expansion of perfect squares in table 3.2.

(iii) A table of sums of products of factors of 45 and 8 can also be compiled (do it!). The correct combination is $5(4) + 9(2) = 20 + 18 = 38$. Therefore,

$$\begin{aligned} 45x^2 + 38xy + 8y^2 &= 45x^2 + 20xy + 18xy + 8y^2 \\ &= (45x^2 + 20xy) + (18xy + 8y^2) \\ &= 5x(9x + 4y) + 2y(9x + 4y) \\ &= (5x + 2y)(9x + 4y) \end{aligned}$$

◈

EXAMPLE 3.5.5. Factorize, by inspection, each of the following quadratic polynomials, which have a positive leading coefficient and a negative constant term:

(i) $51x^2 - 129x - 72$

(ii) $6u^2 + 7u - 20$

Answers:

(i) As before, write down any pair of factors of 72, form all possible pairwise products with the pairs of factors of 51 and *subtract* the results, like this (we only show a few cases):

$1(8) - 51(9) = 8 - 459 = -451$
$51(8) - 1(9) = 408 - 9 = 399$
$3(8) - 17(9) = 24 - 153 = -129$
$17(8) - 3(9) = 136 - 27 = 109$

Because 129 is, without consideration of the sign, the coefficient of x, we factorize the quadratic polynomial by breaking up the middle term according to the third equation of the table above:

$$\begin{aligned} 51x^2 - 129x - 72 &= 51x^2 + 24x - 153x - 72 \\ &= (51x^2 + 24x) - (153x + 72) \\ &= 3x(17x + 8) - 9(17x + 8) \\ &= (3x - 9)(17x + 8). \end{aligned}$$

(ii) Use $2(4) - 3(5) = 8 - 15 = -7$. Then,

$$\begin{aligned} 6u^2 + 7u - 20 &= 6u^2 - 8u + 15u - 20 \\ &= (6u^2 - 8u) + (15u - 20) \\ &= 2u(3u - 4) + 5(3u - 4) \\ &= (2u + 5)(3u - 4). \end{aligned}$$

◈

EXAMPLE 3.5.6. In the case that a quadratic polynomial has a negative leading coefficient, the negative sign can be factorized and then the previous methods can be used. It might also be possible to factorize a constant from all three terms.

(i) $-4x^2 + 17x + 42 = -(4x^2 - 17x - 42) = -(x - 6)(4x + 7)$

(ii) $396y^2 + 69y + 3 = 3(132y^2 + 23y + 1) = 3(11y + 1)(12y + 1)$ ◈

EXAMPLE 3.5.7. A special case that should be recognized immediately is a difference of two squares (see table 3.2).

(i) $49x^2 - 121 = (7x)^2 - (11)^2 = (7x - 11)(7x + 11)$

(ii) $-64u^2 + 100v^2 = -4(16u^2 - 25v^2) = -4(4u - 5v)(4u + 5v)$ ◈

EXAMPLE 3.5.8. The following quadratic polynomials cannot be factorized by inspection because their roots are not rational numbers. What are their factors?

- $x^2 - x + 1$
- $9x^2 + 6x + 4$
- $6x^2 + 7x + 21$ ◈

3.5.4 Solving Quadratic Equations by Factorizing

Quadratic equations were solved by the method of completing the square in section 3.3. However, it is frequently quicker and easier to solve quadratic equations by factorizing the quadratic polynomial by inspection. The principle behind this is that if a product of two factors is zero, then one of the factors must be equal to zero. A few examples will suffice to illustrate this.

EXAMPLE 3.5.9. Find the roots of the quadratic polynomial $2x^2 - 4x - 6$.

Answer: The equation $2x^2 - 4x - 6 = 0$ can be expressed in the factorized form $2(x+1)(x-3) = 0$. This means that $x+1=0$ or $x-3=0$. In other words, the two solutions of the equation are $x=-1$ and $x=3$, that is, the roots are -1 and 3. ◈

EXAMPLE 3.5.10. Find the roots of the quadratic polynomial $11x^2 - 79x + 14$.

Answer: The equation $11x^2 - 79x + 14 = 0$ can be expressed in the factorized form $(11x-2)(x-7) = 0$, so the roots are 2/11 and 7. ◈

Here is a practical problem that involves solving a quadratic equation.

EXAMPLE 3.5.11. The prices of two types of building material differ by 50 cents/m. For a total outlay of \$30, a builder purchases \$15 worth of each material. He then discovers that he has one meter more of the cheaper material than the more expensive material. What is the price per meter of the more expensive material?

Answer: We can denote by x the price (in cents) per meter of the more expensive material. Then the number of meters of the expensive material that the builder purchases is $\frac{1,500}{x}$, and the number of meters of the cheaper material that the builder purchases is $\frac{1,500}{x-50}$. Now, according to the statement of the problem, we can set up the following equation:

$$\frac{1,500}{x} + 1 = \frac{1,500}{x-50}.$$

We multiply each side of the equation by $(x)(x-50)$ and then solve the problem in the following steps:

$$\begin{gathered}
1,500(x-50) + x(x-50) = 1,500x \\
1,500x - 75,000 + x^2 - 50x = 1,500x \\
x^2 - 50x - 75,000 = 0 \\
(x+250)(x-300) = 0 \\
x = -250 \quad \text{or} \quad x = 300.
\end{gathered}$$

The second solution for x is the answer we want. The more expensive material costs \$3 per meter. ◈

3.6 COMPLEX NUMBERS AS MATRICES

The identification of complex numbers with 2×2 matrices provides a way to view complex numbers as concrete objects rather than the mysterious objects they might seem to be. We first need to define matrices and introduce the algebra of 2×2 matrices.

3.6.1 The Algebra of 2×2 Matrices

In the middle of the nineteenth century, the English Mathematician James Joseph Sylvester used the term *matrix* to describe a rectangular array of numbers. The first mathematician to study matrices as elements of an algebraic system was Arthur Cayley, in a paper published in 1858. The theory of matrices developed rapidly and has found many applications in mathematics, physics, engineering, statistics, game theory, and economics.

In this section, we will give the formal definition of a matrix and describe the algebraic operations that can be applied to matrices, including the addition and multiplication of matrices.

DEFINITION 3.6.1. *A rectangular array*

$$\begin{bmatrix} a_{11} & a_{12} & \cdots & a_{1n} \\ a_{21} & a_{22} & \cdots & a_{2n} \\ \vdots & \vdots & \ddots & \vdots \\ a_{m1} & a_{m2} & \cdots & a_{mn} \end{bmatrix}$$

of $m \cdot n$ elements a_{ij}, for $1 \leq i \leq m$ and $1 \leq j \leq n$, arranged in m rows and n columns, is called an $m \times n$ matrix.

REMARK 3.6.1. The elements a_{ij} of a matrix can be chosen from any set. For our purposes, they can be any real numbers.

EXAMPLE 3.6.1. Matrices are usually denoted using uppercase letters, for example,

$$A = \begin{bmatrix} 6 & 3 & -1.1 \\ -2.1 & 4 & 1 \\ 0 & 7 & \pi \end{bmatrix}, B = \begin{bmatrix} 1 & 2 & 3 \\ 4 & 5 & 6 \end{bmatrix}, C = \begin{bmatrix} 5 \\ 4 \\ -2 \end{bmatrix}, \text{ and } D = [\ 15 \quad 22 \quad 0 \quad -4\ ].$$ ◈

According to our definition, A is a 3×3 matrix, B is a 2×3 matrix, C is a 3×1 matrix, and D is a 1×4 matrix. What's more, C is an example of a *column matrix* and D is an example of a *row matrix*.

Any element in a matrix can be referenced using the appropriate row index and the appropriate column index.

EXAMPLE 3.6.2. In the matrix A in example 3.6.1, $a_{23} = 1$; in the matrix B, $b_{12} = 2$; in the matrix C, $c_{31} = -2$; and in the matrix D, $d_{13} = 0$.

Two matrices are said to be equal if and only if they have the same dimensions and the same elements in their corresponding positions.

Any two matrices with the same dimensions can be added by adding the elements in the corresponding positions.

EXAMPLE 3.6.3.

$$\text{If } A = \begin{bmatrix} 1 & 0 & 7.1 \\ -3 & 1.1 & 8 \end{bmatrix} \text{and } B = \begin{bmatrix} 0 & 6 & 11.4 \\ -2 & 6 & 0.1 \end{bmatrix}, \quad \text{then } A+B = \begin{bmatrix} 1 & 6 & 18.5 \\ -5 & 7.1 & 8.1 \end{bmatrix}.$$ ◈

A matrix in which all the elements are equal to zero is called a *zero matrix*. The notation $O_{m \times n}$ can be used for a zero matrix. It is possible for the sum of two matrices to be a zero matrix. In this case, each matrix is the *additive inverse* of the other.

If α is a real number and A is any $m \times n$ matrix, then the *scalar product* αA is the $m \times n$ matrix obtained by multiplying each element of A by α.

EXAMPLE 3.6.4.

$$A = \begin{bmatrix} 2 & 0 \\ 9 & -6 \end{bmatrix}, \ 2A = \begin{bmatrix} 4 & 0 \\ 18 & -12 \end{bmatrix}$$

$$B = \begin{bmatrix} -5 & 3.6 \\ 7 & -2 \\ 0.4 & -1 \end{bmatrix}, \quad (-1)B = \begin{bmatrix} 5 & -3.6 \\ -7 & 2 \\ -0.4 & 1 \end{bmatrix}$$ ◈

REMARK 3.6.2. If B is any matrix, then, instead of $(-1)B$, we can write $-B$ and call it the negative of B. Matrices can now be subtracted in the obvious way, that is, $A - B = A + (-B)$.

The multiplication of matrices is more complicated, and there are different ways to multiply matrices. The usual way to multiply matrices is called the *Cayley product*. We will demonstrate it using 2×2 matrices. It is based on the following formula for multiplying a row matrix and a column matrix:

$$\begin{bmatrix} a_{11} & a_{12} \end{bmatrix} \begin{bmatrix} b_{11} \\ b_{21} \end{bmatrix} = \begin{bmatrix} a_{11}b_{11} + a_{12}b_{21} \end{bmatrix}, \tag{3.7}$$

in which the first element of the row matrix is multiplied by the first element of the column matrix, the second element of the row matrix is multiplied by the second element of the column matrix, and the results are added together. When a 2×2 matrix is multiplied by another 2×2 matrix, the first matrix is regarded as two row matrices, the second matrix is regarded as two column matrices, and the product of the 2×2 matrices is another 2×2 matrix consisting of the four possible products of the two row and two column matrices, that is, if

$$A = \begin{bmatrix} a_{11} & a_{12} \\ a_{21} & a_{22} \end{bmatrix} \quad and \quad B = \begin{bmatrix} b_{11} & b_{12} \\ b_{21} & b_{22} \end{bmatrix},$$

then the product of A and B is

$$AB = \begin{bmatrix} \left(\begin{bmatrix} a_{11} & a_{12} \end{bmatrix} \begin{bmatrix} b_{11} \\ b_{21} \end{bmatrix} \right) & \left(\begin{bmatrix} a_{11} & a_{12} \end{bmatrix} \begin{bmatrix} b_{12} \\ b_{22} \end{bmatrix} \right) \\ \left(\begin{bmatrix} a_{21} & a_{22} \end{bmatrix} \begin{bmatrix} b_{11} \\ b_{21} \end{bmatrix} \right) & \left(\begin{bmatrix} a_{21} & a_{22} \end{bmatrix} \begin{bmatrix} b_{12} \\ b_{22} \end{bmatrix} \right) \end{bmatrix}$$

$$= \begin{bmatrix} a_{11}b_{11} + a_{12}b_{21} & a_{11}b_{12} + a_{12}b_{22} \\ a_{21}b_{11} + a_{22}b_{21} & a_{21}b_{12} + a_{22}b_{22} \end{bmatrix}.$$

The elements of the first row of AB are obtained by multiplying the first row of A, in turn, by the columns of B, and the elements of the second row of AB are obtained by multiplying the second row of A, in turn, by the columns of B.

Multiplying matrices is like patting your head with one hand while rubbing your tummy in a circle with your other hand. Try it—it comes more easily with practice! Look carefully at the following examples. A few important facts about matrix multiplication can be gleaned from them.

EXAMPLE 3.6.5.

(i) $$\begin{bmatrix} 1 & 6 \\ -2 & 3 \end{bmatrix} \begin{bmatrix} 8 & 2 \\ 0 & -1 \end{bmatrix} = \begin{bmatrix} (1)(8)+(6)(0) & (1)(2)+(6)(-1) \\ (-2)(8)+(3)(0) & (-2)(2)+(3)(-1) \end{bmatrix} = \begin{bmatrix} 8 & -4 \\ -16 & -7 \end{bmatrix}$$

(ii) $$\begin{bmatrix} 2 & 0.1 \\ 5 & -3 \end{bmatrix} \begin{bmatrix} 6 & 11 \\ 9 & 2 \end{bmatrix} = \begin{bmatrix} (2)(6)+(0.1)(9) & (2)(11)+(0.1)(2) \\ (5)(6)+(-3)(9) & (5)(11)+(-3)(2) \end{bmatrix} = \begin{bmatrix} 12.9 & 22.2 \\ 3 & 49 \end{bmatrix}$$

(iii) $\begin{bmatrix} 6 & 11 \\ 9 & 2 \end{bmatrix}\begin{bmatrix} 2 & 0.1 \\ 5 & -3 \end{bmatrix} = \begin{bmatrix} (6)(2)+(11)(5) & (6)(0.1)+(11)(-3) \\ (9)(2)+(2)(5) & (9)(0.1)+(2)(-3) \end{bmatrix} = \begin{bmatrix} 67 & -32.4 \\ 28 & -5.1 \end{bmatrix}$

(iv) $\begin{bmatrix} 5 & 0 \\ 0 & 0 \end{bmatrix}\begin{bmatrix} 0 & 0 \\ 0 & 6 \end{bmatrix} = \begin{bmatrix} (5)(0)+(0)(0) & (5)(0)+(0)(6) \\ (0)(0)+(0)(0) & (0)(0)+(0)(6) \end{bmatrix} = \begin{bmatrix} 0 & 0 \\ 0 & 0 \end{bmatrix}$ ◈

REMARK 3.6.3. We learn from (ii) and (iii) above that matrix multiplication is not commutative, that is, if the matrices in these two examples are labeled A and B, then $AB \neq BA$. In example (iv), the product of two matrices is the zero matrix $0_{2\times 2}$, but neither of the matrices being multiplied is a zero matrix. In these two aspects, the algebra of matrices is radically different from the algebra of real numbers.

Some properties of real numbers that do hold true for matrices are the distributive properties and the associativity of multiplication and addition. These properties for 2×2 matrices are stated below as theorems, and the proofs are left as exercises.

THEOREM 3.6.1. *If* A, B, *and* C *are* 2×2 *matrices, then*

(i) $A(B+C) = AB + AC$

(ii) $(B+C)A = BA + CA$

(iii) $A(BC) = (AB)C$

THEOREM 3.6.2. *If* A *and* B *are* 2×2 *matrices and* α *and* β *are real numbers, then*

(i) $(\alpha A)(\beta B) = (\alpha\beta)(AB)$

(ii) $(-A)(-B) = AB$

(iii) $A(\alpha B) = (\alpha A)B = \alpha(AB)$

(iv) $A(-B) = (-A)B = -(AB)$

A special 2×2 matrix is the 2×2 *identity matrix* $I = \begin{bmatrix} 1 & 0 \\ 0 & 1 \end{bmatrix}$. Any 2×2 matrix A multiplied by I_2 remains unchanged, that is, $AI_2 = I_2A = A$.

3.6.2 Complex Numbers as 2×2 Matrices

DEFINITION 3.6.2. *A complex number* z *is a number of the form* $z = x + yi$, *where* x *and* y *are real numbers and* i *has the property* $(i)^2 = -1$. *The number* x *is called the real part of* z *and the number* y *is called the imaginary part of* z.

According to the following important remark, the algebra of complex numbers can be related to the algebra of 2×2 matrices, as we will explain below.

REMARK 3.6.3. We identify the complex number $z = x + yi$ with the matrix $\begin{bmatrix} x & y \\ -y & x \end{bmatrix}$.

If we add (or subtract) two complex numbers, we add (or subtract) their real and imaginary parts.

EXAMPLE 3.6.6. If $z_1 = 2 + 3i$ and $z_2 = -5 + 6i$, then $z_1 + z_2 = -3 + 9i$. ◈

On the other hand, if $z_1 = a + ib$ and $z_2 = c + id$ are identified with the matrices $\begin{bmatrix} a & b \\ -b & a \end{bmatrix}$ and $\begin{bmatrix} c & d \\ -d & c \end{bmatrix}$, respectively, then the sum of the matrices is

$$\begin{bmatrix} a & b \\ -b & a \end{bmatrix} + \begin{bmatrix} c & d \\ -d & c \end{bmatrix} = \begin{bmatrix} a+c & b+d \\ -(b+d) & a+c \end{bmatrix}.$$

The matrix on the right-hand side is the matrix that is identified with the complex number $(a+c)+(b+d)\mathrm{i}$, which is $z_1 + z_2$.

We multiply two complex numbers in the same manner as FOIL, that is, if $z_1 = a + bi$ and $z_1 = c + di$, then

$$\begin{aligned} z_1 z_2 &= ac + adi + bic + bidi \\ &= ac + adi + bci + bdi^2 \\ &= ac + (ad + bc)i + bd(-1) \\ &= ac - bd + (ad + bc)i. \end{aligned}$$

EXAMPLE 3.6.7. If $z_1 = 2 + 3\mathrm{i}$ and $z_2 = -5 + 6\mathrm{i}$, then $z_1 z_2 = -28 - 3\mathrm{i}$. ◈

On the other hand, if we multiply the matrices that are identified above with z_1 and z_2, the result is

$$\begin{bmatrix} a & b \\ -b & a \end{bmatrix}\begin{bmatrix} c & d \\ -d & c \end{bmatrix} = \begin{bmatrix} ac-bd & ad+bc \\ -(ad+bc) & ac-bd \end{bmatrix}.$$

The matrix on the right-hand side is the matrix identified with the complex number $ac - bd + (ad + bc)\mathrm{i}$, which is $z_1 z_2$.

We conclude that the addition and multiplication of the particular 2×2 matrices we have identified with complex numbers exactly replicate the addition and multiplication of the complex numbers. Thus any algebraic calculation that involves addition, subtraction, and multiplication of complex numbers can be carried out with the appropriate 2×2 matrices.

Division of complex numbers (and the identified matrices) can also be defined, but we will not do it here.

EXAMPLE 3.6.8. Because $\mathrm{i} = 0 + 1\mathrm{i}$ and $-1 = -1 + 0\mathrm{i}$, the property $(\mathrm{i})^2 = -1$ has the matrix expression

$$\begin{bmatrix} 0 & 1 \\ -1 & 0 \end{bmatrix}\begin{bmatrix} 0 & 1 \\ -1 & 0 \end{bmatrix} = \begin{bmatrix} -1 & 0 \\ 0 & -1 \end{bmatrix}.$$ ◈

EXAMPLE 3.6.9. It can be verified by means of the quadratic formula that $1 + 2\mathrm{i}$ is a root of the quadratic polynomial $x^2 - 2x + 5$. The matrix expression of this statement is

$$\begin{aligned} &\begin{bmatrix} 1 & 2 \\ -2 & 1 \end{bmatrix}\begin{bmatrix} 1 & 2 \\ -2 & 1 \end{bmatrix} - 2\begin{bmatrix} 1 & 2 \\ -2 & 1 \end{bmatrix} + 5\begin{bmatrix} 1 & 0 \\ 0 & 1 \end{bmatrix} \\ &= \begin{bmatrix} -3 & 4 \\ -4 & -3 \end{bmatrix} + \begin{bmatrix} -2 & -4 \\ 4 & -2 \end{bmatrix} + \begin{bmatrix} 5 & 0 \\ 0 & 5 \end{bmatrix} \\ &= \begin{bmatrix} 0 & 0 \\ 0 & 0 \end{bmatrix}. \end{aligned}$$ ◈

3.7 ROOTS OF POLYNOMIALS

We return to the investigation of polynomials, in particular, the investigation of the properties of the roots of polynomials. Below, we will present a few important theorems that state some general facts about the roots of polynomials.

3.7.1 Factorization Theorems

The brilliant Mathematician Carl Friedrich Gauss (1777–1855), a contemporary of Napoleon Bonaparte, Mozart, and Beethoven, first proved a theorem that is basic to our understanding of polynomials, now known as the *Fundamental Theorem of Algebra*. We will state it, together with its important corollary, the *Complete Factorization Theorem for Polynomials*. We cannot provide a proof of the Fundamental Theorem of Algebra, unfortunately, because this will take us too far into complex function theory and advanced calculus.

REMARK 3.7.1. We can think of a real number as a complex number with zero imaginary part. So we will assume below that the set of complex numbers, sometimes denoted C, includes the set of real numbers.

THEOREM 3.7.1. The Fundamental Theorem of Algebra: *a polynomial of any positive degree with complex coefficients has at least one complex root.*

THEOREM 3.7.2. The Complete Factorization Theorem for Polynomials: *if* $\mathrm{f(x)}$ *is a polynomial of degree* $\mathrm{n}>0$ *with complex coefficients, then there exist* n *complex numbers* $c_1, c_2, \ldots, c_n$, *which are the roots of* $\mathrm{f(x)}$. *This means that*

$$f(x) = a(x-c_1)(x-c_2)\cdots(x-c_n),$$

where a *is the leading coefficient of* $f(x)$.

The Complete Factorization Theorem is a corollary of the Fundamental Theorem of Algebra. The proof of this is left as an exercise. (Hint: apply the Factor Theorem repeatedly.)

REMARK 3.7.2 The roots, $c_1, c_2, \ldots, c_n$, in the Complete Factorization Theorem need not all be different. For example, the polynomial $f(x) = 3x^3 - 9x^2 + 18x - 12 = 3(x-2)^2(x-1)$ has the roots $c_1 = 2$, $c_2 = 2$, and $c_3 = 1$. We say that $c_1 = c_2 = 2$ is a root of *multiplicity* two. In general, some (or all) of the roots $c_1, c_2, \ldots, c_n$, can be real, and some (or all) of the roots can be complex (not real).

It is a fact that the complex roots of a polynomial with real coefficients always come in pairs. In order to explain this, we need the following definition.

DEFINITION 3.6.3. *The conjugate* $\overline{\mathrm{z}}$ *of any complex number* $\mathrm{z = x + yi}$ *is the number obtained by changing the sign in front of the imaginary part, that is* $\overline{\mathrm{z}} = \mathrm{x - yi}$.

The statement of the following remark will be verified in exercise 3.38.

REMARK 3.7.4. If a polynomial $f(x)$ with real coefficients has a complex root c, that is, $f(c) = 0$, then the conjugate $\overline{c}$ is also a root of the polynomial , that is, $f(\overline{c}) = 0$.

EXAMPLE 3.7.1 The complex roots of the quadratic polynomial $x^3 - 3x^2 + 9x + 13$ are $z_1 = 2 + 3\mathrm{i}$ and $z_2 = 2 - 3\mathrm{i}$ (exercise 3.38). Note that the sign in front of the imaginary part of z_2 is the opposite of the sign in front of the imaginary part of z_1, that is, z_2 is the conjugate of z_1. ◈

As a special case, any quadratic polynomial with real coefficients either has two complex (not real) roots, which are conjugates of each other (that is, the quadratic polynomial factorizes as $a(x-c_1)(x-\overline{c}_1)$, where a is a real number and c_1 is a complex number) or it has two real roots. In the first case, the quadratic polynomial is called *irreducible* over the real numbers.

Conversely, any product of the form $a(x-c_1)(x-\overline{c}_1)$, where a is a real number and c_1 is a complex number (not real), can be expanded as an irreducible quadratic polynomial with real coefficients (verify this).

In view of the Complete Factorization Theorem for Polynomials, this leads us to the following conclusion:

THEOREM 3.7.3. *Every polynomial with real coefficients and positive degree can be expressed as a product of linear and irreducible quadratic factors with real coefficients.*

PROOF. According to the Complete Factorization Theorem for Polynomials, any polynomial $f(x)$ with positive degree n can be expressed in the form

$$f(x)=a(x-c_1)(x-\overline{c}_1)\ldots(x-c_j)(x-\overline{c}_j)(x-d_1)(x-d_2)\ldots(x-d_k),$$

where $2j+k=n$, $c_1, c_2, \ldots c_j$ are complex (not real) numbers and $d_1, d_2, \ldots d_k$ are real numbers (some of which can be equal). Here, each product of the form $(x-c_j)(x-\overline{c}_j)$ can be expanded as an irreducible quadratic polynomial with real coefficients, and each factor of the form $(x-d_i)$ is a linear polynomial with real coefficients. □

In the next example, verify that the quadratic factors in the factorization of each polynomial are indeed irreducible quadratic polynomials.

EXAMPLE 3.7.1.

(i) $x^5-9x^3+8x^2-72=(x^2-2x+4)(x-3)(x+3)(x+2)$

(ii) $x^4-2x^3+15x^2-134x+290=(x^2-6x+10)(x^2+4x+29)$ ◈

3.7.2 A Method For Finding the Integer and Rational Roots of a Polynomial

In general, it is difficult to factorize any given polynomial with degree greater than two. Therefore, it helps to have a method for determining whether a given polynomial has any integer or rational roots.

THEOREM 3.7.4. *If the polynomial*

$$f(x)=a_nx^n+a_{n-1}x^{n-1}+\cdots+a_1x+a_0$$

has integer coefficients and, if $\frac{c}{d}$ *is a rational root of* $f(x)$ *such that* c *and* d *have no common factors, then* (i) c *is a factor of the constant term* a_0, *and* (ii) d *is a factor of the leading coefficient* a_n.

PROOF. We suppose that $c \neq 0$ and $d \neq \pm c$ (otherwise, the proof is trivial). The statement that *c*/*d* is a rational root of $f(x)$ means that $f(c/d)=0$. By substitution of *c*/*d* for *x* in the expression given for $f(x)$ this is

$$a_n\left(\frac{c}{d}\right)^n + a_{n-1}\left(\frac{c}{d}\right)^{n-1} + \cdots + a_1\left(\frac{c}{d}\right) + a_0 = 0.$$

If we multiply each side of the equation above by d^n and then add $-a_0d^n$ to each side, the result is

$$a_nc^n + a_{n-1}c^{n-1}d + \cdots + a_1cd^{n-1} = -a_0d^n.$$

Now, c is a factor of the integer that is the left-hand side of the equation; therefore, c must also be a factor of the integer that is the right side of the equation. Because c and d have no common factor (by assumption), we conclude that c must be a factor of a_0. Similarly, a rearrangement of the previous equation leads to

$$a_{n-1}c^{n-1}d + \cdots + a_1cd^{n-1} + a_0d^n = -a_nc^n$$

and, by the same reasoning, d must be a factor of a_n. □

COROLLARY 3.7.1.

(a) *The only possible rational roots of a polynomial with integer coefficients are the rational numbers of the form c/d for which* c *is a factor of the constant term of the polynomial and* d *is a factor of the leading coefficient of the polynomial.*

(b) *The only possible rational roots of a polynomial with integer coefficients and a unit leading coefficient are the integers that are the factors of the constant term of the polynomial.*

EXAMPLE 3.7.2. According to corollary 3.7.1(a), the only possible rational roots of the polynomial $f(x) = 3x^4 + 8x^3 - 2x^2 - 10x + 4$ are ± 1, ± 2, ± 4, $\frac{\pm 1}{3}$, $\frac{\pm 2}{3}$, and $\frac{\pm 4}{3}$. We check all of these possibilities:

$$f(1) = 3, \quad f(2) = 88 \quad f(4) = 1{,}212,$$
$$f(-1) = 7, \quad f(-2) = 0, \quad f(-4) = 268,$$
$$f\left(\frac{1}{3}\right) = 3\left(\frac{1}{3}\right)^4 + 8\left(\frac{1}{3}\right)^3 - 2\left(\frac{1}{3}\right)^2 - 10\left(\frac{1}{3}\right) + 4 = \frac{7}{9},$$
$$f\left(\frac{-1}{3}\right) = 3\left(\frac{-1}{3}\right)^4 + 8\left(\frac{-1}{3}\right)^3 - 2\left(\frac{-1}{3}\right)^2 - 10\left(\frac{-1}{3}\right) + 4 = 4 = \frac{185}{27},$$
$$f\left(\frac{2}{3}\right) = 3\left(\frac{2}{3}\right)^4 + 8\left(\frac{2}{3}\right)^3 - 2\left(\frac{2}{3}\right)^2 - 10\left(\frac{2}{3}\right) + 4 = \frac{-16}{27},$$
$$f\left(\frac{-2}{3}\right) = 3\left(\frac{-2}{3}\right)^4 + 8\left(\frac{-2}{3}\right)^3 - 2\left(\frac{-2}{3}\right)^2 - 10\left(\frac{-2}{3}\right) + 4 = 8,$$
$$f\left(\frac{4}{3}\right) = 3\left(\frac{4}{3}\right)^4 + 8\left(\frac{4}{3}\right)^3 - 2\left(\frac{4}{3}\right)^2 - 10\left(\frac{4}{3}\right) + 4 = \frac{140}{9},$$
$$f\left(\frac{-4}{3}\right) = 3\left(\frac{-4}{3}\right)^4 + 8\left(\frac{-4}{3}\right)^3 - 2\left(\frac{-4}{3}\right)^2 - 10\left(\frac{-4}{3}\right) + 4 = \frac{116}{27}.$$

Therefore, the only rational root is −2. ◈

EXAMPLE 3.7.3. Fully factorize the polynomial $f(x) = 3x^3 + 4x^2 + 74x - 52$.

Answer: According to corollary 3.7.1(a), the only possible rational roots are ±1, ±2, ±4, ±13, ±26, ±52, $\frac{\pm1}{3}$, $\frac{\pm2}{3}$, $\frac{\pm4}{3}$, $\frac{\pm13}{3}$, $\frac{\pm26}{3}$, and $\frac{\pm52}{3}$. It is straightforward to check that $f\left(\frac{2}{3}\right)=0$, and so $x-\frac{2}{3}$ is a factor of $f(x)$. After dividing $f(x)$ by $x-\frac{2}{3}$ (using long or synthetic division), we obtain

$$f(x)=3\left(x-\frac{2}{3}\right)(x^2+2x+26).$$

The roots of $x^2 + 2x + 26$ are $-1\pm5i$, so the full factorization of $f(x)$ is

$$f(x)=(3x-2)(x+1+5i)(x+1-5i).$$ ◈

EXAMPLE 3.7.4. Fully factorize the polynomial $f(x) = x^4 + 2x^3 - 13x^2 - 14x + 24$.

Answer: According to corollary 3.7.1(b), the possible rational roots are ±1, ±2, ±3, ±4, ±6, ±8, ±12, and ±24. Because $f(1) = 0$, $x-1$ is a factor of $f(x)$. After dividing $f(x)$ by $x-1$, we obtain

$$f(x)=(x-1)(x^3+3x^2-10x-24).$$

A root of the cubic polynomial is −2 (a factor of −24). After division by the factor $x+2$, we find that

$$f(x)=(x-1)(x+2)(x^2+x-12).$$

The factorization of the quadratic polynomial can be done by inspection, to produce the full factorization

$$f(x)=(x-1)(x+2)(x-3)(x+4).$$ ◈

3.8 GRAPHS OF POLYNOMIALS

We consider the polynomials which are powers of x, that is, $f(x)=1$ (the identity function), $f(x)=x$, $f(x)=x^2$, $f(x)=x^3$, $f(x)=x^4$, and so on. Their graphs, starting from $y=x$, show a progressive flattening out at the origin, as shown in figure 3.3.

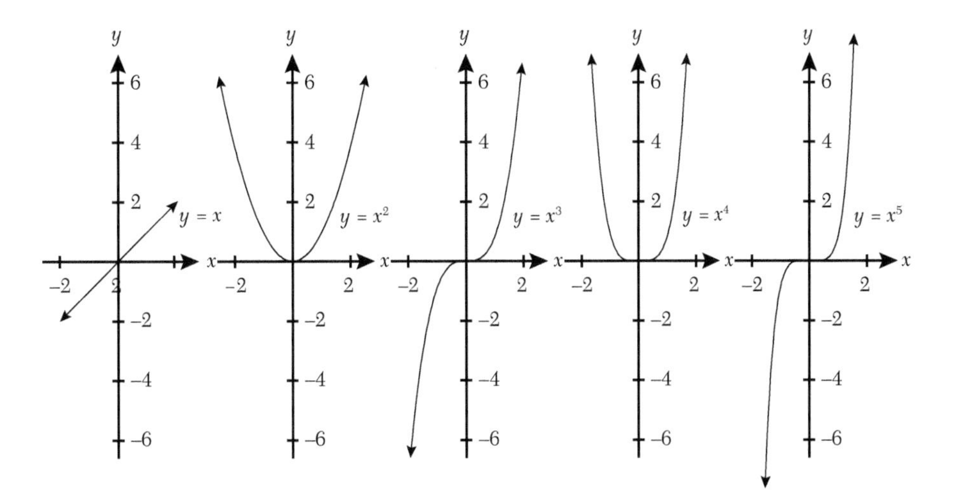

FIGURE 3.3. Basic polynomials.

It is a fact that every polynomial has the property that its graph looks like one of the the graphs above at a small enough scale (small enough resolution). On some graphing calculators, it is possible to plot the graph of a polynomial and "zoom out" until the graph looks like one of the graphs above.

EXAMPLE 3.8.1. Compare the graph of $f(x) = x^5 - 9x^3 + 8x^2 - 72$ at a large scale (high resolution) with the graph at a small scale (low resolution), as shown in figure 3.4. ◈

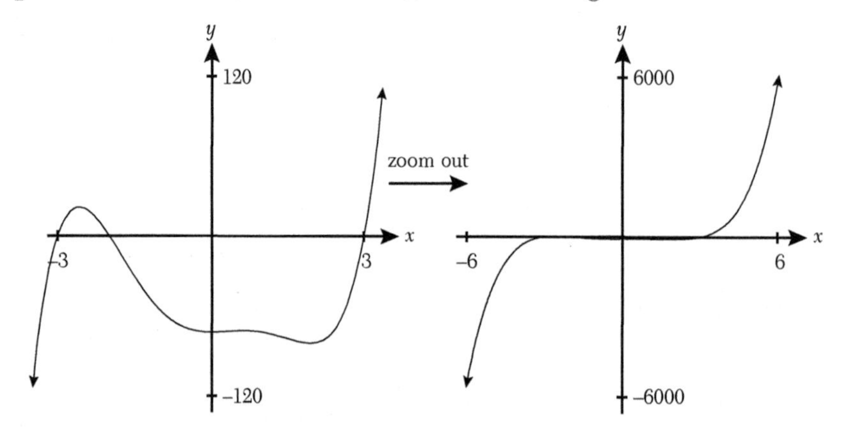

FIGURE 3.4. Zooming out.

REMARK 3.8.1. The number of times the graph of a polynomial crosses the x-axis is at most the degree of the polynomial minus the number of complex roots.

3.9 SOLVING CUBIC, QUARTIC, AND QUINTIC EQUATIONS

This topic is omitted from many textbooks because the methods are difficult and, in any case, nowadays the preferred methods for solving polynomial equations with polynomials of a degree greater than two are computerized, numerical methods. However, a good reason for including this topic here is that the discovery of methods for solving cubic and quartic equations was an important part of the historical development of algebra, and the techniques are a demonstration of the ingenuity that can be employed in solving equations.

3.9.1 Solving Cubic Equations

There are at least two well-known methods for solving cubic equations. Both methods begin with a reduction of the general equation of the form

$$ax^3 + bx^2 + cx + d = 0 \tag{3.8}$$

to an equation of the form

$$X^3 + pX = q, \tag{3.9}$$

where p and q are new constants expressed in terms of a, b, c, and d. This is called a *reduced cubic equation*. The reduction is made by replacing x with $X - \left(\frac{b}{3a}\right)$ in the general cubic equation, that is, formula (3.8), to obtain the equation

$$a\left(X - \frac{b}{3a}\right)^3 + b\left(X - \frac{b}{3a}\right)^2 + c\left(X - \frac{b}{3a}\right) + d = 0.$$

If each of the powers is expanded, then:

$$a\left(X^3-3\left(\frac{b}{3a}\right)X^2+3\left(\frac{b}{3a}\right)^2X-\left(\frac{b}{3a}\right)^3\right)+b\left(X^2-2\left(\frac{b}{3a}\right)X+\left(\frac{b}{3a}\right)^2\right)+c\left(X-\frac{b}{3a}\right)+d=0$$

and this simplifies to

$$aX^3+\left(c-\frac{b^2}{3a}\right)X+\left(d-\frac{bc}{3a}+\frac{2b^3}{27a^2}\right)=0,$$

which can be reformulated as

$$X^3+\left(\frac{c}{a}-\frac{b^2}{3a^2}\right)X=\left(\frac{bc}{3a^2}-\frac{2b^3}{27a^3}-\frac{d}{a}\right),$$

whereupon we make the substitutions

$$p=\left(\frac{c}{a}-\frac{b^2}{3a^2}\right)\text{ and }q=\left(\frac{bc}{3a^2}-\frac{2b^3}{27a^3}-\frac{d}{a}\right)$$

in order to obtain formula (3.9).

Example 3.9.1. In this section, we will find one solution of the cubic equation $2x^3+3x^2-2x+1=0$. First, check for yourself that ± 1 and $\pm\frac{1}{2}$ are not rational roots of the cubic polynomial. According to the method of reduction explained above, x is replaced with $X-\frac{1}{2}$ and p and q are $-\frac{7}{4}$ and $-\frac{5}{2}$, respectively. This means that, if we can find a solution X_0 of the equation

$$X^3-\left(\frac{7}{4}\right)X=-\left(\frac{5}{4}\right),$$

then a solution of the original equation will be $x_0=X_0-\frac{1}{2}$. ◈

A method for solving the reduced equation (formula (3.9)) was published by Girolamo Cardano in his famous work, *Ars Magna*, in 1545. This method, which was probably first discovered by Tartaglia in about 1535, makes use of the identity

$$(u-v)^3+3uv(u-v)=u^3-v^3.$$

If u and v can be chosen in such a way that

$$3uv=p\quad\text{and}\quad u^3-v^3=q \tag{3.10}$$

then a solution for the reduced cubic equation is $X_0=u-v$. The first equation in formula (3.10) can be expressed as $v=\frac{p}{3u}$. If this value for v is substituted in the second equation in formula (3.10), then we obtain

$$(u^3)^2-qu^3-\left(\frac{p}{3}\right)^3=0.$$

This is a quadratic equation in u^3, and one solution (taking the positive sign in the quadratic formula) is

$$u^3 = \frac{q+\sqrt{q^2+4\left(\frac{p}{3}\right)^3}}{2} = \frac{q}{2}+\sqrt{\left(\frac{q}{2}\right)^2+\left(\frac{p}{3}\right)^3}.$$

From this, we also obtain

$$v^3 = u^3 - q = -\frac{q}{2}+\sqrt{\left(\frac{q}{2}\right)^2+\left(\frac{p}{3}\right)^3}.$$

Thus, the solution for the reduced equation is $X_0 = u - v$, that is,

$$\sqrt[3]{\frac{q}{2}+\sqrt{\left(\frac{q}{2}\right)^2+\left(\frac{p}{3}\right)^3}} - \sqrt[3]{-\frac{q}{2}+\sqrt{\left(\frac{q}{2}\right)^2+\left(\frac{p}{3}\right)^3}} \tag{3.11}$$

EXAMPLE 3.9.2. This is a continuation of example 3.9.1. We substitute $p=\frac{-7}{4}$ and $q=\frac{-5}{4}$ in formula (3.11); then

$$X_0 = \sqrt[3]{\frac{1}{24}\sqrt{\frac{332}{3}}-\frac{5}{8}} - \sqrt[3]{\frac{1}{24}\sqrt{\frac{332}{3}}+\frac{5}{8}}$$

and so

$$x_0 = X_0 - \frac{1}{2} \approx -2.0921935.$$

Note that the remaining two solutions for the original equation can be found by dividing $2x^3+3x^2-2x+1$ by $x-x_0$ and then using the quadratic formula to find the roots of the resulting quadratic polynomial. ◈

3.9.2 Solving Quartic Equations

The general quartic equation can be reduced to a quartic equation without a cubic term in the same way that the general cubic equation can be reduced to a cubic equation without a quadratic term. We do not give the details here (check them yourself!). The reduced quartic equation is

$$x^4 + ax^2 + bx + c = 0, \tag{3.12}$$

where a, b, and c are real-valued constants. We assume that $c \neq 0$ and $b \neq 0$. The method that is usually used for solving this is the *method of undetermined coefficients*, given by Descartes in 1637. This method requires the determination of nonzero constants k, s, and t, for which

$$x^4 + ax^2 + bx + c = (x^2 + kx + s)(x^2 - kx + t). \tag{3.13}$$

The two quadratic equations $x^2 + kx + s = 0$ and $x^2 - kx + t = 0$ can then be solved separately to produce four solutions for the quartic equation.

Now expanding the right-hand side of formula (3.13) results in

$$x^4 + ax^2 + bx + c = x^4 + (t - k^2 + s)x^2 + (kt - ks)x + st.$$

Comparison of the coefficients of powers on x of each side leads to the following set of equations relating the coefficients t, k, and s with the coefficients a, b, and c:

$$\begin{aligned} t-k^2+s&=a,\\ kt-ks&=b,\\ st&=c. \end{aligned} \tag{3.14}$$

$$\begin{aligned} t+\frac{c}{t}&=a+k^2,\\ k\left(t-\frac{c}{t}\right)&=b. \end{aligned} \tag{3.15}$$

The equations above can be replaced by two equations by substituting $\frac{c}{t}$ for s in the first two equations:

If we square both sides of the first equation, then we obtain the pair of equations

$$\begin{aligned} \left(t+\frac{c}{t}\right)^2&=(a+k^2)^2,\\ k\left(t-\frac{c}{t}\right)&=b. \end{aligned} \tag{3.16}$$

The reason for doing this is that we can cleverly use the identity

$$\left(t+\frac{c}{t}\right)^2=\left(t-\frac{c}{t}\right)^2+4c$$

to rewrite the left-hand side of the first equation of formula (3.16). We now have the pair of equations

$$\begin{aligned} \left(t-\frac{c}{t}\right)^2&=(a+k^2)^2-4c,\\ k\left(t-\frac{c}{t}\right)&=b. \end{aligned} \tag{3.17}$$

The second equation in formula (3.17) is equivalent to $\left(t-\frac{c}{t}\right)=\frac{b}{k}$, and substituting this in the first equation in formula (3.17) results in

$$\left(\frac{b}{k}\right)^2=(a+k^2)^2-4c.$$

It is convenient to replace k^2 with l and rewrite this equation as a cubic equation in l:

$$l^3+2al^2+(a^2-4c)l-b^2=0 \tag{3.18}$$

By replacing l with $L-\frac{2a}{3}$, this becomes the reduced cubic equation

$$L^3-\left(\frac{a^2}{3}+4c\right)L=\frac{2a^3}{27}-\frac{8ac}{3}+b^2.$$

This cubic equation can be solved using the methods of section 3.9.1. This obtains the required value of l and, by taking the square root of l, the required value for k. Using formula (3.14), we can then solve for t and s in terms of k (do this!) to obtain

$$t=\frac{2ck}{ak+k^3-b} \quad \text{and} \quad s=\frac{c}{t}. \tag{3.19}$$

EXAMPLE 3.9.3. We can solve the quartic equation $x^4 - 6x^2 - 16x - 15 = 0$ by setting $a = -6$, $b = -16$, and $c = -15$ in formula (3.18), which then becomes

$$l^3 - 12l^2 + 96l - 256 = 0.$$

A solution of this equation is $l = 4$. The corresponding value for k is 2 and from formula 3.19 the corresponding values for s and t are 3 and –5, respectively, that is, we have determined that

$$x^4 - 6x^2 - 16x - 15 = (x^2 + 2x + 3)(x^2 - 2x - 5).$$

The roots of $x^2 + 2x + 3$ are $-1 \pm 2i$, and the roots of $x^2 - 2x - 5$ are $1 \pm \sqrt{6}$. These are the four roots of the quartic equation. ◈

3.9.3 Solving Quintic Equations

The quadratic formula (formula (3.3) in section 3.3) is a means to write the roots of a quadratic polynomial in terms of a radical (a square root) that involves the coefficients of the quadratic polynomial. Similarly, formula (3.11) is an expression in radicals (with square and cube roots) for a root of a reduced cubic polynomial in terms of the coefficients of the cubic polynomial. A radical expression for the roots of a (reduced) quartic equation in terms of the coefficients could be given, in principle, but writing it down would be much too cumbersome.

There is, in general, no formula by means of radicals for expressing the roots of a polynomial of degree five (i.e., a quintic polynomial) or higher. This was discovered (after many mathematicians had tried unsuccessfully to find a formula) by the Italian Paolo Ruffini at the beginning of the nineteenth century and also proved independently by the Norwegian Mathematician Niels Henrik Abel in 1824. The determination whether a particular quintic equation is solvable using radicals requires knowledge of *group theory*, a branch of algebra begun by the French teenage Mathematician Évariste Galois, who was born in 1811 and died in 1832 after fighting a duel.

EXERCISES

3.1. Solve for x in each of the following equations (if possible).

(a) $x + 3 = 2x - 7$

(b) $2(7x + 3) = 5(2x - 1)$

(c) $2(3x - 2) = 3(2x + 99)$

(d) $(x + 3)^2 = (x - 1)^2 + 6$

(e) $4(x + 3)^2 - (2x - 1)^2 = 0$

(f) $(x - 7)^2 = 50 - 14x$

3.2. A gambler goes into a casino. At the first table, he doubles his money and then spends \$10 on a martini. At the second table, he triples his money (i.e., he triples the total sum he brings from the first table), and gives the croupier a \$20 tip when he leaves. At the third table, he loses half his money (i.e., half of the total sum he brings from the second table) and then pays \$30 for a cab home. If he has \$95 left over, how much money did he take with him into the casino? (Hint: let x be the amount of money he has when he goes into the casino; then set up an equation involving x according to the given information.)

3.3. Solve for x in each of the following equations.

(a) $\dfrac{1}{x-7} - \dfrac{1}{2x+1} = \dfrac{3}{2x+1}$

(b) $\dfrac{2}{9x-3} + \dfrac{3}{4x+1} = \dfrac{4}{3x-1}$

3.4. Complete the square for each of the following expressions.

(a) $x^2 + 2x$ **(c)** $2x^2 + 5x - 1$

(b) $x^2 - 7x$ **(d)** $-6x^2 - 4x + 2$

3.5. Solve each of the following for x by completing the square.

(a) $x^2 + x = 0$ **(c)** $2x^2 + 5x - 1 = 0$

(b) $x^2 - 7x + 7 = 0$ **(d)** $3x^2 - 7x + 7 = 0$

3.6. Add or subtract the following polynomials, as indicated.

(a) $(x^3 + 2x^2 + 5) + (2x^3 - 3x - 1)$ **(b)** $(2x^4 + 11x^2 - 4x) - (2x^4 + x^3 - 4x - 1)$

3.7. Multiply the polynomials, as indicated.

(a) $(3x^2 + 5)(-2x + 1)$ **(f)** $(1 + 2x)^3$

(b) $(1 - x)(1 + x + x^2 + x^3)$ **(g)** $(1 - 2x)^4$

(c) $(1 - 2x)(1 + 2x + 4x^2 + 8x^3)$ **(h)** $(x + y)^6$

(d) $(1 - 2x)^2$ **(i)** $(x - 2)^6$

(e) $(1 - 2x)(1 + 2x)$ **(j)** $(2 - x)^6$

3.8. A Pythagorean triple is a set of three integers a, b, and c that specify the lengths of the sides of a right triangle, that is, if c is the hypotenuse, then $c^2 = a^2 + b^2$. One example is $a = 3$, $b = 4$, and $c = 5$. There is a method that can generate all possible Pythagorean triples. It works as follows: for any choice of integers m and n, we set $a = m^2 - n^2$, $b = 2mn$, and $c = m^2 + n^2$.

(a) Verify that this assignment of values for a, b, and c determines a Pythagorean triple.

(b) Use a calculator to find three Pythagorean triples that include the number 56.

3.9. Divide $x^4 + 2x^2 + x + 5$ by $x^2 - 3x + 1$ using the method of long division.

3.10. Divide $2x^5 - x^3 + 50x^2 + 8$ by $x - 3$ using first the method of long division and then the method of synthetic division.

3.11. If $f(x) = x^3 - 5x^2 + x - 3$, use the Remainder Theorem to find $f(3)$.

3.12. Using the Factor Theorem, show that $x + 2$ is a factor of $-x^3 - 4x^2 - 3x + 2$.

3.13. Using the Factor Theorem, find a factor of the form $x - c$, where c is an integer, for each of the following polynomials.

(a) $4x^3 + 4x^2 - x - 1$ **(b)** $3x^4 + 8x^3 - 2x^2 - 10x + 4$

3.14. Neatly sketch the graphs of each of the quadratic polynomials in exercise 3.5. Be sure to mark the x- and y-intercepts, the axis of symmetry and the turning point of each parabola.

3.15. Find the equations of the two parabolas shown in figure 3.5.

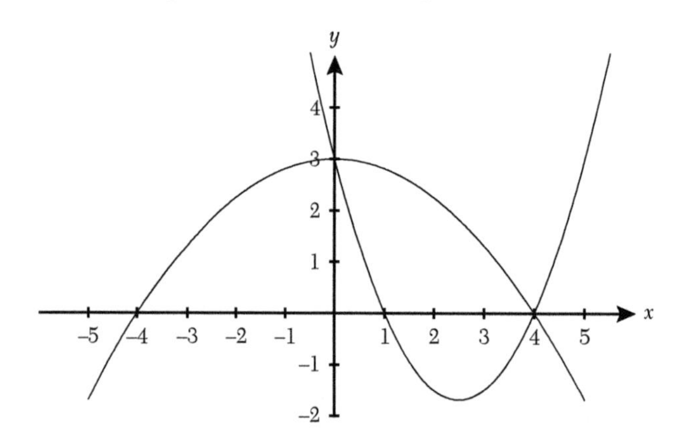

FIGURE 3.5. Two parabolas.

3.16. Prove that the sum of the roots of a quadratic polynomial $ax^2 + bx + c$ is $-\frac{b}{a}$, and the product of the roots is $\frac{c}{a}$.

3.17. Prove that the roots of $m(1-x) = 3 - x^2$ are real but not equal for all real values of m.

3.18. The axis of symmetry of a parabola can be found by taking the average value of the roots. Explain why.

3.19. Factorize the following quadratic polynomials (these are set up to be factorized by grouping).

(a) $14x^2 - 2x + 28x - 4$

(b) $17x^2 + 51x + 6x + 18$

(c) $-132x^2 - 60x + 66x + 30$

3.20. Factorize the following quadratic polynomials by inspection.

(a) $7x^2 + 14x$

(b) $2 - 5y + 2y^2$

(c) $14x^2 + 25x + 9$

(d) $3x^2 + 20x - 63$

(e) $24u^2 - 38u + 15$

(f) $34y^2 + 114y + 36$

(g) $-t^2 + t + 6$

(h) $5s^2 + 5s - 10$

3.21. Factorize the following quadratic polynomials by inspection (these are all perfect squares or a difference of two squares).

(a) $-4x^2 + 8x - 4$

(b) $9a^2 - 66a + 121$

(c) $36u^2 - 361v^2$

(d) $36y^2 + 228y + 361$

(e) $243t^2 - 12$

(f) $9s^2 + 24s + 16$

3.22. Factorize the following polynomials by inspection (if possible).

(a) $169x^2 - 4x$

(b) $4a^2 + 24ab + 36b^2$

(c) $36u^2 - 361v^2$

(d) $2t^2 + 2t + 4$

(e) $54s^2t^2 - 2$

(f) $35y^2 + 13y - 4$

3.23. Solve the following equations for x by factorizing.

(a) $x^2 - x - 2 = 0$

(b) $x^2 - 7x = 0$

(c) $243x^2 - 12 = 0$

(d) $15x^2 - 17x - 4 = 0$

3.24. Solve the following equations for x. Use any method.

(a) $2(x+3)^2 = (x-1)^2 + 4$

(b) $(x-3)(x+2) = (2x-1)(x+1)$

(c) $x(x+1)^2 = (x-1)^3$

(d) $1 = \dfrac{2}{x - \dfrac{2}{x}}$

3.25. Determine the negative value of m for which the equation $-x^2 + 2x - 4 = mx$ has equal roots.

3.26. If $2x^2 + 2ax + 4x = 1 - a$, determine the value of a for which

(a) The sum of the roots is equal to their product.

(b) The roots are numerically equal but have opposite signs.

(c) One of the roots is 0.

3.27. If α and β are the roots of $2x^2 - 3x - 4$, find the value of $\frac{\alpha}{\beta} + \frac{\beta}{\alpha}$ without solving the equation. (Hint: use the formulas for the sum and product of roots from exercise 3.16)

3.28. Suppose that α^2 and β^2 are the roots of the equation $36x^2 - x + 16$. If α and β are both positive, determine a quadratic polynomial with roots α and β, without solving the equation $36x^2 - x + 16 = 0$. (Hint: a quadratic polynomial with roots α and β and unit leading coefficient can be expressed in the form $x^2 - (\alpha + \beta)x + \alpha\beta$.)

3.29. A rectangular closed box with square ends is constructed so that the sum of its breadth (x m) and its length (y m) is equal to 5 m. Its height is equal to its breadth.

(a) Calculate its total surface area in terms of x and y.

(b) If the total surface area of the closed box is 32 m^2, calculate its length and breadth.

3.30. The area of a right triangle is 196, and its hypotenuse is 50. What are the other two sides of the right triangle?

3.31. The sum of two numbers is 28, and the sum of the squares of the two numbers is 554. What are the numbers?

3.32. For a certain triangle, the difference of two sides is 1 unit, the altitude from the third side is 24 units, and the difference of the segments into which the altitude divides the third side is 3 units. What are the sides of the triangle?

3.33. Perform the following matrix and scalar multiplications. (Note: the square of a matrix is the product of a matrix with itself.)

(a) $\begin{bmatrix} 1 & 3 \\ 4 & 1 \end{bmatrix}\begin{bmatrix} -1 & 2 \\ 6 & 0 \end{bmatrix}$

(b) $\begin{bmatrix} 0 & 1 \\ 1 & 0 \end{bmatrix}\begin{bmatrix} 1 & 0 \\ 0 & -1 \end{bmatrix}$

(c) $\frac{1}{15}\begin{bmatrix} 2 & -1 \\ 3 & 6 \end{bmatrix}\begin{bmatrix} 6 & 1 \\ -3 & 2 \end{bmatrix}$

(d) $\begin{bmatrix} 1 & 1 \\ -1 & -1 \end{bmatrix}^2$

(e) $\begin{bmatrix} 6 & 4 \\ -3 & 1 \end{bmatrix}\begin{bmatrix} 1 & -4 \\ 3 & 6 \end{bmatrix}\begin{bmatrix} 11 & -4 \\ 1 & 12 \end{bmatrix}$

(f) $\begin{bmatrix} 6 & 4 \\ -3 & 1 \end{bmatrix}\begin{bmatrix} 11 & -4 \\ 1 & 12 \end{bmatrix}\begin{bmatrix} 1 & -4 \\ 3 & 6 \end{bmatrix}$

(g) $\begin{bmatrix} 0 & 0 \\ 0 & 1 \end{bmatrix}\begin{bmatrix} 0 & 0 \\ 2 & 3 \end{bmatrix}$

(h) $\begin{bmatrix} 0 & 0 \\ 2 & 3 \end{bmatrix}\begin{bmatrix} 0 & 0 \\ 0 & 1 \end{bmatrix}$

3.34. Let $A = \begin{bmatrix} a & b \\ -b & a \end{bmatrix}$ and $B = \begin{bmatrix} c & d \\ -d & c \end{bmatrix}$, for any real numbers a, b, c, and d, where a and b are not both zero and c and d are not both zero. Prove that $AB \neq 0_{2\times2}$ and $BA \neq 0_{2\times2}$.

3.35. The roots of the quadratic polynomial $9x^2 + 1$ are $\pm\frac{i}{3}$. Verify this statement using the appropriate matrices, as was done in example 3.6.8.

3.36. The complex roots of the quadratic polynomial $x^3 - 3x^2 + 9x + 13$ are $2 \pm 3\text{i}$. Verify this statement using the appropriate matrices, as was done in example 3.6.9. What is the other root?

3.37. Use the Fundamental Theorem of Algebra and the Factor Theorem to prove the Complete Factorization Theorem for Polynomials.

3.38. If z is any complex number, prove that $\overline{z^2} = \overline{z}^2$, and if z and w are any complex numbers, prove that $\overline{wz} = \overline{z}\overline{w}$. Note that any real number (regarded as a complex number) is its own conjugate. With this information, prove that, if $f(x)$ is a polynomial with real coefficients, then $\overline{f(c)} = f(\overline{c})$ for any complex number c. Now you should be able to convince yourself that, if a complex number is a root of a polynomial with real coefficients, then its conjugate is also a root.

3.39. Expand the following products of factors into polynomials with real coefficients.

(a) $(2x-1+3\text{i})(2x-1-3\text{i})$

(b) $(2x-\text{i})(2x+\text{i})(3x+2+\text{i})(3x+2-\text{i})$

(c) $(x+1)(x+1+7\text{i})(x+1-7\text{i})$

3.40. Fully factorize the following polynomials.

(a) $24x^3 - 18x^2 - x + 1$

(b) $4x^4 - 53x^2 - 3x - 10$

3.41. Factorize $x^6 + 2x^2 + 1$ into a product of linear and irreducible quadratic factors.

3.42. Fully factorize $x^4 + 5x^2 + 4$.

3.43. Find a solution of the equation $x^3 + 9x^2 - x + 3 = 0$.

3.44. Find a solution of the equation $x^3 + 3x^2 - 5x + 2 = 0$.

3.45. Why did we assume that $b \neq 0$ and $c \neq 0$ at the beginning of section 3.9.3, and why did this allow us to state that k, s, and t were nonzero?

3.46. Find the roots of the quartic equation $x^4 - 17x^2 - 12x - 2$.

3.47. Factorize $x^5 + 4x^3 + x^2 + 4$ into a product of linear and irreducible quadratic factors.

CHAPTER 4

TRIGONOMETRY

4.1 INTRODUCTION

Trigonometry is the starting point for the mathematical description of wavelike phenomena from ripples on a pond to the complicated wave functions that physicists use to describe the states of elementary particles such as protons and electrons. The reason for this, as we will see in section 4.8, is that a sine (or cosine) curve has the shape of a wave.

Trigonometry began with the attempts by early mathematicians to calculate the lengths of chords of circles, which are twice the sine of half of the angle of the chord (see figure 4.1). We know that in the second century BC, the Greek Mathematicians Hipparchus and Menelaus computed tables of lengths of chords. Some early examples of trigonometric identities are found in Hindu works of the fifth century AD. The tangent, secant, and cosecant ratios were introduced by Arabic Mathematicians in the tenth century. The first textbook on trigonometry was published early in the sixteenth century by the German Clergyman and Mathematician Bartholomäus Pitiscus. The science of trigonometry includes planar trigonometry (chapter 4) and spherical trigonometry (chapter 10).

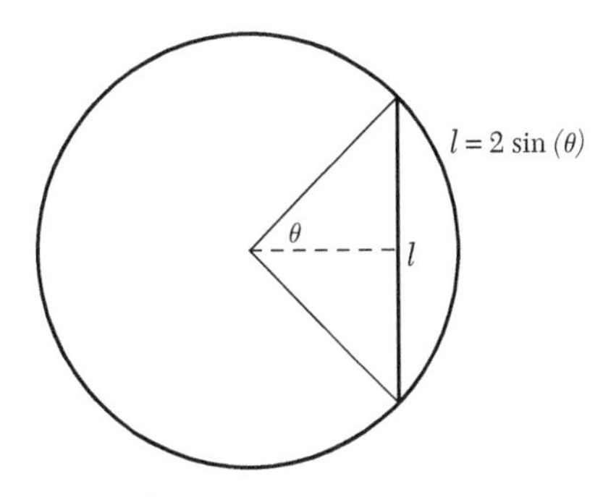

FIGURE 4.1. The length of a chord.

After explaining precisely what is meant by an angle and the radian measure of an angle (section 4.2), we introduce the three basic trigonometric ratios, that is, sine, cosine, and tangent, in section 4.3. The values of these ratios for some special angles are given in section 4.4. Negative angles and periodicity are discussed in section 4.5. The reciprocal trigonometric ratios, that is, cosecant, secant, and cotangent, are introduced in section 4.6, and the cofunction identities are introduced in section 4.7.

A mechanical production of the sine curve is described in detail at the beginning of section 4.8. The basic sine and cosine graphs and some scaling and shifting properties of these graphs are described later. The tangent, cotangent, secant, and cosecant graphs are also given.

We derive the Pythagorean identities and explain the method for solving identities in section 4.9. This is followed by a short section on solving simple trigonometric equations (section 4.10). The addition identities and double- and half-angle identities are derived in sections 4.11 and 4.12, respectively. The sine and cosine rules for solving triangles are introduced in section 4.13.

Components of vectors and the dot product of vectors (a continuation of vectors from chapter 2 are introduced as an application of trigonometry in section 4.14.)

This chapter ends with a final topic on identities (section 4.15). This is an interesting and challenging aspect of learning trigonomety.

Some knowledge of the basic geometry of triangles including right triangles and similar triangles will be helpful for this chapter. These topics are presented in chapter 9.

4.2 ANGLES IN THE CARTESIAN PLANE

An angle in the Cartesian plane is determined by drawing a position vector for any given *reference point* in the plane (except the origin) and a circular arc to indicate the amount by which the position vector is rotated in a counterclockwise direction from the x-axis. If the reference point is not on the x- or y-axis, then the angle is in one of the four quadrants. Figure 4.2 shows an angle α in the first quadrant (corresponding to a reference point P in the first quadrant) and an angle θ finishing in the third quadrant (corresponding to a reference point Q in the third quadrant).

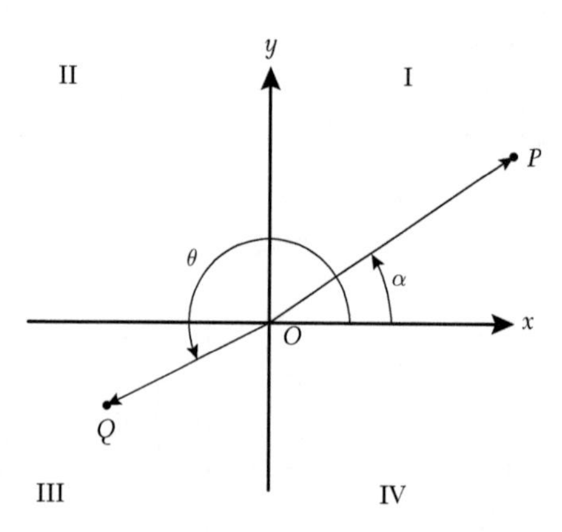

FIGURE 4.2. Quadrants in the Cartesian plane.

An angle in the first quadrant is called *acute*, the second quadrant is *obtuse*, and the third or fourth quadrant is a *reflex angle*.

DEFINITION 4.2.1. *The radian measure of an angle is the length of the arc of the unit circle corresponding to the angle.*

In figure 4.3, the length of the arc of the unit circle corresponding to the angle θ is denoted as l. When it is understood that angles are measured in radians, we can state $\theta = l$. Angles can also be measured in degrees, but radian measure is the unit of measurement in which mathematical formulas (e.g., in calculus) involving trigonometric functions can be expressed in their simplest form.

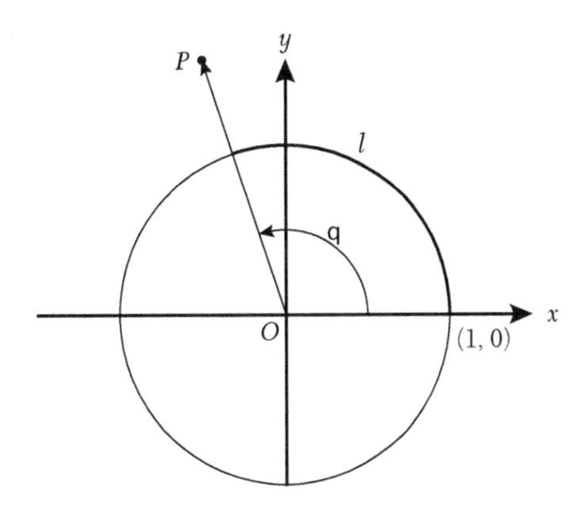

FIGURE 4.3. Radian measure.

It has been known since at least the time of the classical Greek mathematicians that if the radius of a circle is r units, then the circumference (length) of the circle is $2\pi r$ units. In particular, if $r = 1$ (i.e., the circle is a unit circle), then the circumference of the circle is 2π. Therefore, the radian measure of a full angle is 2π (corresponding to 360°) and of any angle is in the same proportion to 2π as its angle is measured in 360°. Figure 4.4 shows four special cases of radian measure.

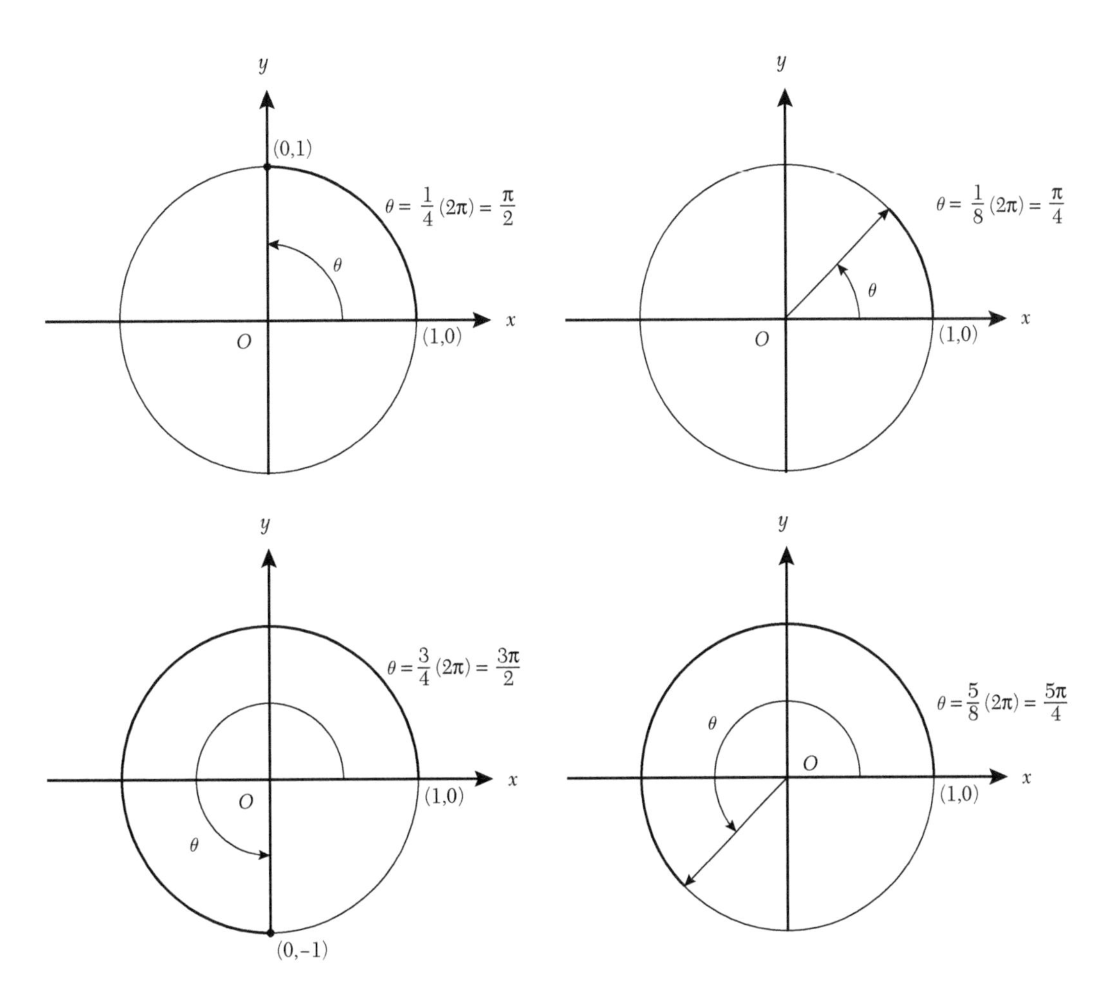

FIGURE 4.4. Some special cases of radian measure.

The formulas for conversion of degrees to radians and vice versa can be derived as follows:

EXAMPLE 4.2.1.

$$60° = 60 \times \frac{\pi}{180} \text{ radians} = \frac{\pi}{3} \text{ radians}, \qquad 30° = 30 \times \frac{\pi}{180} \text{ radians} = \frac{\pi}{6} \text{ radians},$$

$$45° = 45 \times \frac{\pi}{180} \text{ radians} = \frac{\pi}{4} \text{ radians}, \qquad 90° = 90 \times \frac{\pi}{180} \text{ radians} = \frac{\pi}{2} \text{ radians}.$$

◈

4.3 TRIGONOMETRIC RATIOS

DEFINITION 4.3.1. *Corresponding to any reference point* P(a,b) *in the Cartesian plane (except* P(0, 0)) *determining an angle* θ, *we define the trigonometric ratios sine (sin), cosine (cos), and tangent (tan) as follows:*

$$\boxed{\sin(\theta) = \frac{b}{r}, \cos(\theta) = \frac{a}{r}, \text{ and } \tan(\theta) = \frac{b}{a}},$$

where r *is the direct distance of* P *from the origin. If* a = 0 *then* tan(θ) *is undefined.*

REMARK 4.3.1. It is important to realize that, for a fixed value of θ, the values of the trigonometric ratios do not depend on the distance of the reference point P from the origin; that is, the ratios do not depend on the value of r. This is a consequence of the fact that reference points P_1 and P_2 on the same ray from the origin (as shown in figure 4.5) determine similar triangles, and so, by theorem 9.5.14(a), the corresponding ratios of the sides of the triangles are equal.

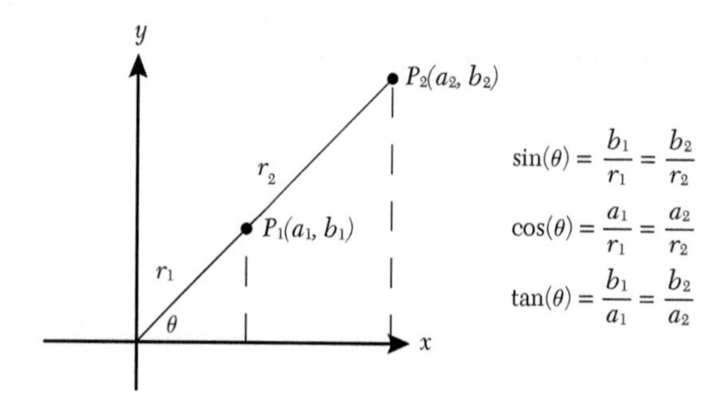

FIGURE 4.5. Trigonometric ratios for similar triangles.

EXAMPLE 4.3.1. In figure 4.6, the sine, cosine, and tangent trigonometric ratios are evaluated for four different angles (one angle in each quadrant). ◈

Furthermore, it can be ascertained from figure 4.6 that the signs of the trigonometric ratios are positive or negative depending on the quadrant to which θ belongs. The sign of each ratio in each quadrant is shown in table 4.1.

TABLE 4.1. The signs of the trigonometric ratios

θ	I	II	III	IV
sin(θ)	+	+	−	−
cos(θ)	+	−	−	+
tan(θ)	+	−	+	−

Table 4.1 can be summarized by means of the *ASTC* diagram shown in figure 4.7: the *A* in the first quadrant means that all trigonometric ratios are positive; the *S* in the second means that sine is positive, and cosine and tangent are negative; the *T* in the third means that tangent is positive, and cosine and sine are negative; and the *C* in the fourth means that cosine is positive, and sine and tangent are negative. This diagram is sometimes memorized using the phrase "*a*ll *s*tudents *t*ake *c*alculus."

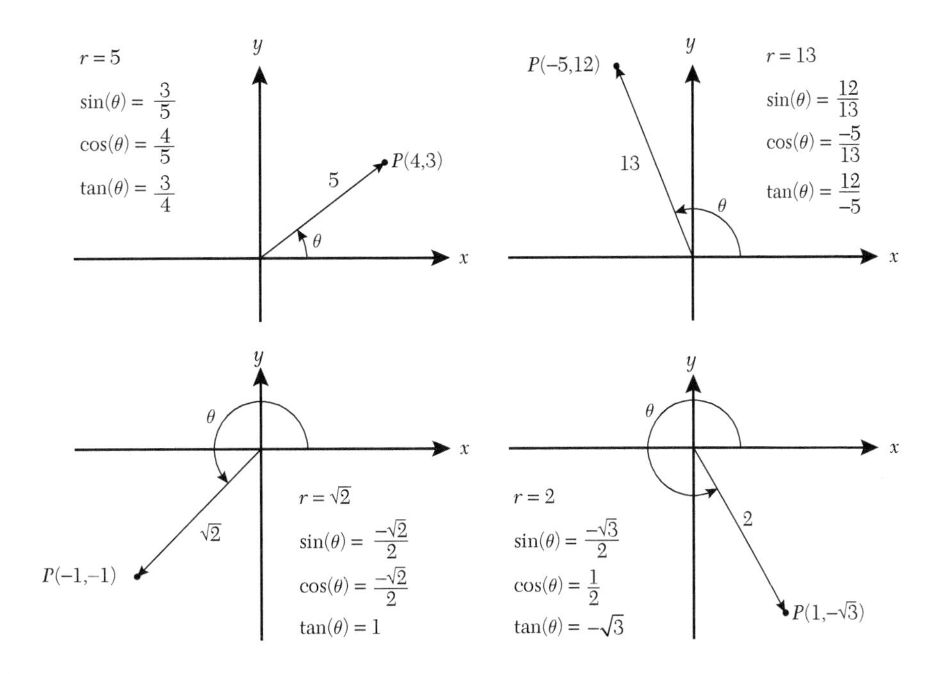

FIGURE 4.6. Examples of trigonometric ratios.

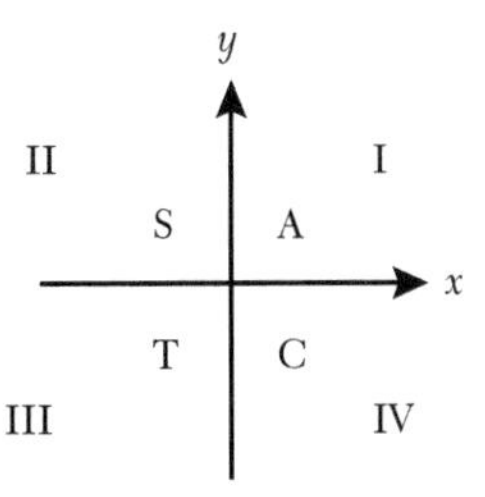

FIGURE 4.7. *ASTC.*

REMARK 4.3.2. It follows from Remark 4.3.1 that any trigonometric ratio can be defined with respect to a reference point that is a unit distance from the origin, that is, a reference point $P(a,b)$ for which $r = \sqrt{a^2 + b^2} = 1$. If the corresponding angle is θ, then

$$\sin(\theta) = \frac{b}{r} = \frac{b}{1} = b, \quad \cos(\theta) = \frac{a}{r} = \frac{a}{1} = a$$

and so, for any point $P(a,b)$ on the unit circle, we can write $P(a,b) = P(\cos(\theta),\ \sin(\theta))$, as shown in figure 4.8.

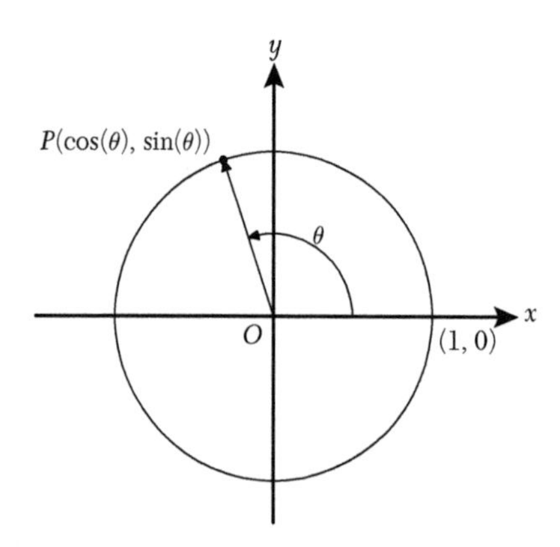

FIGURE 4.8. The definition of sine and cosine ratios.

4.4 SPECIAL ANGLES

There are special cases that can be considered for the position of the reference point *P*. If *P* is on the x- or y-axis, then the y- or x-coordinate is zero, respectively. Figure 4.9 and table 4.2 show five possibilities. Note that the tangent ratio is undefined if $\theta = \frac{\pi}{2}$ or $3\frac{\pi}{2}$. The trigonometric ratios for $\theta = 0$ and 2π coincide because they have the same reference point $P(1,0)$.

TABLE 4.2. Special cases of trigonometric ratios

θ	0	$\frac{\pi}{2}$	π	$\frac{3\pi}{2}$	2π
$\sin(\theta)$	0	1	0	−1	0
$\cos(\theta)$	1	0	−1	0	1
$\tan(\theta)$	0	Undefined	0	Undefined	0

Certain trigonometric ratios occur so frequently in everyday measurements that they warrant special attention. These are the trigonometric ratios of 30°, 45°, and 60°, and the radian measurements of these angles are $\frac{\pi}{6}$, $\frac{\pi}{4}$, and $\frac{\pi}{3}$, respectively. It is helpful to construct the right triangles containing these angles, as shown in figure 4.10. A *right triangle* has one right angle, and the side opposite the right angle is called the *hypotenuse*. In any right triangle, the lengths of the sides satisfy the Pythagorean Theorem (theorem 9.5.8(a)). An isosceles right triangle with short side length equal to 1 and hypotenuse with length $\sqrt{2}$ is shown in the first diagram. We can construct a 60°–30° triangle by dropping a perpendicular from the apex of an equilateral triangle to the base of the triangle in order to produce two congruent right triangles. If the equilateral triangle has side length 2, then the height of the congruent triangles is $\sqrt{3}$ as shown in the second diagram.

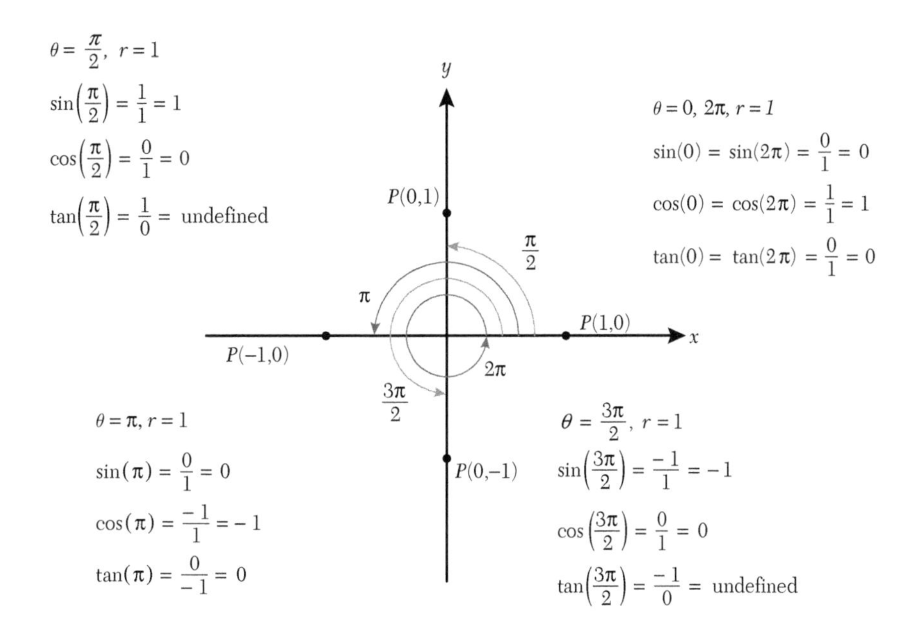

FIGURE 4.9. Special cases of trigonometric ratios.

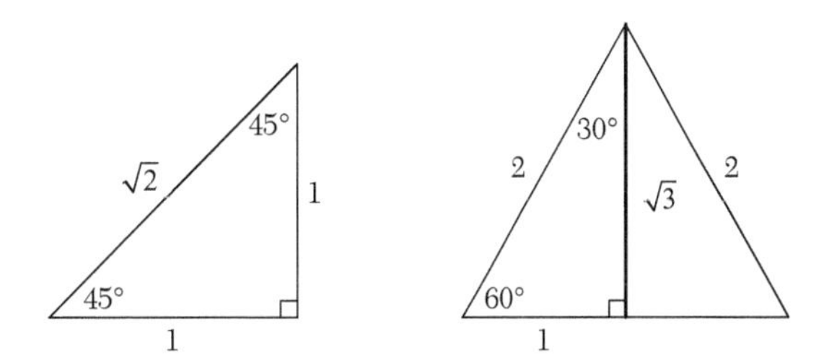

FIGURE 4.10. Special triangles.

If the 45° triangle is scaled by a factor $\frac{1}{\sqrt{2}} = \frac{\sqrt{2}}{2}$, then the length of the hypotenuse will be 1 and the short side lengths will be $\frac{\sqrt{2}}{2}$. Similarly, if the 30°–60° triangle is scaled by a factor $\frac{1}{2}$, then the length of the hypotenuse will be 1, the length of the shortest side will be $\frac{1}{2}$, and the height will be $\frac{\sqrt{3}}{2}$. If each of these scaled triangles is positioned in the Cartesian plane so that the base of the triangle aligns with the x-axis and the hypotenuse coincides with a unit radius from the origin, then the vertex on the unit circle is a coordinate pair from which the trigonometric ratios for $\theta = 45° = \frac{\pi}{4}$ and $\theta = 60° = \frac{\pi}{3}$ or $\theta = 30° = \frac{\pi}{6}$ can be determined. What's more, by reflection of the triangles across the coordinate axes, the trigonometric ratios for certain multiples of $\theta = \frac{\pi}{4}$ and $\theta = 30° = \frac{\pi}{6}$, that is, $\frac{3\pi}{4}, \frac{5\pi}{4}, \frac{7\pi}{4}$ and $\frac{2\pi}{3}, \frac{5\pi}{6}, \frac{7\pi}{6}, \frac{4\pi}{3}, \frac{5\pi}{3}, \frac{11\pi}{6}$, respectively, can also be determined. This is shown in figure 4.11, which, in the context of trigonometry, is sometimes referred to as the "unit circle."

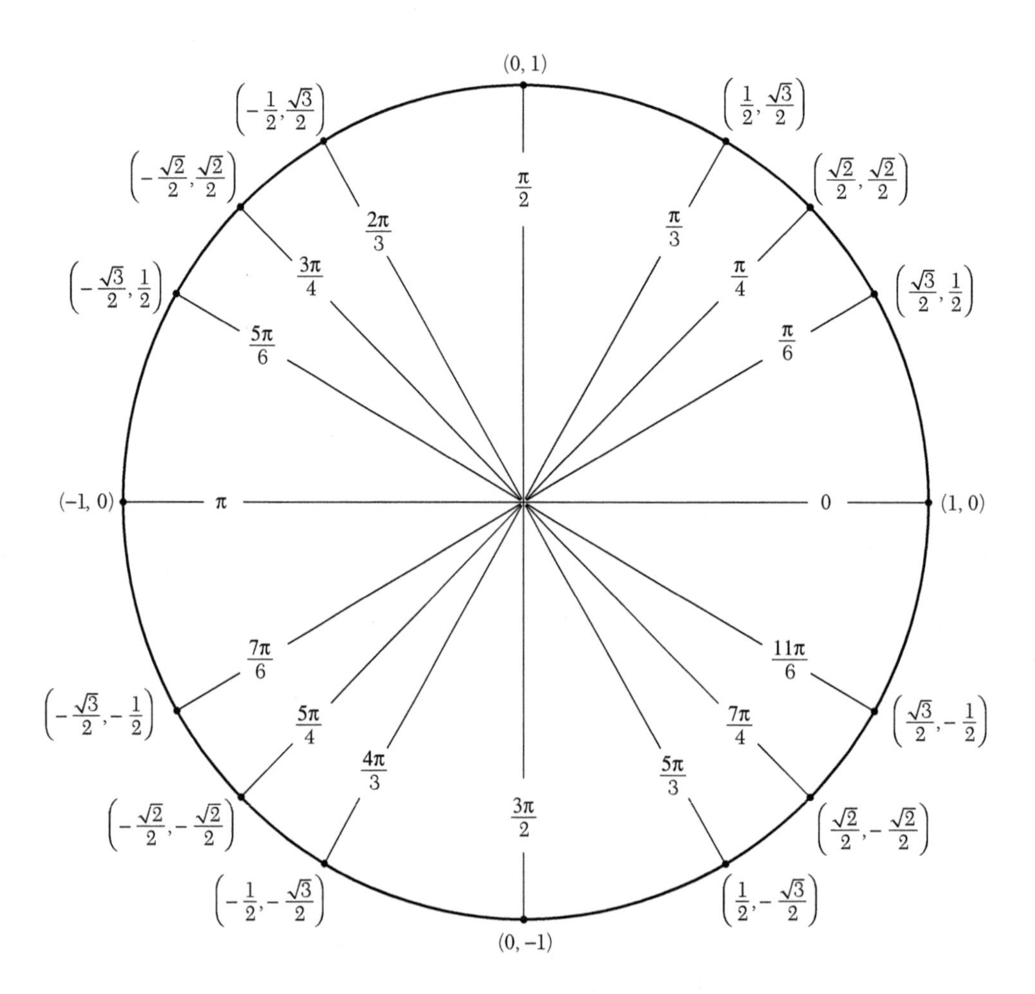

FIGURE 4.11. The unit circle.

EXAMPLE 4.4.1. From figure 4.11 we can read, for example, that $\cos\left(2\frac{\pi}{3}\right)=-\frac{1}{2}$, $\sin\left(\frac{2\pi}{3}\right)=\frac{\sqrt{3}}{2}$, $\cos\left(\frac{3\pi}{4}\right)=\frac{-\sqrt{2}}{2}$, and $\sin\left(\frac{5\pi}{4}\right)=\frac{-\sqrt{2}}{2}$. ◈

4.5 NEGATIVE ANGLES AND PERIODICITY

The definition of trigonometric ratios is unchanged if angles are measured in a clockwise direction rather than a counterclockwise direction; however, an angle measured in a clockwise direction is negative. Figure 4.12 shows the relationship between trigonometric ratios of a positive angle θ in the first quadrant and the corresponding negative angle in the fourth quadrant. (The relationship between a positive angle in the second quadrant and the corresponding negative angle in the third quadrant is similar.) It is clear that the sign of the y-coordinate changes when the sign of any angle changes, and so, the sine and tangent ratios change their sign but the cosine ratio does not change its sign.

EXAMPLE 4.5.1. By reading from the unit circle (figure 4.11), we find that

$$\sin\left(\frac{-\pi}{3}\right)=-\sin\left(\frac{\pi}{3}\right)=\frac{-\sqrt{3}}{2},\ \cos\left(\frac{-\pi}{3}\right)=\cos\left(\frac{\pi}{3}\right)=\frac{1}{2},$$

and

$$\tan\left(\frac{-\pi}{3}\right)=-\tan\left(\frac{\pi}{3}\right)=\left(\frac{-\sqrt{3}}{2}\right)\cdot\left(\frac{2}{1}\right)=-\sqrt{3}.$$

◈

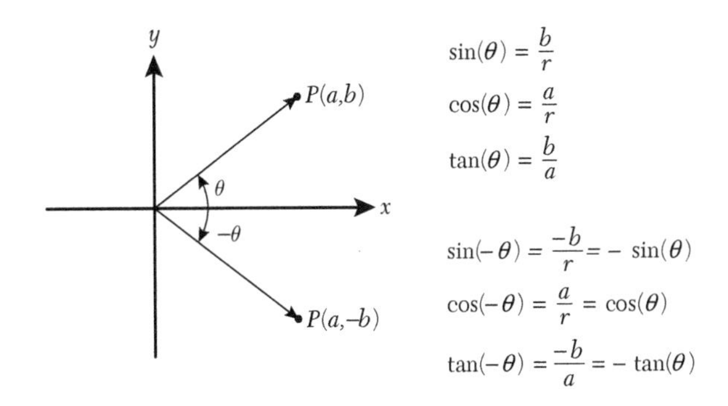

FIGURE 4.12. Negative angles.

It was demonstrated in section 4.4 that $\sin(2\pi)$ is the same trigonometric ratio as $\sin(0)$ because both ratios are determined from the same reference point, namely $P(1,0)$. In general, trigonometric ratios are the same for any two angles that differ by an integer multiple of 2π because the ratios are determined from the same reference point. This property is called *2π-periodicity* of trigonometric ratios. It means that $\sin(\theta+n\cdot 2\pi)=\sin(\theta)$ and $\cos(\theta+n\cdot 2\pi)=\cos(\theta)$ for any value of θ and integer n. The tangent ratio is *π-periodic* because the tangent ratios are equal for two reference points on the same line through the origin in opposite quadrants, that is, $\tan(\theta+n\cdot\pi)=\tan(\theta)$ for any value of θ and integer n.

EXAMPLE 4.5.2. By reading from the unit circle (figure 4.11) again, we find that

$$\sin\left(\frac{13\pi}{4}\right)=\sin\left(\frac{5\pi}{4}+2\pi\right)=\sin\left(\frac{5\pi}{4}\right)=\frac{-\sqrt{2}}{2}$$

and

$$\tan\left(\frac{13\pi}{4}\right)=\tan\left(\frac{\pi}{4}+3\pi\right)=\tan\left(\frac{\pi}{4}\right)=1.$$ ◈

4.6 RECIPROCAL TRIGONOMETRIC RATIOS

DEFINITION 4.6.1. *The trigonometric ratios cosecant (csc), secant (sec), and cotangent (cot) are defined as the reciprocals of the sine, cosine, and tangent ratios, respectively. Therefore, if a reference point $P(a,b)$ is at a distance* r *from the origin, with corresponding angle θ, then*

$$\cos(\theta)=\frac{1}{\sin(\theta)}=\frac{r}{b},\ \sec(\theta)=\frac{1}{\cos(\theta)}=\frac{r}{a},\ \text{and}\ \cot(\theta)=\frac{1}{\tan(\theta)}=\frac{a}{b}.$$

These ratios satisfy the same properties with respect to negative angles as their reciprocals, for example, $\csc(-\theta)=-\csc(\theta)$ and $\sec(-\theta)=\sec(\theta)$, and they have the same periodicity, for example, $\csc(\theta+n\cdot 2\pi)=\csc(\theta)$ and $\cot(\theta+n\cdot\pi)=\cot(\theta)$ for any integer n.

EXAMPLE 4.6.1. From the unit circle (figure 4.11), $\csc\left(\frac{5\pi}{6}\right)=2$, $\sec\left(\frac{5\pi}{6}\right)=\frac{-2}{\sqrt{3}}$, and $\cot\left(\frac{5\pi}{6}\right)=\left(\frac{-\sqrt{3}}{2}\right)\cdot\left(\frac{2}{1}\right)=-\sqrt{3}$. ◈

The following *reciprocal identities* are a consequence of the definitions of the trigonometric ratios (as can be easily verified):

$$\tan(\theta)=\frac{\sin(\theta)}{\cos(\theta)}\ \text{and}\ \cot(\theta)=\frac{\cos(\theta)}{\sin(\theta)}.$$

4.7 COFUNCTION IDENTITIES

The following relationships are known as the *cofunction identities* because each trigonometric ratio is related to its "*co*" ratio:

$$\begin{array}{ll} \sin(\theta)=\cos\left(\frac{\pi}{2}-\theta\right) & \cos(\theta)=\sin\left(\frac{\pi}{2}-\theta\right) \\ \tan(\theta)=\cot\left(\frac{\pi}{2}-\theta\right) & \cot(\theta)=\tan\left(\frac{\pi}{2}-\theta\right) \\ \sec(\theta)=\csc\left(\frac{\pi}{2}-\theta\right) & \csc(\theta)=\sec\left(\frac{\pi}{2}-\theta\right) \end{array} \quad (4.1)$$

These identities can be verified from the symmetry, as shown in figure 4.13 (if θ is in the first quadrant). The general validity of these identities (for any value of θ) can be proved using the identities in section 4.11.

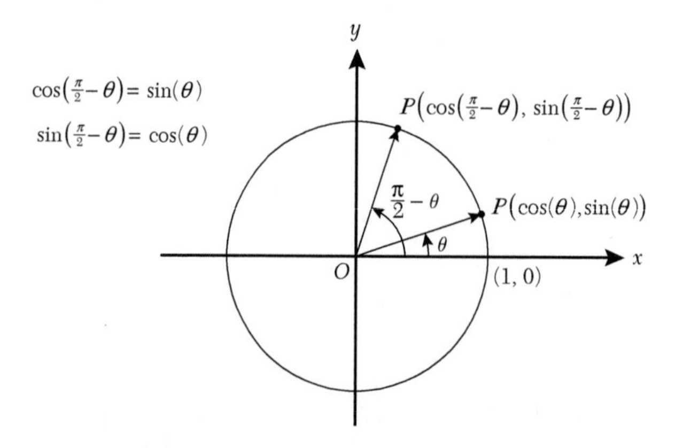

FIGURE 4.13. The cofunction identities.

4.8 TRIGONOMETRIC GRAPHS

In this section, we will demonstrate the generation of a sine graph and, from this, deduce what the cosine, tangent, and other trigonometric graphs should look like.

4.8.1 Generation of a Sine Curve

Suppose that two students, Curtis and Candice, do the following experiment: Curtis pulls a strip of ticker tape through a slot at a constant rate, while Candice turns a disk at a constant rate so that a pen that is attached at a point on the boundary of the disk moves up and down along a groove in the slot. This experiment can actually be performed if the apparatus is available, but it is sufficient for our purposes to think of it as an imaginary experiment. A simplified illustration of the experiment is shown in figure 4.14. The problem is to describe the graph mathematically (wave) that the pen traces out on the ticker tape.

The number L is called the *wave amplitude*, and it is also the radius of the disk. The number d is called the *wave length* or *period*, and it is the horizontal distance from any point on one wave to the equivalent point on the next wave. A *cycle* of the trace is one complete wave, that is, the trace of the pen that results as Candice rotates the disk through one complete revolution. The horizontal dotted line, or *reference line*, is the position of the pen corresponding to the position of the disk when $\theta = 0$. The position of the pen along the reference line can be taken as the value of the independent variable x and the height of the pen above (or depth of the pen below) the reference line can be taken as the value of the dependent variable y. We can assume that $x=0$ and $y=0$ at the starting position of the trace. We will suppose that $y>0$ if the pen is above the reference line and $y<0$ if the pen is below the reference line.

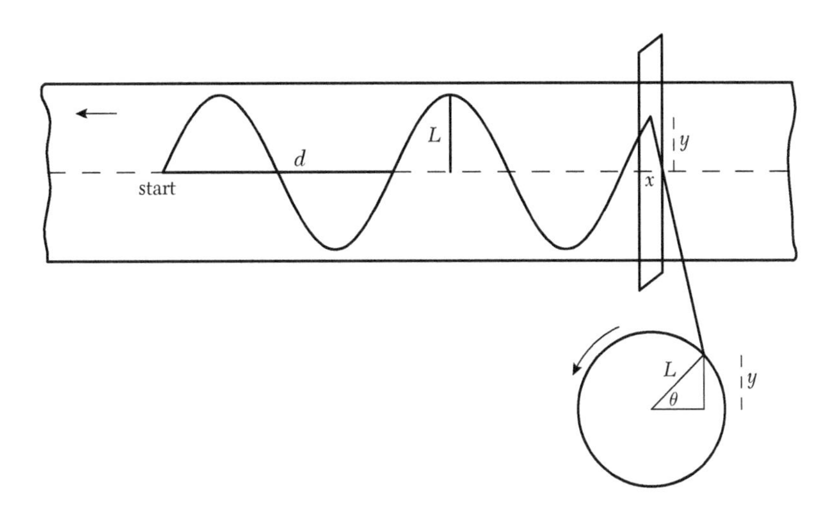

FIGURE 4.14. The generation of a sine curve.

We can deduce from the right triangle shown in the disk that $\frac{y}{L}=\sin(\theta)$, which we can also write as

$$y = L\sin\theta. \tag{4.2}$$

Furthermore, it is logically true that the proportion of x to one complete period d is the same as the proportion of θ to one complete revolution 2π of the disk, that is,

$$\frac{x}{d} = \frac{\theta}{2\pi}.$$

This equation can also be expressed as

$$\theta = \frac{2\pi x}{d}.$$

and substituting this in formula (4.2) produces

$$y = L\sin\left(\frac{2\pi x}{d}\right). \tag{4.3}$$

This equation gives the value of y explicitly in terms of x and tells us that the wave marked out by the trace of the pen is a sine curve.

4.8.2 Sine and Cosine Graphs

If we set $L=1$ and $d=2\pi$ in formula (4.3), then the equation is $y=\sin(x)$ and the corresponding graph (the *sine graph*) is shown in figure 4.15.

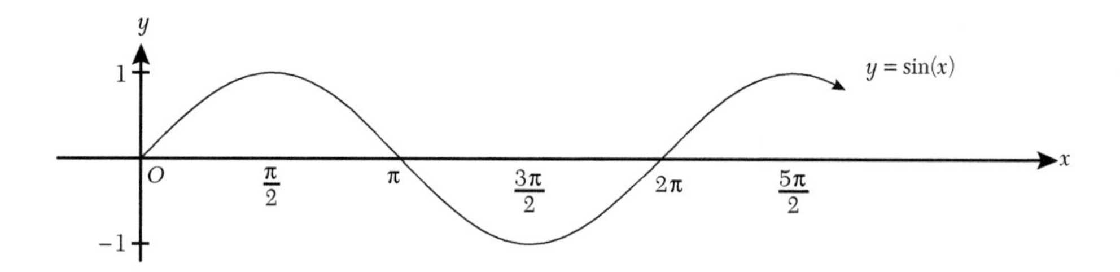

FIGURE 4.15. The sine graph for positive angles.

Note that the sine graph cuts the x-axis at 0, π, and 2π and the sine graph peaks at $x=\left(\frac{\pi}{2}\right)$, $y=1$ and $x=\left(\frac{3\pi}{2}\right)$, $y=-1$. This is consistent with the information in table 4.2. Thus, the period of the sine graph is 2π and its amplitude is 1. Furthermore, the sine graph is positive in the first and second quadrants and negative in the third and fourth quadrants, as indicated in table 4.1. As shown in figure 4.16, the sine graph can also be extended in the negative direction by application of the identity $\sin(-x)=-\sin(x)$.

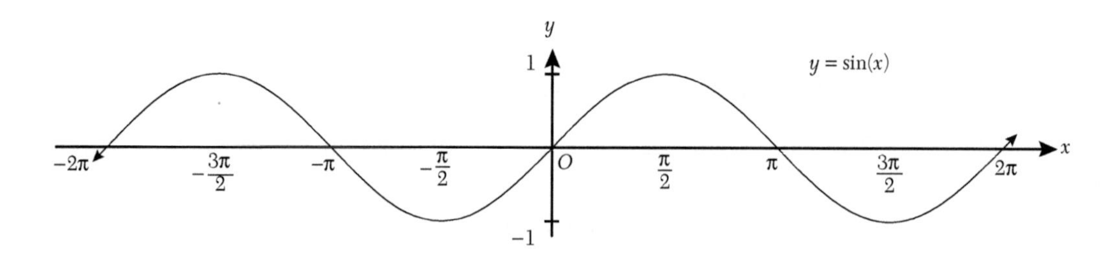

FIGURE 4.16. The sine graph.

The sine graph continues indefinitely in both directions (to the left and right), cuts the x-axis at all integer multiples of π, and peaks at integer multiples of $\frac{\pi}{2}$; that is,

$$\sin(n\pi)=0,\ \sin\left(\frac{\pi}{2}+2n\pi\right)=1,\ \sin\left(\frac{3\pi}{2}+2n\pi\right)=-1,\ \text{for } n\in\mathbb{Z}.$$

The cosine graph can be obtained by an application of the cofunction identity

$$\cos(x)=\sin\left(\frac{\pi}{2}-x\right)=-\sin\left(x-\frac{\pi}{2}\right),$$

which tells us that the cosine graph is a horizontally shifted sine graph, as shown in figure 4.17. (Check that coordinates on the graph correspond with the data in tables 4.1 and 4.2.)

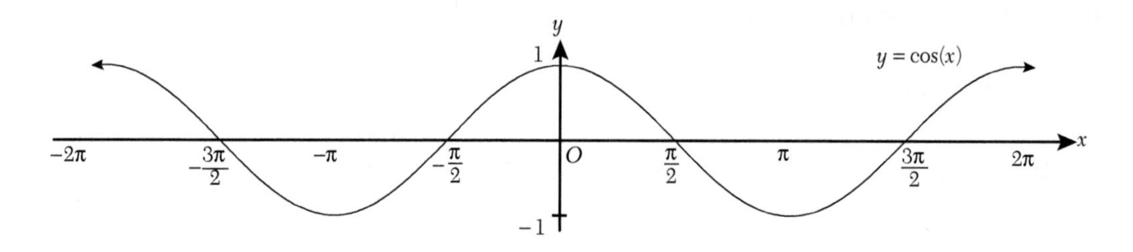

FIGURE 4.17. The cosine graph.

The cosine graph continues indefinitely in both directions (to the left and right), cuts the x-axis at all integer multiples of $\frac{\pi}{2}$, and peaks at integer multiples of π; that is,

$$\cos\left(\frac{n\pi}{2}\right)=0,\ \cos(2n\pi)=1,\ \cos((2n+1)\pi)=-1,\ \text{for } n\in\mathbb{Z}.$$

We say that the cosine graph is $\frac{\pi}{2}$*-out of phase* with the sine graph, because the cosine graph can be obtained by shifting the sine graph to the left by $\frac{\pi}{2}$ units. This property can be expressed as the identity

$$\boxed{\cos(\theta)=\sin\left(\theta+\frac{\pi}{2}\right)}.$$

4.8.3 Scaling and Shifting of the Sine and Cosine Graphs

If we set $L=1$ and $d=\pi$ in formula (4.3), then the equation is $y=\sin(2x)$ and the corresponding graph with period π is shown in figure 4.18.

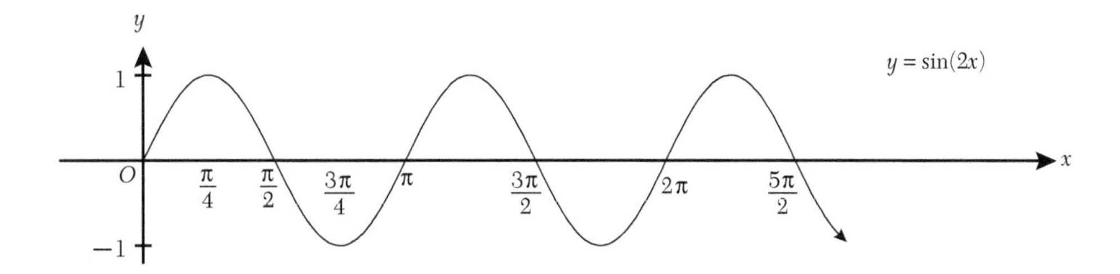

FIGURE 4.18. The sine graph with period π.

This is a *horizontal scaling* of the graph of $y=\sin(x)$.

In general, we make the following remark about the period of a sine graph.

REMARK 4.8.1. The period of the graph $y=L\sin(kx)$, for any real number $k\neq 0$, is $\frac{2\pi}{|k|}$.

EXAMPLE 4.8.1. The period of the graph of $y=-2\sin(\pi x)$ is $\frac{2\pi}{\pi}=2$, and the amplitude is 2. Note that the first peak occurs at $y=-2$ (instead of $y=2$), as shown in figure 4.19. ◈

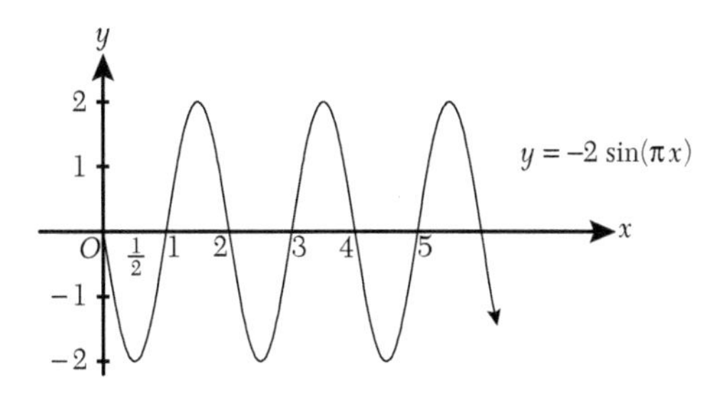

FIGURE 4.19. A sine graph with period 2.

EXAMPLE 4.8.2. The period of the graph of $y=\cos\left(\frac{x}{2}\right)$ is $\frac{2\pi}{1/2}=4\pi$, and the amplitude is 1, as shown in figure 4.20. ◈

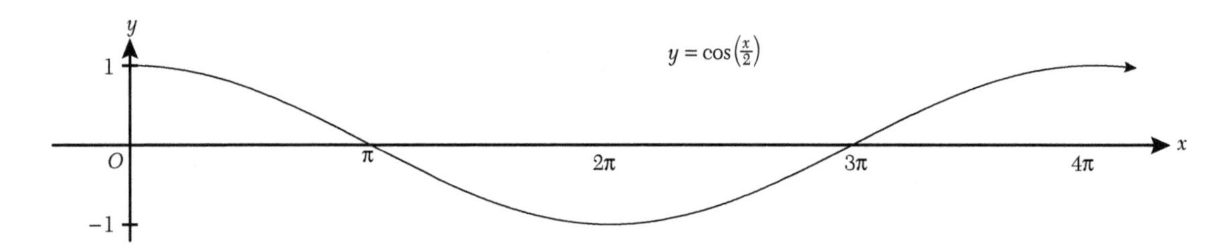

FIGURE 4.20. The cosine graph with period 4π.

In order to draw the graph of an equation $y = \sin(x - a)$ for any real number a, we note that if $x = a$, then $y = 0$; if $x = a + \left(\frac{\pi}{2}\right)$, then $y = 1$; and so on. This means we can take $x = a$ as the starting point on the x-axis for one cycle and label intervals of $\frac{\pi}{2}$ along the x-axis starting from $x = a$. The resulting graph is a *horizontal shift* of the graph of $y = \sin(x)$.

EXAMPLE 4.8.3. One cycle of the graph of $y = \sin\left(x - \left(\frac{\pi}{4}\right)\right)$ starts on the x-axis at $x = \frac{\pi}{4}$, as shown in figure 4.21. ◈

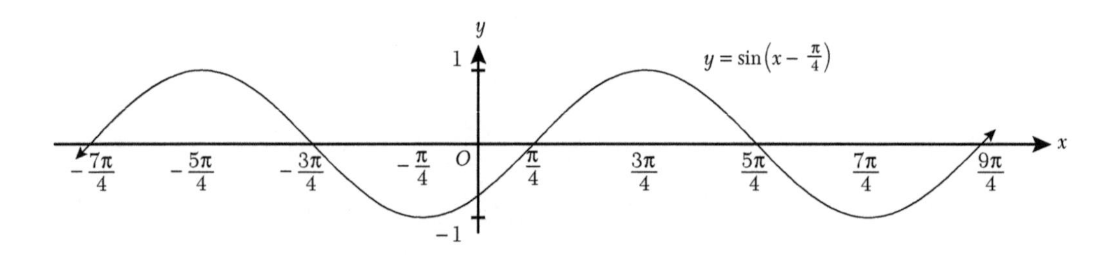

FIGURE 4.21. A shifted sine graph.

The graph of the equation $y = \sin(kx - a)$ or $y = \cos(kx - a)$ will be a combination of a horizontal shift and a horizontal scaling. An equivalent equation is $y = \sin\left(k\left(x - \frac{a}{k}\right)\right)$ or $y = \cos\left(k\left(x - \frac{a}{k}\right)\right)$. The period of the graph is $\frac{2\pi}{|k|}$, and the horizontal shift is to the right or left $\left|\frac{a}{k}\right|$ units, depending on whether the sign of $\frac{a}{k}$ is positive or negative, respectively.

EXAMPLE 4.8.4. The graph of

$$y = \cos\left(-2x - \frac{\pi}{3}\right) = \cos\left(-2\left(x + \frac{\pi}{6}\right)\right) = \cos\left(2\left(x + \frac{\pi}{6}\right)\right)$$

has period π and a shift to the left by $\frac{\pi}{6}$ units, as shown in figure 4.22. ◈

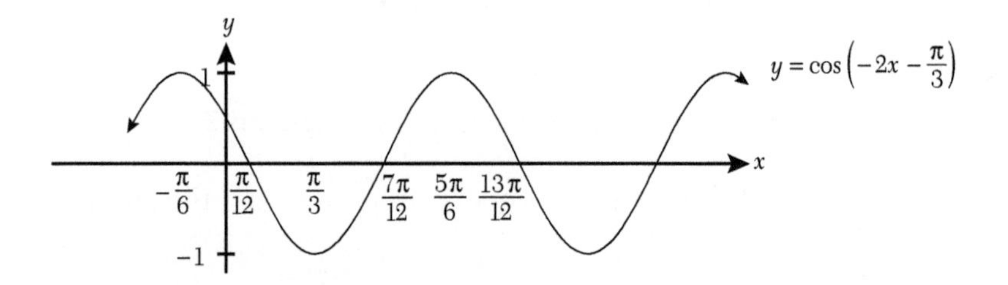

FIGURE 4.22. A shifted cosine graph.

4.8.4 Tangent and Cotangent Graphs

The tangent ratio $\tan(x)$ is undefined at $x = \frac{\pi}{2}$ and $x = \frac{3\pi}{2}$, as shown in table 4.2.

If a coordinate pair (a,b) on the unit circle corresponds to an angle x between 0 and $\frac{\pi}{2}$, then the tangent ratio $\frac{b}{a}$ grows larger and larger as x increases from 0 to $\frac{\pi}{2}$ (because the denominator a grows smaller and smaller, approaching 0, while b increases, approaching 1). This property is indicated in the graph of $y = \tan(x)$ in figure 4.23 by means of a vertical line—called a *vertical asymptote*—perpendicular to the x-axis at $x = \frac{\pi}{2}$. The graph grows closer and closer to the vertical asymptote as x approaches $\frac{\pi}{2}$. Because $\tan(x)$ is periodic with period π, the vertical asymptote repeats at every multiple integer of $\frac{\pi}{2}$. Note that $\tan\left(\frac{\pi}{4}\right) = 1$ and $\tan\left(\frac{-\pi}{4}\right) = -1$ and the sign of the tangent ratio alternates with the quadrants.

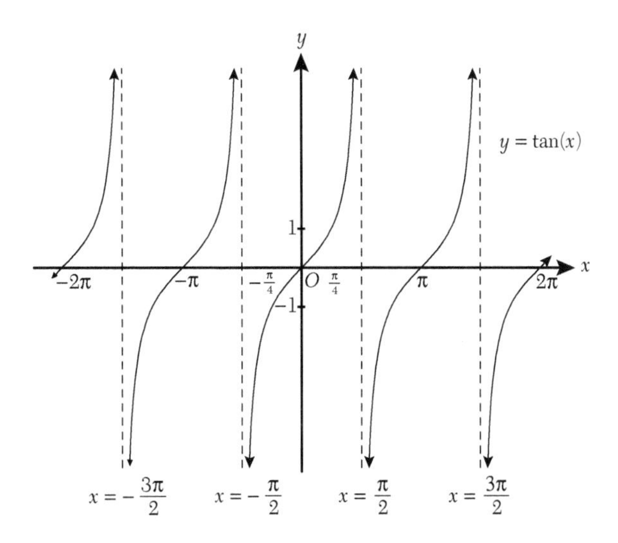

FIGURE 4.23. The tangent graph.

The graph of the cotangent ratio, $\cot(x)$, is undefined if x is any integer multiple of π, because these are the points at which $\tan(x)$ is zero and $\cot(x)$ is the reciprocal of $\tan(x)$. When we draw the graph of $y = \cot(x)$, the vertical asymptotes occur at these values of x. The cotangent ratio is zero at all values of x for which the tangent ratio is undefined. In figure 4.24, one component of the graph of $y = \tan(x)$ is shown by means of a dotted curve and the graph of $\cot(x)$ by means of a solid curve.

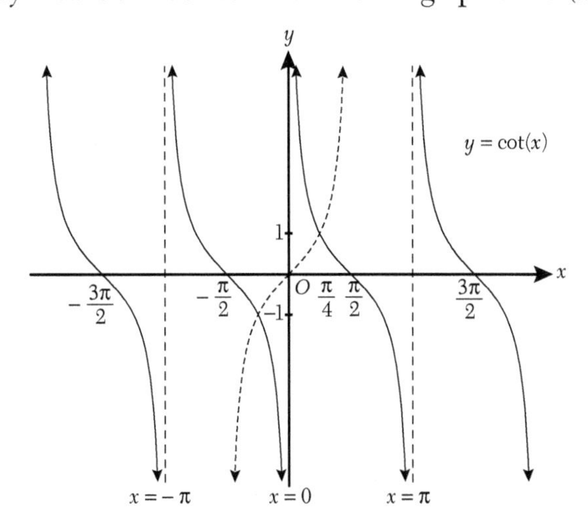

FIGURE 4.24. The cotangent graph.

4.8.5 Cosecant and Secant Graphs

The graphs of $\csc(x)$ and $\sec(x)$ have vertical asymptotes at the values of x where $\sin(x)$ and $\cos(x)$ are zero, respectively. In figures 4.25 and 4.26, the dotted curves are the graphs of $y = \sin(x)$ and $\cos(x)$, and the solid curves are the graphs of $y = \csc(x)$ and $\sec(x)$. The graphs touch at values of x for which $y = 1$ or -1.

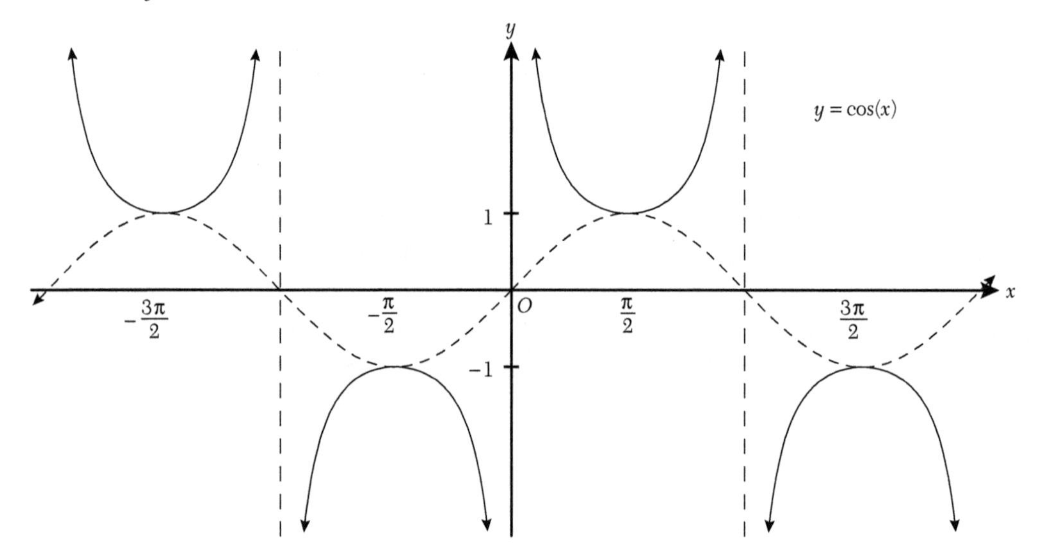

FIGURE 4.25. The cosecant graph.

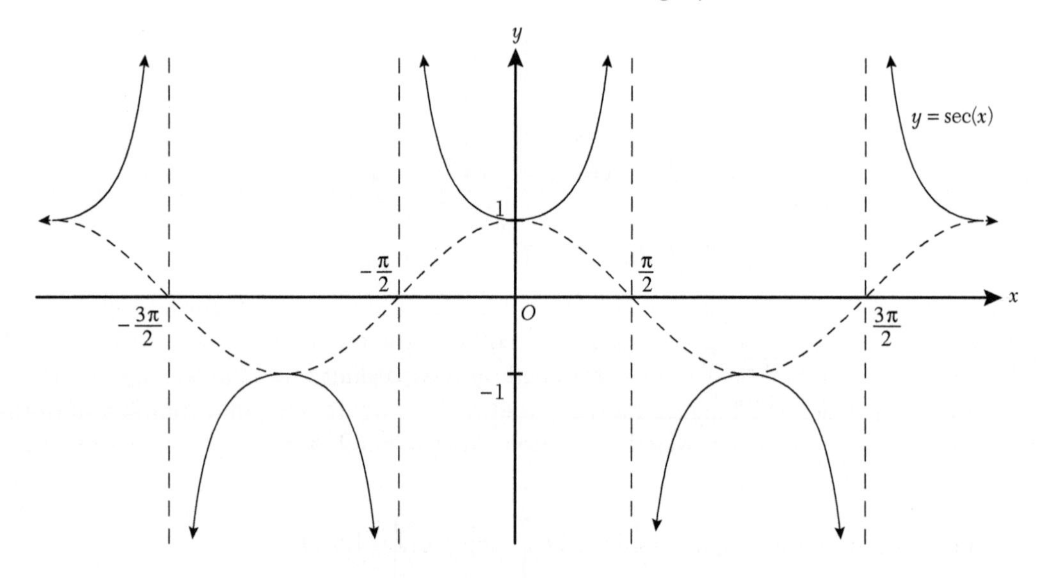

FIGURE 4.26. The secant graph.

4.9 PYTHAGOREAN IDENTITIES

For any point $P(a,b)$ on the unit circle, $a^2 + b^2 = 1$. If θ is the angle corresponding to a reference point $P(a,b)$, then $\cos(\theta) = a$ and $\sin(\theta) = b$. By convention, we write "$\sin^2\theta$" to mean $(\sin(\theta))^2$ and the same for other trigonometric ratios. Thus, another way to write the equation for the unit circle is

$$\boxed{\text{the first Pythagorean identity:} \cos^2\theta + \sin^2\theta = 1} \qquad (4.4)$$

This is a restatement of the Pythagorean Theorem if θ is an angle in the first quadrant, and $a = \cos(\theta)$ and $b = \sin(\theta)$ are regarded as the lengths of the short sides of a right triangle with a hypotenuse of length 1. If we divide both sides of formula (4.4) by $\cos^2\theta$, distribute the denominator on the left-hand side, and rewrite the resulting terms by making use of the reciprocal trigonometric ratios, then formula (4.4) becomes

$$\boxed{\text{the second Pythagorean identity: } 1+\tan^2\theta = \sec^2\theta}. \tag{4.5}$$

In the same way, by dividing both sides of formula (4.4) by $\sin^2(\theta)$, we derive

$$\boxed{\text{the third Pythagorean identity: } 1+\cot^2\theta = \csc^2\theta}. \tag{4.6}$$

Strictly speaking, the second and third Pythagorean identities have been derived only for values of θ for which the terms are defined. In the case of the second Pythagorean identity, for instance, the terms $\tan^2\theta$ and $\sec^2\theta$ are not defined if θ is an integer multiple of $\frac{\pi}{2}$.

In trigonometry, we use known identities to prove or verify more identities. The usual technique for doing this is to denote the left-hand side of the unproven identity as LHS and the right-hand side as RHS. One of the sides, either LHS or RHS, is then restated and manipulated by means of known identities and algebraic methods until it is shown to be equivalent to the other side. Alternatively, both sides can be restated and manipulated until it can be shown that they are both equivalent to some (new) expression. In the following example, each line is a restatement of the previous line by the means explained to the right of each line.

Example 4.9.1. Verify the identity: $\tan(\theta)+\cot(\theta)=\sec^2(\theta)\cot(\theta)$.

Answer:

$$\begin{aligned}
\text{LHS} &= \tan(\theta)+\cot(\theta) \\
&= \tan(\theta)+\frac{1}{\tan(\theta)} && \texttt{reciprocal ratio} \\
&= \frac{1+\tan^2(\theta)}{\tan(\theta)} && \texttt{algebraic addition} \\
&= \frac{\sec^2(\theta)}{\tan(\theta)} && \texttt{second Pythagorean identity} \\
&= \sec^2(\theta)\cot(\theta) && \texttt{reciprocal ratio} \\
&= RHS.
\end{aligned}$$

◈

The next example is an application of FOIL (see section 1.8.1).

Example 4.9.2. Verify the identity: $(\sin(\theta)-\cos(\theta))(\csc(\theta)+\sec(\theta)) = \tan(\theta)-\cot(\theta)$.

Answer:

$$\begin{aligned}
LHS &= (\sin(\theta)-\cos(\theta))(\csc(\theta)+\sec(\theta)) \\
&= \sin(\theta)\csc(\theta)+\sin(\theta)\sec(\theta)-\cos(\theta)\csc(\theta)-\cos(\theta)\sec(\theta) \\
&= 1+\frac{\sin(\theta)}{\cos(\theta)}-\frac{\cos(\theta)}{\sin(\theta)}-1 && \texttt{reciprocal ratios} \\
&= \tan(\theta)-\cot(\theta) && \texttt{reciprocal identities} \\
&= RHS.
\end{aligned}$$

◈

4.10 SOLVING BASIC TRIGONOMETRIC EQUATIONS

Solving trigonometric equations is dealt with in chapter 6; however, here we consider the following basic problem:

EXAMPLE 4.10.1. Solve for $x \in [0, 2\pi]$ in the equation $2\sin(x) = 1$.

Answer: Solving for x means finding the value (or values) for x in the interval for which the equation is a true statement, that is, the values of $x \in [0, 2\pi]$ for which $\sin(x) = \frac{1}{2}$. From an examination of the graphs of $y = \frac{1}{2}$ and $\sin(x)$ on the interval $[0, 2\pi]$ in figure 4.27, we observe that there are exactly two values of x where the graphs intersect. These values are labeled x_1 and x_2 and occur in the first and second quadrants, respectively, that is,

$$0 < x_1 < \frac{\pi}{2} \text{ and } \frac{\pi}{2} < x_2 < \pi.$$

The values for x_1 and x_2 can be read directly from the unit circle (figure 4.11), that is, $x_1 = \frac{\pi}{6}$ and $x_2 = \frac{5\pi}{6}$ because, for these values of the angle, the second coordinate of the corresponding reference point on the unit circle is $\frac{1}{2}$. ◈

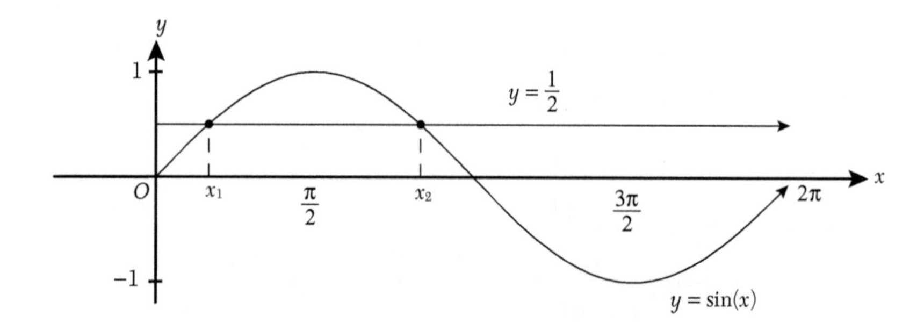

FIGURE 4.27. Solving a trigonometric equation.

EXAMPLE 4.10.2. Solve for θ in the equation $\tan^2(\theta) = 1$, for $\theta \in [0, 2\pi]$.

Answer: The problem is equivalent to solving two separate equations: $\tan(\theta) = 1$ and $\tan(\theta) = -1$, for $\theta \in [0, 2\pi]$. The first equation is true for the two values $\theta_1 = \frac{\pi}{4}$ and $\theta_2 = \frac{5\pi}{4}$, and the second equation is true for the two values $\theta_3 = \frac{3\pi}{4}$ and $\theta_4 = \frac{7\pi}{4}$. These values can be obtained by means of the graphical method we used for the problem above. Thus, the original equation $\tan^2(\theta) = 1$ has the four given solutions $\theta_1, \theta_2, \theta_3$, and θ_4, in the interval $[0, 2\pi]$. ◈

EXAMPLE 4.10.3. Find the general solution of the equation $3\cos(u) = \sqrt{3}\sin(u)$.

Answer: If a number u_0 is a solution of the equation, then we can conclude that $\cos(u_0) \neq 0$ (the sine ratio equals 1 for the angles for which the cosine ratio is zero). Therefore, we can rewrite the equation as $\frac{3}{\sqrt{3}} = \frac{\sin(u)}{\cos(u)}$, which is equivalent to $\sqrt{3} = \tan u$. There is one solution in the first quadrant, which we determine from the unit circle (figure 4.11) to be $u_0 = \frac{\pi}{6}$. Because the tangent ratio is π-periodic, the general solution can be expressed as $u_n = \left(\frac{\pi}{6}\right) + n\pi$, for any integer n. ◈

4.11 ADDITION IDENTITIES

A trigonometric ratio does not distribute through a sum or difference of angles. This is to say, an expression such as $\sin(A+B)$ cannot be evaluated by adding the terms $\sin(A)$ and $\sin(B)$. (Check this by setting A and B both equal to $\frac{\pi}{4}$, for instance.) The correct formula can be derived by starting from an elementary geometric observation, as shown in figure 4.28, where two angles A and B are determined by reference points R and S on the unit circle, respectively. The chord joining R and S is shown by means of a dotted line, and this chord completes the triangle formed by the points R, S, and the origin. In the second diagram in figure 4.28, this triangle is rotated clockwise around the origin, so that the point S rotates to the point $U(1,0)$ and the point R rotates to a point T. The angle determined by the reference point T is $A-B$, and the chord joining T and U has the same length as the chord joining R and S. We can calculate the lengths of the chords RS and TU by means of the distance formula (formula 2.7), as follows:

$$\begin{aligned}(|RS|)^2 &= (\cos(A)-\cos(B))^2+(\sin(A)-\sin(B))^2\\ &= \cos^2(A)-2\cos(A)\cos(B)+\cos^2(B)+\sin^2(A)-2\sin(A)\sin(B)+\sin^2(B)\\ &= 2-2\cos(A)\cos(B)-2\sin(A)\sin(B)\end{aligned}$$

in which the last line is obtained by applying the first Pythagorean identity (formula 4.4) twice, and

$$\begin{aligned}(|TU|)^2 &= (\cos(A-B)-1)^2+(\sin(A-B)-0)^2\\ &= \cos^2(A-B)-2\cos(A-B)+1+\sin^2(A-B)\\ &= 2-2\cos(A-B)\end{aligned}$$

in which the last line is obtained by applying the first Pythagorean identity.

Now, by equating $(|RS|)^2$ and $(|TU|)^2$, we obtain the following *addition identity*:

$$\boxed{\cos(A-B)=\cos(A)\cos(B)+\sin(A)\sin(B)}. \quad (4.7a)$$

In figure 4.28, B is smaller than A, but the identity is true for any values of A and B. What's more, by replacing B with $-B$ in this identity, we obtain another addition identity:

$$\boxed{\cos(A+B)=\cos(A)\cos(B)-\sin(A)\sin(B)}. \quad (4.7b)$$

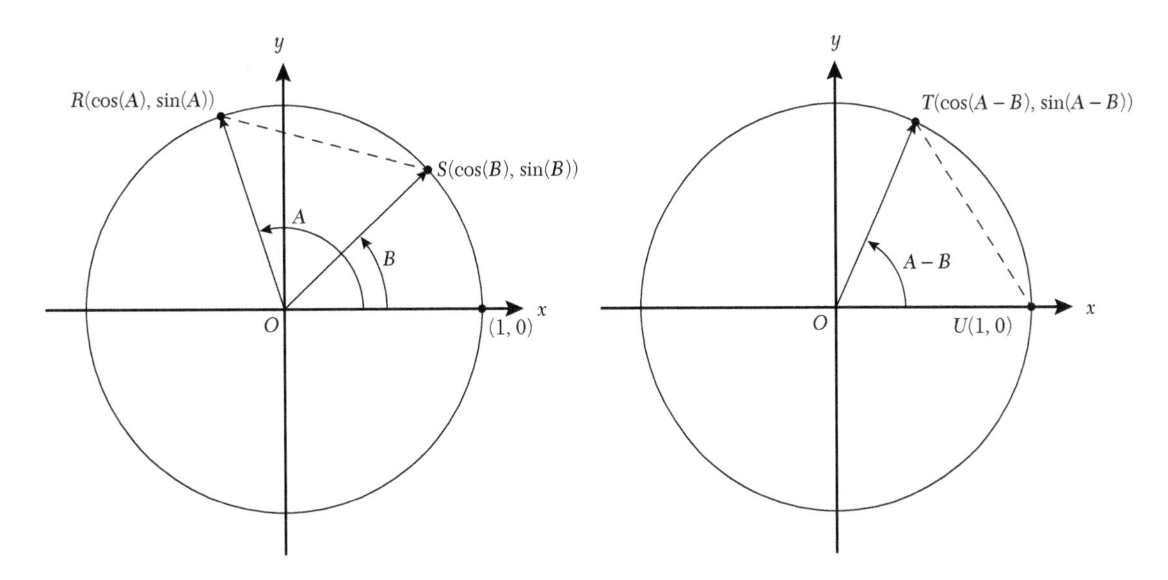

FIGURE 4.28. Derivation of the addition identities.

The following two addition identities for the sine ratio can be derived by means of the cofunction identities (formula 4.1). This is left as an exercise.

$$\boxed{\sin(A+B)=\sin(A)\cos(B)+\cos(A)\sin(B)}, \tag{4.8a}$$

$$\boxed{\sin(A-B)=\sin(A)\cos(B)-\cos(A)\sin(B)}. \tag{4.8b}$$

It is a good idea to memorize these addition identities. It is helpful to remember it in this way: in the second pair of identities, "sin" has the same "sign!"

EXAMPLE 4.11.1. Verify the identity:

$$\boxed{\sin\left(\frac{\pi}{2}+\theta\right)=\cos(\theta)}. \tag{4.9}$$

Answer:

$$\begin{aligned}\sin\left(\frac{\pi}{2}+\theta\right)&=\sin\left(\frac{\pi}{2}\right)\cos(\theta)+\cos\left(\frac{\pi}{2}\right)\sin(\theta)\\&=1\cdot\cos(\theta)+0\cdot\sin(\theta)\\&=\cos(\theta).\end{aligned}$$

◈

Another pair of identities can be proved in a similar way:

$$\boxed{\begin{aligned}\sin(\pi-\theta)&=\sin(\theta)\\\cos(\pi-\theta)&=-\cos(\theta)\end{aligned}}. \tag{4.10}$$

EXAMPLE 4.11.2. Trigonometric ratios of angles that are sums or differences of angles $\frac{\pi}{4}$, $\frac{\pi}{3}$, and $\frac{\pi}{6}$ can be computed using the addition identities; for example,

$$\begin{aligned}\sin\left(\frac{\pi}{12}\right)&=\sin\left(\frac{\pi}{3}-\frac{\pi}{4}\right)\\&=\sin\left(\frac{\pi}{3}\right)\cos\left(\frac{\pi}{4}\right)-\cos\left(\frac{\pi}{3}\right)\sin\left(\frac{\pi}{4}\right)\\&=\frac{\sqrt{3}}{2}\cdot\frac{\sqrt{2}}{2}-\frac{1}{2}\cdot\frac{\sqrt{2}}{2}\\&=\frac{\sqrt{6}-\sqrt{2}}{4}\approx 0.258819.\end{aligned}$$

◈

EXAMPLE 4.11.3. Suppose that α is in the first quadrant and $\sin(\alpha)=\frac{3}{4}$, and that β is in the third quadrant and $\cos(\beta)=\frac{-2}{3}$. Find $\sin(\alpha-\beta)$. Which quadrant does the angle $\alpha-\beta$ belong to?

Answer: In order to use the addition identities, we will need to know the values of $\cos(\alpha)$ and $\sin(\beta)$. According to the first Pythagorean identity,

$$\cos^2\alpha=1-\sin^2\alpha=1-\frac{9}{16}=\frac{7}{16}.$$

Thus, $\cos(\alpha)=\frac{\sqrt{7}}{4}$ because α is in the first quadrant. Similarly, we can determine that $\sin(\beta)=\frac{-\sqrt{5}}{3}$ because β is in the third quadrant.

Now, we have

$$\begin{aligned}\sin(\alpha-\beta)&=\sin(\alpha)\cos(\beta)-\cos(\alpha)\sin(\beta)\\&=\left(\frac{3}{4}\right)\cdot\left(-\frac{2}{3}\right)-\left(\frac{\sqrt{7}}{4}\right)\cdot\left(-\frac{\sqrt{5}}{3}\right)\\&=-\frac{1}{2}+\frac{\sqrt{35}}{12}\\&=\frac{\sqrt{35}-6}{12}.\end{aligned}$$

Because $\sqrt{35}-6<\sqrt{36}-6=0$, we conclude that $\sin(\alpha-\beta)$ is negative. By similar means, we can prove that $\cos(\alpha-\beta)=\frac{(3\sqrt{3}-2\sqrt{7})}{12}$, which is also negative (check this!). Therefore, $\alpha-\beta$ is in the third quadrant (recall the *ASTC* diagram). ◈

There are also addition identies for the tangent ratio. Their derivation is left as an exercise.

$$\boxed{\begin{aligned}\tan(A-B)&=\frac{\tan(A)-\tan(B)}{1+\tan(A)\tan(B)}\\\tan(A+B)&=\frac{\tan(A)+\tan(B)}{1-\tan(A)\tan(B)}\end{aligned}} \quad (4.11)$$

4.12 DOUBLE-ANGLE AND HALF-ANGLE IDENTITIES

A special case of the addition identities is obtained by setting $A=B=\theta$, resulting in the *double-angle identities*:

$$\boxed{\sin(2\theta)=2\sin(\theta)\cos(\theta)}, \quad (4.12)$$

$$\boxed{\cos(2\theta)=\cos^2(\theta)-\sin^2(\theta)}, \quad (4.13)$$

$$\boxed{\tan(2\theta)=\frac{2\tan(\theta)}{1-\tan^2(\theta)}}. \quad (4.14)$$

By making use of the first Pythagorean identity expressed as $\cos^2\theta=1-\sin^2\theta$ or $\sin^2\theta=1-\cos^2\theta$, we can derive two variations of the double-angle identity for the cosine ratio by substitution in formula (4.13). They are

$$\boxed{\cos(2\theta)=1-2\sin^2(\theta)}, \quad (4.15a)$$

$$\boxed{\cos(2\theta)=2\cos^2(\theta)-1}. \quad (4.15b)$$

Consequently, there are three identities for $\cos(2\theta)$ to choose from in any situation.

EXAMPLE 4.12.1. Prove the identity $\frac{\tan(x)}{1+\cos(2x)} = \sin 2x$

Answer: We start with the right hand side of the identity and use formula (4.15b).

$$\begin{aligned}\text{RHS} &= \frac{\sin(2x)}{1+\cos(2x)} \\ &= \frac{2\sin(x)\cos(x)}{1+\left(2\cos^2(x)-1\right)} \\ &= \frac{2\sin(x)\cos(x)}{2\cos(x)\cos(x)} = \frac{\sin(x)}{\cos(x)} = \tan(x) = \text{LHS}.\end{aligned}$$

◈

Formulas (4.15) can be used to derive *half-angle identities* by replacing 2θ with α. Using formula (4.15a), we obtain $\cos(\alpha) = 1 - 2\sin^2\left(\frac{\alpha}{2}\right)$, which implies $\sin^2\left(\frac{\alpha}{2}\right) = \frac{1-\cos(\alpha)}{2}$. The square root can be taken on both sides; however, $\sin\left(\frac{\alpha}{2}\right)$ can be negative, so a negative sign might be needed in front of the square root on the RHS. Thus, the half-angle identity for the sine ratio is

$$\boxed{\sin\left(\frac{\alpha}{2}\right) = \pm\sqrt{\frac{1-\cos(\alpha)}{2}}}. \tag{4.16}$$

Similarly, using formula (4.15b), we can obtain the half-angle identity for the cosine ratio:

$$\boxed{\cos\left(\frac{\alpha}{2}\right) = \pm\sqrt{\frac{1+\cos(\alpha)}{2}}}. \tag{4.17}$$

EXAMPLE 4.12.2. In Example 4.11.2, the value of $\sin\left(\frac{\pi}{12}\right)$ was found by making use of an addition identity. This value can also be computed using formula (4.16):

$$\begin{aligned}\sin\left(\frac{\pi}{12}\right) &= \sin\left(\frac{1}{2}\cdot\frac{\pi}{6}\right) \\ &= \sqrt{\frac{1-\cos\left(\frac{\pi}{6}\right)}{2}} = \sqrt{\frac{1-\frac{\sqrt{3}}{2}}{2}} = \sqrt{\frac{2-\sqrt{3}}{4}}.\end{aligned}$$

Now, this answer does not look the same as the answer in example 4.11.2, which was $\frac{(\sqrt{6}-\sqrt{2})}{4}$. However, by comparing the squares of the answers, it is clear that they are the same:

$$\left(\frac{\sqrt{6}-\sqrt{2}}{4}\right)^2 = \frac{6-2\sqrt{12}+2}{16} = \frac{8-4\sqrt{3}}{16} = \frac{2-\sqrt{3}}{4}.$$

◈

An approximation of the identity in the next example was used by the Indian Mathematician Aryabhata in about 500 AD to construct a table of sines.

EXAMPLE 4.12.3. Prove that

$$\sin((n+1)\alpha) - \sin(n\alpha) = \sin(n\alpha) - \sin((n-1)\alpha) - 4\sin(n\alpha)\sin^2\left(\frac{\alpha}{2}\right).$$

Answer:

$$\begin{aligned}\text{LHS} &= \sin((n+1)\alpha) - \sin(n\alpha)\\ &= \sin(n\alpha)\cos(\alpha) + \cos(n\alpha)\sin(\alpha) - \sin(n\alpha)\\ &= \sin(n\alpha)\big(\cos(\alpha) - 1\big) + \cos(n\alpha)\sin(\alpha)\end{aligned}$$

$$\begin{aligned}\text{RHS} &= \sin(n\alpha) - (\sin(n\alpha)\cos(\alpha) - \cos(n\alpha)\sin(\alpha)) - 4\sin(n\alpha)\left(\frac{1-\cos(\alpha)}{2}\right)\\ &= \sin(n\alpha)(1-\cos(\alpha) - 2(1-\cos(\alpha))) + \cos(n\alpha)\sin(\alpha)\\ &= \sin(n\alpha)\big(\cos(\alpha) - 1\big) + \cos(n\alpha)\sin(\alpha)\end{aligned}$$

The LHS equals the RHS, so the identity is proved. ◈

4.13 SOLVING TRIANGLES

Any triangle has six parts: three sides and three angles. If any three parts of a triangle are given (with at least one of these being the length of a side), then in most cases we can calculate unambiguously what the other parts are by means of the formulas that are derived in this section. In particular, if the triangle is a right triangle, then one angle (the right angle) is automatically given, and if any two other parts are given (with at least one of these being the length of a side), then the trigonometric ratios automatically determine what the other parts are.

4.13.1 Right Triangles

The connection between right triangles and trigonometric ratios has already been indicated a few times in this chapter. Any right triangle can be oriented in the Cartesian plane, so that either of its acute angles coincides with an angle measured in a counterclockwise direction from the x-axis; that is, one vertex is situated at the origin, one side (not the hypotenuse) is aligned with the positive x-axis, and the hypotenuse radiates from the origin to another vertex of the right triangle in the first quadrant. The side along the x-axis is called the *adjacent side*, and the side perpendicular to it is called the *opposite side*, as shown in figure 4.29. If the coordinates of the vertex in the first quadrant are labeled (a,b), then the length of the adjacent side is a, the length of the opposite side is b, and the length of the hypotenuse is $r=\sqrt{a^2+b^2}$. If the angle at the origin is labeled θ, then, according to the definition of the trigonometric ratios, $\sin(\theta)$ is the length of the opposite side divided by the length of the hypotenuse (*opposite over hypotenuse*, for short), $\cos(\theta)$ is the length of the adjacent side divided by the length of the hypotenuse (*adjacent over hypotenuse*, for short), and $\tan(\theta)$ is the length of the opposite side divided by the length of the adjacent side (*opposite over adjacent*, for short). These ratios can be remembered using the mnemonic SOH–CAH–TOA.

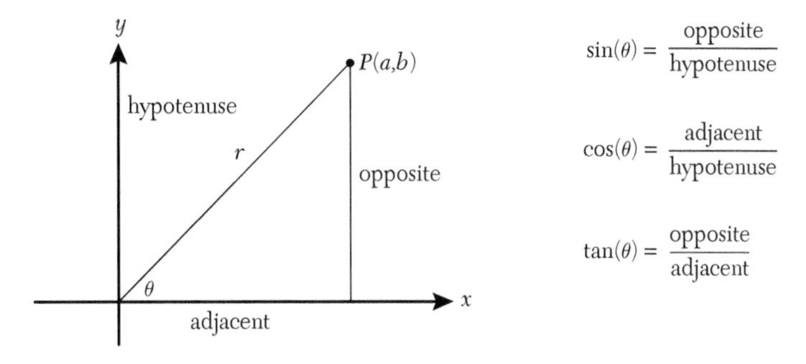

FIGURE 4.29. SOH–CAH–TOA.

EXAMPLE 4.13.1. A well-known right triangle is the “3-4-5” triangle, shown in figure 4.30, in which two angles are labeled α and β, and their trigonometric ratios are shown to the right of the diagram. ◈

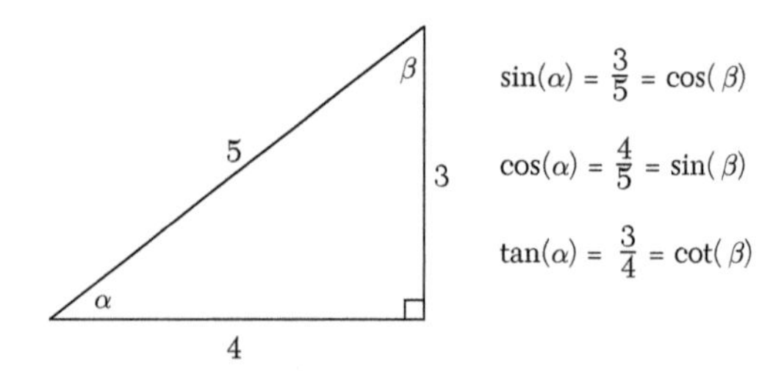

FIGURE 4.30. A 3-4-5 triangle.

Trigonometric ratios are used to compute lengths and angles in all kinds of problem in which right angles occur. Here are some typical examples:

EXAMPLE 4.13.2. The rays of the sun over the top of a flag pole cast a 10 m long shadow and form an angle of 30° with the ground. Find the height of the flag pole.

Answer: The height (h) of the flag pole divided by the length of the shadow (l) is equal to $\tan(30°) = \tan\left(\frac{\pi}{6}\right) = \frac{1}{\sqrt{3}}$, that is, $\frac{h}{10} = \frac{1}{\sqrt{3}}$, which implies that $h = \frac{10}{\sqrt{3}} \approx 5.7735$. ◈

EXAMPLE 4.13.3. In figure 4.31, a triangle has sides a, b, and c, and a line segment drawn from the vertex opposite b is perpendicular to b. Prove that

$$\tan(\theta) = \frac{a\sin(\gamma)}{b - a\cos(\gamma)}.$$

Answer: In the diagram, $b = x + y$, and h is the length of the side adjacent to the two smaller triangles. From the right triangles in the diagram, we can now infer that $\tan(\theta) = \frac{h}{x}$, $\sin(\gamma) = \frac{h}{a}$ (i.e., $h = a\sin(\gamma)$), and $\cos(\gamma) = \frac{y}{a}$ (i.e., $y = a\cos(\gamma)$). Therefore, because $x = b - y$, we have $\tan(\theta) = \frac{a \sin(\gamma)}{b - a(\cos(\gamma)}$. ◈

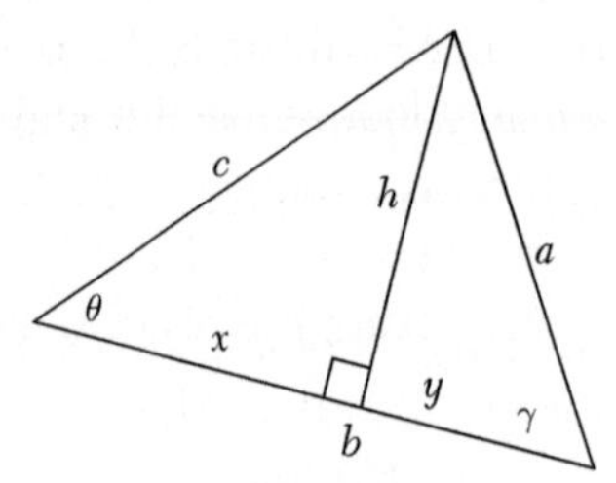

FIGURE 4.31. An application of trigonometric ratios.

4.13.2 The Area Formula and the Sine Rule

Any triangle can be oriented in the Cartesian plane, so that any one of its angles coincides with an angle measured in a counterclockwise direction from the x-axis.

In figure 4.32, a triangle has vertices labeled P, Q, and R, with the vertex P positioned at the origin. The triangle has an obtuse angle at P, but it will not make a difference to the formulas we derive below whether this angle is obtuse. For brevity, we refer to the angle at a vertex by means of the label at the vertex; that is, “$\sin(P)$” means “the sine ratio of the angle measured at P.” The length of a side opposite a vertex is labeled using the same letter as the label at the vertex but in lower case. For example, the length of the side opposite R (i.e., the length of the side coinciding with the x-axis in the diagram) is labeled "r." The coordinates at vertex R are $(q\cos(P),\ q\sin(P))$.

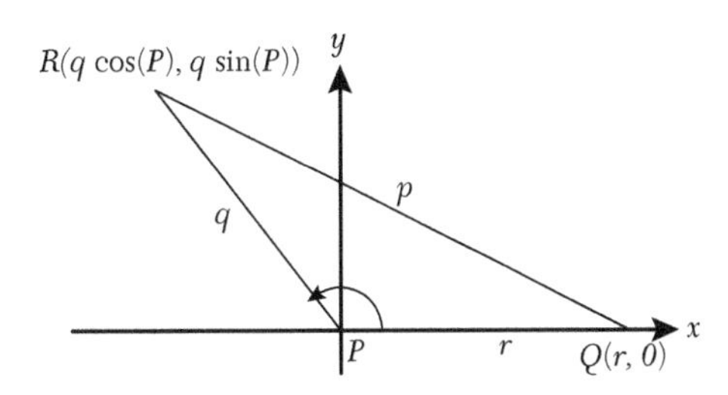

FIGURE 4.32. Derivation of the area rule.

Note that the height of ΔPQR is the second coordinate of R, and the base of the triangle has length r. Therefore, by means of the formula $\frac{1}{2}\times$ base $\times$ height for the area of a triangle, the area of ΔPQR in the diagram is $\frac{1}{2}qr\sin(P)$. Because any of the vertices of ΔPQR could have been situated at the origin, any of the following three formulas can, in fact, be applied to find the area of ΔPQR:

$$\boxed{\text{Area of } \Delta PQR = \frac{1}{2}qr\sin(P) = \frac{1}{2}pr\sin(Q) = \frac{1}{2}pq\sin(R)}.$$

These formulas are just one step away from deriving a useful rule for solving triangles, called the *sine rule*. All we have to do is divide each term above by $\frac{1}{2pqr}$:

$$\boxed{\text{The sine rule}: \frac{\sin(P)}{p} = \frac{\sin(Q)}{q} = \frac{\sin(R)}{r}}. \tag{4.19}$$

There are two situations in which the sine rule is used for solving a triangle. The first is when *two angles* and *the length of any side* are given (as given in the next example), then the sine rule can be used to find the length of either of the remaining sides.

EXAMPLE 4.13.4. If the angles at P and Q in ΔPQR are 30° and 45°, respectively, and side p has length 2, then, to compute the length of side r, we first deduce that the angle at R is 105°, and then use formula (4.19) to write the equation $\frac{r}{\sin(R)} = \frac{p}{\sin(P)}$, which determines that $r = \frac{2\sin(105°)}{\sin(30°)} \approx 3.8637$ (use a calculator). ◈

The second situation is when *the lengths of any two sides* and *a nonincluded angle* are given, then the sine rule can be used to find the length of the remaining side and either of the remaining angles. In this situation, the solution might be ambiguous (meaning that two different triangles can fit the given data), or there might be no solution (no triangle fits the given data). Figure 4.33 shows how an ambiguous solution (case I), unique solution (case II), or no solution (case III) might arise.

In case I, a ΔPQR has an angle of 20° at Q, $r=5$ and $q=3$. The problem is that this can occur in two different ways (as shown by dotted lines), resulting either in a triangle with an obtuse angle or an acute angle at R. In case II, ΔPQR has an angle of 20° at Q, $r=2$ and $q=3$, and this can only occur in one way (in this case, there is no ambiguity). In case III, ΔPQR has an angle of 20° at Q, $r=4$ and $q=1$, and this is impossible because a circle centered at P with radius 1 does not intersect the side p.

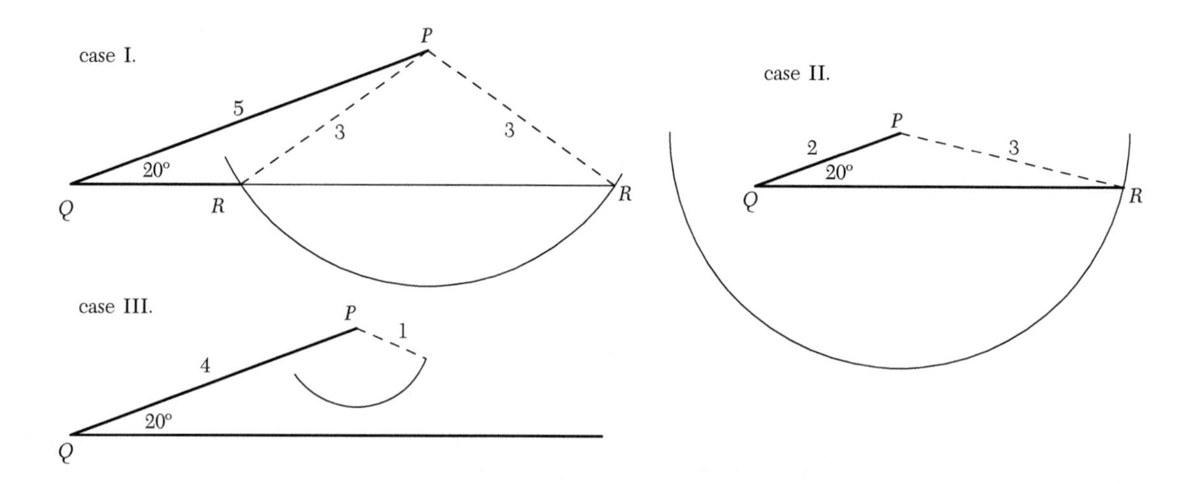

FIGURE 4.33. The cases for the sine rule.

EXAMPLE 4.13.5. Solve ΔPQR in which $Q=50°$, $r=9$, and $q=7$.

Answer: By the sine rule, $\frac{\sin(R)}{9}=\frac{\sin(50°)}{7}$, that is,

$$\sin(R)=\frac{9\sin(50°)}{7}\approx 0.985.$$

This equation has one solution $R_1 \approx 80.04°$ in the first quadrant and another solution $R_2=180°-R_1=99.96°$ in the second quadrant. In the first case, the remaining angle in the triangle is 49.96°, and by another application of the sine rule, the remaining side has length $p\approx 6.996$. In the second case, the remaining angle in the triangle is 30.04, and by another application of the sine rule, the remaining side has length $p\approx 4.575$. Thus, there are two solutions. ◈

The solutions of the following problems employ the sine rule and some other identities derived in this chapter. (Try to solve them yourself before reading the answers.)

Example 4.13.6. In figure 4.34, sides *AB* and *BC* of ΔABC are equal in length (equal to x), and the two angles at *C* are equal. Show that the length of *CD* is $2x$.

Answer: If the magnitude of the angles at *C* is θ, then, because ΔABC is isosceles, the magnitude of the angle at *A* in ΔABC is also θ (according to theorem 9.5.7a), and so, the angle at *B* is $\pi-2\theta$ (because the sum of the angles of a triangle is 180°). Now, we can apply the sine rule to ΔABC to determine that

$$|AC|=\sin(\pi-2\theta)\cdot\frac{x}{\sin(\theta)}=\frac{x\sin(2\theta)}{\sin(\theta)}=\frac{2x\sin(\theta)\cos(\theta)}{\sin(\theta)}=2x\cos(\theta).$$

Because ΔACD is a right triangle, $|CD|=|AC|\sec(\theta)$, and so

$$|CD|=2x\cos(\theta)\sec(\theta)=2x.$$

◈

EXAMPLE 4.13.7. In the second diagram in figure 4.34, *AB* is a vertical tower with its base at point *A*. Two other points *S* and *T* form a triangle with *A* in a horizontal plane that is perpendicular to the tower. Certain angles at the vertices *B*, *T*, and *A* are labeled θ, 2θ, and $90°+\theta$, respectively. (We can suppose that $\theta<30°$.) If the length of *BS* is 2, prove the following: (1) the length of *ST* is 1 and (2) the length of *AT* is $2\cos(2\theta)-1$.

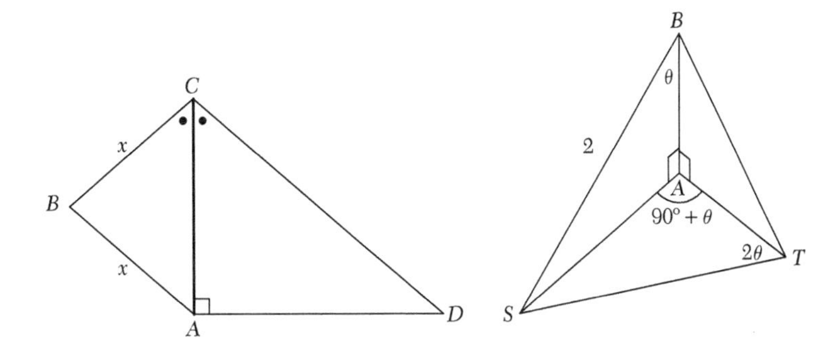

FIGURE 4.34. Problems involving the sine rule.

Answer: (1) Using a trigonometric ratio in the right triangle ABS, we find that $|AS| = 2\sin(\theta)$. Then, by applying the sine rule in ΔAST, we determine that

$$\frac{|ST|}{\sin(90°+\theta)} = \frac{2\sin(\theta)}{\sin(2\theta)}.$$

Because $\sin(90°+\theta) = \cos(\theta)$ and $\sin(2\theta) = 2\sin(\theta)\cos(\theta)$, we determine that $|ST| = 1$. (2) In order to find $|AT|$, note that the angle at S in ΔAST is $90° - 3\theta$, because the sum of the angles in a triangle is 180°. We now apply the sine rule as follows:

$$\frac{|AT|}{\sin(90° - 3\theta)} = \frac{1}{\cos(\theta)}$$

from which we proceed with the following calculation:

$$\begin{aligned}
|AT| &= \frac{\sin(90° - 3\theta)}{\cos(\theta)} = \frac{\cos(3\theta)}{\cos(\theta)} = \frac{\cos(2\theta + \theta)}{\cos(\theta)} \\
&= \frac{\cos(2\theta)\cos(\theta) - 2\sin^2(\theta)\cos(\theta)}{\cos(\theta)} \\
&= \cos(2\theta) - 2\sin^2(\theta) = \cos(2\theta) - 1 + \left(1 - 2\sin^2(\theta)\right) \\
&= \cos(2\theta) - 1 + \cos(2\theta) = 2\cos(2\theta) - 1.
\end{aligned}$$

◈

4.13.3 The Cosine Rule

In the situation in which *the lengths of any two sides* and *the included angle* are given, the *cosine rule* can be used to find the length of the remaining side. The cosine rule can be derived from figure 4.32 by using the distance formula to find the square of the distance between R and Q, that is, the square of the length of p:

$$\begin{aligned}
p^2 &= (q\sin(P) - 0)^2 + (q\cos(P) - r)^2 \\
&= q^2\sin^2(P) + q^2\cos^2(P) - 2qr\cos(P) + r^2 \\
&= q^2\left(\sin^2(P) + \cos^2(P)\right) + r^2 - 2qr\cos(P) \\
&= q^2 + r^2 - 2qr\cos(P).
\end{aligned}$$

In fact, any of the following formulas can be applied to ΔPQR:

$$\boxed{\begin{aligned} \text{The cosine rule:}\quad p^2 &= q^2 + r^2 - 2qr\cos(P) \\ q^2 &= p^2 + r^2 - 2pr\cos(Q) \\ r^2 &= p^2 + r^2 - 2pq\cos(R). \end{aligned}} \tag{4.20}$$

EXAMPLE 4.13.8. An airplane flies at a constant altitude. It flies from a point P at a speed of 700 kmph on a course $S20°W$ for 4 h and then changes to a course $S80°W$ for 5 h at a speed of 800 kmph. After the 9-h period, how far is the plane from P?

Answer: In figure 4.35, the plane is at point Q after 4 h and at point R after 9 h. The triangle formed by points P, Q, and R has an angle of 120° at Q (why?). The distance from P to Q is $4 \times 700 = 2{,}800$ (km) and the distance from Q to R is $5 \times 800 = 4{,}000$ (km). We can determine the distance from P to R by means of the cosine rule. To make the calculation numerically easier, we divide distances by 100:

$$\begin{aligned} |PR|^2 &= |PQ|^2 + |QR|^2 - 2|PQ||PR|\cos(120°) \\ &= 28^2 + 40^2 - 2 \cdot 28 \cdot 40 \cdot (-0.5) = 3{,}504. \end{aligned}$$

Therefore, $|PR| \approx 59.19$, and we conclude that the airplane is approximately 5,919 km from the point P. ◈

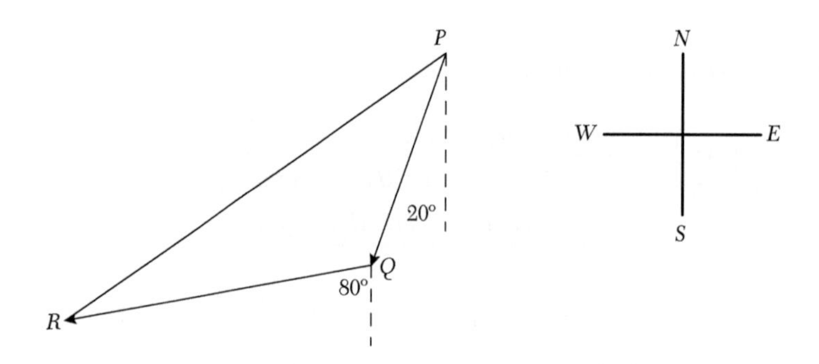

FIGURE 4.35. An application of the cosine rule.

In the next example, the cosine rule is used to find the length of the median of a triangle in terms of the lengths of the sides.

EXAMPLE 4.13.9. The diagram in figure 4.36 shows a median AD of ΔABC. Prove that

$$|AD| = \frac{1}{2}\sqrt{2b^2 + 2c^2 - a^2}.$$

Answer: By an application of the cosine rule to find the length of the median, and by a second application of the cosine rule to find the length of side AB, we obtain the pair of equations:

$$\begin{aligned} |AD|^2 &= b^2 + \left(\frac{a}{2}\right)^2 - 2\left(\frac{a}{2}\right)b\cos(C), \\ c^2 &= b^2 + a^2 - 2ab\cos(C). \end{aligned}$$

If the first equation is multiplied by 2 (that is, each side multiplied by 2) and each side of the second equation is subtracted from the same side of the first equation, then

$$2|AD|^2 - c^2 = b^2 - \frac{a^2}{2}$$

which can be expressed as

$$4|AD|^2 = 2b^2 + 2c^2 - a^2.$$

This leads to the required formula after taking the square root on both sides. ◈

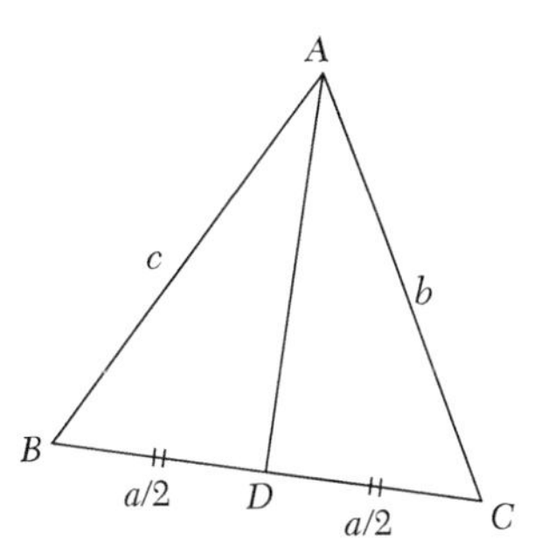

FIGURE 4.36. An application of the cosine rule.

4.14 VECTORS AND TRIGONOMETRY

This section is a continuation of the topic on vectors begun in section 2.6. We will show that the components of a vector can be expressed in terms of trigonometric ratios and explain the geometric interpretation of the dot product.

4.14.1 Components of a Vector

The notion of a component can best be explained by means of a few examples:

EXAMPLE 4.14.1. A kingfisher dives at an angle of 20° to the vertical, at a speed of 3 m/sec, to catch a fish in a dam. The motion of the kingfisher is represented by the velocity vector $\vec{v}$ in figure 4.37. Find the horizontal and vertical components of the velocity vector $\vec{v}$.

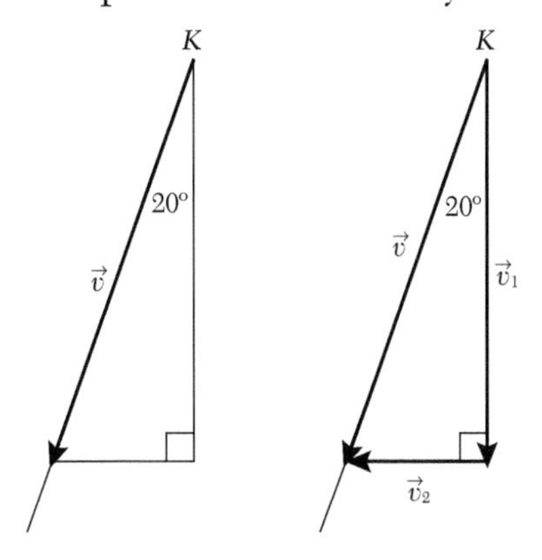

FIGURE 4.37. Components of a vector.

Answer: Figure 4.37 shows a right triangle in which the hypotenuse coincides with the velocity vector $\vec{v}$.

The second diagram shows a vector $\vec{v}_1$ pointing down along the vertical side of the right triangle and another vector $\vec{v}_2$ pointing to the left along the horizontal side. According to the geometric interpretation of vector addition, $\vec{v}_1 + \vec{v}_2 = \vec{v}$. The vectors $\vec{v}_1$ and $\vec{v}_2$ can be called the *vertical and horizontal components* of the vector $\vec{v}$. Recall that the magnitudes (or lengths) of the vectors $\vec{v}$, $\vec{v}_1$, and $\vec{v}_2$ are expressed as $|\vec{v}|$, $|\vec{v}_1|$, and $|\vec{v}_2|$, respectively. From the right triangle, we determine that $\frac{|\vec{v}_1|}{|\vec{v}|} = \cos(20°)$ and $\frac{|\vec{v}_2|}{|\vec{v}|} = \sin(20°)$; that is, the magnitudes of the vertical and horizontal components can be expressed as $|\vec{v}|\cos(20°)$ and $|\vec{v}|\sin(20°)$, respectively. Because $|\vec{v}| = 3$ m/sec, the magnitudes evaluate to $3\cos(20°) \approx 2.819$ m/sec and $3\sin(20°) \approx 1.026$ m/sec, respectively. Therefore, we can describe the vector $\vec{v}_1$ as "2.819 m/sec pointing down," and the vector $\vec{v}_2$ as "1.026 m/sec pointing to the left." ◈

EXAMPLE 4.14.2. A load with mass 50 kg hangs from two taught strings as shown in figure 4.38. Find the tensions (forces) $\vec{T}_1$ and $\vec{T}_2$ in each string.

Answer: The central diagram in figure 4.38 shows the vectors $\vec{T}_1$ and $\vec{T}_2$ pointing along each string counteracting the force $\vec{F}$ due to the load. To the left and right of the central diagram are two smaller diagrams showing the vectors $\vec{T}_1$ and $\vec{T}_2$ expressed as sums of the unknown vertical and horizontal component vectors, that is, $\vec{T}_1 = \alpha_1\vec{i} + \beta_1\vec{j}$ and $\vec{T}_2 = \alpha_2\vec{i} + \beta_2\vec{j}$. In terms of trigonometric ratios,

$$\alpha_1 = -|\vec{T}_1|\cos(52°),\ \beta_1 = |\vec{T}_1|\sin(52°),\ \alpha_2 = |\vec{T}_2|\cos(30°), \text{ and } \beta_2 = |\vec{T}_2|\sin(30°).$$

Thus, we have

$$\begin{aligned} \vec{T}_1 &= -|\vec{T}_1|\cos(52°)\vec{i} + |\vec{T}_1|\sin(52°)\vec{j}, \\ \vec{T}_2 &= |\vec{T}_2|\cos(30°)\vec{i} + |\vec{T}_2|\sin(30°)\vec{j}. \end{aligned} \tag{4.21}$$

If we choose the downward direction as the negative vertical direction, then the force $\vec{F}$ due to gravity acting on the load is $\vec{F} = -50(9.8)\vec{j}$. The combined tensions in the strings exactly counterbalance the load. Therefore,

$$\vec{T}_1 + \vec{T}_2 = -\vec{F} = 490\vec{j}.$$

By substitution of the expressions for $\vec{T}_1$ and $\vec{T}_2$ from formula (4.21), we obtain

$$-|\vec{T}_1|\cos(52°)\vec{i} + |\vec{T}_1|\sin(52°)\vec{j} + |\vec{T}_2|\cos(30°)\vec{i} + |\vec{T}_2|\sin(30°)\vec{j} = 490\vec{j}.$$

Now, by equating coefficients of $\vec{i}$ and $\vec{j}$, we obtain the following pair of *simultaneous equations* in the unknown variables $|\vec{T}_1|$ and $|\vec{T}_2|$:

$$\begin{aligned} -|\vec{T}_1|\cos(52°) + |\vec{T}_2|\cos(30°) &= 0, \\ |\vec{T}_1|\sin(52°) + |\vec{T}_2|\sin(30°) &= 490. \end{aligned}$$

Therefore, the magnitudes of $\vec{T}_1$ and $\vec{T}_2$ (in terms of the unit Newton [N] for measuring the magnitude of a force) are

$$\begin{aligned} |\vec{T}_1| &= \frac{490}{\sin(52°) + \tan(30°)\cos(52°)} \approx 419.5\,N, \\ |\vec{T}_2| &= \frac{|\vec{T}_1|\cos(52°)}{\cos(30°)} \approx 318\,N. \end{aligned}$$

The tensions $\vec{T}_1$ and $\vec{T}_2$ are now obtained by substitution of the values for $|\vec{T}_1|$ and $|\vec{T}_2|$ into formula (4.21):

$$\vec{T}_1 \approx -219.5\vec{i} + 321.5\vec{j} \text{ and } \vec{T}_2 \approx 219.5\vec{i} + 168.5\vec{j}. \quad \diamond$$

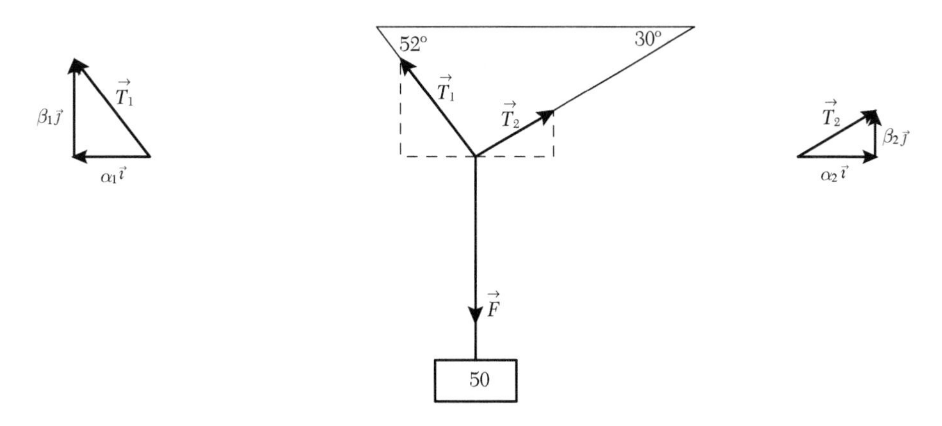

FIGURE 4.38. An application of vector components.

4.14.2 Geometric Interpretation of the Dot Product

The angle between any two vectors is always taken to be the smaller of two positive angles that can be measured if the vectors are joined at their initial points. In particular, if the vectors point in the same direction, then the angle between them is zero, and if the vectors point in opposite directions, then the angle between them is π (180°). The angle between two vectors can never be more than π.

The dot product of two vectors can be interpreted geometrically in terms of the magnitudes of the two vectors and the angle between the vectors as follows: we prove that if θ is the angle between two vectors $\vec{u}$ and $\vec{v}$, then

$$\boxed{\vec{u} \cdot \vec{v} = |\vec{u}|\,|\vec{v}|\cos(\theta)}. \tag{4.22}$$

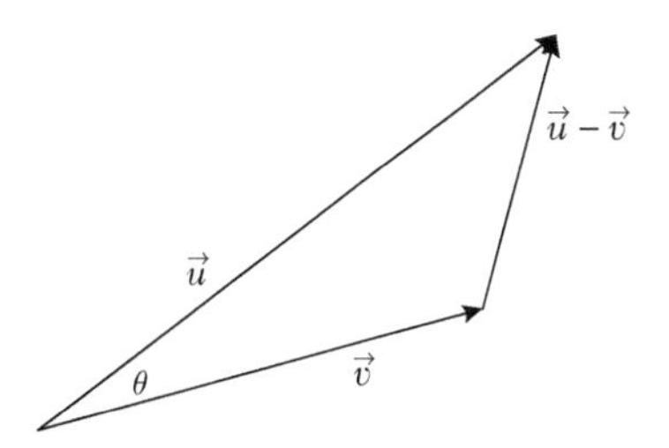

FIGURE 4.39. The angle between vectors.

In figure 4.39, $|\vec{u} - \vec{v}|$ can be computed by an application of the cosine rule:

$$|\vec{u} - \vec{v}|^2 = |\vec{u}|^2 + |\vec{v}|^2 - 2|\vec{u}|\,|\vec{v}|\cos(\theta). \tag{4.23}$$

By means of properties of the dot product, the LHS of formula (4.22) becomes

$$\begin{aligned}|\vec{u}-\vec{v}|^2 &= (\vec{u}-\vec{v})\cdot(\vec{u}-\vec{v}) \\ &= \vec{u}\cdot\vec{u}-2\vec{u}\cdot\vec{v}+\vec{v}\cdot\vec{v} \\ &= |\vec{u}|^2-2\vec{u}\cdot\vec{v}+|\vec{v}|^2.\end{aligned}$$

Therefore, formula (4.23) becomes

$$|\vec{u}|^2-2\vec{u}\cdot\vec{v}+|\vec{v}|^2=|\vec{u}|^2+|\vec{v}|^2-2|\vec{u}||\vec{v}|\cos(\theta),$$

which simplifies to formula (4.22).

Remark 4.14.1. It can be seen as an immediate consequence of formula (4.22) that if two vectors are perpendicular (orthogonal), then the angle between them is $\frac{\pi}{2}$ (90°), and so their dot product is zero. Conversely, if the dot product of two nonzero vectors is zero, then the angle between them must be $\frac{\pi}{2}$.

4.15 MORE IDENTITIES

There are many trigonometric identities. In this chapter, the most basic identities have been provided. In this section, a few more identities will be derived. The starting point will be the addition identities:

$$\begin{aligned}\cos(A-B) &= \cos(A)\cos(B)+\sin(A)\sin(B), \\ \cos(A+B) &= \cos(A)\cos(B)-\sin(A)\sin(B).\end{aligned}$$

If the columns are first added, then subtracted, the resulting equations are

$$\cos(A-B)+\cos(A+B)=2\cos(A)\cos(B), \tag{4.23a}$$

$$\cos(A-B)-\cos(A+B)=2\sin(A)\sin(B). \tag{4.23b}$$

If $A-B$ is replaced by C and $A+B$ replaced by D, then

$$A=\frac{D+C}{2} \quad \text{and} \quad B=\frac{D-C}{2}.$$

The equations in formula (4.23) can now be expressed as the following sum and difference of cosine ratios:

$$\boxed{\begin{aligned}\cos(C)+\cos(D) &= 2\cos\left(\frac{D+C}{2}\right)\cos\left(\frac{D-C}{2}\right), \\ \cos(C)-\cos(D) &= 2\sin\left(\frac{D+C}{2}\right)\sin\left(\frac{D-C}{2}\right)\end{aligned}}. \tag{4.24}$$

A similar pair of identities can be derived in an analogous way for a sum and difference of sine ratios:

$$\boxed{\begin{aligned}\sin(C)+\sin(D) &= 2\sin\left(\frac{C+D}{2}\right)\cos\left(\frac{C-D}{2}\right), \\ \sin(C)-\sin(D) &= 2\cos\left(\frac{C+D}{2}\right)\sin\left(\frac{C-D}{2}\right)\end{aligned}}. \tag{4.25}$$

EXAMPLE 4.15.1. Prove the identity:

$$\frac{\sin(3x)-\sin(x)}{\sin(4x)} = \frac{1}{2\cos(x)}.$$

Answer:

$$\begin{aligned}
\text{LHS} &= \frac{\sin(3x)-\sin(x)}{\sin(4x)} \\
&= \frac{2\cos\left(\frac{3x+x}{2}\right)\sin\left(\frac{3x-x}{2}\right)}{2\sin(2x)\cos(2x)} \\
&= \frac{\sin(x)}{\sin(2x)} = \frac{\sin(x)}{2\sin(x)\cos(x)} = \frac{1}{2\cos(x)} = \text{RHS}
\end{aligned}$$

◈

EXERCISES

4.1. Convert the following angles in radians to degrees. The answer should be rounded off to three decimal digits.

(a) $\frac{7\pi}{4}$

(b) $\frac{13\pi}{6}$

(c) $\frac{11\pi}{7}$

(d) $\frac{\pi}{16}$

(e) 1

(f) 3

4.2. Convert the following angles in degrees to radians.

(a) $150°$

(b) $390°$

(c) $1°$

(d) $6°$

(e) $\left(\frac{150}{\pi}\right)^{\circ}$

(f) $\left(\frac{720}{\pi}\right)^{\circ}$

4.3. Answer the following questions.

(a) If $\tan(\theta) < 0$ and $\csc(\theta) > 0$ for some angle θ, to which quadrant does θ belong?

(b) If $\sec(\alpha) > 0$ and $\cot(\alpha) < 0$ for some angle α, to which quadrant does α belong?

4.4. Find the value of $\csc(\theta)$ if θ belongs to the first quadrant and $\cot(\theta) = \frac{11}{6}$.

4.5. Given that $\sin(s) = -\frac{3}{8}$ and s is in the third quadrant, find the values of $\cos(s)$ and $\tan(s)$.

4.6. Find the exact value of each of the following trigonometric ratios.

(a) $\sec\left(\frac{\pi}{3}\right)$

(b) $\tan\left(\frac{\pi}{3}\right)$

(c) $\sin(135°)$

(d) $\cos(135°)$

(e) $\cot(210°)$

(f) $\sin(210°)$

(g) $\csc\left(\frac{5\pi}{3}\right)$

(h) $\cos\left(\frac{5}{3}\right)$

4.7. Find the exact value of each of the following trigonometric ratios.

(a) $\tan\left(\frac{4\pi}{3}\right)$

(b) $\cot\left(\frac{11\pi}{6}\right)$

(c) $\sin\left(-\frac{12\pi}{3}\right)$

(d) $\cos\left(\frac{2\pi}{3}\right)$

(e) $\csc\left(\frac{7\pi}{4}\right)$

(f) $\tan\left(\frac{11\pi}{4}\right)$

(g) $\sec(\pi)$

(h) $\sec\left(-\frac{5\pi}{6}\right)$

4.8. What are the values of u and v in each of the diagrams in figure 4.40?

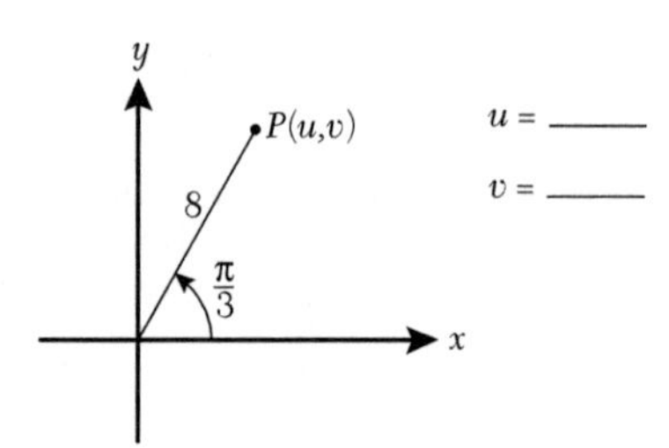

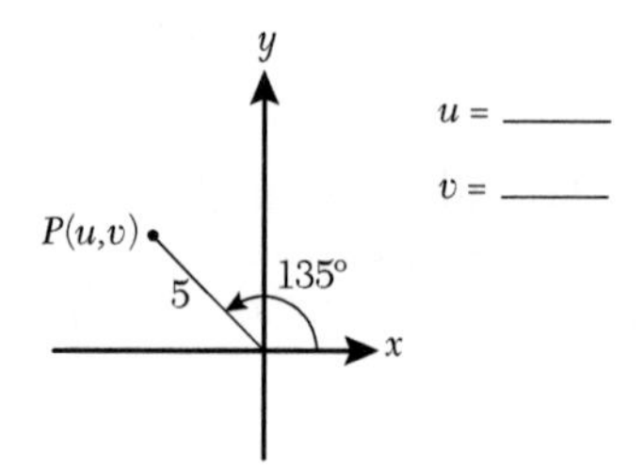

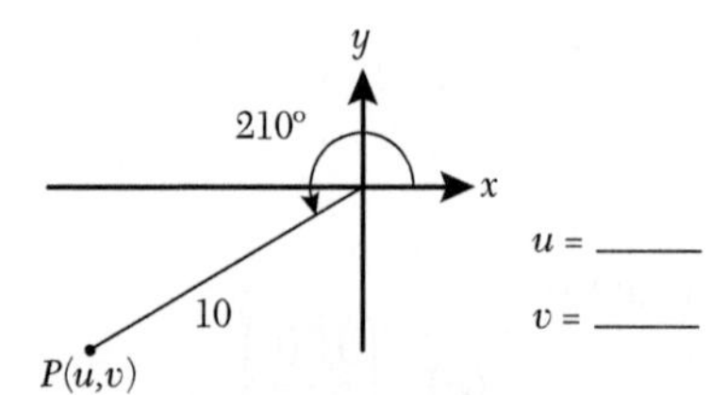

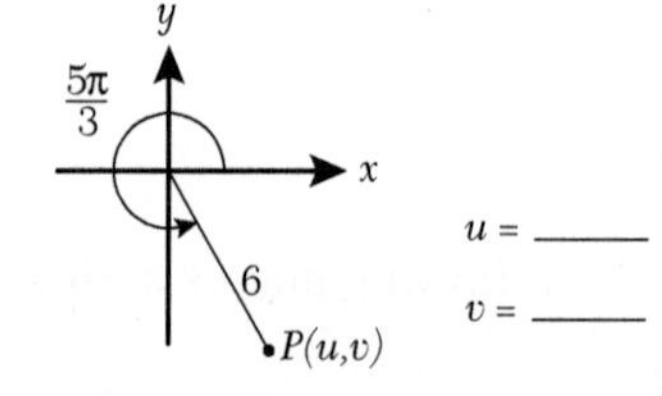

FIGURE 4.40. Find the values of u and v.

4.9. What is the magnitude of θ in each of the diagrams in figure 4.41?

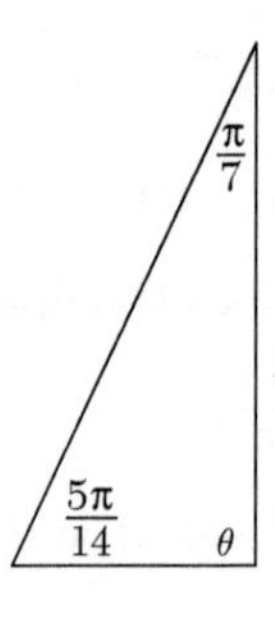

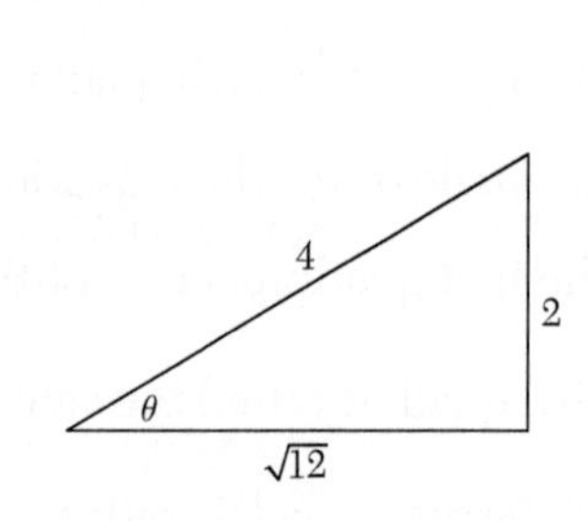

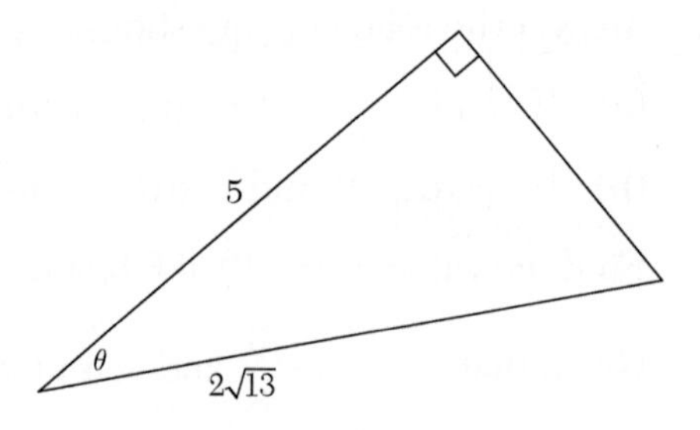

FIGURE 4.41. Find the magnitude of θ.

4.10. Suppose that ΔABC is a right triangle with a right angle at C, and the sides a, b, and c (opposite the vertices A, B, and C, respectively) form a geometric sequence (assume that $a < b < c$). Find the magnitude of the angle at B. (Hint: a, b, and c form a geometric sequence if and only if $\frac{c}{b} = \frac{b}{a}$. A calculator will be needed to get the answer.)

4.11. Solve for $\theta \in [0, 2\pi]$ in each of the following equations:

(a) $\cos(\theta) = \frac{1}{2}$

(b) $2\sin(\theta) = -\sqrt{2}$

(c) $\sin(2\theta) = \cos(2\theta)$

(d) $\sin^2\theta = \cos^2\theta$

4.12. Prove the cofunction identities (formula (4.1)) using formulas (4.7) and (4.8).

4.13. Prove the identity:

$$\frac{\sec(\alpha)+\sec(\alpha)\sin(\alpha)}{1-\sin^2\alpha} = \frac{1}{\cos(\alpha)-\cos(\alpha)\sin(\alpha)}$$

(Hint: $1-x^2 = (1-x)(1+x)$.)

4.14. Prove the identity:

$$\cot(\theta)+\tan(\theta) = \frac{1}{\cos(\theta)\sin(\theta)}$$

4.15. Carefully sketch the graphs of the following equations. Label the peaks of the graphs and label the intercepts on the axes. Show at least one full period for each graph.

(a) $y = \sin\left(\frac{2x}{3}\right)$

(b) $y = \cos(\pi(x-1))$

(c) $y = 1.5\sin(-x)$

(d) $y = 0.5\cos(0.5x)$

(e) $y = \frac{\tan(x)}{2}$

(f) $y = -\sec(x)$

4.16. Simplify the following expressions by making use of the first Pythagorean identity (formula (4.4)):

(a) $(\sin(u)+\cos(u))^2 - 2\sin(u)\cos(u)$

(b) $(\sin(-x)-\cos(-x))^2 - 1$

4.17. Prove the addition identities for the sine ratio (formula (4.8)). (Hint: make use of the cofunction identities.)

4.18. Prove the addition identities for the tangent ratio (formula (4.11)).

4.19. Use the addition identities or half-angle identities to evaluate the following ratios:

(a) $\cos\left(\frac{\pi}{12}\right)$

(b) $\cos\left(\frac{3\pi}{8}\right)$

(c) $\sin\left(\frac{\pi}{12}\right)$

(d) $\sin\left(\frac{3\pi}{8}\right)$

(e) $\cos\left(\frac{7\pi}{24}\right)$

(f) $\sin\left(\frac{7\pi}{12}\right)$

4.20. For each of the following sets of values, solve for triangle ABC. Be sure to find all possible solutions (triangles).

(a) $a = 15$, $b = 25$, and $A = 45°$

(b) $a = 4$, $b = 3$, and $B = 30°$

(c) $a = 15$, $b = 10$, and $A = 45°$

4.21. If $\tan(\theta) = \frac{1}{4}$, what is $\tan(2\theta)$?

4.22. A lighthouse is built on the edge of a vertical cliff that is 20 m high. From a point 70 m from the base of the cliff, the angle of elevation to the top of the lighthouse is 30°. How tall is the lighthouse?

4.23. Prove, for any triangle PQR with sides p, q, and r, that

$$\frac{q^2 + r^2 - p^2}{pr} = 2\sin(Q)\cot(P).$$

4.24. Prove that the area of any triangle PQR is $\dfrac{p^2 \sin(Q)\sin(R)}{2\sin(P)}$.

4.25. Two observers, 1 mile apart on level ground, notice an unidentified flying object (UFO) hovering above a radio station directly between them. The angles of elevation from each observer to the UFO are 25° and 45°, respectively. At what altitude is the UFO hovering?

4.26. Derive the half-angle identity

$$\tan\left(\frac{\alpha}{2}\right) = \frac{\sin(\alpha)}{1+\cos(\alpha)}$$

from the half-angle identities for the sine and cosine ratios.

4.27. Kyle watches a kite ahead of him at an angle of 45° to the horizontal (level ground), while Nathan, standing in front of Kyle, watches the same kite at an angle of 60° to the horizontal (see figure 4.42). The direct distance from Kyle to the kite is $10\sqrt{2}$ m. How far is Kyle standing behind Nathan?

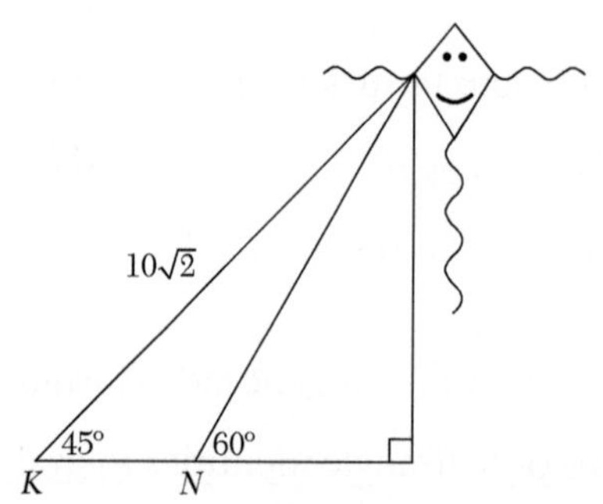

FIGURE 4.42. Kyle and Nathan watching a kite.

4.28. Figure 4.43 shows a vertical tower AB with its base at point A. Two other points S and T form a triangle with A in a horizontal plane that is perpendicular to the tower. The angle of elevation of B from S is equal to θ; that is, $B\hat{S}A = \theta$ and $S\hat{A}T = 2\theta$. The sides AT and ST are of equal length x. Prove the following:

(a) The length of AS is $2x\cos(2\theta)$,

(b) The length of AB is
$2x\tan(\theta)\cos(2\theta)$.

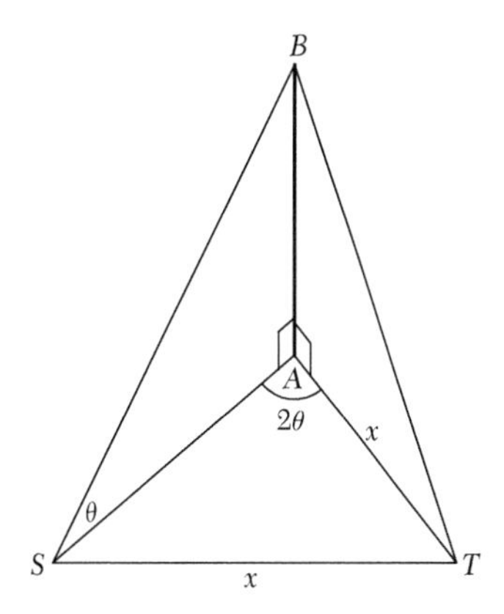

FIGURE 4.43. Diagram for exercise 4.30.

4.29. In figure 4.44, if the three angles at P that are marked equal are equal to θ, show that $|AC| = 2|PC|\sin(\theta)$; hence, or otherwise, show that $\frac{|AC|}{|CD|} = 2\cos(2\theta)$.

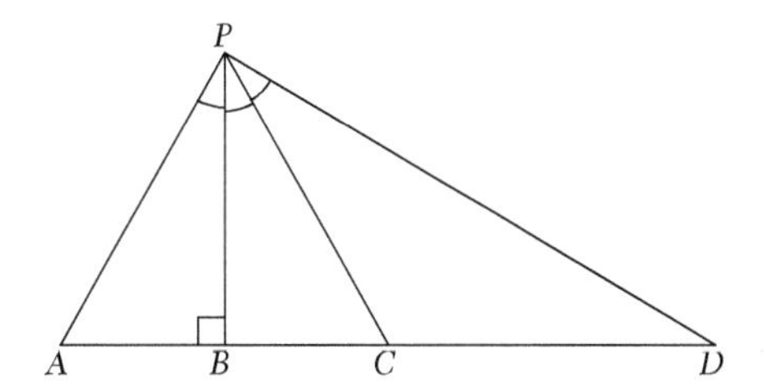

FIGURE 4.44. Diagram for exercise 4.32.

4.30. Prove the identity:

$$\frac{\cos(x)+\cos(3x)}{\sin(x)+\sin(3x)} = \frac{1}{2}\left(\cot(x)-\tan(x)\right).$$

4.31. Prove the identity:

$$\sin(2x)+\sin(4x)+\sin(6x) = 4\sin(3x)\cos(2x)\cos(x).$$

4.32. A golf caddy pulls a bag of golf clubs along a level green with a force of 500 N exerted at an angle of 42° to the horizontal. Find the horizontal and vertical components of the force.

4.33. In figure 4.45, PQ and ST are vertical towers. From a point R, in the same horizontal plane as Q and T, the angles of elevation to P and S are ϕ and 2ϕ, respectively. $S\hat{Q}R = 90° + \phi$ and $S\hat{R}Q = 90° - 2\phi$. If $|QR| = a$, express $|PQ|$ and $|ST|$ each in terms of a and a trigonometric ratio of ϕ. Thus prove that $\frac{|ST|}{|PQ|} = \frac{2\cos^2(\phi)}{\tan(\phi)}$. Calculate $\frac{|ST|}{|PQ|}$ if $\phi = 30°$. Leave your answer in radical notation.

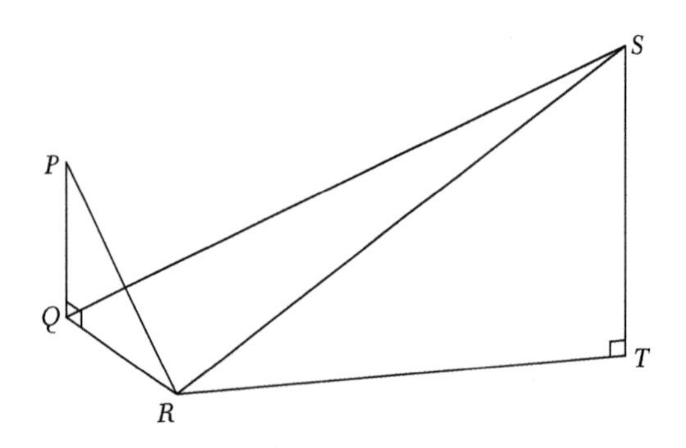

FIGURE 4.45. Diagram for exercise 4.36.

4.34. Two forces $\vec{F_1}$ and $\vec{F_2}$ with magnitudes 8 and 10 N, respectively, act on an object at the origin (shown in figure 4.46). Find the resultant force $\vec{F}$ acting at the origin, that is, find the magnitude $|\vec{F}|$ and determine the direction of $\vec{F}$ by finding the value of θ.

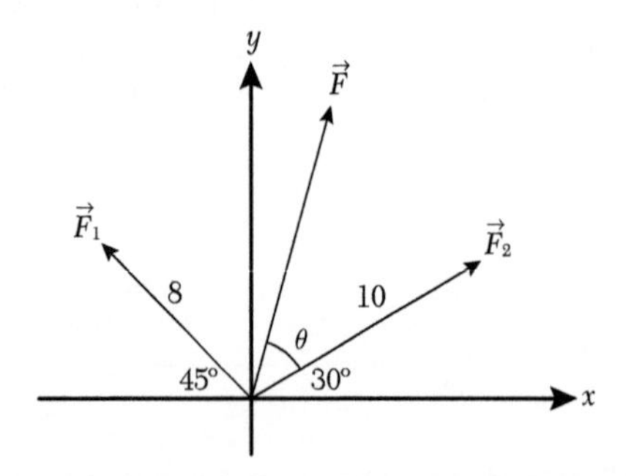

FIGURE 4.46. Diagram for exercise 4.37.

4.35. Determine whether the following pairs of vectors are orthogonal, parallel, or neither.

(a) $\vec{u} = \langle -1,2 \rangle, \vec{v} = \langle 4,2 \rangle$

(b) $\vec{u} = \langle -1,2 \rangle, \vec{v} = \langle -2,4 \rangle$

(c) $\vec{u} = \langle 1,-2 \rangle, \ \vec{v} = \langle -2,4 \rangle$

(d) $\vec{u} = \langle -1,-2 \rangle, \ \vec{v} = \langle 4,2 \rangle$

CHAPTER 5

FUNCTIONS

5.1 INTRODUCTION

Calculus can be regarded as the science of curves. A basic problem relating to curves is measuring the area enclosed by a particular curve. The Greek Mathematician Archimedes, who lived in Sicily in the third century BC, found a way to measure the area enclosed by a parabola. This is explained in his famous treatise *Quadrature of the Parabola*. In another treatise, called *Spiral Lines*, he proved that a spiral line encloses one-third of the area enclosed by the circle surrounding it. These treatises, together with other treatises and books that he wrote, can be taken as the beginning of calculus.

A curve can most conveniently be defined as the graph of a function. In the modern presentation of calculus, as we will see in chapters 7 and 8 of this book, the operations of calculus are applied to functions.

In this chapter, we begin with the set theoretic definition of a function that has been the preferred definition of a function since the early twentieth century. In fact, we see in section 5.2 that a function is a special type of relation and that a relation is a pairing of elements of two sets called the domain and codomain. If the codomain of a function is $\mathbb{R}$ (the set of real numbers), then the function is called a real-valued function. Examples of real-valued functions that have already been studied in earlier chapters of this book (although not explicitly mentioned as such) are polynomials and trigonometric functions.

Most of us are familiar with the graphing capabilities of pocket calculators and computers. In section 5.3, we explain how the graph of a function can be generated from a table of values, and this can, in fact, be used to generate the graph of an equation, an example of which is given in section 5.3.1. A simple test, described in section 5.3.2, called the vertical line test, can be used to decide whether any given graph can be considered the graph of a real-valued function.

The types of real-valued function that we introduce in this chapter are the absolute value function (in section 5.4), exponential functions (in section 5.5), root functions (in section 5.7), rational functions (in section 5.6), and logarithmic functions (in section 5.13.2). Piecewise defined functions, introduced in section 5.8, are a means to draw more complicated graphs and also, as we will see in chapter 7 a means to understand the left and right limits.

Functions can also be regarded as elements of an *algebra of functions*, that is, as explained in section 5.10, functions can be added, subtracted, divided, and multiplied together. In other words, it is possible to operate on functions with the same operations that are performed with real numbers. Functions can also be composed with each other as a way of linking functions together, so that the output of one function is the input for another function.

Transformations of functions, including vertical and horizontal shifts, vertical and horizontal scaling, and reflections across the axes, are introduced in section 5.11.

There arise situations where we would like to describe figures (think of spirals, loops, and flowers) in the plane that cannot be described as the graphs of real-valued functions. Vector-valued functions, introduced in section 5.12, differ from real-valued functions in that a value in the domain is mapped to a vector (a pair of real numbers) rather than a single real number. Thus, the graph of a vector-valued function (which is called a trajectory) may contain loops and self-intersections.

This chapter ends with an introduction to inverse functions and the logarithmic function and inverse trigonometric functions as examples of the construction of inverse functions by reflections of the graphs of one-to-one functions about the line $y = x$.

5.2 RELATIONS AND FUNCTIONS

The definition of a relation always involves two sets called A and B.

DEFINITION 5.2.1. *A relation, which we may call* r, *is a set of coordinate pairs, where the first member of each coordinate pair is any element of a set* A *and the second member is any element of a set* B.

Thus, each element of the relation is a pairing of any element of A with any element of B. It is easy to make up examples from the world we live in.

EXAMPLE 5.2.1. The set A is {shark, whale, penguin, tuna, porpoise, turtle, albatross}, set B is {Atlantic ocean, Pacific Ocean, Antarctic Ocean, Arctic Ocean, Gulf of Mexico, Bering Straight, South China Sea, Amazon River}, and the relation r is {(shark, Pacific Ocean), (shark, Gulf of Mexico), (shark, Amazon River), (whale, Pacific Ocean), (penguin, Antarctic Ocean), (tuna, Atlantic Ocean), (porpoise, Bering Straight), (turtle, Gulf of Mexico)}. A diagrammatic representation of the relation is shown in figure 5.1. ◈

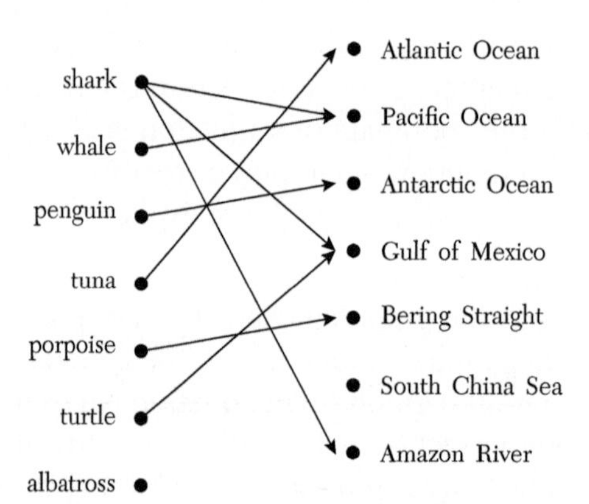

FIGURE 5.1. A diagrammatic representation of a relation.

Some observations need to be made about the relation defined in this example and the properties of a relation that are permissible, in general. First, an element of A may be paired with more than one element of B (e.g., "shark" is paired with three elements of B). Second, different elements of A may be paired with the same element of B (e.g., "shark" and "turtle" are both paired with "Gulf of Mexico"). Third, some elements of A ("albatross") may not be paired with any element of B, and finally, some elements of B ("South China Sea") may not be paired with any element of A.

The *domain* of the relation r is the subset of A consisting of those elements of A that are paired with elements of B. The *range* of the relation r is the subset of B consisting of those elements of B that are paired with elements of A. Thus, the domain of r is the set {shark, whale, penguin, tuna, porpoise, turtle} and the range of r is the set {Atlantic Ocean, Pacific Ocean, Antarctic Ocean, Arctic

Ocean, Gulf of Mexico, Bering Straight, Amazon River}. The order in which the elements of sets A and B, the domain, and range are listed does not matter.

REMARK 5.2.1. Any selection of points (coordinate pairs) in the Cartesian plane determines a relation because we can identify the x axis with set A and the y axis with set B. Evidently, any relation whose domain and range are both subsets of $\mathbb{R}$ can be identified with a collection of points in the Cartesian plane.

EXAMPLE 5.2.2. Plot the coordinate pairs given by the relation:

$$r = \{(3,-1),(2,0),(2,1),(1,1),(1,0),(1,-1),(0,0)\}$$

in the Cartesian plane. What is the domain of r? What is the range of r? ◈

A special type of relation is one in which *every* element of set A is paired with *only one* element of set B. This type of relation is called a *function*. Note that the domain of a function coincides with set A. Set B is called the *codomain*. The definition of a function does not exclude the possibility of an element of B being paired with more than one element of A. The definition of the range of a function is the same as the definition of the range of a relation. A function is usually labeled f (in the same way that a relation is usually labeled r). Here is the formal definition of a function:

DEFINITION 5.2.2. *A function is a rule that assigns to each element* x *in a set* A, *called the domain, exactly one element* f(x) *in a set* B, *called the codomain.*

REMARK 5.2.2. In the definition of a function f, the letter x is a variable. This means that x is representative of any element of the domain of f. While the letters f and x are typically used for the name and variable in the definition of a function, other letters, or names, could be used. The letters g and h are typically used for the name of a function, and the letters r, s, t, u, v, w, y, and z are also typically used for the name of the variable.

EXAMPLE 5.2.3. Using the same sets A and B as in example 5.2.1 above, an example of a function is {(shark, Pacific Ocean), (whale, Pacific Ocean), (penguin, Antarctic Ocean), (tuna, Atlantic Ocean), (porpoise, Bering Straight), (turtle, Gulf of Mexico), (albatross, Bering Straight)}. Figure 5.2 is a diagrammatic representation of this function. ◈

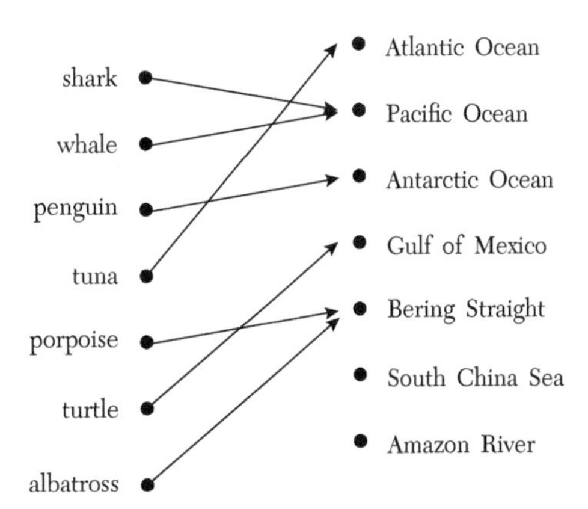

FIGURE 5.2. A diagrammatic representation of a function.

REMARK 5.2.3. A useful analogy of a function is the machine analogy. The machine takes an input x, performs some kind of operation on the input, and gives an output that we call $f(x)$. This is shown in figure 5.3.

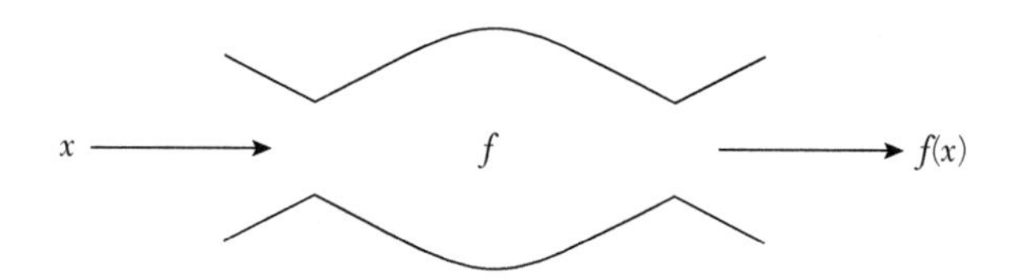

FIGURE 5.3. The machine analogy of a function.

REMARK 5.2.4. In this textbook, the domain of a function will always be a subset of $\mathbb{R}$ and the codomain will either be $\mathbb{R}$, in which case the function is real-valued, or the set of two-dimensional vectors (with real-valued components), in which case the function is vector-valued.

REMARK 5.2.5. Two functions are the same if and only if they have the same domain, codomain, range, and pairing of elements of the domain with elements of the codomain. (The order in which the elements of the domain, codomain, and range are listed does not matter.)

EXAMPLE 5.2.4. If we modify the function defined in the previous example, by choosing a different codomain, for example, B = {Atlantic Ocean, Pacific Ocean, Antarctic Ocean, Gulf of Mexico, Bering Straight, South China Sea} (i.e., Amazon River is excluded) and, all else being the same, we regard the functions as different (unequal) functions. ◈

REMARK 5.2.6. If a function $f(x)$ is given, the domain may, or may not, be stated explicitly as part of the definition of the function. In the latter case, it should be assumed that the domain is the largest set of real numbers for which the function is defined. In other words, the domain is the set of all real values x for which $f(x)$ is a real number.

In the next two examples, the domain of the function is stated explicitly as the set A and the codomain is the set of real numbers.

EXAMPLE 5.2.5. If $f(x)=3x$ and $A=\mathbb{N}$, then the domain is the set of natural numbers, and the range is the set of all natural numbers that are a multiple of 3, which we can denote as $3\mathbb{N}$. ◈

EXAMPLE 5.2.6. If $g(x)=x^2$ and $A=[0,4]$, then the range is the interval [0, 16]. ◈

In the following four examples, the domain is not stated explicitly.

EXAMPLE 5.2.7. $f(t)=\frac{1}{2(t^2+t)}$. This function is defined for any value of t for which the denominator is not zero. Therefore, the only real numbers that should be excluded from the domain are 0 and −1. In set notation, the domain of f is:

$$A=\{t\in\mathbb{R}\,|\,t\neq 0, t\neq -1\}.$$

Using interval notation, the domain of f is a union of three open intervals:

$$A=(-\infty,-1)\cup(-1,0)\cup(0,\infty).$$

◈

EXAMPLE 5.2.8. $h(y)=\sqrt{y}$ This function is defined as long as y is non-negative.

$$A=\{y\in\mathbb{R}\,|\,y\geq 0\}=[0,\infty)$$

◈

EXAMPLE 5.2.9. $f(x)=\frac{1}{\sqrt{x}}$ is defined as long as x is positive.

$$A=\{x\in\mathbb{R}\,|\,x>0\}=(0,\infty)$$

◈

EXAMPLE 5.2.10. $f(s)=\sqrt{4-3s-s^2}$ is defined as long as $4-3s-s^2$ is non-negative, that is

$$\begin{aligned} A &= \{s \in \mathbb{R} \mid 4-3s-s^2 \geq 0\} \\ &= \{s \in \mathbb{R} \mid (4+s)(1-s) \geq 0\}. \end{aligned}$$

It is helpful to draw a number line to solve the inequality $(4+s)(1-s)\geq 0$ (i.e., to find all values of s for which the inequality is a true statement), as shown in the diagram below. Because the factor $4+s$ changes sign at $s=-4$ and the factor $1-s$ changes sign at $s=1$, the sign of the product $(4+s)(1-s)$ will change sign at $s=-4$ and $s=1$, while its sign will be constant on each of the intervals $(-\infty,-4)$, $(-4,1)$, and $(1,\infty)$. A test point in the interval $(-4,1)$, for example, $s=0$, determines that the product is positive in this interval (because $(4+0)(1-0)=4>0$), and, consequently, the product must be negative in the intervals $(-\infty,-4)$ and $(1,\infty)$. Therefore, we can express the domain of $f(s)$ as:

$$A=\{s \in \mathbb{R} \mid -4 \leq s \leq 1\}=[-4,1].$$

◈

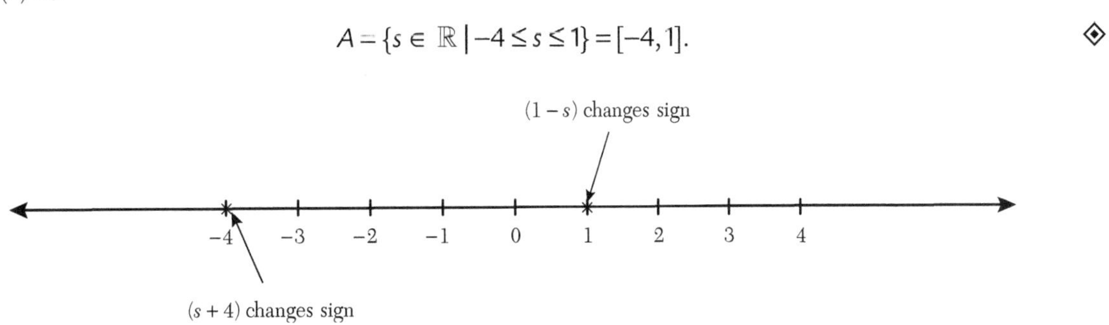

FIGURE 5.4. Solving an inequality on the number line.

In the next example we mention some types of function that were introduced in chapters 3 and 4.

EXAMPLE 5.2.11. A linear function is of the form $f(x)=mx+c$, where m and c are real numbers; a quadratic function is of the form $f(x)=ax^2+bx+c$, where a, b, and c are real numbers; a cubic function is of the form $f(x)=ax^3+bx^2+cx+d$, where a, b, c, and d are real numbers; and so on. In general, a polynomial function is a polynomial expression, as defined in section 3.4. The domain of any polynomial function is the set of real numbers, unless specified otherwise. The basic trigonometric functions are $f(x)=\sin(x)$, the sine function; $f(x)=\cos(x)$, the cosine function; and $f(x)=\tan(x)$, the tangent function (or tan function, for short). The sine and cosine functions have the same domain, that is, the set of real numbers, and the domain of the tan function is the union of intervals of the form $\left(n\pi-\frac{\pi}{2}, n\pi+\frac{\pi}{2}\right)$, where n is any integer. The reciprocal trigonometric functions, that is, the cosecant, secant, and cotangent functions, can be defined similarly. We can also define functions using a mixture of polynomial and trigonometric terms, for example, $f(x)=\sin^3(x)-5\cos(2x)+x^2$. ◈

The *evaluation* of a real-valued function involves substitution of any real number (belonging to the domain of the function) for x and simplification of the resulting expression.

EXAMPLE 5.2.12. If, $f(x)=\sin^3(x)-5\cos(2x)+x^2$, then:

$$\begin{aligned} f\left(\frac{\pi}{4}\right) = \sin^3\left(\frac{\pi}{4}\right) - 5\cos\left(\frac{2\pi}{4}\right) + \left(\frac{\pi}{4}\right)^2 &= \frac{\sqrt{2}}{4} - 5(0) + \frac{\pi^2}{16} \\ &= \frac{(4\sqrt{2}+\pi^2)}{16} \end{aligned}$$

◈

5.3 VISUALIZING FUNCTIONS

Diagrams or graphs can help us visualize the properties of a function. Most students are familiar with graphing/plotting software on pocket calculators or computers, but how does the software generate graphs that look like smooth curves? It is a visual trick. A computer-generated curve is, in fact, composed of many straight-line segments that join together and seemingly blend into a smooth curve. These line segments join coordinate pairs that are determined from a *table of values* generated by the software according to the definition of the function.

EXAMPLE 5.3.1. Table 5.1 gives the values of the function $f(x) = x^3$ at half-integer points from −3 to 3. The coordinate pairs given in the table are (−3, −27), (−2.5, −15.625), (−2, −8), and so on. As a comparison, we first generate a graph by plotting the coordinate pairs with integer values of x only, as shown in the first diagram in figure 5.5, and then generate a graph using all of the coordinate pairs, as shown in the second diagram. Clearly, the jagged edges in the second graph are less pronounced. ◈

TABLE 5.1. Table of values for $y = x^3$

x	−3	−2.5	−2	−1.5	−1	−0.5	0	0.5	1	1.5	2	2.5	3
$y = x^3$	−27	−15.625	−8	−3.375	−1	−0.125	0	0.125	1	3.375	8	15.625	27

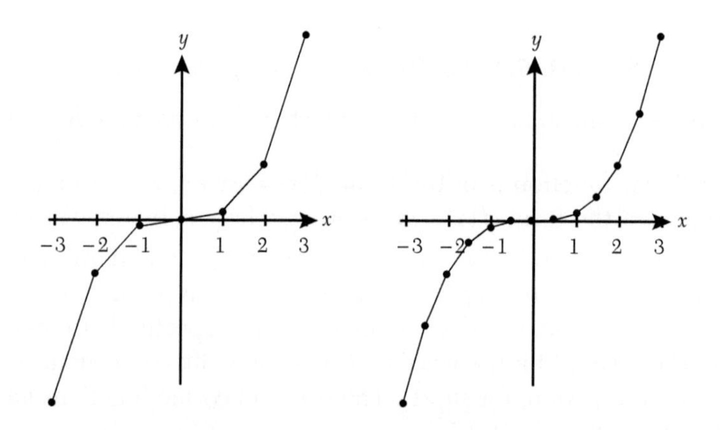

FIGURE 5.5. Generating the graph of a cubic function.

It is important to remember that the appearance of the graph of a function depends on the domain of the function. A dramatic illustration of this is the graph of the function $f(x) = \sin(x)$, with the set of natural numbers chosen for the domain. This graph, shown in figure 5.6 (up to $x = 280$), looks very different from the familiar graph of the sine function in which the set of real numbers is the domain (as shown in figure 4.16).

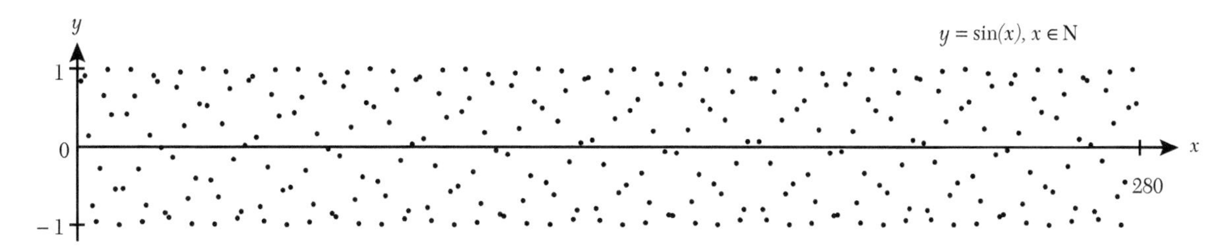

FIGURE 5.6. The sine function plotted at positive integers.

5.3.1 Graphs of Equations

The graph of an equation is generated from the *solution set* of the equation. This is the set of all coordinate pairs for which the equation is a true statement.

EXAMPLE 5.3.2. The solution set of the equation $\sqrt[3]{x^2} + \sqrt[3]{y^2} = 4$ is the set S of all coordinate pairs (x, y) for which the equation is a true statement, that is

$$S = \{(x, y) \mid \sqrt[3]{x^2} + \sqrt[3]{y^2} = 4 \text{ is true}\}$$

Some coordinate pairs that belong to the solution set S are $(\pm 8, 0), (0, \pm 8)$, $(\pm 3\sqrt{3}, \pm 1)$, and $(\pm 1, \pm 3\sqrt{3})$. We can generate an approximate graph of S by plotting these twelve points, as shown in the first diagram of figure 5.7 below. In the second diagram, a "smooth" curve is generated using 256 points. This graph is called an astroid. (Note that $3\sqrt{3} \approx 5.196$.) ◈

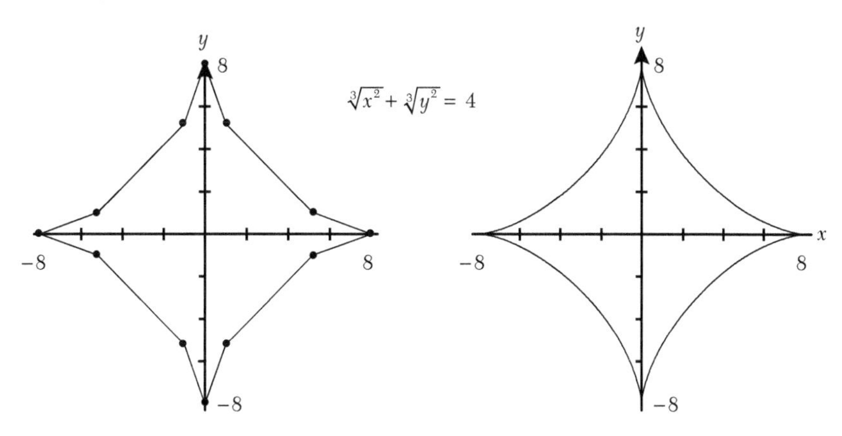

FIGURE 5.7. Generating the graph of an astroid.

5.3.2 The Vertical Line Test

The graph of an equation is not necessarily the graph of a function. A visual test called the *vertical line test* is used to tell when a graph is the graph of a function (like the astroid above). According to this test, if any vertical line cuts the graph at most once, then the graph is the graph of a function. Stated otherwise, if it is possible to draw a vertical line that cuts the graph at two or more points, then the graph is not the graph of a function. Figure 5.8 below contains some diagrams that illustrate this. Three of the graphs are the graphs of functions (the two semicircles and the parabola) and three are not (the circle, the sideways parabola, and the astroid).

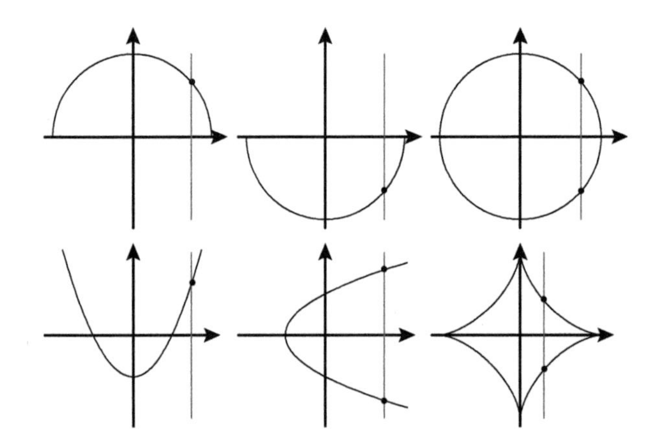

FIGURE 5.8. The vertical line test.

5.4 THE ABSOLUTE VALUE FUNCTION

The absolute value of a real number was introduced in section 1.6 Here, we introduce the *absolute value function*: $f(x) = |x|$, with the domain being the set of real numbers. The graph of $f(x)$ in the Cartesian plane is in the shape of a "V," with the point of the "V" at the origin. The graph can be modified, for example, $f(x) = 2|x|$ gives a steeper "V" and $f(x) = |x| - 1$ shifts the "V" down by one unit as shown in figure 5.9 below.

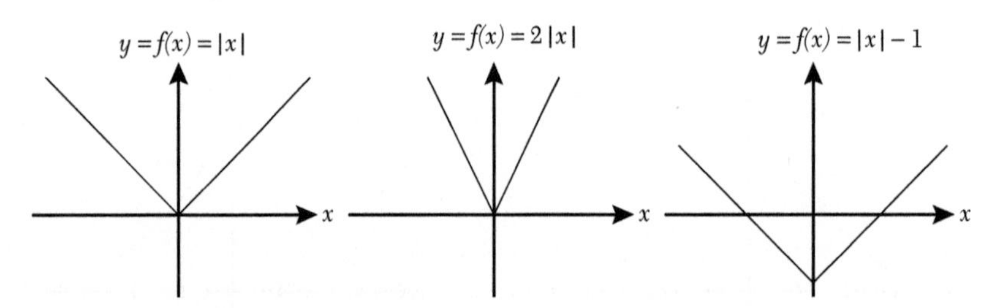

FIGURE 5.9. Graphs of the absolute value function.

5.5 EXPONENTIAL FUNCTIONS

We are not yet in a position to define a function $f(x) = a^x$ with the set of real numbers as the domain because we have not yet defined the meaning of an exponent that is not a natural number. (In section 1.9, we stated that an exponent is a symbol for repeated multiplication if the exponent is a natural number.) In section 5.5.3, we will explain the meaning of a^x if x is a rational number, and in section 5.5.2, we will explain the meaning of a^x if x is an irrational number. In order to simplify matters, we assume that $a > 0$.

5.5.1 Fractional Exponents

We are guided by the laws for exponents in table 1.6 in section 1.14, where it was supposed that m and n were integers. If we suppose instead that m and n were rational numbers, that is, if $m = \frac{1}{2}$ and $n = 2$, for example, then according to the second law for exponents,

$$(a^{\frac{1}{2}})^2 = a^{\left(\frac{1}{2}\right)\cdot 2} = a^1 = a.$$

We are led to define $a^{1/2}$ as $\sqrt{a}$. Similarly, if $m = \frac{1}{3}$ and $n = 3$, for example, then

$$(a^{\frac{1}{3}})^3 = a^{\left(\frac{1}{3}\right)\cdot 3} = a^1 = a.$$

and so we are led to define $a^{1/3}$ as $\sqrt[3]{a}$. In general, if k is any natural number and n is any integer, then

$$\boxed{a^{\frac{1}{k}} = \sqrt[k]{a}} \text{ and } \boxed{a^{\frac{n}{k}} = \left(a^{\frac{1}{k}}\right)^n = \left(\sqrt[k]{a}\right)^n = \left(a^n\right)^{\frac{1}{k}} = \sqrt[k]{a^n}}.$$

The following example demonstrates how expressions with fractional exponents can be evaluated or simplified.

EXAMPLE 5.3.3.

(i) $64^{\frac{3}{2}} = (\sqrt{64})^3 = 8^3 = (2^3)^3 = 2^9 = 512$

(ii) $64^{\frac{-3}{2}} = (\sqrt{64})^{-3} = \dfrac{1}{8^3} = \dfrac{1}{512}$

(iii) $243^{\frac{4}{5}} = (\sqrt[5]{243})^4 = 3^4 = 81$

(iv) $4^{\frac{2}{3}} \cdot 4^{\frac{1}{4}} = 4^{\frac{2}{3}+\frac{1}{4}} = (2^2)^{\frac{11}{12}} = 2^{\frac{11}{6}} = 2 \cdot 2^{\frac{5}{6}} = 2 \cdot \sqrt[6]{32}$ ◈

5.5.2 Irrational Exponents

It is difficult, at this stage of the book, to justify why numbers such as 2^{π}, $7^{\sqrt{2}}$, 4^e, π^e, or 2^{π^e}, that is, numbers formed with irrational exponents, should be regarded as real numbers. For now, a good enough justification is that any irrational exponent can be replaced with a rational exponent (a fraction) that approximates the irrational exponent as closely as needed, simply by truncating the infinite decimal expansion of the irrational number. For example, because an approximate value for e is $\frac{3{,}020}{1{,}111}$ (see section 1.12.3), the number 4^e can be approximated by $4^{\frac{3{,}020}{1{,}111}} = \sqrt[1{,}111]{4^{3{,}020}} \approx 43.308$. If you compute 4^e on a pocket calculator, it will give a decimal number very close to this value.

Because exponents have now been defined for real numbers (integer, rational, and irrational exponents), the laws for exponents given in table 1.6 can be generalized to the case where the integers m and n are the real numbers x and y, respectively, as shown in table 5.2.

5.5.3 The Graphs of Exponential Functions

If $a > 0$, we can now take the domain of the function $f(x) = a^x$ to be the set of real numbers. This is the *exponential function*. We can set $a = 2$ and draw the graph of $y = f(x)$ in the Cartesian plane. As always, we can make a table of values for the function, using some integer values for x.

TABLE 5.3. Table of values for $y = 2^x$

x	-3	-2	-1	0	1	2	3
$y = f(x) = 2^x$	$\frac{1}{8}$	$\frac{1}{4}$	$\frac{1}{2}$	1	2	4	8

In the first diagram in figure 5.10, we show an approximate graph of $f(x)=2^x$, using the seven coordinate pairs in table 5.3, and in the second diagram, we show the graph of $f(x)=2^x$ as a smooth curve.

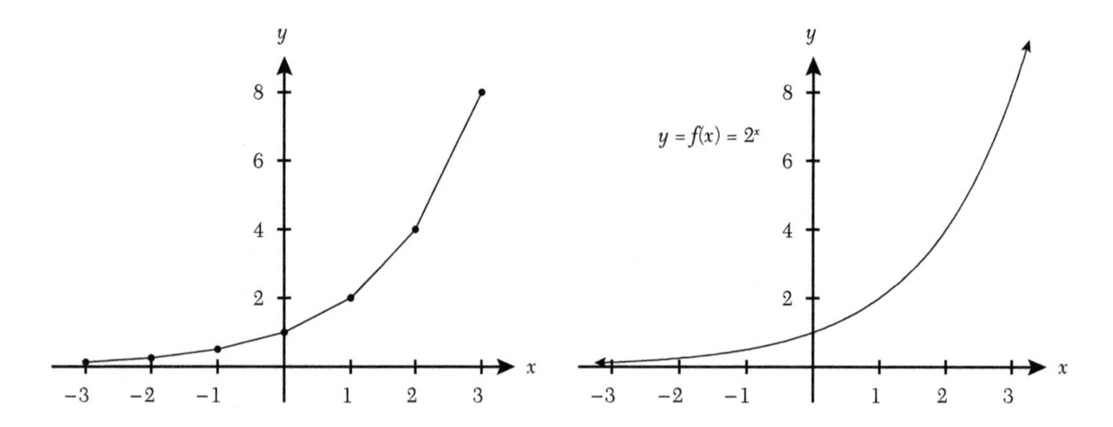

FIGURE 5.10. Generating the graph of an exponential function.

In general, if $0 < a < 1$, the graph of the exponential function $f(x)=a^x$ increases rapidly to the right and decreases rapidly to the left of the y axis. However, if $0 < a < 1$, then the graph increases rapidly to the left and decreases rapidly to the right of the y axis (why?). If $a=1$, then the graph is a constant function (the identity function) because $1^x=1$ for all values of x (why?). Note that the graph of $f(x)=a^x$ intersects the y axis at $y=1$.

TABLE 5.2. Laws for exponents

(I)	$a^x \cdot a^y = a^{x+y}$	Exponents with the same base can be added
(II)	$\left(a^x\right)^y = a^{xy}$	A power of an exponent becomes a product of exponents
(III)	$(a \cdot b)^x = a^x b^x$	An exponent distributes through a product
(IV)	$\left(\dfrac{a}{b}\right)^x = \dfrac{a^x}{b^x}$	An exponent distributes through a quotient
(V)	$\dfrac{a^x}{a^y} = a^{x-y}$	An exponent in the denominator changes sign in the numerator
(VI)	$\dfrac{a^x}{a^y} = \dfrac{1}{a^{y-x}}$	An exponent in the numerator changes sign in the denominator

5.6 RATIONAL FUNCTIONS

DEFINITION 5.6.1. *A rational function is a function of the form* $f(x)=\frac{p(x)}{q(x)}$, *where* $p(x)$ *and* $q(x)$ *are polynomials. We also refer to an expression of the form* $\frac{p(x)}{q(x)}$, *where* $p(x)$ *and* $q(x)$ *are polynomials, as a rational expression.*

EXAMPLE 5.6.1. A basic example of a rational function is $f(x)=\frac{1}{x}$ (let $p(x)=1$ and $q(x)=x$). The domain of $f(x)$ is $\mathbb{R}\setminus\{0\}$, that is, the set of real numbers, excluding zero. The graph of is $f(x)=\frac{1}{x}$ is a hyperbola, as shown in figure 5.11, in which the y axis is a vertical asymptote. This means that the graph of $y=\frac{1}{x}$ grows closer and closer to the y axis as the value of x gets closer and closer to zero, while the corresponding y value increases toward infinity or minus infinity depending whether x approaches zero through positive or negative values, respectively. ◈

In this chapter, we will suppose that the polynomials $p(x)$ and $q(x)$ in the definition of a rational function are polynomials with real coefficients. With this in mind, we recall theorem 3.7.3, which states that every polynomial with real coefficients and positive degree can be factorized as a product of linear and irreducible quadratic factors with real coefficients. If the factorizations of $p(x)$ and $q(x)$ have no linear or irreducible quadratic factors in common, then $\frac{p(x)}{q(x)}$ is a rational expression in *simplest form*. If $p(x)$ and $q(x)$ do have linear or irreducible quadratic factors in common, then all of these common factors can be canceled across the division sign, resulting in a new rational expression, which is then in simplest form.

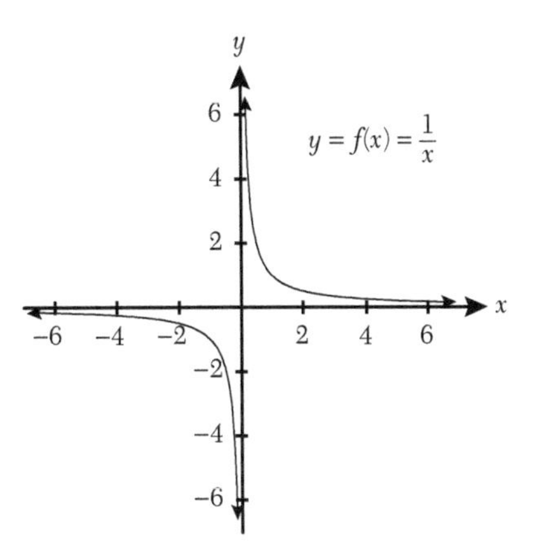

FIGURE 5.11. The graph of the rational function $y=\frac{1}{x}$.

EXAMPLE 5.6.2. Consider the rational expression $\frac{x^3-x^2-x-2}{x^2-4}$. By means of the methods described in section 3.7, we find the factorizations $x^3-x^2-x-2=(x-2)(x^2+x+1)$ and $x^2-4=(x-2)(x+2)$. We therefore obtain the following simplification of the rational expression by cancellation of the common factor $(x-2)$:

$$\frac{x^3-x^2-x-2}{x^2-4}=\frac{(x-2)(x^2+x+1)}{(x-2)(x+2)}=\frac{x^2+x+1}{x+2}.$$

The rational expression $\frac{x^2+x+1}{x+2}$ is in simplest form. ◈

REMARK 5.6.1. The functions $f(x)=\frac{x^3-x^2-x-2}{x^2-4}$ and $g(x)=\frac{x^2+x+1}{x+2}$ are not the same, even though the rational expression that defines the function g can be obtained from the rational expression that defines the function f by cancellation of a common factor. To find why f and g are not the same function, observe that the domain of f is $\{x \in \mathbb{R} \mid x \neq 2, x \neq -2\}$, whereas the domain of g is $\{x \in \mathbb{R} \mid x \neq -2\}$. Indeed, according to the PEMDAS rule introduced in section 1.10, $f(2) = \frac{0}{0}$, which is not defined (not a real number), whereas $g(2) = \frac{7}{4}$. Thus, it is clear that $x = 2$ is in the domain of g but not in the domain of f.

The graphs of rational functions will be discussed in detail in chapter 7 For the time being, it is enough to know that the graph of a rational function $f(x) = \frac{p(x)}{q(x)}$ has vertical asymptotes at those, and only those, values of x for which the rational expression in simplest form, obtained by simplification of $\frac{p(x)}{q(x)}$, is not defined. These values of x are precisely the values x_i of the linear factors $(x - x_i)$ belonging to the factorization of $q(x)$ that remains after all common factors of $p(x)$ and $q(x)$ have been canceled.

Example 5.6.3. The rational function $f(x) = \frac{x^3 - x^2 - x - 2}{x^2 - 4}$ has a vertical asymptote at $x = -2$ (and this is the only vertical asymptote). The graphs of $f(x)$ and $g(x) = \frac{x^2 + x + 1}{x + 2}$ are shown in figure 5.12 below. The graph of f differs from the graph of g only at $x = 2$. We draw an open circle above $x = 2$ on the graph of $y = f(x)$ to indicate that $x = 2$ is not in the domain of f. The vertical asymptote through $x = -2$ is shown as a dotted line. ◈

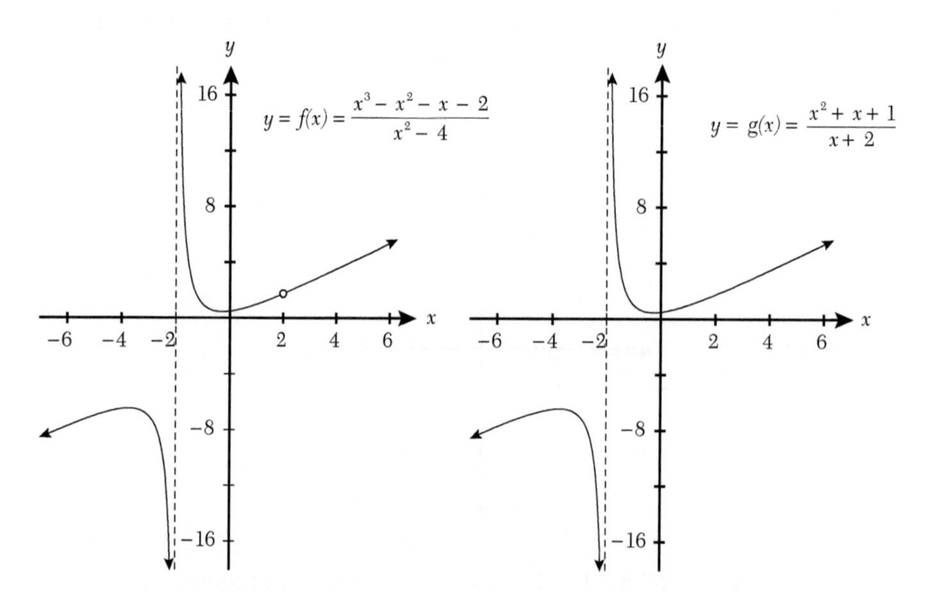

FIGURE 5.12. Graphs of two rational functions that differ at one point.

5.7 ROOT FUNCTIONS

Definition 5.7.1. *Root functions are functions of the form* $f(x) = \sqrt[n]{x}$. *If n is even, then the domain is the set of non-negative real numbers. If n is odd, then the domain is the set of real numbers.*

The graphs of $f(x) = \sqrt[n]{x}$ for $n = 2$ and $n = 3$ are shown in the first and second diagrams, respectively, in figure 5.13 below.

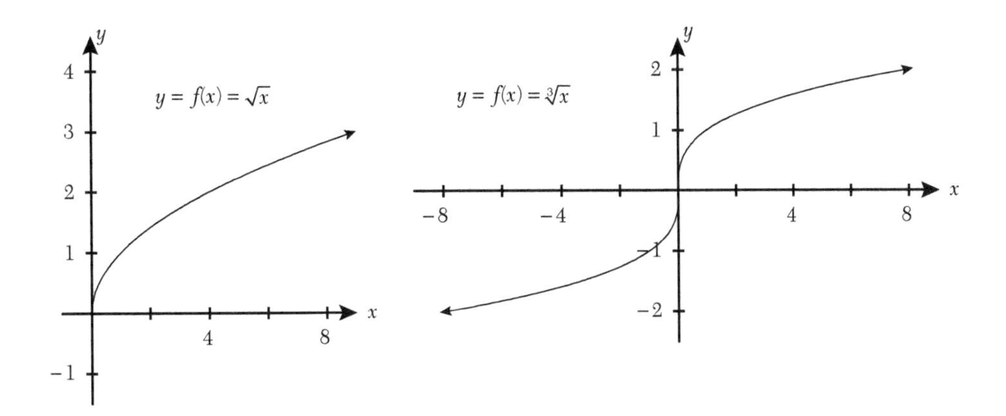

FIGURE 5.13. Even and odd root functions.

In general, if n is even, then the graph of $f(x)=\sqrt[n]{x}$ will have the same shape as the graph of $f(x)=\sqrt{x}$, and if n is odd, then the graph of $f(x)=\sqrt[n]{x}$ will have the same shape as the graph of $f(x)=\sqrt[3]{x}$.

5.8 PIECEWISE DEFINED FUNCTIONS

We now introduce the notion of a *piecewise defined function*.

DEFINITION 5.8.1. *A piecewise defined function is a function with different definitions on different intervals of the real line.*

If the separate definitions of piecewise defined functions are familiar, as in the examples below, then it is possible to draw a graph corresponding to each definition.

If the graphs on two adjacent intervals do not join at the common end point of the two intervals, then the value of the piecewise defined function is indicated by placing a bullet on the end point of the graph that is across the correct y value and an open circle on the end point of the graph that is not the function value. This can be understood with the help of the following example.

EXAMPLE 5.8.1. We define the piecewise defined function

$$f(x)=\begin{cases} 1-x & \text{if } x\leq 1 \\ x^2 & \text{if } x>1 \end{cases}$$

The graph of $y=f(x)$ is shown in figure 5.14. The location of the bullet in the graph determines that $f(1)=0$. ◈

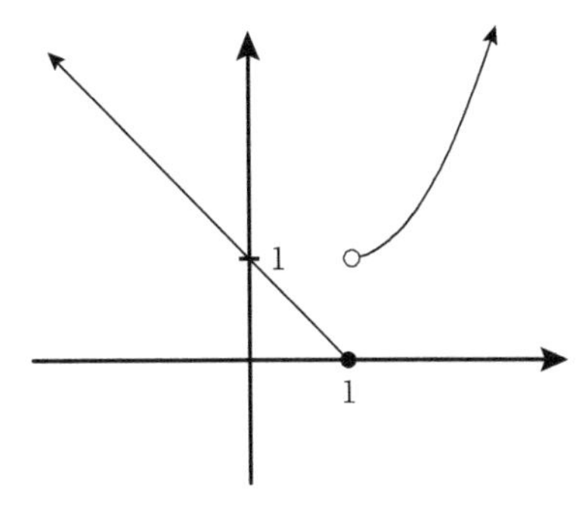

FIGURE 5.14. A piecewise defined function.

A familiar function that can be expressed in terms of a piecewise definition is the absolute value function.

EXAMPLE 5.8.2. The graph of

$$f(x) = |x| = \begin{cases} -x & \text{if } x \le 0 \\ x & \text{if } x \ge 0 \end{cases}$$

can be drawn by erasing the dotted portions of the lines $y = x$ and $y = x$, as shown in the first diagram in figure 5.15 below, which leaves behind the graph of the absolute value function. Similarly, the graph of

$$f(x) = x|x| = \begin{cases} x(-x) = -x^2 & \text{if } x \le 0 \\ x(x) = x^2 & \text{if } x \ge 0 \end{cases}$$

can be drawn by erasing the dotted portions of the parabolas shown in the second diagram of figure 5.15. ◈

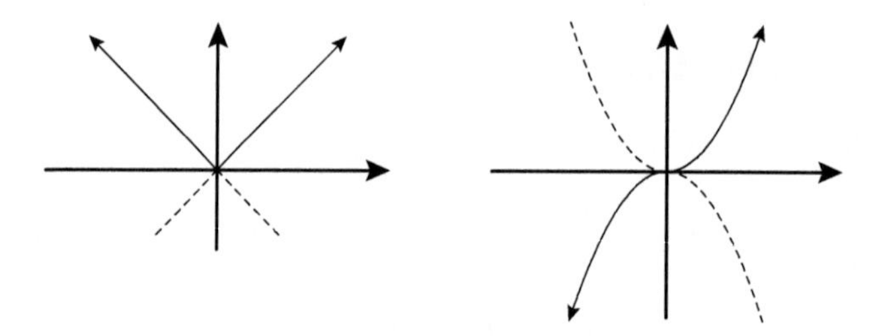

FIGURE 5.15. Piecewise defined functions.

5.9 SYMMETRY OF FUNCTIONS

The graphs of some functions have the property of being symmetric across the axes. We identify two types of symmetry.

DEFINITION 5.9.1. *If the graph of a function folds onto itself over the y axis, then the function is called an even function; if the graph of a function matches itself when it is reflected both across the* y *axis and across the* x *axis, then the function is called an odd function.*

REMARK 5.9.1. At should be understoond from definition 5.9.1 that in order to talk about a function being even or odd, its domain, as a subset of the number line, should be symmetric about the origin, that is, if x is an element of the domain, then $-x$ is also an element of the domain.

EXAMPLE 5.9.1. In figure 5.16, the graphs on the left are the graphs of even functions, and the graphs on the right are the graphs of odd functions. ◈

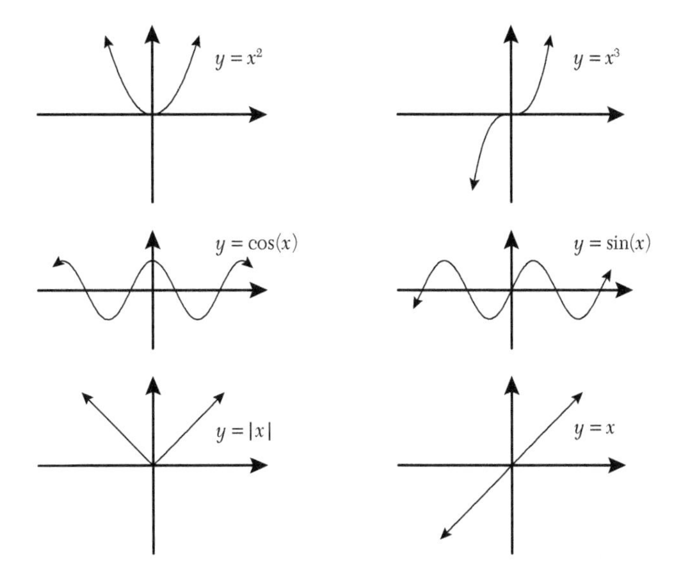

FIGURE 5.16. Even and odd functions.

REMARK 5.9.2. A function is an even function if and only if it has the algebraic property $f(-x) = f(x)$ for any value x in the domain of f, and a function is an odd function if and only if, it has the algebraic property $f(-x) = -f(x)$ for any value x in the domain of f.

EXAMPLE 5.9.2. $f(x) = x + x^5$ is an odd function because

$$f(-x) = (-x) + (-x)^5 = -x - x^5 = -(x + x^5) = -f(x)$$

◈

5.10 OPERATIONS ON FUNCTIONS

Functions can be regarded as elements of an algebraic system, meaning that we can add, subtract, multiply, and divide them. We can also operate on functions by forming compositions of functions.

5.10.1 The Algebra of Functions

DEFINITION 5.10.1. *Suppose that* f *and* g *are functions with domains* A *and* B, *respectively, then we can create the new functions.*

$$f + g, \quad f - g, \quad fg, \quad f / g,$$

defined by

$$
\begin{aligned}
(f+g)(x) &= f(x) + g(x), && \text{for } x \in A \cap B \\
(f-g)(x) &= f(x) - g(x), && \text{for } x \in A \cap B \\
(fg)(x) &= f(x) \cdot g(x), && \text{for } x \in A \cap B \\
\left(\frac{f}{g}\right)(x) &= \frac{f(x)}{g(x)}, && \text{for } x \in A \cap B / \{x \in \mathbb{R} \mid g(x) = 0\}.
\end{aligned}
$$

The next example demonstrates that it is possible to prove statements about the properties of functions, in general (i.e., statements that do not require specification of any particular functions).

EXAMPLE 5.10.1. Prove that a sum of odd functions is an odd function.

Answer: Suppose that f and g are odd functions, then for a value of x in the domain of both functions,

$$\begin{aligned}(f+g)(-x) &= f(-x)+g(-x) && \text{sum of functions}\\ &= -f(x)-g(x) && \text{property of odd functions}\\ &= -(f(x)+g(x)) && \text{factorizing } -1\\ &= -(f+g)(x) && \text{sum of functions}\end{aligned}$$

As we have proved that $(f+g)(-x) = -(f+g)(x)$, we conclude that $f+g$ is an odd function. ◈

5.10.2 Compositions of Functions

DEFINITION 5.10.2. *If the output from one function is taken to be the input for another function, then this forms a composition of the functions. If the first function is* g *and the second function is* f, *then the composition is denoted as* $f \circ g$.

We can use the machine analogy again (figure 5.17 below), with g taking an input x and giving an output u and f taking u as its input and giving an output y. The composition $f \circ g$ takes x as its input and gives y as its output.

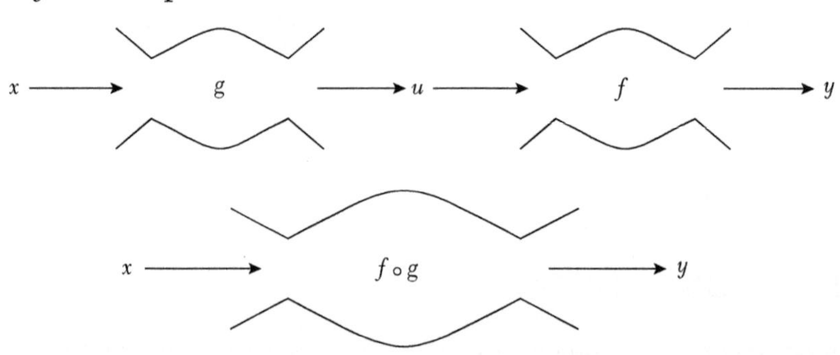

FIGURE 5.17. Machine analogy of the composition of two functions.

It is a familiar experience for us to hook up two machines, where the first machine provides data or some other kind of input to the second machine. For example, a computer sends a bit stream to a printer when a "print" command is executed from the keyboard of a computer. If the printer does not print as expected, then the problem could be with the printer or with the computer (or the keyboard). A simpler example is a hose pipe connected to a faucet. If we turn on the faucet to water a flower bed, then water might not come out of the hose if there is a kink in the hose. However, if there is no kink in the hose, then we would check to see whether water is actually coming out of the faucet. The principle is the same with mathematical functions. We cannot hope for a composition $f \circ g$ of functions to give an output if the input x is not in the domain of the first function g. Another way to put it is that the domain of the composition $f \circ g$ should be a subset of the domain of g (i.e., any value excluded from the domain of the first function should also be excluded from the domain of the composition).

EXAMPLE 5.10.2. If $g(x) = \frac{x-1}{x+1}$ and $f(u) = \frac{1}{u}$, then we can obtain the expression for $f \circ g$:

$$\begin{aligned}(f\circ g)(x) &= f(g(x))\\ &= f\left(\frac{x-1}{x+1}\right)\\ &= \frac{x+1}{x-1}.\end{aligned}$$

The domain of $f \circ g$ should exclude $x = 1$ (because the denominator of $\frac{x+1}{x-1}$ should not be zero) and also exclude $x = -1$ (because this value is not in the domain of g). Therefore, the domain of $f \circ g$ is:

$$\{x \in \mathbb{R} \mid x \neq 1, x \neq -1\}.$$ ◈

EXAMPLE 5.10.3. If $h(x) = \frac{1}{x^2}$ and $f(x) = \sqrt{x}$, then we can form the composition in two ways:

$$(f \circ h)(x) = f(h(x)) = f\left(\frac{1}{x^2}\right) = \sqrt{\frac{1}{x^2}} = \frac{1}{\sqrt{x^2}} = \frac{1}{|x|}$$

$$(h \circ f)(x) = h(f(x)) = h\left(\sqrt{x}\right) = \frac{1}{\left(\sqrt{x}\right)^2} = \frac{1}{x}.$$

The domain of $f \circ h$ is $\{x \in \mathbb{R} \mid x \neq 0\}$ and the domain of $h \circ f$ is $\{x \in \mathbb{R} \mid x > 0\}$. ◈

5.11 TRANSFORMATIONS OF FUNCTIONS

Functions can be transformed in various ways. The transformations below result in a vertical or horizontal shift of the graph of the function, or a vertical or horizontal stretch or compression of the graph of the function. Graphs can also be reflected across the y axis or the x axis.

5.11.1 Vertical and Horizontal Shifts

DEFINITION 5.11.1. *Suppose that* $c > 0$ *and* $f(x)$ *is any real-valued function, then the graphs of the following equations are related to the graph of* $y = f(x)$ *by means of a shift up, down, left, or right.*

- $y = f(x) + c$ *(graph shifts up c units)*
- $y = f(x) - c$ *(graph shifts down c units)*
- $y = f(x + c)$ *(graph shifts to the left c units)*
- $y = f(x - c)$ *(graph shifts to the right c units)*

EXAMPLE 5.11.1. If $f(x) = |x|$ and $g(x) = f(x-1) = |x-1|$, then the effect of the transformation can be understood from table 5.4, which shows the values of $f(x)$ and $g(x)$ for some integer values of x. The graphs are shown in figure 5.18. ◈

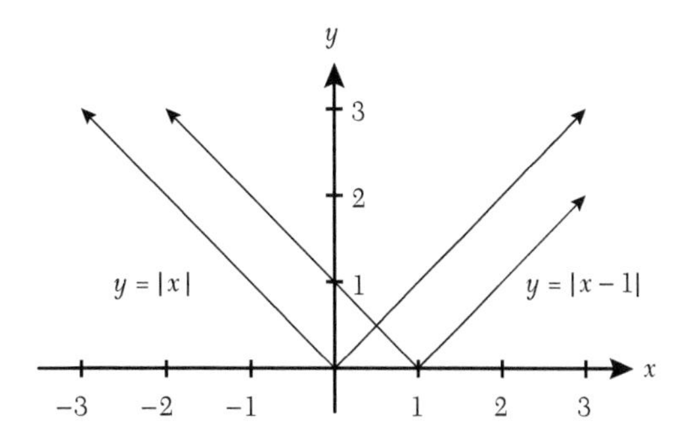

FIGURE 5.18. The shifting of an absolute value graph.

TABLE 5.4. Table of values for $y = g(x) = |x-1|$

x	−3	−2	−1	0	1	2	3
$f(x) = \|x\|$	3	2	1	0	1	2	3
$g(x) = \|x-1\|$	4	3	2	1	0	1	2

5.11.2 Vertical and Horizontal Scaling

Definition 5.11.2. *Suppose that* $c > 1$ *and* f(x) *is any real-valued function, then the graphs of the following equations are related to the graph of* $y = f(x)$ *by means of a vertical or horizontal stretching or compression of the graph.*

- $y = cf(x)$ *(graph stretches vertically)*
- $y = \dfrac{f(x)}{c}$ *(graph compresses vertically)*
- $y = f(cx)$ *(graph compresses horizontally)*
- $y = f\left(\dfrac{x}{c}\right)$ *(graph stretches horizontally)*

Examples of the scaling of cosine and sine graphs were given in section 4.8.3 There are more examples for other types of functions in the exercises at the end of this chapter.

5.11.3 Reflections Across the Axes

Definition 5.11.3. *If* f(x) *is any real-valued function, then the graph of* $\mathrm{y} = \mathrm{f(x)}$ *can reflect across the* y *axis or the* x *axis.*

- $y = -f(x)$ *(graph reflects across the* x *axis)*
- $y = f(-x)$ *(graph reflects across the* y *axis)*

Example 5.11.2. Let $f(x) = 2^x$. The graph in figure 5.19 below, shows the graph of $y = f(x)$ together with its reflection across the y axis (i.e., the graph of $y = f(-x) = 2^{-x}$). ◈

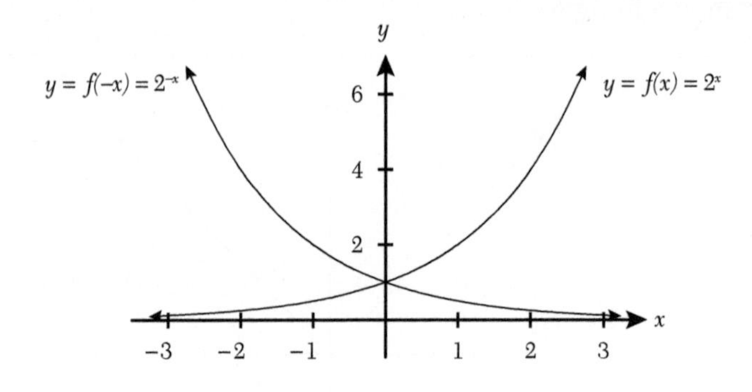

FIGURE 5.19. Reflection of an exponential graph across the *y* axis.

5.12 VECTOR-VALUED FUNCTIONS

A graph that fails the vertical line test is not the graph of a real-valued function, but it could be the graph of a vector-valued function. We will explain below what is meant by a vector-valued function. The graph of a vector-valued function is a *curve* that, together with a specified direction, is called a *directed curve* or a *trajectory*.

DEFINITION 5.12.1. *A vector-valued function is defined in component form, or in terms of the standard basis vectors, as*

$$\vec{r}(t) = \langle x(t), y(t)\rangle \text{ for } a \le t \le b \quad \text{or} \quad \vec{r}(t) = x(t)\vec{i} + y(t)\vec{j} \text{ for } a \le t \le b$$

where x(t) *and* y(t) *are real-valued functions of a variable* t *(called a parameter),* a *is a real number or* $-\infty$, *and b is a real number or* ∞.

DEFINITION 5.12.2. *The trajectory of a vector-valued function* $\vec{r}(t)$ *is the curve*

$$\{(x(t), y(t)) \mid a \le t \le b\}$$

directed with increasing values of t.

REMARK 5.12.1. The trajectory of a vector-valued function with two components, as defined earlier, is a directed curve in the Euclidean plane (i.e., two-dimensional space). If a vector-valued function is defined with three components, then its trajectory is a directed curve in three-dimensional space.

EXAMPLE 5.12.1. Some values for the vector-valued function $\vec{r}(t) = \langle t^2 - 2t, t+1\rangle$, that is, $x(t) = t^2 - 2t$ and $y(t) = t + 1$, are computed in table 5.5 below. The terminal point of the position vector for any of the vectors in the table is a point on the trajectory of $\vec{r}(t) = \langle x(t), y(t)\rangle$, as shown in figure 5.20 below. The trajectory looks like a parabola turned on its side. That's because it is a parabola! (The reason for this is given at the end of section 5.12.3.) ◈

TABLE 5.5. Table of values for $\vec{r}(t) = \langle t^2 - 2t, t+1\rangle$

t	-2	-1	0	1	2	3	4
$x(t)$	8	3	0	-1	0	3	8
$y(t)$	-1	0	1	2	3	4	5
$\langle x(t), y(t)\rangle$	$\langle 8,-1\rangle$	$\langle 3,0\rangle$	$\langle 0,1\rangle$	$\langle -1,2\rangle$	$\langle 0,3\rangle$	$\langle 3,4\rangle$	$\langle 8,5\rangle$

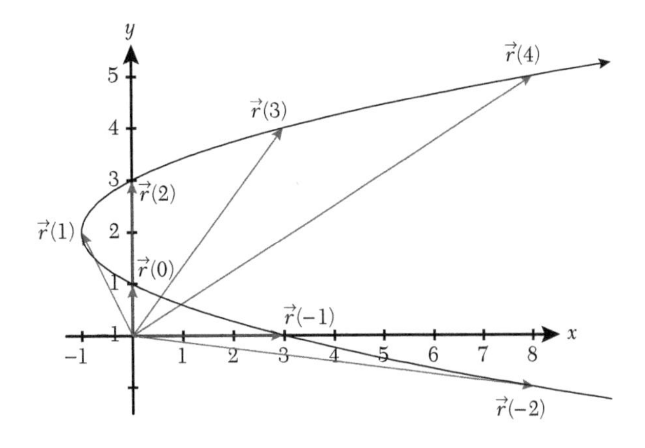

FIGURE 5.20. The trajectory of the vector-valued function $\vec{r}(t) = \langle t^2 - 2t, t+1\rangle$.

Some familiar graphs, for example, a circle and a line, can also be described as the trajectories of vector-valued functions.

5.12.1 The Vector-Valued Function for a Circle

If $\vec{r}$ is the position vector for a point on a circle centered at the origin with radius R, that is, the circle $x^2 + y^2 = R^2$, then

$$\frac{x}{|\vec{r}|} = \frac{x}{R} = \cos\theta, \quad \text{and} \quad \frac{y}{|\vec{r}|} = \frac{y}{R} = \sin\theta$$

where θ is the angle in a counterclockwise direction from the x axis to the vector $\vec{r}$. (Refer to figure 5.21.) By solving for x and y above, and regarding θ as a parameter, the vector-valued function for a circle can be described as

$$\vec{r}(\theta) = \langle R\cos\theta, R\sin\theta \rangle,$$

where $-\infty < \theta < \infty$. As θ increases from 0, the radius vector $r(\theta)$ winds around the circle in a counterclockwise direction, completing one revolution for each increment in the value of θ by 2π. Similarly, as θ decreases, the radius vector $r(\theta)$ winds around the circle in a clockwise direction, completing one revolution for each decrement in the value of θ by 2π.

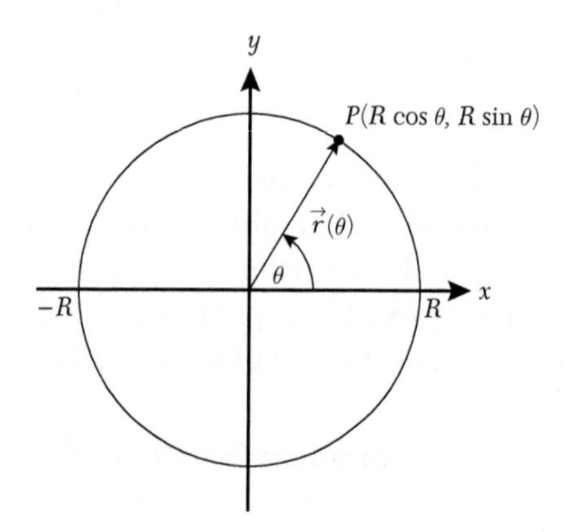

FIGURE 5.21. The vector equation for a circle.

5.12.2 The Vector-Valued Function for a Line

Recall that the Cartesian equation for a line can be determined when the slope m and a point $P(a,b)$ on the line are given. The equation for the line is then $y - b = m(x - a)$.

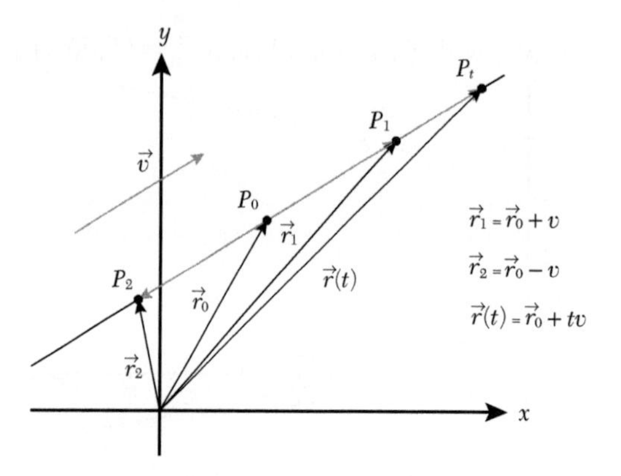

FIGURE 5.22. The vector equation for a line.

Alternatively, as we now demonstrate by figure 5.22, a line can be determined when a vector $\vec{v}$ parallel to the line and a position vector $\vec{r}_0$ for any point on the line are given. We see that P_0 is the terminal point of $\vec{r}_0$ and any other point (e.g., points P_1, P_2, and P_t in the diagram) on the line can be reached by adding an appropriate scalar multiple of v to $\vec{r}_0$. Generally, if the scalar multiple of $\vec{v}$ is denoted as $t\vec{v}$, then the position vector for a point on the line can be described as $\vec{r}_0 + t\vec{v}$. Thus, we regard t as the parameter for the vector-valued function $\vec{r}(t)$ defined by

$$\vec{r}(t) = \vec{r}_0 + t\vec{v}.$$

If $\vec{r}_0$ and $\vec{v}$ are expressed in component form as $\vec{r}_0 = \langle x_0, y_0 \rangle$ and $\vec{v} = \langle a, b \rangle$, then $\vec{r}(t)$ can also be expressed in component form as

$$\vec{r}(t) = \langle x_0, y_0 \rangle + t\langle a, b \rangle = \langle x_0 + ta, y_0 + tb \rangle.$$

That is, $\vec{r}(t) = \langle x(t), y(t) \rangle$, where

$$x(t) = x_0 + ta$$
$$y(t) = y_0 + tb$$

These two equations are called the *parametric equations* for the line.

EXAMPLE 5.12.2. Find a vector equation and parametric equations for the line passing through the points $P(3,-2)$ and $Q(5,7)$.

Answer: A vector parallel to the line is the directed line segment joining P to Q. Thus, in terms of the notation above, $\vec{v} = \langle 5-3, 7-(-2) \rangle = \langle 2,9 \rangle$. A position vector for a point on the line can be either the position vector for P or the position vector for Q, so we may set $\vec{r}_0 = \langle 3,-2 \rangle$ (the position vector for P). Therefore,

$$\vec{r}(t) = \vec{r}_0 + t\vec{v} = \langle 3,-2 \rangle + t\langle 2,9 \rangle = \langle 3+2t, -2+9t \rangle$$

and the parametric equations are

$$x(t) = 3 + 2t$$
$$y(t) = -2 + 9t$$

◈

REMARK 5.12.2. The expression of the vector and parametric equations for a line depends on the choice of parameter. In example 3.12.2, if t is replaced by $s+1$, then the parametric equations for the line (in terms of parameter s) are

$$x(s) = 5 + 2s$$
$$y(s) = 7 + 9s.$$

These are the equations that result from the choice $\vec{r}_0 = \langle 5,7 \rangle$ (the position vector for Q).

It should be no surprise that it is possible to convert the vector equation, or parametric equations, into the familiar Euclidean equation for a line.

EXAMPLE 5.12.3. This is a continuation of example 5.12.2. If the parametric equations are expressed as

$$x = 3 + 2t$$
$$y = -2 + 9t$$

then by solving for t in each equation,

$$t = \frac{x-3}{2} \text{ and } t = \frac{y+2}{9}$$

the parameter t can be eliminated by setting the expressions for t equal to each other, that is

$$\frac{y+2}{9} = \frac{x-3}{2}$$

which simplifies to

$$y = \frac{9x}{2} - \frac{31}{2}$$

This is the Cartesian equation for the line. ◈

5.12.3 Exploring Vector-Valued Functions

The trajectories of vector-valued functions can form spirals and loops, or they can be transformations of the graphs of familiar real-valued functions. Let's look at a few examples.

EXAMPLE 5.12.4. A vector-valued function with a spiral trajectory is obtained by multiplying the vector-valued function for a unit circle by the parameter value. For instance, the trajectory of $\vec{r}(\theta) = \langle \cos(2\pi\theta), \sin(2\pi\theta) \rangle$ is contained in the unit circle, with one complete revolution of the circle for every increase of θ by one unit. Now, if $\vec{s}(\theta) = \theta\vec{r}(\theta) = \langle \theta\cos(2\pi\theta), \theta\sin(2\pi\theta) \rangle$, for $\theta \geq 0$, then the distance from the origin to $\vec{s}(\theta)$ is always equal to θ (check this using the distance formula). This means the trajectory starts at the origin (with $\theta = 0$) and follows a circular path that gets farther and farther from the origin, as shown in figure 5.23 below.

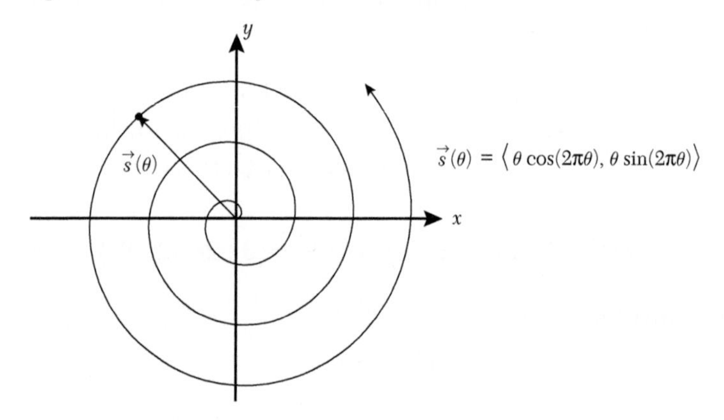

FIGURE 5.23. The vector function for a counter-clockwise spiral.

EXAMPLE 5.12.5. If a vector-valued function has trigonometric functions with different periods in its first and second components, then its trajectory forms loops around the origin. For example, if $\vec{r}(\theta) = \langle \cos(2\theta), \sin(3\theta) \rangle$, for $\theta \geq 0$, then the trajectory is a repeating pattern of intersecting loops passing through the origin, as shown in figure 5.24. ◈

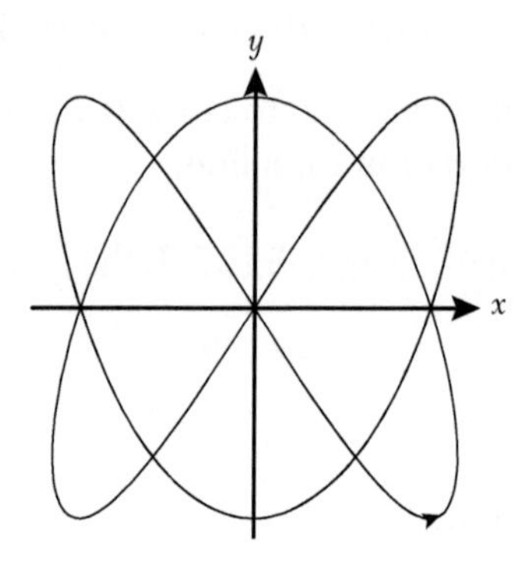

FIGURE 5.24. A trajectory with loops.

EXAMPLE 5.12.6. This is a continuation of example 5.12.1. It can be verified by elimination of the parameter t that the trajectory of

$$\vec{r}(t) = \langle t^2 - 2t, t + 1 \rangle$$

is a parabola. If we write

$$x = t^2 - 2t \text{ and } y = t + 1$$

then in the expression for x, t can be replaced by $y - 1$, resulting in the expression $x = (y-1)^2 - 2(y-1)$. This simplifies to $x = y^2 - 4y + 3$, which is the equation for a parabola with the usual roles of x and y interchanged. ◈

5.13 INVERSE FUNCTIONS

A function maps every element of its domain to an element of its codomain. As mentioned earlier, the elements of the codomain that are targeted in this way form a subset of the codomain called the range of the function. Some functions have the property that every element in the range is targeted only once, meaning that there is a unique element in the domain that maps to it. Such a function is called *invertible*.

DEFINITION 5.13.1. *If the mapping from the domain to the range of a function can be reversed, that is, if we can take the range to be the domain of a new function, then we call this an inverse function.*

In this section, we are going to investigate the property of invertibility and introduce the notation for an inverse function.

5.13.1 The Inverse of a Point

The notion of invertability of a function can more easily be understood if we know what is meant by the *inverse of a point* in the Cartesian plane. Lets say we select a point $P(D,C)$ with $0 < D < C$. Then, P is in the first quadrant and lies above the line $y = x$. What can we mean by the inverse of the point P? Because points on the x-axis are mapped to points on the y-axis, D is mapped to C. Our intention now is to reverse the mapping, so we want to map C to D. This can be indicated by means of another point, say $Q(C,D)$, which has D and C switched. The point Q is also in the first quadrant and lies below the line $y = x$. Figure 5.25 shows how the points P and Q relate geometrically to the line $y = x$ (the dotted line). Because of the congruence of the triangle in the diagram, P and Q are equidistant from the line $y = x$ and the line joining P to Q is perpendicular to the line $y = x$. For these two reasons, Q is called the *reflection* of P across the line $y = x$.

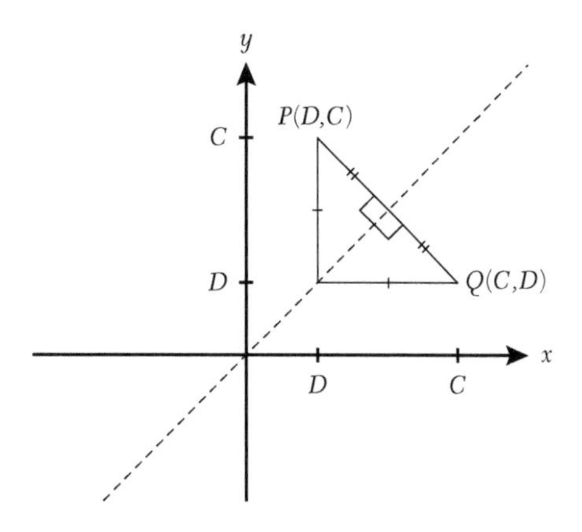

FIGURE 5.25. The inverse of a point in the plane.

5.13.2 Logarithmic Functions

As will be explained fully in section 5.13.3 below, an example of a function, that is, invertible is an exponential.

Definition 5.13.2. *The inverse of an exponential function is called a logarithmic function.*

In this section, we will demonstrate the construction of logarithmic functions and explore their properties.

If we suppose that P in figure 5.25 is a point on the graph of an exponential function, as shown in figure 5.26, then every point reflects to a point across the line $y = x$ in the same way that P reflects to Q. The reflected graph obtained in this way is the graph of a logarithmic function. Because the equation for an exponential graph is $y = a^x$, where $a > 0$ and $a \neq 1$, and P is a point on the graph, the value of C in terms of D is $C = a^D$. Now, if we want to express the value of D in terms of C, we use the notation $D = \log_a C$. Then $Q(C,D)$ is a point on the graph of $y = \log_a x$, called a *logarithmic graph*. The logarithmic function $f(x) = \log_a x$ is defined for positive values of x (its graph lies entirely to the right of the y axis). The number a is called the *base* of the logarithm. Note that $\log_a a^D = \log_a C = D$ and $a^{\log_a C} = a^D = C$. These are called *cancellation equations* and are true for any positive value of C and any value of D.

In summary, we have

$$y = a^x \leftrightarrow x = \log_a y, \quad y > 0 \text{ (inverse property)}$$
$$\log_a a^x = x, \quad x \in \mathbb{R} \text{ and } a^{\log_a x} = x, \quad x > 0 \text{ (cancellationequations)} \qquad (5.1) \text{ and } (5.2)$$

In section 5.13.3, we will demonstrate that exponential and logarithmic graphs pass the *horizontal line test*. This property can be stated algebraically: if c_1 and c_2 are positive real numbers and d_1 and d_2 are real numbers, then (supposing that $a > 0$ and $a \neq 1$)

$$(\log_a c_1 = \log_a c_2 \text{ iff } c_1 = c_2, \quad a^{d_1} = a^{d_2} \text{ iff } d_1 = d_2) \qquad (5.3)$$

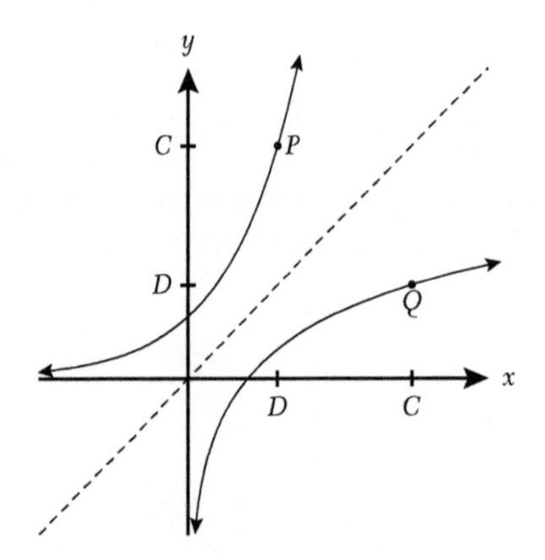

FIGURE 5.26. The logarithmic function.

(The abbreviation iff is being used for "if and only if.")

We list the laws for logarithms in table 5.6 below. They are a consequence of formulas (5.1) to (5.3) and the laws for exponents (see table 5.2).

TABLE 5.6. Laws for logarithms

(I)	$\log_a xy = \log_a x + \log_a y$	A log of a product becomes a sum of logs.
(II)	$\log_a\left(\dfrac{x}{y}\right) = \log_a x - \log_a y$	A log of a quotient becomes a difference of logs.
(III)	$\log_a x^r = r\log_a x$	A log of an exponent becomes a product.

PROOF OF (I). If $c = \log_a xy$, then (using the second cancellation equation)

$$a^c = a^{\log_a xy} = xy = \left(a^{\log_a x}\right)\left(a^{\log_a y}\right) = a^{\log_a x + \log_a y}$$

This means that $c = \log_a x + \log_a y$, which was to be proved.

The properties (II) and (III) can be proved similarly. □

EXAMPLE 5.13.1. According to the cancellation equations and the laws for logarithms,

(i) $\log_2 64 = \log_2 2^6 = 6.$

(ii) $\log_3 \dfrac{1}{81} = \log_3 3^{-4} = -4.$

(iii) $\log_{10} 3.75 + \log_{10} \dfrac{8}{3} = \log_{10} \dfrac{30}{8} + \log_{10} \dfrac{8}{3} = \log_{10}\left(\dfrac{30}{8}\right)\left(\dfrac{8}{3}\right) = \log_{10} 10 = 1.$ ◈

EXAMPLE 5.13.2. Solve for x in the equation $\log_4\left(\log_3\left(\log_2 x\right)\right) = 1.$

Answer: We start by raising each side as a power of 4 and then apply the second cancellation equation.

$$4^{\log_4(\log_3(\log_2 x))} = 4^1$$
$$\log_3\left(\log_2 x\right) = 4$$

We repeat the process by raising each side as a power of 3 and using the inverse property to obtain the final answer.

$$3^{\log_3(\log_2 x)} = 3^4$$
$$\log_2 x = 3^4$$
$$x = 2^{81}$$

◈

A frequently occurring base of the logarithm is the number e, introduced in section 1.12.3. The notation $\log_e$ is abbreviated as "ln". This is called the *natural logarithm*. Here are the properties of the natural logarithm.

$$y = e^x \leftrightarrow x = \ln y, \quad y > 0 \text{ (inverse property)}$$
$$\ln e^x = x, \quad x \in \mathbb{R} \text{ and } e^{\ln x} = x, \quad x > 0 \text{ (cancellation equations)} \qquad (5.4) \text{ and } (5.5)$$

The graphs of some logarithmic functions are shown in figure 5.27. All of the graphs pass through the x axis at $x = 1$. The laws for natural logarithms, as a special case of the laws for logarithms in table 5.6, are stated in table 5.7.

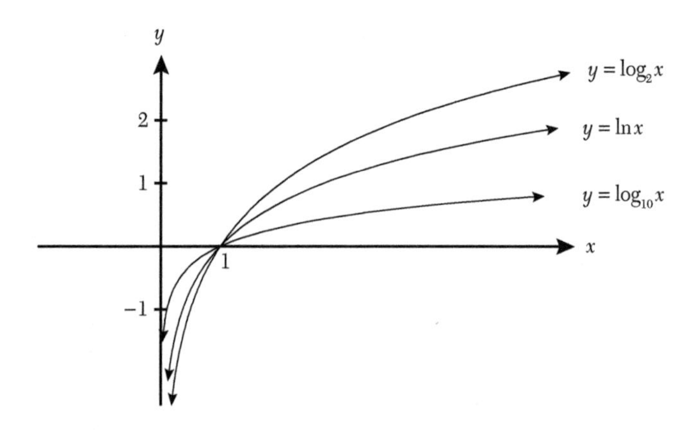

FIGURE 5.27. Graphs of logarithmic functions.

TABLE 5.7. Laws for natural logarithms

(I)	$\ln xy = \ln x + \ln y$	A log of a product becomes a sum of logs
(II)	$\ln\left(\dfrac{x}{y}\right) = \ln x - \ln y$	A log of a quotient becomes a difference of logs
(III)	$\ln x^r = r \ln x$	A log of an exponent becomes a product

More advanced methods of solving equations will be a topic in chapter 6. The next example is a foretaste of this.

Example 5.13.3. Solve for x in the equation $\ln(x+1)+\ln(x-1)=2$.

Answer: Property (I) of the laws for logarithms can be used to rewrite the left-hand side. After that the inverse property and ordinary algebra can be used to solve for x:

$$\ln(x+1)(x-1)=2$$
$$(x+1)(x-1)=e^2$$
$$x^2-1=e^2$$
$$x=\sqrt{e^2+1}$$

The positive root was taken in the last step because the terms $\ln(x+1)$ and $\ln(x-1)$ are not defined if $x=-\sqrt{e^2+1}$. ◈

A *change of base formula* is useful in many situations and is easy to derive. The equation $x=\log_a b$ is equivalent to $a^x=b$, and if we can take the logarithm of both sides with any choice of base, that is, $\log_c a^x = \log_c b$, then, by property (III) of the laws for logarithms, $x\log_c a = \log_c b$. Therefore, we derive the formula

$$\boxed{\log_a b = \frac{\log_c b}{\log_c a}} \tag{5.6}$$

where b is a positive real number, and a and c are positive real numbers not equal to 1.

EXAMPLE 5.13.4. Use the change of base formula to evaluate $\log_4 8$.

Answer:

$$\log_4 8 = \frac{\log_2 8}{\log_2 4} = \frac{3}{2}. \qquad \diamond$$

Logarithms provide a useful measuring scale, called a *logarithmic scale*, for measurement over a very large range of possible values. A unit increase in a logarithmic scale corresponds to an increase by a factor equal to the base chosen for the logarithm. Human sensory perception and many natural processes manifest according to a logarithmic scale. Regarding our sense of hearing is that, for example, equal ratios of frequencies are perceived as equal differences of pitch. Another example of a logarithmic scale is devised by Charles Richter in 1935 for measuring earthquakes. According to this scale, the magnitude of an earthquake is determined by taking the logarithm of the amplitude of waves measured by a seismograph. A two-point increase in this scale corresponds to a thousand times more energy released by an earthquake. For example, the tremendous earthquake that hit Japan in 2011 measured 9.0 on the Richter scale, and so released a thousand times more energy than the devastating earthquake that shook Haiti in 2010, which measured 7.0 on the Richter scale.

5.13.3 The Inversion of One-to-One Functions

A real-valued function $f(x)$ is a *one-to-one function* if and only if the graph of $y = f(x)$ passes the *horizontal line test*. A graph passes the horizontal line test if and only if any horizontal line in the Cartesian plane cuts the graph of $y = f(x)$ at most once. An exponential function, for example, passes the horizontal line test, but a quadratic function, for example, does not, as shown in figure 5.28 below.

A horizontal line reflected across the line $y = x$ becomes a vertical line, and, therefore, the reflected graph of a one-to-one function passes the vertical line test. Recall that this is a test that determines whether a graph is the graph of a function. We conclude that the reflected graph of a one-to-one function is the graph of a function. Therefore, a one-to-one function $f(x)$ is invertible, and the reflection of its graph about the line $y = x$ is the graph of its inverse function. The notation we use for its inverse function is $f^{-1}(x)$.

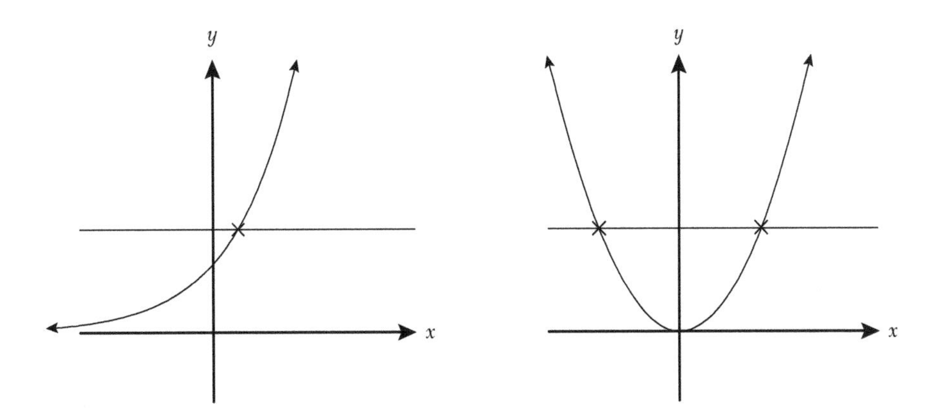

FIGURE 5.28. The horizontal line test.

EXAMPLE 5.13.5. If $f(x) = a^x$, where $a > 0$ and $a \neq 1$, then $f^{-1}(x) = \log_a x$. ◈

EXAMPLE 5.13.6. The function $f(x) = x^3 + 2$ is a one-to-one function (its graph is the cubic graph shifted up two units); therefore, it has an inverse function $f^{-1}(x)$. The graphs of $y = f(x) = x^3 + 2$ and $y = f^{-1}(x)$ are shown in figure 5.29. ◈

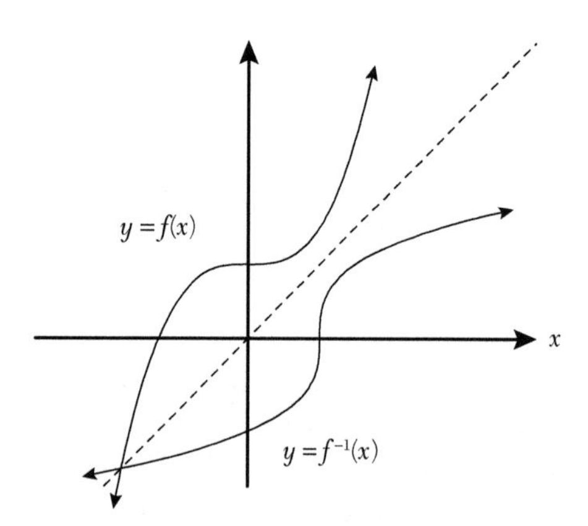

FIGURE 5.29. The inverse of a cubic function.

Given a one-to-one function $f(x)$, there is the following method by which the explicit expression for its inverse function $f^{-1}(x)$ can be found:

(i) Write the equation $x = f(y)$, that is, interchange x and y in the equation $y = f(x)$, using the explicit expression for $f(x)$;

(ii) then solve for y (if possible). The resulting expression is for $f^{-1}(x)$ (if an explicit expression can be found).

EXAMPLE 5.13.7. If $f(x) = x^3 + 2$, then, according to the method above, an expression for the inverse function $f^{-1}(x)$ can be found by solving the equation $x = y^3 + 2$ for y. The solution is $y = \sqrt[3]{x-2}$ (check this). Therefore, $f^{-1}(x) = \sqrt[3]{x-2}$, and the graph of $y = f^{-1}(x)$ is the graph of the cube root function shifted two units to the right (the reflected graph shown in figure 5.29). ◈

EXAMPLE 5.13.8. Real-valued functions that are strictly increasing or decreasing are one-to-one functions. We give a precise definition of these types of functions in the next section. ◈

5.13.4 Increasing and Decreasing Functions

In the following definition, we will restrict our attention to intervals in the domain of a function. We also make a distinction between *increasing* and *strictly increasing*, and *decreasing* and *strictly decreasing*.

DEFINITION 5.13.3.

- *A function is* increasing *on an interval* (I) *if* $f(x_1) \le f(x_2)$ *whenever* $x_1 < x_2$ *in* (I).
- *A function is* decreasing *on an interval* (I) *if* $f(x_1) \ge f(x_2)$ *whenever* $x_1 < x_2$ *in* (I).
- *A function is* strictly increasing *on an interval* (I) *if* $f(x_1) < f(x_2)$ *whenever* $x_1 < x_2$ *in* (I).
- *A function is* strictly decreasing *on an interval* (I) *if* $f(x_1) > f(x_2)$ *whenever* $x_1 < x_2$ *in* (I).

Some functions that are strictly increasing on their domains are the root functions $f(x) = \sqrt{x}$, $f(x) = \sqrt[3]{x}$, etc.; odd powers of x, that is, $f(x) = x$, $f(x) = x^3$, etc.; and exponential functions a^x, if $a > 1$. The trigonometric function $f(x) = \tan(x)$ is strictly increasing on every interval $\left((2n-1)\frac{\pi}{2}, (2n+1)\frac{\pi}{2}\right)$, where n is any integer. The absolute value function and even powers of x, that is, $f(x) = x^2$, $f(x) = x^4$, etc. are strictly decreasing on the interval $(-\infty,0]$ and strictly increasing on the interval $[0,\infty)$. Any constant function, according to our definition, is both an increasing and a decreasing function (but not strictly increasing or strictly decreasing).

If a function $f(x)$ is (strictly) increasing on an interval, then its negative, that is, $-f(x)$, is (strictly) decreasing on the same interval, and vice versa.

5.13.5 Inverse Trigonometric Functions

The trigonometric functions are not one-to-one functions (their graphs fail the horizontal line test); however, if their domains are suitably restricted, then the corresponding graphs do pass the horizontal line test. In particular, if $f(x)=\sin(x)$ for $\frac{-\pi}{2}\le x\le\frac{\pi}{2}$, then the graph of $y=f(x)$ passes the horizontal line test, that is, f is a one-to-one function (see the first diagram of figure 5.30). The graph of its corresponding inverse function (the *inverse sine function* or *arcsine function*) is also shown in the first diagram as the reflection across the line $y=x$. We use the notations $\arcsin(x)$ or $\sin^{-1}x$ for the inverse sine function. The graph of $y=\sin^{-1}x$ is shown again in the third diagram. Similarly, if $f(x)=\cos(x)$ for $0\le x\le\pi$, then the graph of $y=f(x)$ passes the horizontal line test; that is, f is a one-to-one function. The graph of its corresponding inverse function (the *inverse cosine function* or *arccosine function*) is shown in the second diagram of figure 5.30 as the reflection across the line $y=x$. We use the notations $\arccos(x)$ or $\cos^{-1}x$ for the inverse cosine function. The graph of $y=\cos^{-1}x$ is shown again in the fourth diagram in figure 5.30.

The *inverse tangent function* (or *arctan function*) is defined by restricting the tangent function to the interval $\left(-\frac{\pi}{2},\frac{\pi}{2}\right)$ and reflecting across the line $y=x$, as shown in figure 5.31. Note that the vertical asymptotes $x=\frac{-\pi}{2}$ and $x=\frac{\pi}{2}$ for the tangent function become the horizontal asymptotes $y=\frac{-\pi}{2}$ and $y=\frac{\pi}{2}$ for the arctan function.

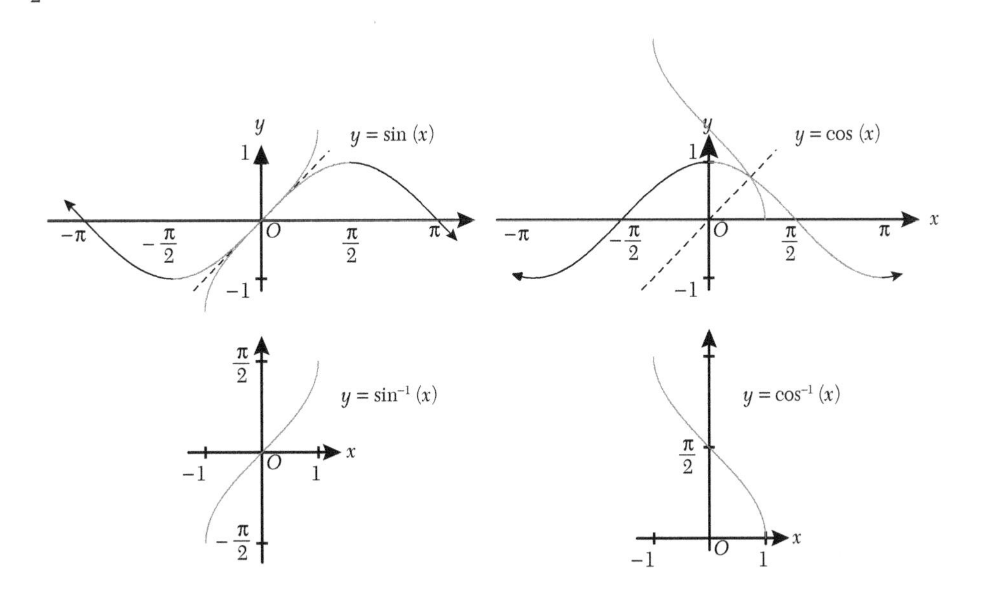

FIGURE 5.30. The inverse sine and inverse cosine functions.

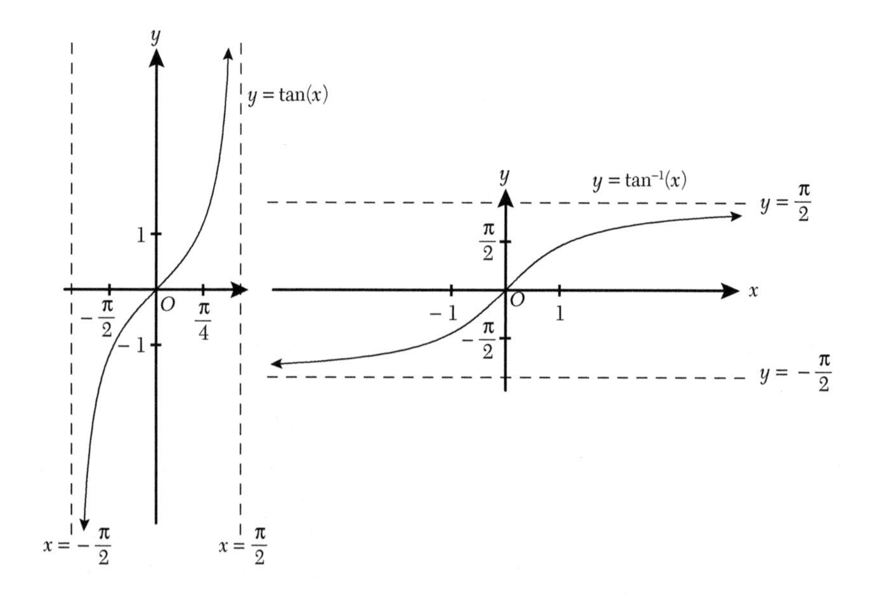

FIGURE 5.31. The inverse tangent function.

It is also possible to define inverse secant, cosecant, and cotangent functions, but we will not do so here.

In the examples below it will be helpful to refer to table 5.8, which lists the domain and range of each of the inverse trigonometric functions. Any answer obtained from a calculation involving an inverse trigonometric function should be in accordance with the information in table 5.8; that is, any answer must be in the correct range.

TABLE 5.8. Domain and range of the inverse trigonometric functions

	$\sin^{-1}$	$\cos^{-1}$	$\tan^{-1}$
Domain	$[-1,1]$	$[-1,1]$	$(-\infty,\infty)$
Range	$\left[-\frac{\pi}{2},\frac{\pi}{2}\right]$	$[0,\pi]$	$\left(-\frac{\pi}{2},\frac{\pi}{2}\right)$

Example 5.13.9.

(i) $\sin^{-1}\left(-\frac{1}{2}\right)=-\frac{\pi}{6}\ \left(\text{because } \sin\left(\frac{\pi}{6}\right)=-\frac{1}{2}\right)$

(ii) $\tan^{-1}(1)=\frac{\pi}{4}\ \left(\text{because } \tan\left(\frac{\pi}{4}\right)=1\right)$

(iii) $\sin^{-1}\left(\sin\left(\frac{\pi}{3}\right)\right)=\frac{\pi}{3}$

(iv) $\sin^{-1}\left(\sin\left(\frac{7\pi}{3}\right)\right)=\sin^{-1}\left(\sin\left(\frac{\pi}{3}\right)\right)=\frac{\pi}{3}$

(Watch out! The answer is not $\frac{7\pi}{3}$ because this is not in the range of the arcsine function.)

(v) $\sin\left(\sin^{-1}\left(\dfrac{7}{11}\right)\right) = \dfrac{7}{11}$

(vi) $\sin^{-1}\left(\sin\left(\dfrac{3\pi}{4}\right)\right) = \sin^{-1}\left(\sin\left(\pi - \dfrac{3\pi}{4}\right)\right) = \sin^{-1}\left(\sin\left(\dfrac{\pi}{4}\right)\right) = \dfrac{\pi}{4}$

(We used the first identity in formula (4.10) to get an answer in the range of the arcsine function.) ◈

In the following example, keep in mind that trigonometric ratios are ratios of the sides of right triangles. Thus, if $x > 0$, then $\tan^{-1} x$, for example, can be regarded as an angle in a right triangle with opposite side x, adjacent side 1, and, by the Pythagorean Theorem, hypotenuse $\sqrt{1+x^2}$. (If $x < 0$, then $\tan^{-1} x$ is a negative angle, and we place the corresponding triangle in the fourth quadrant.) It is also helpful to know the identity

$$\sin(\cos^{1} x) = \cos(\sin^{-1} x) = \sqrt{1-x^2}, \quad \text{for } -1 \le x \le 1.$$

Example 5.13.10. Simplify the expressions:

(i) $\cos(\tan^{-1} x)$

(ii) $\cos\left(\sin^{-1}\left(\dfrac{3}{5}\right)\right)$

(iii) $\cos\left(\sin^{-1}\left(\dfrac{1}{4}\right) + \sin^{-1}\left(\dfrac{3}{4}\right)\right)$

Answers:

(i) As explained above, $\cos(\tan^{-1} x) = \dfrac{1}{\sqrt{1+x^2}}$.

(ii) Because $\sin^{-1}\left(\dfrac{3}{5}\right)$ is an angle in a 3-4-5 triangle, $\cos\left(\sin^{-1}\left(\dfrac{3}{5}\right)\right) = \dfrac{4}{5}$.

(iii) Lets $A = \sin^{-1}\left(\dfrac{1}{4}\right)$ and $B = \sin^{-1}\left(\dfrac{3}{4}\right)$, then A is an angle in a right triangle with adjacent side $\sqrt{15}$ and hypotenuse 4, and B is an angle in a right triangle with adjacent side $\sqrt{7}$ and hypotenuse 4. Now we expand using formula (4.7b)

$$\begin{aligned}\cos(A+B) &= \cos(A)\cos(B) - \sin(A)\sin(B)\\ &= \frac{\sqrt{15}}{4}\cdot\frac{\sqrt{7}}{4} - \frac{1}{4}\cdot\frac{3}{4}\\ &= \frac{\sqrt{105}-3}{16}\end{aligned}$$

◈

EXERCISES

5.1. Decide whether each of the diagrams in figure 5.32 determines a relation. In each case, the elements of set A are in the first column, and the elements of set B are in the second column. An arrow points from the first element of a coordinate pair to the second element of the coordinate pair.

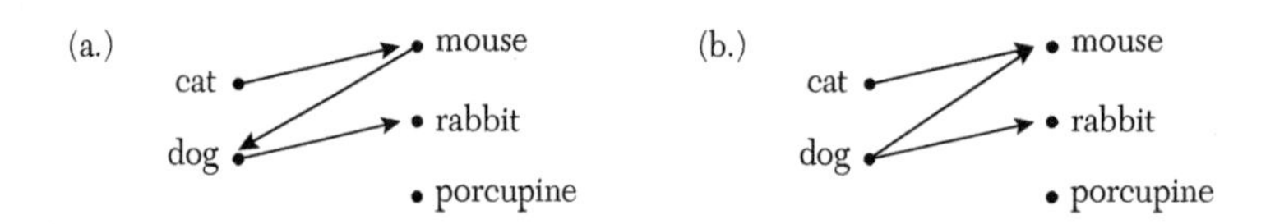

FIGURE 5.32.

5.2. Plot each of the following relations in the Cartesian plane. What is the range of each relation?

(a) $\{(0,0.5),(0,1.5),(-0.5,1),(0.5,1),(-2,0),(0,0),(2,0),(0,-1),(-1,-2),(1,-2)\}$

(b) $\{(-2,2),(-2,0),(-1,0),(-1,-2),(0,-2),(0,-1),(2,-1)\}$

5.3. If C = {mouse, cat, dog} and D = {mouse, rabbit, porcupine}, which of the following is a relation with domain in C and range in D?

(a) {(mouse,dog), (mouse,porcupine), (mouse,rabbit), (cat,rabbit)}

(b) {(mouse,dog), (mouse,porcupine), (mouse,rabbit), (cat,rabbit)}

(c) {(dog,mouse), (porcupine,mouse), (rabbit,mouse), (rabbit,cat)}

(d) {(mouse,rabbit)}

5.4. Which of the following relations is a function? Find the domain and range of each function.

(a) $\{(0,1),(1,2),(2,3),(3,4),(4,5),(5,0)\}$

(b) $\{(-2,-2),(-1,-2),(0,0),(1,1),(2,1)\}$

(c) $\{(-2,-2),(-2,-1),(0,0),(1,1),(1,2)\}$

(d) $\{(-2,0),(0,2),(2,0),(0,-2)\}$

5.5. Which of the following sets cannot be the description of a function?

(a) $\{f(1)=-1,\ f(-1)=1,\ f(0)=-1\}$

(b) $\{f(-1)=-1,\ f(1)=1,\ f(0)=0\}$

(c) $\{f(0)=-1,\ f(1)=1,\ f(0)=1\}$

(d) $\{f(1)=0,\ f(-1)=0,\ f(0)=0\}$

(e) $\{f(1)=f(-1),\ f(-1)=f(0),\ f(0)=1\}$

5.6. If $A=\{-3,-1,0,1,3\}$ and $B=\{-3,0,3\}$, which of the following sets define a function with domain A and codomain B?

(a) $\{(-3,-3),(-1,-3),(0,0),(1,-3),(3,3)\}$

(b) $\{(3,-3),(1,-3),(0,0),(-1,3),(-3,3)\}$

(c) $\{(3,1),(1,-1),(0,0),(-1,1),(-3,1)\}$

(d) $\{(-3,1),(0,0),(3,1)\}$

(e) $\{(-3,0),(-1,0),(0,0),(1,0),(3,0)\}$

5.7. In which of the following definitions of pairs of functions (i) and (ii) are f and g the same function?

(a) (i) domain = $\{1,2,3,4\}$, codomain = $\{r,s,t,u\}$, $f=\{(1,r),(2,u),(3,t),(4,s)\}$

(ii) domain = $\{4,3,2,1\}$, codomain = $\{r,s,t,u\}$, $g=\{(4,s),(3,t),(2,u),(1,r)\}$

(b) (i) domain = $\{1,2,3,4\}$, codomain = $\{r,s,t,u,w\}$, $f=\{(1,u),(2,u),(3,t),(4,r)\}$

(ii) domain = $\{1,2,3,4\}$, codomain= $\{r,s,t,u,x\}$, $g=\{(1,u),(2,u),(3,t),(4,r)\}$

(c) (i) domain = $\{1,2,3,4\}$, codomain = $\{r,s,t\}$, $f=\{(1,r),(2,s),(3,t),(4,t)\}$

(ii) domain = $\{1,2,3,4\}$, codomain = $\{r,s,t\}$, $g=\{(1,r),(2,s),(3,s),(4,t)\}$

5.8. Find the domains of the following functions.

(a) $f(x)=\dfrac{1}{x+3}$

(b) $g(x)=\dfrac{-x}{(2x-1)(x+4)}$

(c) $f(t)=\dfrac{5}{3t^2+22t+7}$

(d) $h(u)=\dfrac{11}{\sqrt{2-u}}$

(e) $h(u)=\dfrac{1}{2-\sqrt{u}}$

(f) $f(s)=\dfrac{s^2+s-2}{s^2-s-12}$

(g) $h(r)=\sqrt{r^2-14r+33}$

(h) $f(z)=\dfrac{3}{\sqrt{z^2-14z+33}}$

5.9. Evaluate the following functions at each of the indicated values of x.

(a) $f(x)=2x^3+x^2+3$, at $x=-2,-1,0,1,2$

(b) $g(x)=\sin^2(x)-\cos^2(x)+x$, at $x=-\dfrac{\pi}{2}, \dfrac{-\pi}{3}, 0, \dfrac{\pi}{3}, \dfrac{\pi}{2}$

5.10. Sketch a graph of the function $f(x)=x+x^3$ by plotting points from a table of values using some integer values of x and then some integer and half-integer values of x. Connect the points with straight-line segments.

5.11. Sketch a graph of the equation $2xy=1+y^2$, $x,y>0$, by plotting points from a table of values obtained by solving the equation with some positive integer values substituted for y and then solving the equation with some positive integer values substituted for x. Connect the points with straight-line segments.

5.12. Use the vertical line test to decide whether or not each of the graphs in figure 5.33 below is the graph of a function.

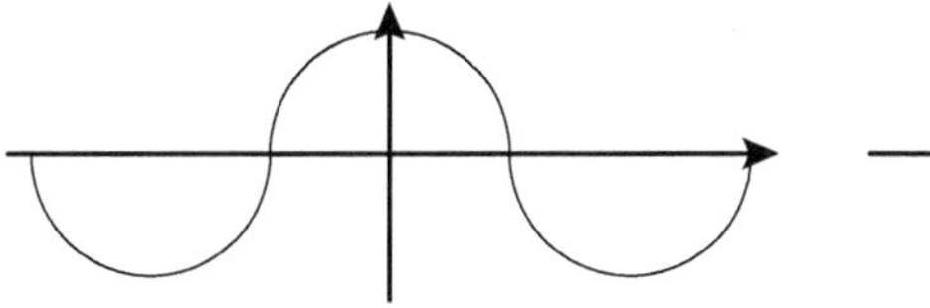
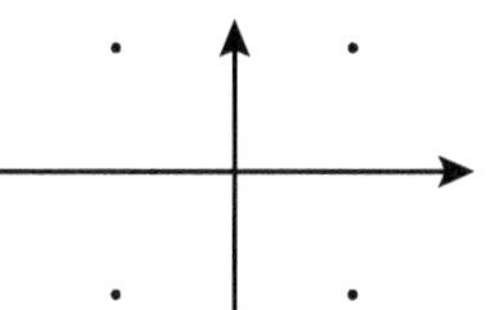
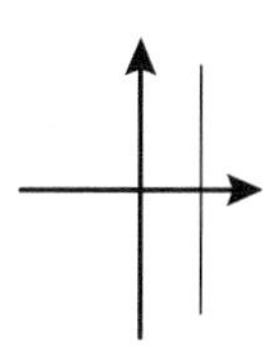

FIGURE 5.33.

5.13. Sketch the graph of $f(x)=x^2$ with domain A, for each of the following choices of A.

(a) $A=[-2,2]$

(b) $A=[-3,-2]\cup[-1,0]\cup[1,3]$

(c) $A=\{-3,-1.5,0,1,2\}$

(d) $A=[0,3]$

(e) $A=\mathbf{Z}\cap[-3,3]$

5.14. Let $f(x)=x\sin x$. Answer the following questions:

(a) What is the domain of f?

(b) If n is an integer, what is the value of $f(n\pi)$?

(c) If n is a non-negative integer, what is the value of $f\left(\dfrac{(4n+1)\pi}{2}\right)$?

(d) If n is a positive integer, what is the value of $f\left(\frac{(4n-1)\pi}{2}\right)$?

(e) Based on your answers from (a) to (d) above, sketch the graph of $y = f(x)$, for $x \geq 0$.

5.15. Try to explain the behavior of the graph of $f(x) = \sin(x)$, where the domain is the set of natural numbers, that is, the graph shown in figure 5.6.

5.16. Draw the graphs of $f(x) = |x-1|$ and $g(x) = |x+1|$ on the same set of axes. (Hint: start by making a table of values.)

5.17. Simplify the following expressions.

(a) $343^{\frac{2}{3}}$

(b) $343^{\frac{2}{-3}}$

(c) $216^{\frac{1}{3}}$

(d) $64^{\frac{1}{3}} \cdot 64^{\frac{-1}{4}}$

(e) $(64^{\frac{1}{2}})^{\frac{1}{3}}$

(f) $7^{\frac{1}{3}} - 7^{\frac{-4}{3}}$

5.18. Evaluate the following functions at each of the indicated values of x.

(a) $f(x) = \left(\frac{1}{4}\right)^x$, at $x = -3, -2, -1, 0, 1, 2, 3$

(b) $g(x) = 2^x + \left(\frac{1}{2}\right)^x$, at $x = -3, -2, -1, 0, 1, 2, 3$

5.19. Draw the graphs of $f(x) = 2^x$, $g(x) = 3^x$, and $h(x) = 10^x$ on the same set of axes.

5.20. Draw the graphs of $f(x) = \left(\frac{1}{3}\right)^x$ and $g(x) = 3^x$ on the same set of axes.

5.21. Draw the graphs of the following functions.

(a) $f(x) = 3^{x+1}$

(b) $f(x) = 3 \cdot 2^x$

(c) $g(x) = 5^x - 1$

(d) $g(x) = 2^x + 3^x$

(e) $h(x) = e^x$

(f) $h(x) = 2^x + 2^{-x}$

5.22. Draw the graphs of the following functions. You might need to start with a table of values.

(a) $f(x) = \dfrac{1}{x+1}$

(b) $f(x) = \dfrac{1}{(x+1)^2}$

(c) $f(x) = \dfrac{1}{x-1}$

(d) $f(x) = \dfrac{1}{(x-1)^3}$

(e) $g(x) = \dfrac{1}{x-1}$

(f) $g(x) = \dfrac{1}{x^2+1}$

(g) $h(x) = \dfrac{x^2}{x^2+1}$

(h) $h(x) = \dfrac{1-x}{x}$

(i) $h(x) = \dfrac{-x^2}{x^2+1}$

5.23. Determine the simplest form of the following rational expressions.

(a) $\dfrac{x^2+3x+2}{x^2+x-2}$

(b) $\dfrac{x^3+2x^2-16x+4}{x^2-2x+1}$

(c) $\dfrac{4x^3+2x^2+2x+14}{x^3-4x^2+4x-4}$

(d) $\dfrac{66x^2-7x-3}{42x^2-17x-4}$

(e) $\dfrac{1+x^3}{1+x-x^3-x^4}$

(f) $\dfrac{27-3x^4}{3\sqrt{3}-(3-\sqrt{3})x-x^2}$

5.24. Sketch the graphs of the following functions. What is the domain of each function?

(a) $f(x)=\dfrac{x+2}{x^2+3x+2}$

(b) $f(x)=\dfrac{x-1}{x^2-1}$

(c) $g(x)=\dfrac{x}{x^2-2x}$

(d) $g(x)=\dfrac{x^2}{x}$

(e) $h(x)=\dfrac{x+1}{x^3+1}$

(f) $h(x)=\dfrac{x^3+1}{x+1}$

5.25. Sketch the graphs of the following functions.

(a) $f(x)=\sqrt[4]{x}$

(b) $f(x)=\sqrt[5]{x}$

5.26. Sketch the graphs of the following piecewise defined functions.

(a) $f(x)=\begin{cases} x+2 & \text{if } x\leq -2 \\ \sin\left(\dfrac{\pi x}{2}\right) & \text{if } -2<x<2 \\ x+2 & \text{if } x\geq 2 \end{cases}$

(b) $f(x)=\begin{cases} \cos(\pi x) & \text{if } x<-2 \\ x & \text{if } -2\leq x\leq 2 \\ \sin(2\pi x) & \text{if } x>2 \end{cases}$

(c) $f(x)=\begin{cases} \sqrt{-x} & \text{if } x\leq 0 \\ \dfrac{x}{2} & \text{if } 0<x<2 \\ \sqrt{x-2} & \text{if } x\geq 2 \end{cases}$

5.27. Make use of the piecewise definition of the absolute value function to sketch the graphs of the following functions. (Hint: refer to example 5.8.2)

(a) $f(x)=|x|+x$

(b) $f(x)=\sin(|x|)$

(c) $f(x)=\dfrac{|x|}{x}$

(d) $f(x)=\sin\left(\dfrac{\pi|x|}{x}\right)$

5.28. Decide whether each of the following functions is even, odd, or neither. An algebraic test can be used.

(a) $f(x)=x^2+x^4$

(b) $f(x)=x+\sin(x)$

(c) $g(t)=\sin(|t|)$

(d) $g(t)=\cos(t)+\sin(t)$

(e) $h(y)=y\cos(y)$

(f) $h(y)=|y|\sin(y)$

(g) $k(z)=\dfrac{z+z^3}{1+z^2}$

(h) $k(z)=\dfrac{z+z^3}{1+z^3}$

(i) $l(w)=e^{w+w^3}$

(j) $l(w)=2^w+2^{-w}$

(k) $l(w)=2^w-2^{-w}$

5.29. If $f(x)=\sqrt{x}$ and $g(x)=\dfrac{1}{x-1}$, find the domains of the functions $f+g$, $f\cdot g$, $\frac{f}{g}$, and $\frac{g}{f}$. Evaluate each of the following expressions.

(a) $(f+g)(4)$

(b) $f\cdot g(16)$

(c) $\dfrac{f}{g}(9)$

(d) $\dfrac{f(4)+g(9)}{g(4)+f(9)}$

(e) $\dfrac{f}{g}(9)+\dfrac{g}{f}(6)$

(f) $\dfrac{f\cdot g(4)}{f\cdot g(9)}$

5.30. If $f(x)=2^x$ and $g(x)=3^x$, evaluate each of the following expressions.

(a) $(f+g)(-1)$

(b) $f\cdot g(3)$

(c) $\dfrac{f}{g}(1)$

(d) $\dfrac{f}{g}(2)+\dfrac{g}{f}(-2)$

5.31. If $g(t)=t^2$ and $h(t)=|t|$, (i) plot the graph of $y=\dfrac{g}{h}(t)$.

5.32. If $g(z)=\dfrac{z}{z+1}$ and $h(z)=\dfrac{z+1}{z+2}$, find the domains of the functions $g+h$, $g\cdot h$, $\dfrac{h}{g}$, and $\dfrac{g}{h}$. Evaluate each of the following expressions (if possible).

(a) $(g-h)(4)$

(b) $g\cdot h(1)$

(c) $g\cdot h(-1)$

(d) $\dfrac{g}{h}(5)+\dfrac{h}{g}(-5)$

5.33. Prove carefully that the product of **any** pair of odd functions is an even function and that the absolute value of **any** odd function is an even function. (Your proof should be a general proof, that is, a proof not involving any particular functions.)

5.34. If $f(x)=\dfrac{2}{x}$, find the simplest expressions for the following functions and find their domains.

(a) $f\circ f(x)$

(b) $f\circ f\circ f(x)$

(c) $f\circ f\circ f\circ f(x)$

5.35. If $f(x)=\dfrac{1}{x}+1$ and $g(x)=\sqrt{x-1}$, find the simplest expressions for the following functions and find their domains.

(a) $f\circ g(x)$

(b) $g\circ f(x)$

(c) $g\circ g(x)$

5.36. Let $f(t)=t^3+1$, $g(t)=t-1$, and $h(t)=\sqrt[3]{t}-1$. Find the simplest expressions for the functions (1) $f\circ g(t)$, (2) $g\circ f(t)$, and (3) $f\circ h(t)$ and find their domains.

5.37. If $f(u)=2^u$, $g(v)=\dfrac{1}{v}$, determine the following values.

(a) $f\circ g(-2)$

(b) $f\circ g(1/2)$

(c) $g\circ f\circ g(1/3)$

(d) $f\circ g\circ f(-3)$

(e) $f\circ f\circ f(2)$

(f) $g\circ g\circ g(-1)$

5.38. Sketch the graph of each of the following functions as a vertical or horizontal shift of the graph of function that you are familiar with. Be sure to label the intercepts of the graph with the axes.

(a) $f(x) = |x-3|$

(b) $f(x) = e^{x-2}$

(c) $f(x) = 4^{-x} - 4$

(d) $g(t) = \sin(\pi t - \pi)$

(e) $g(t) = 1 + \cos(t - \pi)$

(f) $g(t) = t^2 + 2t + 2$

(g) $h(u) = \sqrt{u+1} - 1$

(h) $h(u) = \dfrac{1}{u+1}$

(i) $h(u) = \dfrac{u+1}{u+2}$

5.39. Sketch the graph of each of the following functions as a vertical stretching or compression of the graph of a function you are familiar with. Be sure to label the intercepts of the graph with the axes.

(a) $f(x) = 2|x-3|$

(b) $f(x) = 4 \cdot 4^{-x}$

(c) $g(t) = \dfrac{\tan(t)}{2}$

(d) $g(t) = 4t^2 + 8t + 4$

(e) $h(u) = 3\sqrt{u+1}$

(f) $h(u) = \dfrac{1}{2u+2}$

5.40. Sketch the graph of each of the following functions as a horizontal stretching or compression of the graph of a function you are familiar with. Be sure to label the intercepts of the graph with the axes.

(a) $f(x) = \sin(2x)$

(b) $f(x) = \cos\left(\dfrac{x}{3}\right)$

(c) $g(t) = \tan\left(\dfrac{t}{2}\right)$

(d) $g(t) = \dfrac{t^2}{4} + t + 2$

(e) $h(u) = \sqrt{3u+1}$

(f) $h(u) = \dfrac{1}{2u+1}$

5.41. Sketch the graph of the each of the following functions as a reflection over the x or y axes of the graph of a function you are familiar with. Be sure to label the intercepts of the graph with the axes.

(a) $f(x) = |3-x|$

(b) $f(x) = 4^{-x+1}$

(c) $g(t) = -\sec(t)$

(d) $g(t) = \sin(2\pi - t)$

(e) $h(u) = -\sqrt{1-u}$

(f) $h(u) = \dfrac{1}{2-u}$

5.42. Sketch the graph of each of the following functions as a combination of transformations of the graph of a function that you are familiar with. Be sure to label the intercepts of the graph with the axes.

(a) $f(x) = 1 - |3-x|$

(b) $f(x) = 2 \cdot 4^{2-x}$

(c) $g(t) = 1 + \tan\left(t + \dfrac{\pi}{4}\right)$

(d) $g(t) = 1 - \sin\left(\dfrac{\pi}{4} - 2t\right)$

(e) $h(u) = \sqrt{1-3u}$

(f) $h(u) = 1 + \dfrac{1}{2-4u}$

5.43. Figure 5.34 shows the graph of $y=\sin(\pi x)-\frac{1}{2}$, $y=(x-h)^2-k$, for some constants k and h, and the graph of $y=mx+c$, for some constants m and c. Based on the information given in the graph determine the values of the constants k, h, m, and c. You will need to do a few calculations. (The axis of symmetry of the parabolas is shown with a dotted line, and the line and parabola have an intersection point at $x=\frac{19}{10}$.)

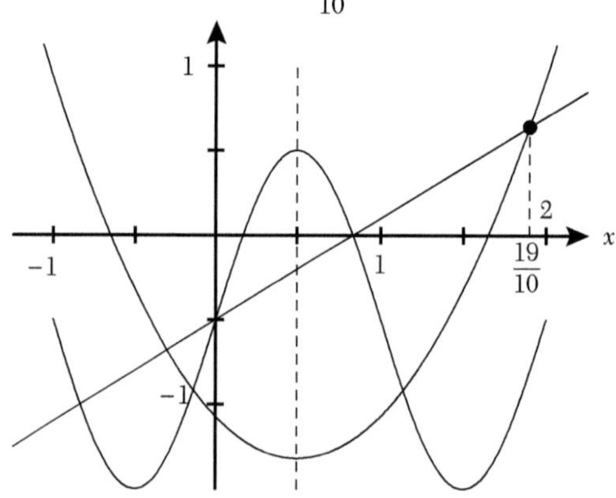

FIGURE 5.34.

5.44. Sketch the graphs of the following piecewise defined functions.

(a) $f(x)=\begin{cases} 2^{x+1} & \text{if } x<-1 \\ 1 & \text{if } -1\le x\le 1 \\ 2^{-x+1} & \text{if } x>1 \end{cases}$

(c) $f(x)=\begin{cases} |2x+1| & \text{if } -2\le x\le 0 \\ |2x-1| & \text{if } 0\le x\le 2 \end{cases}$

(b) $f(x)=\begin{cases} \sqrt{2+x} & \text{if } -2\le x<0 \\ 0 & \text{if } x=0 \\ -\sqrt{2-x} & \text{if } 0<x\le 2 \end{cases}$

(d) $f(x)=\begin{cases} \frac{1}{x}-1 & \text{if } -4\le x\le -1 \\ 2x & \text{if } -1<x<1 \\ \frac{1}{x}+1 & \text{if } 1\le x\le 4 \end{cases}$

5.45. Sketch the trajectory of the following vector-valued function:

$$\vec{r}(t)=\langle t+\cos(\pi t), \sin(\pi t)\rangle$$

where $0\le t\le 2$. Indicate the direction of the trajectory for increasing values of t. It will help to construct a table of values.

5.46. Suppose that $x(t)=t^2$ and $y(t)=\sin\left(\frac{\pi t}{2}\right)$, then the vector-valued function $\vec{r}(t)=\langle x(t), y(t)\rangle$ determines the trajectory of an Asian tiger mosquito, for $0\le t\le 4$. Compute a table of values for some values of t and sketch the trajectory of the mosquito.

5.47. Sketch the trajectory of the following vector-valued function.

$$\vec{r}(\theta)=\langle 2\cos(-\theta),\ 2\sin(-\theta)\rangle$$

where $-\infty<\theta<\infty$. Indicate the direction of the trajectory for increasing values of θ.

5.48. Find (a) a vector-valued function, (b) parametric equations, and (c) a Cartesian equation for the line that is parallel to the vector $\langle -2,6\rangle$ and passes through the point $P(-4,5)$.

5.49. The trajectory of an ice skater starting at the center of an ice rink is given by the vector-valued function $\vec{r}(t)=\langle x(t),y(t)\rangle$ as t increases from $t=0$ to $t=8$, with $x(t)$ and $y(t)$ given by the formulas below. Compute a table of values for some values of t and sketch the skater's trajectory.

$$x(t)=\frac{t\cos(\pi t/2)}{2}, \quad y(t)=\frac{t\sin(\pi t/2)}{2}$$

5.50. A charged particle in a magnetic chamber moves with coordinates $\vec{r}(t)=\langle x(t),y(t)\rangle$ as t increases from $t=0$, with $x(t)$ and $y(t)$ given as $x(t)=2+\cos(t)$ and $y(t)=3+\sin(t)$. Compute a table of values for some values of t and sketch the trajectory of the particle.

5.51. Restate the following equations according to the inverse property for logarithms; that is, translate each logarithmic statement into exponential form and restate each exponential statement into logarithmic form.

(a) $\log_5 125=3$

(b) $2^{-7}=\frac{1}{128}$

(c) $\log_{16} 2=\frac{1}{4}$

(d) $7^3=343$

(e) $\frac{1}{64}^{-\frac{1}{3}}=4$

(f) $\log_{\frac{3}{5}}\frac{25}{9}=-2$

5.52. Evaluate each logarithm.

(a) $\log_5 625$

(b) $\log_{16}\frac{1}{8}$

(c) $\log_9\frac{1}{9}$

(d) $\log_9 1$

(e) $\log_3(\log_{27}3)$

(f) $\log_{25}\frac{8}{125}$

(g) $\log_3\sqrt{81}$

(h) $\log_4\frac{1}{\sqrt{256}}$

5.53. Prove (II) and (III) of the laws for logarithms (table 5.6).

5.54. Use the laws for logarithms to simplify the following expressions.

(a) $\log_5 625+\log_3 81$

(b) $\log_{15}\frac{1}{5}+\log_{15}75$

(c) $\log_3 33-\log_3 99$

(d) $\log_2 6-\log_2 3+\log_2\sqrt{2}$

(e) $\log_{10}4.25-\log_{10}17+\log_{10}0.4$

(f) $\log_2 8^{11}$

(g) $\log_5 343-\log_5 35^3$

(h) $\log_4\sqrt{32}+\log_4 2$

5.55. Solve for x.

(a) $2^{\log_2 x} = 21$

(b) $3^x = \sqrt{3^3}$

(c) $\log_5 5^x = 2$

(d) $2^{3x-7} = 64$

(e) $\log_2 x = -3$

(f) $\log_3(6x+4) = 3$

(g) $\log_3(\log_8(\log_2 x)) = -1$

(h) $\log_2(\log_3(\log_4 x)) = 0$

(i) $\log_{10} x + \log_{10}(x+3) = 1$

(j) $8 = 2^{5x} \cdot 4^{x^2}$

(k) $2 = \log_3(x+2) + \log_3(x-2)$

(l) $\ln x + \ln(2x) = 1$

5.56. Use the change of base formula to simplify the following logarithms.

(a) $\log_9 27$

(b) $\log_6 8 + \log_2 \sqrt[3]{36}$

(c) $\dfrac{\log_{12} 8}{\log_8 12}$

5.57. Simplify the expression $\dfrac{\dfrac{e^x e^{\ln x}}{x} - e^x e^{2\ln x}}{e^{2x}(1+x)}$.

5.58. Sketch the graphs of the following functions and sketch the reflections of the graphs about the line $y = x$. Which functions are invertible?

(a) $f(x) = e^x - 1$

(b) $f(x) = \sqrt{x+1}$

(c) $f(x) = x^2 - 2$

(d) $f(x) = \log_2 x$

(e) $f(x) = \dfrac{1}{x}$

(f) $f(x) = \dfrac{1}{x+1}$

(g) $f(x) = \cos(x)$

(h) $f(x) = \cos(x)$ for $0 \le x \le \pi$

5.59. Find an explicit expression for the inverse function of each of the functions in the previous exercise. State the domain of the inverse function in each case.

5.60. Identify the intervals on which the function $f(x) = -x^2 - 7x + 8$ is increasing or decreasing.

5.61. In which intervals is $f(x) = \sin(x)$ increasing?

5.62. Answer True or False for each of the following statements. You may assume that all functions are real-valued functions.

(1) If f and g are increasing functions on an interval I, then $f+g$ is an increasing function on the interval (I).

(2) If f and g are strictly decreasing functions on an interval I, then $-f-g$ is a strictly decreasing function on the interval (I).

(3) The function $f(x) = x^3 - x$ is increasing on the interval $[-1,1]$.

(4) If $f(x)$ is an increasing function on an interval I, then $x+f(x)$ is a strictly increasing function on the interval (I).

(5) If $f(x)$ is an increasing function on an interval $[3,7]$, then $f(7)\geq f(3)$.

(6) If $f(x)$ is a strictly increasing function on the interval $[3,7]$, then $f(7)\geq f(3)$.

(7) If $f(x)$ is a decreasing function on an interval $[0,4]$, then $x^3-f(x)$ is a strictly increasing function on the interval $[1,3]$.

5.63. Find the exact value of each of the following expressions.

(a) $\sin^{-1}(-1)$

(b) $\cos^{-1}(-1)$

(c) $\tan^{-1}(\sqrt{3})$

(d) $\sin^{-1}(\sin(2))$

(e) $\cos^{-1}\left(\cos\left(\frac{\pi}{4}\right)\right)$

(f) $\tan^{-1}\left(\tan\left(\frac{3\pi}{4}\right)\right)$

(g) $\cos(\cos^{-1}(-0.7))$

(h) $\sin(2\cos^{-1}(-1))$

(i) $\sin\left(2\sin^{-1}\left(\frac{\sqrt{3}}{2}\right)\right)$

(j) $\cos\left(2\sin^{-1}\left(\frac{\sqrt{3}}{2}\right)\right)$

(k) $\sin\left(\cos^{-1}\left(\frac{1}{4}\right)+\cos^{-1}\left(\frac{3}{4}\right)\right)$

(l) $\sin\left(\cos^{-1}\left(-\frac{2}{3}\right)+\cos^{-1}\left(\frac{1}{3}\right)\right)$

5.64. Simplify the following expressions.

(a) $\sin(\sin^{-1}(x))$

(b) $\sin(\sin^{-1}(2x))$

(c) $\sin(\cos^{-1}(2x))$

(d) $\tan(\cos^{-1}(x))$

(e) $\sin(\tan^{-1}(y))$

(f) $\sin(2\cos^{-1}(-y))$

(g) $\sin\left(2\sin^{-1}\left(\frac{\sqrt{3y}}{2}\right)\right)$

(h) $\cos\left(2\sin^{-1}\left(\frac{y}{2}\right)\right)$

CHAPTER 6

TECHNIQUES OF ALGEBRA

6.1 INTRODUCTION

This chapter will bring the student up to the level of skill needed in algebra in order to be able to solve problems in calculus. Section 6.2 deals with the algebra of rational expressions, that is, adding, subtracting, multiplying, dividing, and simplifying rational expressions. This is followed by section 6.3 on algebra with rational exponents and radicals (the algebra of expressions involving square and cube roots, for example).

The heart of this chapter is section 6.4 which is an overview of the methods for solving equations involving all the types of expression we have considered in this book. These include polynomials, rational expressions, the absolute value, exponents, radicals, logarithms, and trigonometric expressions. In section 6.5, students will get their first exposure to the formulas for partial fractions. This will not only give them more practice with the manipulation of rational expressions but also prepare them for the topic of integration of rational expressions that they will do later in their calculus curriculum (not in this book).

Finally, section 6.6 deals with the methods of solving inequalities, where, again, the examples involve all the types of expression mentioned above. At the end, there are some demonstrations of the solutions of two-variable inequalities as shaded regions in the Cartesian plane.

6.2 THE ALGEBRA OF RATIONAL EXPRESSIONS

We will demonstrate in this section how to add, subtract, multiply, and divide rational expressions in order to produce new rational expressions. For this reason, we can talk about the algebra of rational expressions. A rational expression of one variable is of the form $\frac{p(x)}{q(x)}$ where $p(x)$ and $q(x)$ are polynomials in one variable, a rational expression of two variables is of the form $\frac{p(x,y)}{q(x,y)}$ where $p(x,y)$ and $q(x,y)$ are polynomials in two variables, and so on. Note that we can regard any polynomial as a rational expression with a unit in its denominator.

In this section, we will only work with rational expressions in one variable. We will also suppose that the polynomials have real coefficients and so can, in principle, be factored into a product of linear and irreducible quadratic factors. Students will recall that a rational expression $\frac{p(x)}{q(x)}$ is in simplified form if $p(x)$ and $q(x)$ have no common factors.

We have explained the procedure for simplifying rational expressions in section 5.6. We will use this technique in some of the examples below.

6.2.1 Multiplying and Dividing Rational Expressions

Multiplying rational expressions is as simple as multiplying ordinary fractions; that is, numerators and denominators are multiplied together. The resulting fraction can be simplified by canceling common factors in the numerator and denominator.

EXAMPLE 6.2.1. Write the following product of rational expressions as a single rational expression in simplified form:

$$\frac{x}{x^2-x-2}\cdot\frac{2}{(x-2)^2}\cdot\frac{1-x^2}{x^5+x^4+x^3}.$$

Answer: If the denominators in the first and third rational expressions are factorized, then the product of the numerators divided by the product of the denominators is

$$\frac{2x(1-x)(1+x)}{(x+1)(x-2)^3x^3(x^2+x+1)}=\frac{2(1-x)}{(x-2)^3x^2(x^2+x+1)}.$$ ◈

Division by a rational expression, as with ordinary fractions, is a case of multiplying by the reciprocal of the rational expression.

EXAMPLE 6.2.2. Write the following ratio of rational expressions as a single rational expression:

$$\frac{x+3}{2x-6}\div\frac{x^3+27}{x^2-2x-3}.$$

Answer: If all the polynomials are factorized and the division is expressed as multiplication by the reciprocal, then the result is

$$\frac{x+3}{2(x-3)}\cdot\frac{(x-3)(x+1)}{(x+3)(x^2-3x+9)}=\frac{(x+1)}{2(x^2-3x+9)}.$$ ◈

6.2.2 Adding and Subtracting Rational Expressions

The basic idea for adding rational expressions was explained in example 3.2.4, where the rational expressions $\frac{1}{4x-2}$ and $\frac{-3}{x+5}$ were added by finding their lowest common denominator (LCD); that is, the polynomial $2(2x-1)(x+5)$. In general, if the denominators of the rational expressions to be added are not given in factorized form, then it helps (if possible) to find all the linear and irreducible quadratic factors of the denominators by factorizing the denominators. It might be possible to simplify the rational expressions before adding them.

The LCD is obtained by identifying all of the linear or irreducible quadratic factors that occur in all of the denominators and taking the product of all of these factors, with the multiplicity of each factor chosen according to the highest multiplicity that occurs in any of the denominators.

Subtracting a rational expression is the same as adding its negative, so we need not say more about this.

EXAMPLE 6.2.3. Perform the addition of the rational expressions, as indicated. Write the answer as a single rational expression:

$$\frac{x}{x^2-x-2}+\frac{2}{(x-2)^2}+\frac{1-x^2}{x^5+x^4+x^3}.$$

Answer: The denominators of the first and third expressions factorize as $(x-2)(x+1)$ and $x^3(x^2+x+1)$, respectively. Thus, in all of the denominators, we can identify the linear factors $(x+1)$, $(x-2)$, and x (with the highest multiplicities for each being 1, 2, and 3, respectively) and the irreducible factor (x^2+x+1) (with multiplicity 1); so the LCD is

$$(x+1)(x-2)^2x^3(x^2+x+1).$$

We now multiply and divide each rational expression above by the product of missing factors needed to make up the LCD in its denominator, resulting in this sum of fractions:

$$\frac{x(x-2)x^3(x^2+x+1)}{(x+1)(x-2)^2x^3(x^2+x+1)}+\frac{2(x+1)x^3(x^2+x+1)}{(x+1)(x-2)^2x^3(x^2+x+1)}+\frac{(1-x^2)(x+1)(x-2)^2}{(x+1)(x-2)^2x^3(x^2+x+1)}.$$

Because all the fractions are expressed with the same denominator, the numerators can be added, resulting in the rational expression

$$\frac{x(x-2)x^3(x^2+x+1)+2(x+1)x^3(x^2+x+1)+(1-x^2)(x+1)(x-2)^2}{(x+1)(x-2)^2x^3(x^2+x+1)}.$$

Each of the three terms in the numerator can be expanded into a sum of powers of x and all of these can be added to give the final answer:

$$\frac{x}{x^2-x-2}+\frac{2}{(x-2)^2}+\frac{1-x^2}{x^5+x^4+x^3}=\frac{4-7x^2+3x^3+5x^4+2x^5+x^6+x^7}{(x+1)(x-2)^2x^3(x^2+x+1)}$$

which cannot be simplified (check this!). The denominator can be left in factorized form. ◈

Example 6.2.4. Add the rational expressions, as indicated. Write the answer as a single rational expression:

$$\frac{x^3+x^2-8x-12}{x^4-3x^3-4x^2+12x}+\frac{1}{(x+2)^2}.$$

Answer: By the factor theorem (theorem 3.4.2), $(x-3)$ is a factor of $x^3+x^2-8x-12$ and then, by long division or synthetic division, we find that $x^3+x^2-8x-12=(x+2)^2(x-3)$. Next, $x^4-3x^3-4x^2+12x=x(x^3-3x^2-4x+12)$ and the cubic polynomial can be factorized by grouping, with the result

$$x(x^3-3x^2-4x+12)=x(x-3)(x^2-4)=x(x-3)(x-2)(x+2).$$

Thus, after cancellation of common factors, the sum is

$$\frac{(x+2)}{x(x-2)}+\frac{1}{(x+2)^2}$$

which we now want to express as a single rational expression. Because the LCD is $x(x-2)(x+2)^2$, the sum can be expressed as

$$\frac{(x+2)^3}{x(x-2)(x+2)^2}+\frac{x(x-2)}{x(x-2)(x+2)^2}=\frac{x^3+7x^2+10x+8}{x(x-2)(x+2)^2}.$$ ◈

Remark 6.2.1. Rational expressions can also be expressed using negative exponents, for example, $x(1+x)^{-2}$ means the same thing as $\frac{x}{(1+x)^2}$. Some exercises involving negative exponents are included at the end of this chapter.

6.3 ALGEBRA WITH RATIONAL EXPONENTS

Fractional exponents were introduced in section 5.5.1. We now go a step further and introduce algebraic expressions that involve fractional exponents. In table 6.1, we give the products of some binomials and trinomials with terms containing fractional exponents.

TABLE 6.1. Products involving rational exponents

Product	Name
$(x^{\frac{1}{2}}+y^{\frac{1}{2}})(x^{\frac{1}{2}}-y^{\frac{1}{2}})=x-y$	Difference of two squares
$(x^{\frac{1}{2}}+y^{\frac{1}{2}})^2=x+2x^{\frac{1}{2}}y^{\frac{1}{2}}+y$	Perfect square trinomial
$(x^{\frac{1}{2}}-y^{\frac{1}{2}})^2=x-2x^{\frac{1}{2}}y^{\frac{1}{2}}+y$	Perfect square trinomial
$(x^{\frac{1}{3}}+y^{\frac{1}{3}})(x^{\frac{2}{3}}-x^{\frac{1}{3}}y^{\frac{1}{3}}+y^{\frac{2}{3}})=x+y$	Sum of two cubes
$(x^{\frac{1}{3}}-y^{\frac{1}{3}})(x^{\frac{2}{3}}+x^{\frac{1}{3}}y^{\frac{1}{3}}+y^{\frac{2}{3}})=x-y$	Difference of two cubes
$(x^{\frac{1}{3}}+y^{\frac{1}{3}})^3=x+3x^{\frac{2}{3}}y^{\frac{1}{3}}+3x^{\frac{1}{3}}y^{\frac{2}{3}}+y$	Perfect cube
$(x^{\frac{1}{3}}-y^{\frac{1}{3}})^3=x-3x^{\frac{2}{3}}y^{\frac{1}{3}}+3x^{\frac{1}{3}}y^{\frac{2}{3}}-y$	Perfect cube

The terms with fractional exponents in table 6.1 can also be expressed as radicals, as demonstrated in table 6.2.

TABLE 6.2. Products involving radicals

Product	Name
$(\sqrt{x}+\sqrt{y})(\sqrt{x}-\sqrt{y})=x-y$	Difference of two squares
$(\sqrt{x}+\sqrt{y})^2=x+2\sqrt{xy}+y$	Perfect square trinomial
$(\sqrt{x}-\sqrt{y})^2=x-2\sqrt{xy}+y$	Perfect square trinomial
$(\sqrt[3]{x}+\sqrt[3]{y})(\sqrt[3]{x^2}-\sqrt[3]{xy}+\sqrt[3]{y^2})=x+y$	Sum of two cubes
$(\sqrt[3]{x}-\sqrt[3]{y})(\sqrt[3]{x^2}+\sqrt[3]{xy}+\sqrt[3]{y^2})=x-y$	Difference of two cubes
$(\sqrt[3]{x}+\sqrt[3]{y})^3=x+3\sqrt[3]{x^2}y+3\sqrt[3]{xy^2}+y$	Perfect cube
$(\sqrt[3]{x}-\sqrt[3]{y})^3=x-3\sqrt[3]{x^2y}+3\sqrt[3]{xy^2}-y$	Perfect cube

Expressions involving fractional exponents can be multiplied, divided, added, or subtracted in the same manner as was demonstrated above for rational expressions.

Example 6.3.1. Add the expressions with fractional exponents, as indicated. Write the answer as a single fraction without negative exponents.

$$3(2x+5)^{\frac{1}{2}} - x(2x+5)^{\frac{-1}{2}}.$$

Answer:

$$\begin{aligned}
3(2x+5)^{\frac{1}{2}} - x(2x+5)^{\frac{-1}{2}} &= 3(2x+5)^{\frac{1}{2}} - \frac{x}{(2x+5)^{\frac{1}{2}}} \\
&= \frac{3\left((2x+5)^{\frac{1}{2}}\right)^2 - x}{(2x+5)^{\frac{1}{2}}} \\
&= \frac{3(2x+5)-x}{(2x+5)^{\frac{1}{2}}} \\
&= \frac{5(x+3)}{(2x+5)^{\frac{1}{2}}}.
\end{aligned}$$

◈

Example 6.3.2. Add the following terms with fractional exponents, as indicated. Write the answer as a single fraction with a rational denominator.

$$\frac{3}{x^{\frac{3}{2}}-1} - \frac{2}{x^{\frac{1}{2}}+1} + \frac{x^{\frac{1}{2}}}{x-1}.$$

Answer: We will begin by rationalizing the denominator of each term. This can be done with the help of the formulas in table 6.1. In particular, note that, if the denominator of the first term is multiplied by $x^{\frac{3}{2}}+1$, then the product is x^3-1, and if the denominator of the second term is multiplied by $x^{\frac{1}{2}}-1$, then the product is $x-1$. The third term can be left as it is. Of course, we cannot change the terms so we multiply and divide the first and second terms by the factors $x^{\frac{3}{2}}+1$ and $x^{\frac{1}{2}}-1$, respectively. We proceed to add the terms:

$$\begin{aligned}
&\frac{3}{(x^{\frac{3}{2}}-1)}\frac{(x^{\frac{3}{2}}+1)}{(x^{\frac{3}{2}}+1)} - \frac{2}{(x^{\frac{1}{2}}+1)}\frac{(x^{\frac{1}{2}}-1)}{(x^{\frac{1}{2}}-1)} - \frac{x^{\frac{1}{2}}}{1-x} \\
&= \frac{3(x^{\frac{3}{2}}+1)}{x^3-1} - \frac{2(x^{\frac{1}{2}}-1)}{x-1} + \frac{x^{\frac{1}{2}}}{x-1} \\
&= \frac{3(x^{\frac{3}{2}}+1)}{x^3-1} + \frac{2-x^{\frac{1}{2}}}{x-1} \\
&= \frac{3(x^{\frac{3}{2}}+1)}{(x-1)(x^2+x+1)} + \frac{(2-x^{\frac{1}{2}})(x^2+x+1)}{(x-1)(x^2+x+1)} \\
&= \frac{x^{\frac{5}{2}}+2x^2+2x^{\frac{3}{2}}+2x-x^{\frac{1}{2}}+5}{x^3-1}.
\end{aligned}$$

◈

EXAMPLE 6.3.3. Add the terms involving radicals, as indicated. Write the answer as a single fraction with a rational denominator.

$$\frac{\sqrt[3]{x}+1}{\sqrt[3]{x^2}+\sqrt[3]{x}+1}-\frac{\sqrt[3]{x^2}+1}{x+1}.$$

Answer: For this problem refer table 6.2. Note that the denominator of the first term can be rationalized by multiplying by $\frac{\sqrt[3]{x}-1}{\sqrt[3]{x}-1}$, thus

$$\begin{aligned}\frac{(\sqrt[3]{x}+1)}{(\sqrt[3]{x^2}+\sqrt[3]{x}+1)}\frac{(\sqrt[3]{x}-1)}{(\sqrt[3]{x}-1)}-\frac{\sqrt[3]{x^2}+1}{x+1}&=\frac{\sqrt[3]{x^2}-1}{x-1}-\frac{\sqrt[3]{x^2}+1}{x+1}\\&=\frac{(x+1)(\sqrt[3]{x^2}-1)}{(x+1)(x-1)}-\frac{(x-1)(\sqrt[3]{x^2}+1)}{(x-1)(x+1)}\\&=\frac{2\sqrt[3]{x^2}-2x}{x^2-1}.\end{aligned}$$

◈

6.4 SOLVING EQUATIONS

The objective of this section is, by means of twelve examples, to give an overview of methods that can be used to solve equations that involve all the expressions that have been introduced so far in this book, including the absolute value, radicals, exponents, and logarithms. All but one of the examples will require that a quadratic equation be solved at some stage. It is important to check whether the solutions obtained at the end are solutions for the original equation because some terms of the original equation might not be defined if the "solutions" are plugged in and, if this is the case, they need to be excluded.

It is best to learn the methods for solving equations that can also be applied to solving inequalities. (It helps to develop good habits from the start.) For example, if rational expressions occur somewhere in an equation, then it is tempting to multiply all terms of the equation by a factor that will get rid of the rational expressions. While this does lead to the correct solution when solving an equation, it is not a good method for solving an inequality because multiplying all of the terms by a factor will require the direction of the inequality to change if the factor is negative (and this will depend on the value of the variable), which is likely to cause a great deal of confusion!

Students who would like to see more examples of solving equations can consult an algebra textbook, where each type of example would be treated in detail in a separate section.

EXAMPLE 6.4.1. Solve for x in the equation

$$(3x-2)(4x+3)=5x(2x+1).$$

Answer: After the expansion of terms on each side of the equation, all the terms can be brought to the left-hand side of the equation, resulting in a quadratic equation that can be solved by factorizing, as explained in section 3.5.4:

$$\begin{aligned}12x^2+x-6&=10x^2+5x\\2x^2-4x-6&=0\\2(x-3)(x+1)&=0.\end{aligned}$$

Therefore, the solutions are $x=3$ and $x=-1$.

◈

Example 6.4.2. Solve for x in the equation

$$x^2+\frac{1}{x^2}-9\left(x-\frac{1}{x}\right)=12.$$

Answer: A trick is to make use of the identity $x^2+\frac{1}{x^2}=\left(x-\frac{1}{x}\right)^2+2.$ (Verify it.) The equation can thus be expressed in the following form, resembling a quadratic equation:

$$\left(x-\frac{1}{x}\right)^2-9\left(x-\frac{1}{x}\right)-10=0.$$

To simplify this problem, we make the identification $V\equiv\left(x-\frac{1}{x}\right)$. In terms of the new variable V, the equation becomes an ordinary quadratic equation that can be factorized as:

$$V^2-9V-10=0$$

$$(V-10)(V+1)=0.$$

The solutions for this equation are $V=10$ and $V=-1$. In terms of the variable x, this gives the following two equations:

$$x-\frac{1}{x}=10 \quad\text{and}\quad x-\frac{1}{x}=-1.$$

We can rewrite each expression as a rational expression equal to 0:

$$\frac{x^2-10x-1}{x}=0 \quad\text{and}\quad \frac{x^2+x-1}{x}=0.$$

A basic property of fractions to remember in this situation is that a fraction is 0 if and only if its numerator is equal to 0. We thus obtain all solutions for each equation by setting the numerator of the rational expression in each equation equal to 0, resulting in two quadratic equations:

$$x^2-10x-1=0 \quad\text{and}\quad x^2+x-1=0.$$

The solutions for the first and second equation are $x=5\pm\sqrt{26}$ and $x=\frac{-1\pm\sqrt{5}}{2}$. These are all the solutions for the original equation. ◈

Example 6.4.3. If the product of two real numbers is $\frac{1}{3}$ and their reciprocals differ by $\frac{1}{2}$, find the numbers.

Answer: We will denote the two real numbers as x and y. According to the statement of the problem, $x\cdot y=\frac{1}{3}$ and $\frac{1}{x}-\frac{1}{y}=\frac{1}{2}$. We would like to solve an equation that involves a single variable, which can be either x or y, so we solve the first equation for y. This gives $y=\frac{1}{3x}$. By making use of this value for y in the second equation, the latter becomes

$$\frac{1}{x}-3x=\frac{1}{2}.$$

If all the terms are brought to the left side of the equation and the terms are added as rational expressions, then the equation becomes:

$$\frac{-6x^2 - x + 2}{2x} = 0.$$

Once again, a solution is possible if and only if the numerator is equal to zero. Therefore, we factor the numerator to obtain the equation

$$(3x + 2)(2x - 1) = 0.$$

The two solutions for x that we get from this equation are $x = -\frac{2}{3}$ and $x = \frac{1}{2}$, and the corresponding values for y (which we get from the equation $x \cdot y = \frac{1}{3}$) are $y = -\frac{1}{2}$ and $y = \frac{2}{3}$. We can therefore express the final solution as the two pairs of values $\left(-\frac{2}{3}, -\frac{1}{2}\right)$ and $\left(\frac{1}{2}, \frac{2}{3}\right)$. ◈

Example 6.4.4. A blue creepy crawly cleans a swimming pool in x number of hours, and a red creepy crawly takes 3 h longer. If, working together, the two creepy crawlies clean the swimming pool in 3 h, 36 min, find the value of x.

Answer: The blue creepy crawly cleans the portion $\frac{1}{x}$ of the swimming pool in 1 h, and the red creepy crawly cleans the portion $\frac{1}{x+3}$ of the swimming pool in 1 h. This means that, working together, they clean the portion $\frac{1}{x} + \frac{1}{x+3}$ of the swimming pool in 1 h. Because 3 h, 36 min can be expressed as $\frac{18}{5}$ h, we can use the given information to write the equation

$$\frac{18}{5}\left(\frac{1}{x} + \frac{1}{x+3}\right) = 1.$$

Now, we solve for x:

$$\begin{aligned} \frac{x+3}{x(x+3)} + \frac{x}{x(x+3)} - \frac{5}{18} &= 0 \\ \frac{18(2x+3) - 5x(x+3)}{18x(x+3)} &= 0 \\ \frac{-5x^2 + 21x + 54}{18x(x+3)} &= 0 \\ 5x^2 - 21x - 54 &= 0 \\ (5x+9)(x-6) &= 0. \end{aligned}$$

The negative solution $-\frac{9}{5}$ can be discarded, leaving the positive solution $x = 6$ as the solution to the problem: it takes the blue creepy crawly 6 h to clean the swimming pool. ◈

Example 6.4.5. Solve for x in the equation

$$\frac{2x+7}{6x^2 - x - 15} = \frac{4x-3}{15x^2 - 19x - 10} - \frac{3x+4}{10x^2 + 19x + 6}.$$

Answer: The first step is to factorize all of the denominators:

$$\frac{2x+7}{(3x-5)(2x+3)} = \frac{4x-3}{(5x+2)(3x-5)} - \frac{3x+4}{(5x+2)(2x+3)}.$$

Each of the factors $(5x+2)$, $(3x-5)$, and $(2x+3)$ occurs with multiplicity one, so the LCD is their product. Therefore

$$\frac{(2x+7)(5x+2)}{(5x+2)(2x+3)(3x-5)}-\frac{(4x-3)(2x+3)}{(5x+2)(2x+3)(3x-5)}+\frac{(3x+4)(3x-5)}{(5x+2)(2x+3)(3x-5)}=0.$$

We now add the terms on the left-hand side of the equation:

$$\frac{(2x+7)(5x+2)-(4x-3)(2x+3)+(3x+4)(3x-5)}{(5x+2)(2x+3)(3x-5)}=0.$$

and expand all the products in the numerator to get

$$\frac{11x^2+30x+3}{(5x+2)(2x+3)(3x-5)}=0.$$

Setting the numerator equal to zero and applying the quadratic formula results in the two solutions

$$x=\frac{-15\pm\sqrt{192}}{11}.$$ ◈

EXAMPLE 6.4.6. Solve for x in the equation

$$|(x+2)(3x-1)|=10.$$

Answer: When solving an equation involving an absolute value like the one above, there are always two possibilities:

$$(x+2)(3x-1)=10 \quad \text{or} \quad (x+2)(3x-1)=-10,$$

which is equivalent to

$$3x^2+5x-12=0 \quad \text{or} \quad 3x^2+5x+8=0.$$

The solutions for the first equation are $x=\frac{4}{3}$ and $x=-3$ and the second equation are $x=\frac{-5\pm\sqrt{-71}}{6}=\frac{-5\pm i\sqrt{71}}{6}$. These are all the solutions for the original equation. ◈

EXAMPLE 6.4.7. Solve for x in the equation

$$5^{x+1}-5^{x-1}=120.$$

Answer: A common factor for the terms on the left-hand side of the equation is 5^x. The algebra that follows is straightforward:

$$\begin{aligned} 5^x\left(5-\frac{1}{5}\right)&=120 \\ 5^x\left(\frac{24}{5}\right)&=120 \\ 5^x&=5^2 \\ x&=2. \end{aligned}$$ ◈

Example 6.4.8. Solve for x in the equation

$$4x^{\frac{1}{3}} - 6x^{-\frac{1}{3}} = 5.$$

Answer: This is equivalent to the equation

$$4x^{\frac{1}{3}} - \frac{6}{x^{\frac{1}{3}}} - 5 = 0,$$

which, in turn, is equivalent to

$$\frac{4\left(x^{\frac{1}{3}}\right)^2 - 5x^{\frac{1}{3}} - 6}{x^{\frac{1}{3}}} = 0.$$

By setting $V = x^{\frac{1}{3}}$ and the numerator equal to zero, we obtain a quadratic equation

$$4V^2 - 5V - 6 = 0$$

In factorized form, this is

$$(4V - 3)(V + 2) = 0.$$

Because the two solutions are $V = \frac{3}{4}$ and $V = -2$, we get two equations $x^{\frac{1}{3}} = \frac{3}{4}$ and $x^{\frac{1}{3}} = -2$ that we need to solve for x. Finally, the two solutions for the problem are $x = \frac{27}{64}$ and $x = -8$. ◈

In most cases, the approach to solving an equation in which polynomials occur inside a radical is to take a power of both sides of the equation that eliminates one or more of the radicals. The procedure can be repeated until there are no radicals left in the expression, and the equation can then be solved. In the next example, the square is taken on both sides, and this needs to be done twice.

Example 6.4.9. Solve for x in the equation

$$\sqrt{x+6} + \sqrt{x-1} = 7.$$

Answer:

$$\begin{aligned}
\sqrt{x+6} + \sqrt{x-1} &= 7 \\
\sqrt{x+6} &= 7 - \sqrt{x-1} \\
x + 6 &= 49 - 14\sqrt{x-1} + x - 1 \\
14\sqrt{x-1} &= 42 \\
\sqrt{x-1} &= 3 \\
x - 1 &= 9 \\
x &= 10.
\end{aligned}$$

It's a good idea to check the solution:

$$\sqrt{10+6} + \sqrt{10-1} = \sqrt{16} + \sqrt{9} = 7.$$

◈

Equations involving logarithms very often reduce to rational expressions, as in the next example.

EXAMPLE 6.4.10. Solve for x in the equation

$$2\log(x+1)-\log(x+2)=0.$$

Answer:

$$2\log(x+1)-\log(x+2)=0$$
$$\log\frac{(x+1)^2}{x+2}=0.$$

By the inverse property of logarithms (formula (6.1)), this is equivalent to

$$\frac{(x+1)^2}{x+2}=1$$
$$\frac{x^2+2x+1}{x+2}-\frac{x+2}{x+2}=0$$
$$\frac{x^2+x-1}{x+2}=0$$
$$x^2+x-1=0$$
$$x=-\frac{1}{2}\pm\frac{\sqrt{5}}{2}.$$

Once again, it's a good idea to check both solutions. Taking the positive sign in front of the square root gives a solution for the equation:

$$2\log\left(-\frac{1}{2}+\frac{\sqrt{5}}{2}+1\right)-\log\left(-\frac{1}{2}+\frac{\sqrt{5}}{2}+2\right)=\log\left(\frac{1}{2}+\frac{\sqrt{5}}{2}\right)^2-\log\left(\frac{3}{2}+\frac{\sqrt{5}}{2}\right)$$
$$=\log\left(\frac{1}{4}+\frac{\sqrt{5}}{2}+\frac{5}{4}\right)-\log\left(\frac{3}{2}+\frac{\sqrt{5}}{2}\right)$$
$$=0.$$

However, taking the negative sign in front of the square root does not give a solution because $2\log\left(-\frac{1}{2}-\frac{\sqrt{5}}{2}+1\right)$ simplifies to $2\log\left(\frac{1}{2}-\frac{\sqrt{5}}{2}\right)$, which is not defined. (A logarithm is not defined for negative numbers.) ◈

Solving trigonometric equations like the equation in the next example is a skill that involves the correct application of trigonometric identities in order to arrive at an equation that can be solved using the algebraic methods learned so far (e.g., the method of factorizing to solve quadratic equations). If there are cosine and sine terms involved, then the Pythagorean identity $\cos^2(\theta)+\sin^2(\theta)=1$ would typically be used to rewrite the equation so that all the terms are powers of $\sin(\theta)$ or $\cos(\theta)$.

EXAMPLE 6.4.11. Solve for x in the equation

$$\sqrt{3}\cos(x)-\sin(x)=1.$$

Answer:

$$\sqrt{3}\cos(x) = 1 + \sin(x)$$
$$(\sqrt{3}\cos(x))^2 = (1 + \sin(x))^2$$
$$3\cos^2(x) = 1 + 2\sin(x) + \sin^2(x)$$
$$3(1 - \sin^2(x)) = 1 + 2\sin(x) + \sin^2(x)$$
$$4\sin^2(x) + 2\sin(x) - 2 = 0$$
$$2(2\sin(x) - 1)(\sin(x) + 1) = 0$$

$$\sin(x) = \frac{1}{2} \quad \text{or} \quad \sin(x) = -1$$

$$x = \frac{\pi}{6} + 2n\pi \quad \text{or} \quad x = \frac{5\pi}{6} + 2n\pi \quad \text{or} \quad x = \frac{3\pi}{2} + 2n\pi$$

where n is any integer. As there is a step in which both sides of the equation are squared, we need to check whether these are all solutions of the original equation. The solutions $x = \frac{\pi}{6} + 2n\pi$ and $x = \frac{3\pi}{2} + 2n\pi$ are indeed valid solutions but $x = \frac{5\pi}{6} + 2n\pi$ are not solutions (recall that the cosine ratio is negative in the second quadrant). ◈

The trigonometric equation that we solved in the previous example actually belongs to a class of trigonometric equations that we will solve in the next example, using a different method.

EXAMPLE 6.4.12. Solve for x in the equation

$$a\cos(x) + b\sin(x) = c.$$

where a, b, and c are real numbers such that $a \neq 0$, $b > 0$, and $|c| \leq \sqrt{a^2 + b^2}$

Answer: Note that if we multiply both sides of the equation in example 6.4.11 by -1, then it is an equation of the type above, with $a = -\sqrt{3}$, $b = 1$, and $c = -1$. The first step in solving the equation above is to divide both sides by $\sqrt{a^2 + b^2}$:

$$\left(\frac{a}{\sqrt{a^2 + b^2}}\right)\cos(x) + \left(\frac{b}{\sqrt{a^2 + b^2}}\right)\sin(x) = \frac{c}{\sqrt{a^2 + b^2}}.$$

The reason for writing the equation this way is that we can express $\frac{a}{\sqrt{a^2 + b^2}}$ as the cosine of some undetermined angle θ. Indeed, it is helpful to draw a right-angled triangle with the short sides labeled a and b and the hypotenuse labeled $\sqrt{a^2 + b^2}$ and θ placed so that $\cos(\theta) = \frac{a}{\sqrt{a^2 + b^2}}$ and $\sin(\theta) = \frac{b}{\sqrt{a^2 + b^2}}$. (If $a > 0$ then θ will be in the first quadrant and $a < 0$ then θ will be in the second quadrant.) We replace the coefficients of $\cos(x)$ and $\sin(x)$ on the left-hand side of the equation by $\cos(\theta)$ and $\sin(\theta)$, respectively, so that the equation becomes

$$\cos(\theta)\cos(x) + \sin(\theta)\sin(x) = \frac{c}{\sqrt{a^2 + b^2}}.$$

By means of the trigonometric identity in formula (4.7a), we can write the equation above as

$$\cos(\theta - x) = \frac{c}{\sqrt{a^2+b^2}}.$$

We write the solution for this equation as follows.

$$\theta - x = \cos^{-1}\left(\frac{c}{\sqrt{a^2+b^2}}\right) + 2n\pi \quad \text{or} \quad \theta - x = -\cos^{-1}\left(\frac{c}{\sqrt{a^2+b^2}}\right) + 2n\pi$$

where n is any integer. Because we can replace θ with $\cos^{-1}\left(\frac{a}{\sqrt{a^2+b^2}}\right)$, the final answer can be expressed as:

$$x = \cos^{-1}\left(\frac{a}{\sqrt{a^2+b^2}}\right) - \cos^{-1}\left(\frac{c}{\sqrt{a^2+b^2}}\right) + 2n\pi$$

$$\text{or} \quad x = \cos^{-1}\left(\frac{c}{\sqrt{a^2+b^2}}\right) + \cos^{-1}\left(\frac{c}{\sqrt{a^2+b^2}}\right) + 2n\pi$$

where n is any integer. Note that if we now make the substitutions $a = -\sqrt{3}$, $b = 1$, and $c = -1$, then we get the solution for exercise 6.4.11:

$$x = \cos^{-1}\left(-\frac{\sqrt{3}}{2}\right) - \cos^{-1}\left(-\frac{1}{2}\right) + 2n\pi = \frac{5\pi}{6} - \frac{2\pi}{3} + 2n\pi = \frac{\pi}{6} + 2n\pi$$

$$\text{or} \quad x = \cos^{-1}\left(-\frac{\sqrt{3}}{2}\right) + \cos^{-1}\left(-\frac{1}{2}\right) + 2n\pi = \frac{5\pi}{6} + \frac{2\pi}{3} + 2n\pi = \frac{3\pi}{2} + 2n\pi.$$

◈

6.5 PARTIAL FRACTIONS

Imagination and creativity are required to find ways to reverse processes in mathematics. In this section on partial fractions, we are going to discover methods for reversing the addition of rational expressions. A simple example of adding rational expressions is

$$1 - \frac{2}{x+2} = \frac{x+2}{x+2} - \frac{2}{x+2} = \frac{x+2-2}{x+2} = \frac{x}{x+2}.$$

If we write the equations in reverse order, then we obtain

$$\frac{x}{x+2} = \frac{x+2-2}{x+2} = \frac{x+2}{x+2} - \frac{2}{x+2} = 1 - \frac{2}{x+2}.$$

By means of this reversed calculation, we write $\frac{x}{x+2}$ as $1 - \frac{2}{x+2}$. A situation in which it is useful to do this is when we need to draw the graph of $y = \frac{x}{x+2}$. It might not be immediately obvious what this graph should look like, but in the alternative form $y = 1 - \frac{2}{x+2}$ we know immediately that, if we start with the graph of $y=\frac{2}{x}$, then we can shift this graph two units to the left, reflect the graph over the x-axis and shift the graph up one unit to finally obtain the graph we want.

A more general form of the calculation above is:

$$\frac{x}{x+a}=\frac{x+a-a}{x+a}=\frac{x+a}{x+a}-\frac{a}{x+a}=1-\frac{a}{x+a}.$$

Here, a can be any real number, but it is easier to let a be any positive number and instead use the following calculation if "a" above is negative.

$$\frac{x}{x-a}=\frac{x-a+a}{x-a}=\frac{x-a}{x-a}+\frac{a}{x-a}=1+\frac{a}{x-a}.$$

It is very helpful to memorize the pair of equations that we have derived:

$$\boxed{\frac{x}{x+a}=1-\frac{a}{x+a} \quad \text{and} \quad \frac{x}{x-a}=1+\frac{a}{x-a}, \quad \text{where } a>0}. \tag{6.1}$$

Example 6.5.1.

(i) $\frac{x}{x+7}=1-\frac{7}{x+7}$

(ii) $\frac{x}{x-6}=1+\frac{6}{x-6}$ ◈

A calculation that is not as easy to reverse is shown in the next example.

Example 6.5.2.

$$\begin{aligned}\frac{1}{2}\frac{1}{x-5}-\frac{1}{2}\frac{1}{x-3}&=\frac{1}{2}\frac{(x-3)}{(x-5)(x-3)}-\frac{1}{2}\frac{(x-5)}{(x-5)(x-3)}\\&=\frac{1}{2}\frac{(x-3)-(x-5)}{(x-5)(x-3)}\\&=\frac{1}{2}\frac{2}{(x-5)(x-3)}\\&=\frac{1}{(x-5)(x-3)}\end{aligned}$$ ◈

This motivates that we write the general equation

$$\frac{1}{(x-a)(x-b)}=\frac{A}{x-a}+\frac{B}{x-b}. \tag{6.2}$$

where a and b are any real numbers with $a \neq b$, and then figure out what A and B should be. The way to do this is not difficult: we begin by adding the rational expressions on the right-hand side of the equation:

$$\begin{aligned}\frac{1}{(x-a)(x-b)} &= \frac{A}{x-a} + \frac{B}{x-b} \\ &= \frac{A(x-b)}{(x-a)(x-b)} + \frac{B(x-a)}{(x-a)(x-b)} \\ &= \frac{A(x-b)+B(x-a)}{(x-a)(x-b)} \\ &= \frac{Ax-Ab+Bx-Ba}{(x-a)(x-b)} \\ &= \frac{(A+B)x-Ab-Ba}{(x-a)(x-b)}.\end{aligned}$$

Now the rational expressions on each side of the equation are equal to each other and have the same denominator. Therefore, it must be true that the numerators on each side of the equation are equal. Hence,

$$1 = (A+B)x - Ab - Ba.$$

This equation might look confusing. What it actually says is that the polynomial $0x+1$ on the left-hand side of the equation is equal to the polynomial $(A+B)x - Ab - Ba$ on the right-hand side. Now, it is a basic fact that two polynomials $p(x)$ and $q(x)$ are the same polynomials if and only if the corresponding coefficients of the powers of x are all equal. Thus, the equation above is true if and only if:

$$0 = A+B \quad \text{and} \quad 1 = -Ab - Ba.$$

According to the first equation, $A = -B$. If this is substituted in the second equation, then:

$$1 = -(-B)b - Ba = B(b-a).$$

In other words, $B = \frac{1}{b-a}$. From this we also get $A = \frac{1}{a-b}$. Therefore, we have solved for A and B in terms of a and b, and these expressions for A and B can be replaced in formula (6.2), that is,

$$\boxed{\frac{1}{(x-a)(x-b)} = \frac{1}{(a-b)}\frac{1}{(x-a)} + \frac{1}{(b-a)}\frac{1}{(x-b)}}. \tag{6.3}$$

where $a \neq b$. A good way to memorize formula (6.3) is to note that the coefficient of the term $\frac{1}{x-a}$ on the right-hand side of the equation can be obtained by deleting the factor $(x-a)$ in the denominator of the expression on the left-hand side and replacing x with a in the adjacent factor $(x-b)$ (to obtain the coefficient $\frac{1}{a-b}$). Similarly, the coefficient of the term $\frac{1}{x-b}$ on the right-hand side can be obtained by deleting the factor $(x-b)$ in the denominator of the expression on the left-hand side of the equation and replacing x with b in the adjacent factor $(x-a)$ (to obtain the coefficient $\frac{1}{b-a}$). The terms $\frac{1}{(a-b)}\frac{1}{(x-a)}$ and $\frac{1}{(b-a)}\frac{1}{(x-b)}$ in the equation above are called *partial fractions* of the rational expression $\frac{1}{(x-a)(x-b)}$.

EXAMPLE 6.5.3.

(i) $$\frac{1}{(x-5)(x-3)} = \frac{1}{2}\frac{1}{(x-5)} - \frac{1}{2}\frac{1}{(x-3)}$$

(ii) $$\frac{1}{(x-4)(x-3)} = \frac{1}{(x-4)} - \frac{1}{(x-3)}$$

(iii) $$\frac{1}{(x+6)(x-3)} = -\frac{1}{9}\frac{1}{(x+6)} + \frac{1}{9}\frac{1}{(x-3)}$$

(iv) $$\frac{6}{(x+6)(x+3)} = 6\left(-\frac{1}{3}\frac{1}{(x+6)} + \frac{1}{3}\frac{1}{(x+3)}\right) = -\frac{2}{(x+6)} + \frac{2}{(x+3)}$$

(v) $$\frac{1}{(10-x)(x+6)} = \frac{-1}{(x-10)(x+6)} = \frac{-1}{16}\left(\frac{1}{x-10} - \frac{1}{x+6}\right) = \frac{1}{16}\left(\frac{1}{10-x} + \frac{1}{x+6}\right)$$ ◈

In the following calculation, we will use the method above to find the partial fractions in the more general case where the linear factors in the denominator are of the form $(kx-a)$ and $(lx-b)$, where k, l, a, and b are real numbers with $k \neq 0$, $l \neq 0$, and $\frac{a}{k} \neq \frac{b}{l}$.

$$\frac{1}{(kx-a)(lx-b)} = \frac{1}{kl}\left(\frac{1}{\left(x-\frac{a}{k}\right)\left(x-\frac{b}{l}\right)}\right)$$

$$= \frac{1}{kl}\left(\frac{1}{\left(\frac{a}{k}-\frac{b}{l}\right)\left(x-\frac{a}{k}\right)} + \frac{1}{\left(\frac{b}{l}-\frac{a}{k}\right)\left(x-\frac{b}{l}\right)}\right)$$

$$= \frac{1}{\left(l\left(\frac{a}{k}\right)-b\right)(kx-a)} + \frac{1}{\left(k\left(\frac{b}{l}\right)-a\right)(lx-b)}.$$

The interpretation of this result is that the coefficient of $\frac{1}{(kx-a)}$ on the right-hand side is obtained by deleting the factor $(kx-a)$ in the denominator on the left-hand side and substituting the root of this linear factor, that is, $\frac{a}{k}$ for x in the remaining part of the expression. The coefficient of $\frac{1}{(kx-a)}$ on the right-hand side can be obtained similarly. The following notation can be used:

$$\frac{1}{(kx-a)(lx-b)} = \frac{A}{(kx-a)} + \frac{B}{(lx-b)},$$
$$\text{where } A = \left.\frac{1}{(lx-b)}\right|_{x=\frac{a}{k}},\quad B = \left.\frac{1}{(kx-a)}\right|_{x=\frac{b}{l}},\quad \text{and } k, l \neq 0, \frac{a}{k} \neq \frac{b}{l} \tag{6.4}$$

The vertical bar indicates that a substitution is made for x.

EXAMPLE 6.5.4.

(i) $$\frac{1}{(3x-1)(2x-1)}=\frac{1}{\left(2\left(\frac{1}{3}\right)-1\right)}\frac{1}{(3x-1)}+\frac{1}{\left(3\left(\frac{1}{2}\right)-1\right)}\frac{1}{(2x-1)}=\frac{-3}{(3x-1)}+\frac{2}{(2x-1)}$$

(ii) $$\frac{1}{(3x-1)(2x+1)}=\frac{1}{\left(2\left(\frac{1}{3}\right)+1\right)}\frac{1}{(3x-1)}+\frac{1}{\left(3\left(\frac{-1}{2}\right)-1\right)}\frac{1}{(2x+1)}=\frac{3}{5(3x-1)}-\frac{2}{5(2x+1)}$$

(iii) $$\frac{10}{(1+x)(1-4x)}=\frac{10}{(1-4(-1))}\frac{1}{(1+x)}+\frac{10}{\left(1+\left(\frac{1}{4}\right)\right)}\frac{1}{(1-4x)}=\frac{2}{(1+x)}+\frac{8}{(1-4x)}$$ ◈

This method of finding the partial fractions of a rational expression with a product of two linear factors in the denominator and a constant in the numerator can be applied to find the partial fractions of a rational expression with a product of *three* linear factors in the denominator and a constant in the numerator. The idea is to group terms and then apply the method twice:

$$\begin{aligned}\frac{1}{(x-a)(x-b)(x-c)}&=\left(\frac{1}{(x-a)(x-b)}\right)\frac{1}{(x-c)}\\&=\left(\frac{1}{(a-b)}\frac{1}{(x-a)}+\frac{1}{(b-a)}\frac{1}{(x-b)}\right)\frac{1}{(x-c)}\\&=\frac{1}{(a-b)}\frac{1}{(x-a)(x-c)}+\frac{1}{(b-a)}\frac{1}{(x-b)(x-c)}\\&=\frac{1}{(a-b)}\left(\frac{1}{(a-c)}\frac{1}{(x-a)}+\frac{1}{(c-a)}\frac{1}{(x-c)}\right)+\frac{1}{(b-a)}\left(\frac{1}{(b-c)}\frac{1}{(x-b)}+\frac{1}{(c-b)}\frac{1}{(x-c)}\right)\\&=\frac{1}{(a-b)}\frac{1}{(a-c)}\frac{1}{(x-a)}+\frac{1}{(b-a)}\frac{1}{(b-c)}\frac{1}{(x-b)}+\frac{1}{(a-b)}\left(\frac{1}{(c-a)}-\frac{1}{(c-b)}\right)\frac{1}{(x-c)}\\&=\frac{1}{(a-b)}\frac{1}{(a-c)}\frac{1}{(x-a)}+\frac{1}{(b-a)}\frac{1}{(b-c)}\frac{1}{(x-b)}+\frac{1}{(a-b)}\frac{(c-b)-(c-a)}{(c-a)(c-b)}\frac{1}{(x-c)}\\&=\frac{1}{(a-b)}\frac{1}{(a-c)}\frac{1}{(x-a)}+\frac{1}{(b-a)}\frac{1}{(b-c)}\frac{1}{(x-b)}+\frac{1}{(c-a)(c-b)}\frac{1}{(x-c)}.\end{aligned}$$

Thus, we have derived the formula

$$\boxed{\frac{1}{(x-a)(x-b)(x-c)}=\frac{1}{(a-b)}\frac{1}{(a-c)}\frac{1}{(x-a)}+\frac{1}{(b-a)}\frac{1}{(b-c)}\frac{1}{(x-b)}+\frac{1}{(c-a)(c-b)}\frac{1}{(x-c)}}. \tag{6.5}$$

where $a \neq b \neq c \neq a$. Note that the shortcut method explained above for finding the coefficients of the partial fractions when there are two linear terms in the denominator can also be applied when there are three linear terms in the denominator. That is, to find the coefficient of the term $\frac{1}{(x-a)}$ on the right-hand side of the formula above, delete the factor $(x-a)$ in the denominator of the expression on the left-hand side of the equation and replace x with a in the remaining part of the expression (to obtain the coefficient $\frac{1}{(a-b)(a-c)}$). The other two coefficients can be obtained similarly.

Furthermore, there is also the more general formula (which we will not derive), which is a generalization of formula (6.5):

$$\frac{1}{(kx-a)(lx-b)(mx-c)}=\frac{A}{(kx-a)}+\frac{B}{(lx-b)}+\frac{C}{(mx-c)},$$

$$\text{where}\quad A=\left.\frac{1}{(lx-b)(mx-c)}\right|_{x=\frac{a}{k}},\quad B=\left.\frac{1}{(kx-a)(mx-c)}\right|_{x=\frac{b}{l}},\quad \text{and}\quad C=\left.\frac{1}{(kx-a)(lx-b)}\right|_{x=\frac{c}{m}}. \tag{6.6}$$

$$a\neq b\neq c\neq a,\quad \frac{a}{k}\neq\frac{b}{l}\neq\frac{c}{m}\neq\frac{a}{k}$$

EXAMPLE 6.5.5.

(i) $$\frac{1}{(x-1)(x-2)(x-3)}$$

$$=\frac{1}{(1-2)(1-3)}\frac{1}{(x-1)}+\frac{1}{(2-1)(2-3)}\frac{1}{(x-2)}+\frac{1}{(3-1)(3-2)}\frac{1}{(x-3)}$$

$$=\frac{1}{2}\frac{1}{(x-1)}-\frac{1}{(x-2)}+\frac{1}{2}\frac{1}{(x-3)}$$

(ii) $$\frac{1}{(x-1)(x+2)(x-3)}$$

$$=\frac{1}{(1+2)(1-3)}\frac{1}{(x-1)}+\frac{1}{(-2-1)(-2-3)}\frac{1}{(x+2)}+\frac{1}{(3-1)(3+2)}\frac{1}{(x-3)}$$

$$=-\frac{1}{6}\frac{1}{(x-1)}+\frac{1}{15}\frac{1}{(x+2)}+\frac{1}{10}\frac{1}{(x-3)}$$

(iii) $$\frac{65}{(2x-1)(3x+1)(x-4)}$$

$$=\frac{65}{\left(3\left(\frac{1}{2}\right)+1\right)\left(\frac{1}{2}-4\right)}\frac{1}{(2x-1)}+\frac{65}{\left(2\left(-\frac{1}{3}\right)-1\right)\left(-\frac{1}{3}-4\right)}\frac{1}{(3x+1)}+\frac{65}{(2(4)-1)(3(4)+1)}\frac{1}{(x-4)}$$

$$=-\frac{52}{7}\frac{1}{(2x-1)}+\frac{9}{(3x+1)}+\frac{5}{7}\frac{1}{(x-4)}$$

◈

The next type of rational expression that we consider involves a linear factor in the numerator and a product of two linear factors in the denominator. We find the partial fractions by grouping the factors and making use of the formulas that have already been derived. Here is how this works in the simplest case:

$$\frac{x}{(x-a)(x-b)}=\left(\frac{x}{x-a}\right)\frac{1}{(x-b)}=\left(1+\frac{a}{x-a}\right)\frac{1}{(x-b)}=\frac{1}{(x-b)}+\frac{a}{(x-a)(x-b)}$$

$$=\frac{1}{(x-b)}+\frac{a}{(a-b)}\frac{1}{(x-a)}+\frac{a}{(b-a)}\frac{1}{(x-b)}$$

$$=\left(1+\frac{a}{b-a}\right)\frac{1}{(x-b)}+\frac{a}{(a-b)}\frac{1}{(x-a)}$$

$$=\frac{b}{(b-a)}\frac{1}{(x-b)}+\frac{a}{(a-b)}\frac{1}{(x-a)}.$$

where $a \neq b$. In the general case, the coefficients of the partial fractions are:

$$\boxed{\begin{array}{l} \dfrac{p(x)}{(kx-a)(lx-b)} = \dfrac{A}{(kx-a)} + \dfrac{B}{(lx-b)}, \\ \text{where } A = \left.\dfrac{p(x)}{(lx-b)}\right|_{x=\frac{a}{k}}, B = \left.\dfrac{p(x)}{(kx-a)}\right|_{x=\frac{b}{l}} \quad k, l \neq 0, \ \dfrac{a}{k} \neq \dfrac{b}{l}, \text{ and } p(x) \text{ is any linear factor} \end{array}}. \tag{6.7}$$

EXAMPLE 6.5.6.

(i) $\dfrac{2x+1}{(x-1)(x-2)} = \dfrac{2(1)+1}{(1-2)}\dfrac{1}{(x-1)} + \dfrac{2(2)+1}{(2-1)}\dfrac{1}{(x-2)} = -\dfrac{3}{(x-1)} + \dfrac{5}{(x-2)}$

(ii) $\dfrac{3x-4}{(2x-5)(x+6)} = \dfrac{3\left(\frac{5}{2}\right)-4}{\left(\frac{5}{2}+6\right)}\dfrac{1}{(2x-5)} + \dfrac{3(-6)-4}{(2(-6)-5)}\dfrac{1}{(x+6)} = \dfrac{7}{17}\dfrac{1}{(2x-5)} + \dfrac{22}{17}\dfrac{1}{(x+6)}$ ◈

We also consider rational expressions that involve a quadratic factor in the numerator and a product of three linear factors in the denominator. Again, we will find the partial fractions by grouping the factors and making use of the formulas that have already been derived. In the simplest case, this is:

$$\begin{aligned}
&\frac{x^2}{(x-a)(x-b)(x-c)} \\
&= \left(\frac{x}{(x-a)(x-b)}\right)\frac{x}{(x-c)} = \left(\frac{a}{(a-b)}\frac{1}{(x-a)} + \frac{b}{(b-a)}\frac{1}{(x-b)}\right)\frac{x}{(x-c)} \\
&= \frac{a}{(a-b)}\frac{x}{(x-a)(x-c)} + \frac{b}{(b-a)x(x-b)(x-c)} \\
&= \frac{a}{(a-b)}\left(\frac{a}{(a-c)}\frac{1}{(x-a)} + \frac{c}{(c-a)}\frac{1}{(x-c)}\right) + \frac{b}{(b-a)}\left(\frac{b}{(b-c)}\frac{1}{(x-b)} + \frac{c}{(c-b)}\frac{1}{(x-c)}\right) \\
&= \frac{a^2}{(a-b)(a-c)}\frac{1}{(x-a)} + \frac{b^2}{(b-a)(b-c)}\frac{1}{(x-b)} + \frac{c}{(a-b)}\left(\frac{a}{c-a} + \frac{b}{b-c}\right)\frac{1}{(x-c)} \\
&= \frac{a^2}{(a-b)(a-c)}\frac{1}{(x-a)} + \frac{b^2}{(b-a)(b-c)}\frac{1}{(x-b)} + \frac{c^2}{(c-a)(c-b)}\frac{1}{(x-c)}, \qquad \text{where } a \neq b \neq c \neq a.
\end{aligned}$$

By now, it should be no surprise that the formula for the coefficients of the partial fractions, in the general case, is

$$\boxed{\begin{array}{l} \dfrac{p(x)}{(kx-a)(lx-b)(mx-c)} = \dfrac{A}{(kx-a)} + \dfrac{B}{(lx-b)} + \dfrac{C}{(mx-c)} \\ \text{where } \ A = \left.\dfrac{p(x)}{(lx-b)(mx-c)}\right|_{x=\frac{a}{k}}, \ B = \left.\dfrac{p(x)}{(kx-a)(mx-c)}\right|_{x=\frac{b}{l}}, \text{ and} \\ C = \left.\dfrac{p(x)}{(kx-a)(lx-b)}\right|_{x=\frac{c}{m}} \quad k \neq l \neq m \neq 0, \ \dfrac{a}{k} \neq \dfrac{b}{l} \neq \dfrac{c}{m} \neq \dfrac{a}{k}, \text{ and } p(x) \text{ is any quadratic factor.} \end{array}}$$

EXAMPLE 6.5.7.

$$\frac{x^2+x+1}{(2x-3)(x+1)(x+2)}=\frac{\left(\frac{3}{2}\right)^2+\frac{3}{2}+1}{\left(\frac{3}{2}+1\right)\left(\frac{3}{2}+2\right)}\frac{1}{(2x-3)}+\frac{(-1)^2+(-1)+1}{(2(-1)-3)(-1+2)}\frac{1}{(x+1)}+\frac{(-2)^2+(-2)+1}{(2(-2)-3)(-2+1)}\frac{1}{(x+2)}$$

$$=\frac{19}{35}\frac{1}{(2x-3)}-\frac{1}{5}\frac{1}{(x+1)}+\frac{3}{7}\frac{1}{(x+2)}$$

◈

There are expressions equivalent to those we have derived for rational expressions with products of four or more different linear factors in the denominator. For these, and for the cases that were considered above, the degree of the numerator is supposed to be less than the degree of the denominator (expanded as a polynomial). In the case that the numerator of a rational expression has the same or larger degree than the degree of the denominator, then the numerator should first be divided by the denominator using long division.

EXAMPLE 6.5.8. Find the partial fractions of $\frac{2x^2+1}{(3x-2)(x+1)}$

Answer: The expanded form of the denominator is $3x^2+x-2$. By means of long division, we find that

$$\frac{2x^2+1}{3x^2+x-2}=\frac{2}{3}+\frac{-2x+7}{3(3x^2+x-2)}=\frac{2}{3}+\frac{-2x+7}{3(3x-2)(x+1)}.$$

Now, by means of the formulas above, we can find the partial fractions:

$$\frac{2x^2+1}{(3x-2)(x+1)}=\frac{2}{3}+\frac{-2\left(\frac{2}{3}\right)+7}{3\left(\frac{2}{3}+1\right)}\frac{1}{(3x-2)}+\frac{-2(-1)+7}{3(3(-1)-2)}\frac{1}{(x+1)}$$

$$=\frac{2}{3}+\frac{17}{15}\frac{1}{(3x-2)}-\frac{3}{5}\frac{1}{(x+1)}.$$

◈

The methods that have been used in this section can be used to find the partial fractions of rational expressions when some of the linear factors in the denominator are repeated, for example, rational expressions of the type $\frac{1}{(x-a)^2(x-b)}$, where $a\neq b$. However, in order to find the partial fractions of rational expressions of the most general type, for example, when some of the factors in the denominator are irreducible quadratic expressions, an algebraic technique that involves solving systems of equations is preferable. We will not discuss this here.

6.6 INEQUALITIES

In example 5.2.10, the inequality $4-3s-s^2\geq 0$ was solved to find the domain of the function $f(s)=\sqrt{4-3s-s^2}$. The method that was used to solve this inequality is basically the method that can be used to solve any inequality that involves polynomial or rational expressions. In this section, we are going to practice this method of solving inequalities. To begin with, we need to see how an inequality can be simplified.

6.6.1 Simplifying Inequalities

If x is a real variable (i.e., x may be any real number), then an inequality such as $2x+3<9$ represents all real numbers less than 3. The reason is that if a real number less than 3 is substituted for x (e.g., $x=1$), then the inequality is true ($2(1)+3<9$ is a true statement because $5<9$); on the other hand, if a real number greater than or equal to 3 is substituted for x (e.g., $x=5$), then the inequality is false ($2(5)+3<9$ is a false statement because $13 \not< 9$). To see precisely why $x=3$ is the cutoff point, we create an equivalent inequality (i.e., an inequality with the same solution) by subtracting 3 from both sides of the inequality to obtain $2x<6$ and then dividing both sides of the inequality by 2 to obtain $x<3$.

Performing the same algebraic operation on both sides of an inequality to produce an equivalent inequality is the basic method for simplifying inequalities. However, we need to be careful when multiplying or dividing both sides of an inequality by a negative number. We know, for instance, that $-5<-2$ is true but changing the sign on both sides, which is the same as multiplying both sides by -1, produces $5<2$, which is false. Similarly, multiplying both sides by -12, for example, produces $60<24$, which is also false. What we have to do when we multiply or divide an inequality by a negative number is change the direction of the inequality. For example, an inequality that is equivalent to $-2x+14<10-6x$ is the inequality $x-7>3x-5$ (divide both sides by -2).

Example 6.6.1. The inequality $1-x^4<2-x^2-3x^4$ can be expressed in the factorized form

$$(1+x^2)(1-x^2)<(1+x^2)(2-3x^2).$$

Because the factor $1+x^2$ is positive for all values of x, we can divide both sides of the inequality by $1+x^2$ to obtain the equivalent inequality $(1-x^2)<(2-3x^2)$. If we now add $3x^2$ to both sides and subtract 1 from both sides, we obtain the equivalent inequality $2x^2<1$. This, in turn, is equivalent to $x^2<\frac{1}{2}$. ◈

Example 6.6.2. The inequality $-2x^3+5x^2-8x+3\le -x^3+2x^2-3x$ can be expressed in the factorized form

$$(2x-1)(-x^2+2x-3)\le x(-x^2+2x-3).$$

Because the factor $-x^2+2x-3$ is negative for all values of x (because the equation $0=-x^2+2x-3$ has imaginary roots and the graph of $y=-x^2+2x-3$ is a downward pointing parabola), we obtain the equivalent inequality $2x-1\ge x$ by dividing both sides by $-x^2+2x-3$ and changing the direction of the inequality. This can be further simplified to the inequality $x\ge 1$. ◈

6.6.2 Solving Inequalities

In this section, we will look at inequalities involving one real variable (x), beginning with inequalities involving polynomials and then moving onto inequalities involving rational expressions, roots, absolute values, and, finally, other kinds of expression (exponents, logarithms, and trigonometric ratios).

When solving an inequality involving polynomials, the first objective, as with solving an equation involving polynomials, is to move all the terms to one side of the inequality (so that zero is on the other side) and then to factorize the resulting expression as a product of linear and irreducible quadratic factors.

The next step is to determine the sign of this product on each of the intervals (of the number line) determined by the roots of the linear factors. A test point chosen from any one of the intervals can be used to determine the sign of the product on that interval. The sign (of the product) on all the other intervals can then be determined by using the fact that it changes across any root corresponding to a linear factor raised to an odd power. The solution for the inequality is then the union of the intervals on which the inquality is a true statement. The examples below will make this clear.

EXAMPLE 6.6.3. Solve the inequality

$$2x+1<3-x.$$

Answer: The solution is $x<\frac{2}{3}$, which can alternatively be expressed by stating that x is an element of the interval $\left(-\infty,\frac{2}{3}\right)$, that is, $x\in\left(-\infty,\frac{2}{3}\right)$. ◈

EXAMPLE 6.6.4. Solve the inequality

$$11x^2+13x+2<0.$$

Answer: In factorized form, the inequality is $(11x+2)(x+1)<0$. The roots of the linear factors are $x=-\frac{2}{11}$ and $x=-1$. These two roots divide the number line into the intervals $(-\infty,-1)$, $\left(-1,-\frac{2}{11}\right)$, and $\left(-\frac{2}{11},\infty\right)$. At $x=0$ (a test point in the interval $\left(-\frac{2}{11},\infty\right)$), the sign of the product $(11x+2)(x+1)$ is positive. Therefore, the sign of the product $(11x+2)(x+1)$ will be positive on the interval $\left(-\frac{2}{11},\infty\right)$, negative on the adjacent interval $\left(-1,-\frac{2}{11}\right)$, and positive on the interval $(-\infty,-1)$. The final solution can then be expressed by means of the double inequality $-1<x<-\frac{2}{11}$ or $x\in\left(-1,-\frac{2}{11}\right)$. ◈

EXAMPLE 6.6.5. Solve the inequality

$$x^3\geq x^2+5x+3.$$

Answer: This inequality is equivalent to $x^3-x^2-5x-3\geq 0$. In factorized form, this is

$$(x+1)^2(x-3)\geq 0.$$

The roots of the linear factors are $x=-1$ and $x=3$. Again, we can use $x=0$ as a test point to determine that the product $(x+1)^2(x-3)$ is negative on the interval $(-1,3)$. Because there is no sign change at $x=-1$ (the linear factor $(x+1)$ is squared), the product $(x+1)^2(x-3)$ is also negative on the interval $(-\infty,-1)$. However, the product $(x+1)^2(x-3)$ will be positive on the interval $(3,\infty)$. Thus, the inequality is solved if $x=-1$ (where the product is zero) *or* if $x\geq 3$, that is, $x\in\{-1\}\cup[3,\infty)$. (The square bracket means that 3 is included in the interval.) ◈

EXAMPLE 6.6.6. Solve the inequality

$$x^5+x^4-x^3-x^2-x+1<0.$$

Answer: By the factor theorem, $(x-1)$ is a factor of the left-hand side, and by long division or synthetic division we obtain the factorization

$$\begin{aligned}x^5+x^4-x^3-x^2-x+1&=(x-1)(x^4+2x^3+x^2-1)\\&=(x-1)((x^2+x)^2-1)\\&=(x-1)(x^2+x+1)(x^2+x-1)\end{aligned}$$

and so the inequality can be expressed as $(x-1)(x^2+x+1)(x^2+x-1)<0$. The factor x^2+x+1 is irreducible (the roots are imaginary); however, the factor x^2+x-1 has the real roots $x=-\frac{1}{2}-\frac{\sqrt{5}}{2}$ and $x=-\frac{1}{2}+\frac{\sqrt{5}}{2}$. Therefore, the intervals we need to consider for the solution for the inequality are $\left(-\infty,-\frac{1}{2}-\frac{\sqrt{5}}{2}\right)$, $\left(-\frac{1}{2}-\frac{\sqrt{5}}{2},-\frac{1}{2}+\frac{\sqrt{5}}{2}\right)$, $\left(-\frac{1}{2}+\frac{\sqrt{5}}{2},1\right)$, and $(1,\infty)$. It is helpful to label these intervals I_1, I_2, I_3, and I_4, respectively. (Mathematicians use the " $:=$ " symbol when naming or defining something, so in future we will use the notation $I_1:=\left(-\infty,-\frac{1}{2}-\frac{\sqrt{5}}{2}\right)$, etc., for naming intervals.) The product

$$(x-1)(x^2+x+1)(x^2+x-1)$$

is positive in the interval I_2 (check $x=0$) and negative in the adjacent intervals I_1 and I_3. Therefore, the solution for the inequality is $x\in I_1$ or $x\in I_3$, that is, $x\in\left(-\infty,-\frac{1}{2}-\frac{\sqrt{5}}{2}\right)\cup\left(-\frac{1}{2}+\frac{\sqrt{5}}{2},1\right)$. ◈

When an inequality involves rational expressions, the first objective, as before, is to obtain all the terms on one side of the inequality (and zero on the other side) and then to add the terms to obtain a single rational expression in simplest form. The numerator and denominator of this rational expression can then be factorized separately into a product of linear and irreducible quadratic factors. Thereafter, the principle for solving the inequality is exactly the same as explained for polynomials above: odd powers of linear factors in the numerator or denominator change their sign at their roots while irreducible quadratic factors in the numerator or denominator do not change their sign and even powers of linear factors are positive except at their roots (where they are zero).

EXAMPLE 6.6.7. Solve the inequality

$$\frac{1}{x}<x.$$

Answer: This is equivalent to $\frac{1}{x}-x<0$, which can be expressed as

$$\frac{(1-x)(1+x)}{x}<0.$$

The roots of the linear factors are $x=-1$, $x=0$, and $x=1$. These roots divide the number line into the intervals $(-\infty,-1)$, $(-1,0)$, $(0,1)$, and $(1,\infty)$. A test point (e.g., $x=2$) determines that the rational expression is negative in the interval $(1,\infty)$. The solution for the inequality is therefore $x\in(-1,0)\cup(1,\infty)$. ◈

EXAMPLE 6.6.8. Solve the inequality

$$\frac{1}{x-1}<\frac{1}{x+1}+\frac{1}{7-x}.$$

Answer: This is equivalent to

$$\frac{(x+1)(7-x)-(x-1)(7-x)-(x^2-1)}{(x-1)(x+1)(7-x)}<0$$

which simplifies to

$$\frac{(x+5)(x-3)}{(1-x)(1+x)(7-x)}<0.$$

The roots of the linear factors are $x=-5,\ -1,\ 1,\ 3,$ and 7, which divide the number line into the intervals $(-\infty,-5)$, $(-5,-1)$, $(-1,1)$, $(1,3)$, $(3,7)$, and $(7,\infty)$. A test point (e.g., $x=0$) determines that the rational expression is negative in the interval $(-1,1)$. Therefore, the final solution is $x\in(-\infty,-5)\cup(-1,1)\cup(3,7)$. ◈

Example 6.6.9. Solve the inequality

$$\frac{1}{x^2+2x-7}\geq\frac{1}{x^2+2x+7}.$$

Answer: This is equivalent to

$$\frac{(x^2+2x+7)-(x^2+2x-7)}{(x^2+2x-7)(x^2+2x+7)}\geq 0$$

which simplifies to

$$\frac{14}{(x^2+2x-7)(x^2+2x+7)}\geq 0.$$

The factor (x^2+2x+7) is irreducible and (x^2+2x-7) has roots $x=-1\pm\sqrt{8}$. Thus, the intervals we consider are $I_1:=(-\infty,-1-\sqrt{8})$, $I_2:=(-1-\sqrt{8},-1+\sqrt{8})$, and $I_3:=(-1+\sqrt{8},\infty)$. The test point $x=0$ determines that the rational expression on the left-hand side of the inequality above is negative in the interval I_2. Therefore, the solution for the inequality is $x\in(-\infty,-1-\sqrt{8})\cup(-1+\sqrt{8},\infty)$. ◈

Inequalities involving root terms can be solved in a manner similar to solving equations with root terms, as long as the following facts are taken into careful consideration (assume that k is a natural number, $k>1$, and a and b are real numbers):

(i) If k is odd, then $\sqrt[k]{a}<\sqrt[k]{b}$ if and only if $a<b$.

(ii) If k is even, and $a>0$ and $b>0$, then $\sqrt[k]{a}<\sqrt[k]{b}$ if and only if $a<b$.

Example 6.6.10. Solve the inequality

$$\sqrt{x+6}<x.$$

Answer: The root is not defined if $x<-6$, so we make the restriction $x\geq-6$. Because a square root is always nonnegative, the inequality will be false if $-6\leq x<0$ and also false if $x=0$. Now, if $x>0$, then $\sqrt{x+6}<x$ is true if and only if $x+6<x^2$ is true (by (ii) above). The previous inequality is equivalent to $(x-3)(x+2)>0$. Therefore the final solution is $x\in(3,\infty)$. ◈

Example 6.6.11. Solve the inequality

$$1-2\sqrt{2x-1}<12-x.$$

Answer: The inequality is equivalent to $x-11<2\sqrt{2x-1}$. The root is not defined if $x<\frac{1}{2}$, so we make the restriction $x\geq\frac{1}{2}$. Because a square root is always nonnegative the inequality will be true if $\frac{1}{2}\leq x\leq 11$. If we suppose that $x>11$, then the inequality is true if $(x-11)^2<4(2x-1)$. This is equivalent to $(x-25)(x-5)<0$, which is true if $11<x<25$ (because we are supposing that $x>11$). The final solution is $x\in\left[\frac{1}{2},25\right)$. ◈

An inequality involving an absolute value should be interpreted either as a double inequality or pair of inequalities, depending on the direction of the inequality. For example, if k is a positive number and $p(x)$ is an algebraic expression (a polynomial or rational expression, for example), then $|p(x)|<k$ is equivalent to the double inequality $-k<p(x)<k$, and $|p(x)|>k$ is equivalent to the pair of inequalities $p(x)>k$ or $p(x)<-k$. The double inequality can be solved by treating it as two inequalities and taking the intersection of their solutions as the final solution. The pair of inequalities can be solved by taking the union of their solutions as the final solution.

Example 6.6.12. Solve the inequality

$$|2x+1|<11.$$

Answer: This is equivalent to the double inequality $-11<2x+1<11$. We will solve each of the inequalities $-11<2x+1$ and $2x+1<11$, separately. The solution for the first inequality is $-6<x$ and the second inequality is $x<5$. Thus, the final solution is $x\in(-6,5)$. ◈

Example 6.6.13. Solve the inequality

$$|x^2+x-12|>8.$$

Answer: There are two cases to consider here. We first consider the case $x^2+x-12>8$, which is equivalent to $(x+5)(x-4)>0$. The solution for this case is $x<-5$ or $x>4$. The second case, we need to consider is $x^2+x-12<-8$, which is equivalent to $x^2+x-4<0$. The roots of x^2+x-4 are $x=\frac{-1\pm\sqrt{17}}{2}$, and so the solution for this case is $\frac{-1-\sqrt{17}}{2}<x<\frac{-1+\sqrt{17}}{2}$. Note that $-5<\frac{-1-\sqrt{17}}{2}<\frac{-1+\sqrt{17}}{2}<4$. We can conclude, therefore, that the final solution is

$$x\in(-\infty,-5)\cup\left(\frac{-1-\sqrt{17}}{2},\frac{-1+\sqrt{17}}{2}\right)\cup(4,\infty).$$ ◈

Another way to solve inequalities involving absolute values, especially if there is more than one absolute value, is to break up the problem into cases depending on whether the expressions inside the absolute values are positive or negative. The two cases here are $|p(x)|=p(x)$ if $p(x)\geq 0$ and $|p(x)|=-p(x)$ if $p(x)\leq 0$.

Example 6.6.14. Solve the inequality

$$|3x+1|-|x+4|<1.$$

Answer: The four cases we need to consider are:

Case (i) $3x+1\geq 0$ and $x+4\geq 0$;

Case (ii) $3x+1<0$ and $x+4\geq 0$;

Case (iii) $3x+1\geq 0$ and $x+4<0$;

Case (iv) $3x+1<0$ and $x+4<0$.

Case (i) is equivalent to $x\geq -\frac{1}{3}$ and $x\geq -4$. These two inequalities combine as $x\geq -\frac{1}{3}$, and, in this case, the original inequality becomes $(3x+1)-(x+4)<1$. This simplifies to $x<2$, so the solution for this case is $-\frac{1}{3}\leq x<2$.

Case (ii) is equivalent to $x<-\frac{1}{3}$ and $x \geq -4$. These two inequalities combine as $-4 \leq x < -\frac{1}{3}$, and the original inequality becomes $-(3x+1)-(x+4)<1$. This simplifies to $x>-\frac{3}{2}$, so the solution for this case is $-\frac{3}{2}<x<-\frac{1}{3}$.

Case (iii) is equivalent to $x \geq -\frac{1}{3}$ and $x<-4$. These two inequalities have no intersection, so there is no inequality to solve for this case.

Case (iv) is equivalent to $x<-\frac{1}{3}$ and $x<-4$. These two inequalities combine as $x<-4$, and the original inequality becomes $-(3x+1)+(x+4)<1$. This simplifies to $x>1$, so there is no solution for this case (because we have the restriction $x<-4$).

Therefore, by combining cases (i) and (ii) we obtain the final solution $x \in \left(-\frac{3}{2}, 2\right)$. ◈

The only types of problem involving exponential, trigonometric, and logarithmic expressions that we will consider in this section are those that can be solved using the same methods used to solve inequalities that involve polynomial or rational expressions. So, essentially, nothing new is being introduced now. Knowledge of the graphs of exponential, trigonometric, and logarithmic expressions comes in very useful!

Example 6.6.15. Solve the inequality

$$2^x - 2^{-x+1} - 1 > 0.$$

Answer: We represent 2^x as X and then write the inequality as $X-\frac{2}{X}-1>0$, which is equivalent to $\frac{X^2-X-2}{X}>0$, or, in factorized form, $\frac{(X-2)(X+1)}{X}>0$. We may suppose that $X>0$, as it represents 2^x, which is positive for all values of x. Therefore, the solution for the inequality is $X>2$ (we discard the other possibility, $X<-1$). In terms of x, this is $2^x>2$, which is satisfied if and only if $x>1$, that is, $x \in (1,\infty)$. ◈

Example 6.6.16. Solve the inequality

$$\cos(x)\sin(x) < \frac{1}{4}.$$

Answer: We multiply both sides of the inequality by 2 and make use of the identity $\sin(2\theta) \equiv 2\sin(\theta)\cos(\theta)$. The inequality then becomes $\sin(2x)<\frac{1}{2}$. With reference to the graph of the sine function, we can deduce that the inequality is solved if $2x$ belongs to any interval of the form $\left(\frac{5\pi}{6}+2n\pi, \frac{13\pi}{6}+2n\pi\right)$, where n is any integer. This in turn means that $x \in \left(\frac{5\pi}{12}+n\pi, \frac{13\pi}{12}+n\pi\right)$, where n is any integer. ◈

Example 6.6.17. Solve the inequality

$$(\log_3(x))^2 + \log_3(x) - 2 > 0.$$

Answer: In factorized form, this is $(\log_3(x)+2)(\log_3(x)-1)>0$, and so the solution is $\log_3(x)<-2$ or $\log_3(x)>1$. The first of these inequalities is equivalent to $0<x<3^{-2}$, which is the same as $0<x<\frac{1}{9}$ (these are the values of x for which the graph $y=\log_3(x)$ is below the horizontal line $y=-2$), and the second of these inequalities is equivalent to $x>3^1$, which is the same as $x>3$ (these are the values of x for which the graph $y=\log_3(x)$ is above the horizontal line $y=1$). The solution for the problem, therefore, is $x\in(0,\frac{1}{9})\cup(3,\infty)$. ◈

6.6.3 Two-Variable Inequalities

A solution for an inequality of the form

$$p(x,y)<q(x,y),$$

where $p(x,y)$ and $q(x,y)$ are expressions in terms of variables x and y, is a pair of real numbers (a,b), such that if $x=a$ and $y=b$, then the inequality is a true statement. The solution set for the inequality is the set of all such solutions. It can be represented as the following set S:

$$S=\{(a,b)\mid p(a,b)<q(a,b)\}.$$

Each solution (a,b) can be regarded as a pair of coordinates in the Cartesian plane, and the set S can be regarded as a subset of the Cartesian plane. Therefore, in all of the examples below, the set S can be represented by shading a region of the Cartesian plane.

It is difficult, in general, to find the solution for an inequality involving two variables. However, if the inequality relates to a familiar graph, then the solution can be obtained by shading the region above or below the graph.

Example 6.6.18. Consider the inequality

$$x<y.$$

The solution set S for this inequality is the set of all coordinate pairs (a,b), where a and b are real numbers and $a<b$. In the same way that the solution set of the equation $x=y$ is represented in the Cartesian plane as a line through the origin, the solution set S for the inequality $x<y$ can be represented in the Cartesian plane as the region above this line (because every point directly above any point on this line has the same x coordinate but larger y coordinate). In figure 6.1, this is the shaded region. The line $y=x$ is shown as a dotted line because it is not included in the set S. ◈

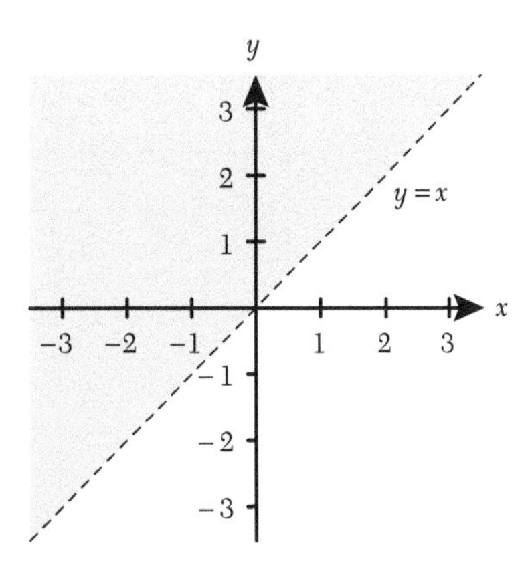

FIGURE 6.1. The graphical representation of the solution set for $x<y$.

Example 6.6.19. The solution sets for the inequalities $y\le x^2$, $x^2+y^2<9$ and the double inequality $x-1\le y<x+1$ are the shaded regions in the first, second, and third diagrams in figure 6.2, respectively. Note that, in the third diagram, the lower boundary line for the shaded region is shown as a solid line because equality is included in the lower inequality $x-1\le y$, whereas the upper boundary

line for the shaded region is shown as a dotted line because the upper inequality $y < x + 1$ is a strict inequality. ◈

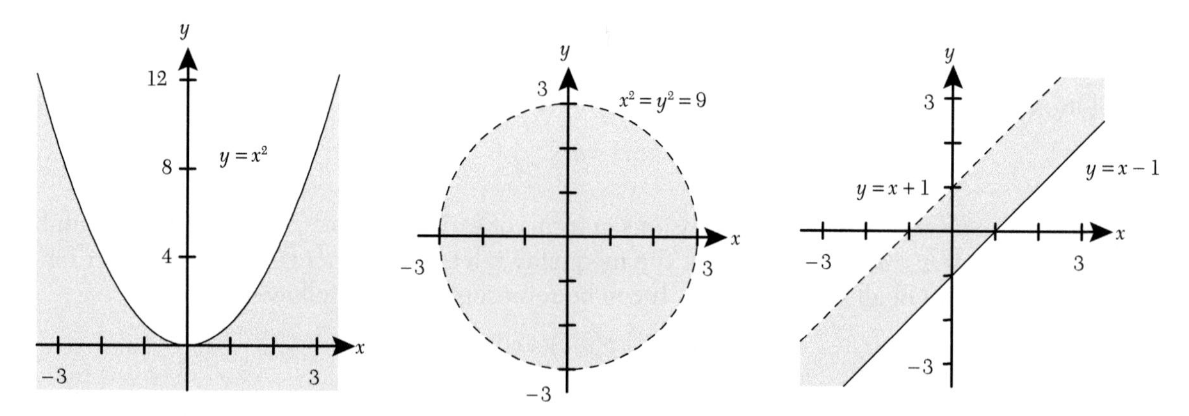

FIGURE 6.2. Diagram for example 6.6.19.

We can also regard the shaded region in the third diagram in figure 6.2 as a representation of the solution set for the *system of inequalities* $\{y < x + 1, y \geq x - 1\}$, that is

$$S = \{(a,b) \mid b < a + 1\} \cap \{(a,b) \mid b \geq a - 1\}$$

We now give a few more examples of systems of inequalities and their corresponding shaded regions.

Example 6.6.20. The shaded region representing the solution set for the system of inequalities

$$\{y > -2x^2 - x + 6,\ y \leq 2x - 3,\ x \leq 2.5\}$$

is shown in figure 6.3. ◈

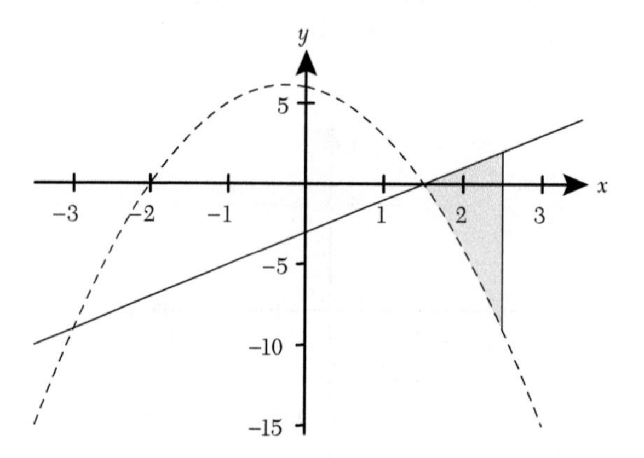

FIGURE 6.3. Diagram for example 6.6.20.

Example 6.6.21. The shaded region representing the solution set for the system of inequalities

$$\{x^2 + y^2 \leq 9,\ xy \leq 1,\ y - x < 0\}$$

is shown in figure 6.4. (Note that inequality $xy \leq 1$ is represented by the region between the two components of the graph of the hyperbola $y = \frac{1}{x}$.) ◈

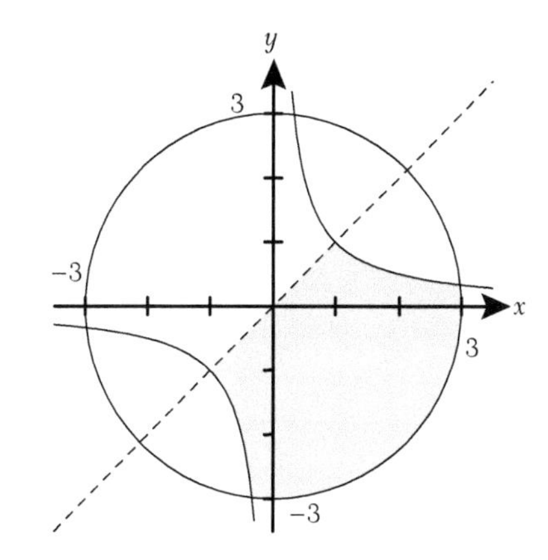

FIGURE 6.4. Diagram for example 6.6.21.

EXAMPLE 6.6.22. The shaded region representing the solution set for the system of inequalities

$$\{x \le 1 - y^2, x \ge y^2 - 1\}$$

is shown in figure 6.5. It's a good idea to check that a test point in the region that has been shaded verifies the inequalities. In this case, the origin $(0,0)$ can be used a test point. Because both inequalities are true if $x=0$ and $y=0$, the correct region is shaded. ◈

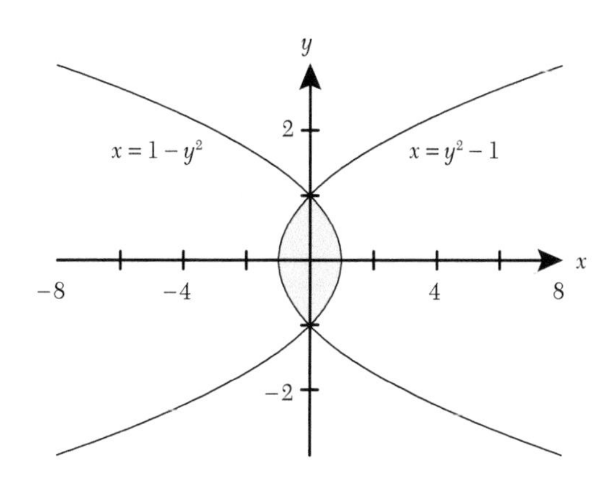

FIGURE 6.5. Diagram for example 6.6.22.

EXERCISES

6.1. Perform the indicated product or division of rational expressions. Write each answer as a single, simplified rational expression.

(a) $\dfrac{4x^3+4x^2+x}{x-1} \cdot \dfrac{x^2-1}{1-4x^2}$

(b) $\dfrac{t^3+t^2+121t+121}{t^4-14641} \cdot \dfrac{t^2+9t-22}{t+2}$

(c) $\dfrac{r^2-r-42}{r^2-r-20} \div \dfrac{r^2-5r-14}{r^2+r-12}$

(d) $\dfrac{y^4+64}{y^4+4y^3+6y^2+4y+1} \div \dfrac{y^2-2y-24}{y+1}$

6.2. Perform the indicated operation (addition, subtraction, multiplication, or division) of rational expressions. Write each answer as a single, simplified rational expression.

(a) $\dfrac{16}{x^2-16}+\dfrac{4x}{x-4}$

(b) $\dfrac{16}{x^2-16}-\dfrac{4x}{x-4}$

(c) $\dfrac{t}{t-3}-\dfrac{6t}{t^2-9}+\dfrac{7}{t+3}$

(d) $\dfrac{r+1}{r^2+5r}+\dfrac{r-1}{r^3+5r^2}+\dfrac{1}{r(r+5)^2}$

(e) $\dfrac{y}{6y^2+y+1}+\dfrac{2}{y+2}+1$

(f) $\dfrac{\frac{2}{x}-\frac{x}{2}}{\frac{3}{x}+\frac{x}{3}}$

(g) $\dfrac{\frac{1}{(3+t)^3}-\frac{1}{(3+t)^2}}{3+t}$

(h) $\dfrac{7r^2-6r+1}{\frac{r}{r+1}+\frac{1}{7r-7}}$

6.3. Perform the indicated operation of rational expressions. Write each answer as a single, simplified rational expression without negative exponents.

(a) $\dfrac{x+x^{-1}}{x^{-2}}$

(b) $1+y-(1-y)^{-1}$

(c) $1+t-2t(1+t)^{-1}$

(d) $\left(2r-6(r+2)^{-1}\right)\cdot\left(1+3(r-1)^{-1}\right)$

(e) $\left(3-2(2s+1)^{-1}+10(s-1)^{-1}\right)\div\left(1-6(2s+1)^{-1}+2(s+1)^{-1}\right)$

6.4. Factorize the following expressions so that the resulting factors contain only square roots, cube roots, and squares of cube roots.

(a) $1-x^{\frac{2}{3}}$

(b) $6x^{\frac{2}{3}}-x^{\frac{1}{3}}-1$

(c) $6x+20\sqrt{x}+14$

(d) $x+6\sqrt[3]{x^2}+12\sqrt{x}+8$

(e) $1-x^{\frac{1}{2}}+x^{\frac{1}{3}}-x^{\frac{5}{6}}+x^{\frac{2}{3}}-x^{\frac{7}{6}}$ (Hint: factor by grouping in pairs.)

6.5. Add or subtract the expressions, as indicated. Write the answer as a single fraction without negative exponents.

(a) $4(x+2)^{\frac{1}{3}}-16(x+2)^{-\frac{2}{3}}$

(b) $\dfrac{(2w+3)^{\frac{1}{2}}}{(w-1)^{\frac{1}{4}}}-\dfrac{(2w+3)^{-\frac{1}{2}}}{(w-1)^{-\frac{3}{4}}}$

(c) $\dfrac{(z+1)^{\frac{1}{2}}}{(z-1)^{-\frac{1}{2}}}+\dfrac{(z+1)^{-\frac{1}{2}}}{(z-1)^{\frac{1}{2}}}$

6.6. Add or subtract the expressions, as indicated. Write the answer as a single fraction and rationalize the denominator.

(a) $\dfrac{1}{(1-\sqrt{x})(2+\sqrt{x})}+\dfrac{1}{(1+\sqrt{x})(2-\sqrt{x})}$

(b) $\dfrac{1}{6y+20\sqrt{y}+14}-\dfrac{1}{6y-20\sqrt{y}+14}$

(c) $\dfrac{3}{t^{\frac{3}{2}}+1}-\dfrac{2}{t^{\frac{1}{2}}-1}+\dfrac{t^{\frac{1}{2}}}{1+t}$

(d) $\dfrac{\sqrt[3]{w+1}}{\sqrt[3]{w^2}-\sqrt[3]{w+1}}+\dfrac{\sqrt[3]{w^2+1}}{w-1}$

6.7. Simplify the following rational expressions.

(a) $\dfrac{x^2 - 3x - 54}{x^2 + 10x + 24}$

(b) $\dfrac{r^3 - 8}{2r^4 - 32}$

(c) $\dfrac{3t^2 - 21t + 30}{15t^2 - 10t - 40}$

6.8. Solve each equation for x.

(a) $15(x^2 + 1) = 34x$

(b) $(x + 1)(x - 2) = (2x - 1)(x + 2)$

(c) $(x + 1)(x - 2) = 2(x - 5)(x - 4)$

6.9. Solve each equation for t.

(a) $(t + 3)^2 + 2(t + 3) - 3 = 0$

(b) $\left(t + \dfrac{3}{t} - 1\right)\left(t + \dfrac{3}{t} + 1\right) = 15$

(c) $(t^2 + t - 5)(t^2 + t - 7) = 15$

(d) $(t + 1)^2 = (t + 2)^2$

(e) $4(t + 2) + \dfrac{4}{(t + 2)} = 1$

(f) $t^2 + \dfrac{1}{t^2} - 11\left(t - \dfrac{1}{t}\right) = 14$

(g) $\dfrac{t^4 + 1}{t^2} + 7\left(\dfrac{t^2 + 1}{t}\right) - 16 = 0$

6.10. Two numbers differ by 2 and their reciprocals differ by $\frac{1}{60}$. Find the two numbers.

6.11. The product of two consecutive numbers added to the square of their sum is 151. Find the numbers.

6.12. If the squares of two numbers add up to 5, and three times one of the numbers plus the other number equals 5, what are the numbers?

6.13. A blue hose can fill a swimming pool in x h and a red hose takes 6 h longer. Running together, they can fill the swimming pool in 4 h. Find x.

6.14. Solve each equation for r.

(a) $\dfrac{r + 1}{r + 4} = \dfrac{2(r - 4)}{r - 2}$

(b) $\dfrac{r}{2r - 1} + \dfrac{r + 5}{r + 1} = 3$

(c) $\dfrac{5}{r - 2} - \dfrac{3}{r - 1} = \dfrac{1}{r - 3}$

(d) $\dfrac{14}{r + 2} - \dfrac{1}{r - 4} = 1$

(e) $\dfrac{r + 1}{r + 1} = \dfrac{r - 1}{r - 1}$

6.15. Solve each equation for y.

(a) $\dfrac{2y^2}{y - 3} + \dfrac{4y - 6}{y + 3} = \dfrac{108}{y^2 - 9}$

(b) $\left(\dfrac{y}{y + 2}\right)^2 - \dfrac{2y}{y + 2} - 16 = 0$

(c) $\dfrac{3y + 1}{6y^2 - 11y + 3} - \dfrac{2y - 5}{8y^2 - 2y - 15} = \dfrac{5y + 3}{12y^2 + 11y - 5}$

(d) $\dfrac{y^2 + 4}{y + 2} = \dfrac{y - 2}{y^2 + 11}$

6.16. Solve each equation for v.

(a) $|v-2|=v+4$

(b) $|v-2|=v^2+1$

(c) $2|v^2-2|=3-v^2$

(d) $|v|^2-2|v|+1=0$

(e) $|(v-4)(v+2)|=7$

(f) $|(v-3)(v-1)|=v-1$

(g) $|(v-3)(v-1)|=(v-3)(v-1)$

6.17. Solve each equation for x.

(a) $3^x3^{x-1}=\sqrt{3}$

(b) $9^x=\dfrac{1}{3}\sqrt{3}$

(c) $8\cdot 7^x=392$

(d) $6^{x+1}-6^{x-1}=210$

(e) $7^{2x}-6\cdot 7^x-7=0$

(f) $\dfrac{5^x+1}{5^x+4}=\dfrac{2(5^x-4)}{5^x-2}$

(g) $\left(\dfrac{1}{2}\right)^{x^2-9}=4^{x+3}$

6.18. Solve each equation for w.

(a) $w^{\frac{1}{3}}=-2$

(b) $(w-4)^{\frac{2}{3}}=2$

(c) $(w^{\frac{1}{3}}-1)^{\frac{2}{3}}=2$

(d) $2w^{\frac{4}{5}}=\dfrac{32}{81}$

(e) $4w^{\frac{2}{3}}-6w^{-\frac{2}{3}}=5$

(f) $6w-20\sqrt{w}+14=0.$

(g) $w^{-1}+w^{-\frac{1}{2}}=\dfrac{3}{4}$

6.19. Solve each equation for s.

(a) $s=3+\sqrt{5s-9}$

(b) $s+\sqrt{5s+19}=-1$

(c) $\sqrt{s-3}+\sqrt{s+2}=5$

(d) $2\sqrt{3-s}=3\sqrt{2-s}$

(e) $\sqrt{(3-s)(1-s)}=\sqrt{s-1}$

(f) $\sqrt{20-s}=\sqrt{9-s}+3$

(g) $\sqrt{7s+11}=1+\sqrt{5s+6}$

(h) $\sqrt{1+4\sqrt{s}}=1+\sqrt{s}$

(i) $\sqrt{s+2}-\sqrt{s-2}=\sqrt{2s}$

(j) $\sqrt{3+s}+\sqrt{s}=\dfrac{5}{\sqrt{s}}$

(k) $\sqrt[4]{2s^2+3}=s$

6.20. Solve each equation for v.

(a) $\log_2 v=8$

(b) $\log_3(v-2)=-1$

(c) $\log 256+\log v=\log 1{,}024$

(d) $\log v=1-\log(v+9)$

(e) $\ln(v^2+1)=1$

(f) $\log v=\log 2v$

(g) $\log_2 3v=\log_2 3+\log_2 v$

(h) $\log v^2=\log(6-v)$

(i) $2\log v=\log(6-v)$

(j) $\log v=-\log(v+3)$

(k) $\log_6(3v+4)=2+\log_6(v-3)$

(l) $\log_2\sqrt{v}=\sqrt{\log_2 v}$

(m) $\log v\log(v+1)\log(v+2)=0$

6.21. Solve each equation for k.

(a) $\ln(k^2+1)=1+\ln 2k$

(b) $\log_2(k+3)+\log_2\frac{1}{k}=\log_2 128$

(c) $2\log_8(2k+3)=\log_8(k+1)$

(d) $\log(k-3)-\log(3k-10)-\log\frac{1}{k}=0$

(e) $\log(9k+10)^2=2+\log(2k+5)+\log(k+2)$

(f) $2\log(9k+10)=2+\log(2k+5)+\log(k+2)$

6.22. Solve each equation for x.

(a) $\cos(2x)+\sin(x)=1$

(b) $\sin(3x)\cos(2x)-\cos(3x)\sin(2x)=\frac{\sqrt{2}}{2}$

(c) $\sin(4x)-2\sin(2x)=0$

(d) $\sec^2(x)+3\tan(x)-11=0$

(e) $\cos(\pi-x)+\sin\left(x-\frac{\pi}{2}\right)=1$

6.23. Solve each equation for θ.

(a) $2\cos(\theta)+2\sin(\theta)=\sqrt{6}$

(b) $12\sin^2(\theta)-17\sin(\theta)+6=0$

(c) $3\cos^2(\theta)-2\sin(\theta)-1=0$

6.24. Solve each equation for α.

(a) $-12\cos(\alpha)+5\sin(\alpha)=10.$

(b) $3\cos(\alpha)-4\sin(\alpha)=2.$

(c) $3\cos(\alpha)-\sqrt{3}\sin(\alpha)=2.$

6.25. **(a)** Use the trigonometric identity $\cos(2\theta)=2\cos^2(\theta)-1$ to prove that

$$2\cos\left(\frac{\pi}{8}\right)=\sqrt{2+\sqrt{2}} \quad \text{and} \quad 2\sin\left(\frac{\pi}{8}\right)=\sqrt{2-\sqrt{2}}$$

(b) Solve the trigonometric equation for θ in the interval $[0,2\pi]$:

$$\sin(2\theta)-\sqrt{2-\sqrt{2}}\cos(\theta)-\sqrt{2+\sqrt{2}}\sin(\theta)+\frac{\sqrt{2}}{2}=0$$

(Hint: $(x-a)(y-b)=xy-ay-by+ab$.)

6.26. **(a)** Use trigonometric identities to prove that

$$\sin(2x)+\sin(4x)+\sin(6x)=4\sin(3x)\cos(2x)\cos(x)$$

(b) Solve the trigonometric equation for x in the interval $[0,2\pi]$:

$$\sin(2x)+\sin(4x)+\sin(6x)=2\cos(2x)\cos(x)$$

6.27. Find the partial fractions of the following rational expressions.

(a) $\dfrac{x}{x+4}$

(b) $\dfrac{x}{x-4}$

(c) $\dfrac{x+3}{x+6}$

(d) $\dfrac{x+3}{x-6}$

(e) $\dfrac{2x}{2x+7}$

(f) $\dfrac{x}{2x-7}$

(g) $\dfrac{x+1}{2x+1}$

(h) $\dfrac{x-1}{2x-1}$

6.28. Find the partial fractions of the following rational expressions.

(a) $\dfrac{1}{(t-4)(t-2)}$

(b) $\dfrac{1}{(t+4)(t+2)}$

(c) $\dfrac{3}{(t+3)(t-3)}$

(d) $\dfrac{1}{3(t-2)(t+1)}$

(e) $\dfrac{2}{(2t+7)(t-1)}$

(f) $\dfrac{1}{7(t-7)(t+1)}$

(g) $\dfrac{1}{(3t+4)(3t-4)}$

(h) $\dfrac{1}{(5t-1)(1-3t)}$

6.29. Find the partial fractions of the following rational expressions.

(a) $\dfrac{y}{(y-4)(y-2)}$

(b) $\dfrac{y}{(y+4)(y+2)}$

(c) $\dfrac{3y}{(y+3)(y-3)}$

(d) $\dfrac{y}{3(y-2)(y+1)}$

(e) $\dfrac{2y}{(2y+7)(y-1)}$

(f) $\dfrac{y+17}{(y-7)(y+1)}$

(g) $\dfrac{2y+1}{(3y+4)(3y-4)}$

(h) $\dfrac{12y+5}{(5y-1)(1-3y)}$

6.30. Find the partial fractions of the following rational expressions.

(a) $\dfrac{1}{(w-4)(w-2)(w+3)}$

(b) $\dfrac{1}{(w+4)(w+2)(w-2)}$

(c) $\dfrac{3w}{(w+3)(w-3)(w+1)}$

(d) $\dfrac{w}{3(w-2)(w+1)(w-8)}$

(e) $\dfrac{2w}{(2w+7)(w-1)(w+4)}$

(f) $\dfrac{w^2+17w}{(w-7)(w+1)}$

(g) $\dfrac{w^2+2w+1}{(3w+4)(3w-4)(w+1)}$

(h) $\dfrac{2w^2+2w+1}{(3w+4)(3w-4)(w+1)}$

6.31. Find the partial fractions of the following rational expressions.

(a) $\dfrac{r^2-r+3}{(r-4)(r-2)}$

(b) $\dfrac{r^2+3r+1}{(r+2)(r-2)}$

(c) $\dfrac{(r-3)(r-1)}{(r+3)(r+1)}$

(d) $\dfrac{r^3+2r+1}{(r-2)(r+1)(r-8)}$

(e) $\dfrac{r^3+1}{(r-2)(r+1)(r-1)}$

6.32. If $a \neq b$, then the partial fractions of $\frac{1}{(x-a)^2(x-b)}$ are of the form

$$\frac{A_1}{(x-a)}+\frac{A_2}{(x-a)^2}+\frac{B}{(x-b)}$$

for constants A_1, A_2, and B. For example,

$$\frac{1}{(x-1)^2(x-2)}=-\frac{1}{(x-1)}-\frac{1}{(x-1)^2}+\frac{1}{(x-2)}$$

Find the partial fractions of the following rational expressions. (Hint: start by grouping the factors.)

(a) $\dfrac{1}{(x-4)^2(x-2)}$

(b) $\dfrac{1}{(x-4)(x-2)^2}$

(c) $\dfrac{1}{(x+4)^2(x+2)}$

6.33. Solve the following inequalities.

(a) $6x-7>x+2$

(b) $2x^2-9x-35\leq 0$

(c) $(x+1)(x+2)^2(x+3)^3(x+4)^4>0$

(d) $(x+1)(x-2)<2(x-5)(x-4)$

(e) $6x^3+2\leq 3x+11x^2$

(f) $(x^2+x+2)(x^2-x+2)(x^2-6)<0$

(g) $(x^4+3x^2+4)(x^2-6)>0$

6.34. Solve the inequality $x^4+x^3-x^2-2x-2\geq 0$.
(Hint: rewrite the polynomial as $x^4+x^3+x^2-2x^2-2x-2$ and factor by grouping.)

6.35. Solve the following inequalities.

(a) $\dfrac{u+9}{u+6}\leq 3$

(b) $\dfrac{1}{u^2}<u^2$

(c) $\dfrac{3-u}{1+u}<u$

(d) $\dfrac{3-u}{u^2-16}\geq 0$

(e) $\dfrac{-4}{2-u}<1+u$

(f) $\dfrac{(3u+1)(1-u)^2}{u^3(2+u)}<0$

(g) $\dfrac{3u+1}{u^2-9u-22u}<0$

(h) $\dfrac{1}{2u^2-20u-22u}<\dfrac{1}{u^2+14u+33}$

(i) $\dfrac{1}{u+2}+\dfrac{1}{u-2}\geq\dfrac{1}{u-4}$

(j) $\dfrac{1}{u^3+2u+3}>\dfrac{1}{2-u^2-u^3}$

6.36. Solve the following inequalities.

(a) $\sqrt{y} \leq y$

(b) $2\sqrt{3-y}+1>y$

(c) $\sqrt{7y+2} \leq 6-y$

(d) $\dfrac{1}{\sqrt{y+3}}<y+1$

6.37. Solve the following inequalities.

(a) $|1-2v|-3<0$

(b) $|1-2v|-3>0$

(c) $|12v+13| \leq 156$

(d) $|1-2v|+6<3v$

(e) $|3-v| \geq 4v-2$

(f) $|4v-2|<3-v$

(g) $|3-v| \leq 3-v$

(h) $|(4-v)(3-v)| \leq 2$

(i) $|(4-v)(3-v)|<5-v$

(j) $|(4-v)(3-v)|<\dfrac{4-v}{4}$

(k) $|v+3|+|4v-1|<1$

(l) $|v+3|-|4v-1|<1$

6.38. Solve the following inequalities.

(a) $(3^x-1)^2>4$

(b) $(3^x-1)^2>4$ (express your answer as a logarithm)

(c) $4^x-2^x>0$

(d) $\log_2(4^x-2^x)>0$ (express your answer as a logarithm)

(e) $12\sin^2(\theta)-17\sin(\theta)+6<0$ (express your answers in terms of $\sin^{-1}$)

(f) $\sin(2x)+\sin(x)>0$

(g) $(\log_3(x))^2+3\log_3(x)+2<0$

(h) $\log_3(x)>\log_4(x)$

(i) $\ln(x)-\dfrac{2}{\ln(x)}>1$

6.39. Shade the regions in the Cartesian plane representing the solution set for each of the following inequalities.

(a) $y \leq 3x-4$

(b) $x<3y-4$

(c) $y \geq x^2-2x+4$

(d) $x>y^2-2y+4$

(e) $x \leq 2^y$

(f) $x<2^y$

6.40. Shade the regions in the Cartesian plane representing the solution set for each of the following double inequalities.

(a) $0<y<x$

(b) $0<y<2$

(c) $-4<y<1-2x$

(d) $0<y<\sin(x)$

(e) $-\sqrt{x+1}<y<\sqrt{x+1}$

(f) $-\dfrac{1}{x}<y<\dfrac{1}{x}$

(g) $x^2-2x<y<2x-x^2$

(h) $x^2+2x-1<y<1+2x-x^2$

(i) $\sqrt{4-x^2}<y \leq |x-1|-1$

6.41. Shade the regions in the Cartesian plane representing the solution set for each of the following systems of inequalities.

(a) $\{y \geq \sqrt{16-x^2}, y \geq |-x+4|, -4 < x < 4\}$

(b) $\{y < 2^x, y > -2^x, x > 0\}$

(c) $\{y < \log_3(x), y > \log_4(x), x > 64\}$

(d) $\{y < x+1, y > x-1, y < 1-x, y > -1-x\}$

(e) $\{y < -\frac{x}{2}+2, y > x-4, y > -x-2\}$

(f) $\{x^2+2x-1 < y < 1+2x-x^2, y > 2x\}$

(g) $\{x^2+y^2 \leq 1, x^2+2x+y^2 \leq 0, y > 0\}$

(h) $\{y < \sqrt{x+1}, y < \sqrt{x-1}, y > x^2-1\}$

(i) $\{y > \sqrt{x+1}, y < \sqrt{x-1}, y > x^2-1\}$

(j) $\{x > \frac{1}{y}, x > y, x < 4\}$

6.42. Figure 6.6 shows a circle centered at the origin with radius equal to 4, the graph of $y = -\frac{1}{2}|x-1|$, and a parabola that intersects the x-axis at $x = -2$ and $x = 6$ and the y-axis at $y = 4$.

(a) Determine the equation of the circle

(b) Determine the equation of the parabola

(c) Express the shaded area as a representation of the solution set of a system of inequalities

(d) Calculate the domain of the shaded area (i.e., all x values for points in the shaded area)

(e) Calculate the range of the shaded area (i.e., all y values for points in the shaded area).

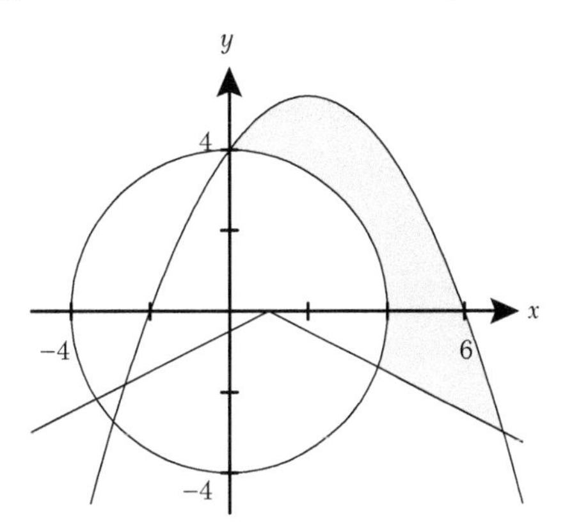

FIGURE 6.6. Diagram for exercise 6.42.

6.43. The two inequalities $y < \frac{1}{x}$ and $xy < 1$ do not have the same solution set (check this!). Sketch the shaded regions representing the solution sets for these two inequalities. (Recall that the graph of the hyperbola $y = \frac{1}{x}$ has two components.)

CHAPTER 7

LIMITS

7.1 INTRODUCTION

The concept of a limiting value of a function plays an important role in calculus, because the formal definition of the derivative of a function at a point in its domain can be expressed as the limiting value of a particular expression involving the function.

The meaning of "limit" in mathematics is more subtle than that in everyday speech. A speed limit that applies on a highway is a speed that motorists may not exceed. The meaning of "limit" in mathematics is similar to that in the following sentence: "In the *minute to win it* competition the contestant was pushed to the limit of his abilities." Thus, a "limit" in mathematics is something (like a number or geometrical figure) that is approached and might or might not be reached.

It is in keeping with the historical approach of this book to begin with the method of exhaustion as an example of an occurrence of a limit in mathematics, as this is the method that Archimedes and other Greek mathematicians of his time used to calculate approximate values of certain areas, for example, the area of a disk. In section 7.3, the concept of a limit is explained carefully using number sequences without giving the completely rigorous treatment (involving ε arguments) that are given in more advanced textbooks. Students of this book will probably not benefit from such a theoretical approach at this stage.

The notion of the left or right limit of a function, introduced simplistically (by reading from a graph) in section 7.4, leads to the definition of continuity of a function in section 7.5. The property of continuity is important because many theorems about functions, for example, the Intermediate Value Theorem (in section 7.7), apply only to functions that are continuous.

Most of the skills that students need to learn in this chapter are introduced in sections 7.6 and 7.8. They are the algebraic skills required for calculating limits. The algebraic methods that have been explained in chapters 3 and 6, such as expanding and factorizing expressions, come into play here.

The study of the graphs of rational functions is one of the topics in this book. In section 7.9, the different kinds of behavior of the graph of a rational function near a vertical asymptote are investigated. This involves the behavior of a rational function $f(x)$ as x approaches a vertical asymptote from the left or the right.

This chapter ends with a demonstration (in section 7.10) of the way the Squeeze Theorem and rules for limits are used to evaluate certain types of limits involving trigonometric ratios. The evaluation of these limits will make it possible to compute the derivatives of trigonometric functions in chapter 8.

7.2 THE METHOD OF EXHAUSTION

The areas of certain geometrical shapes (e.g., a disk) can be calculated as closely as desired by approximating the shape with polygons and computing the areas of the polygons. This is called the *method of exhaustion*. The area of the geometrical shape can be regarded as the limit of the areas of polygons that approximate the shape in a more and more refined way.

EXAMPLE 7.2.1. If a regular polygon is inscribed in a disk (see figure 7.1), then the area of the polygon approximates the area of disk. (Recall that in a regular polygons all the sides have the same length, and they all meet at the same angle.) Such inscribed polygons fill out more and more of the disk as their number of sides increases (i.e., as their sides get shorter and shorter). Figure 7.1 shows an equilateral triangle, regular hexagon, and regular dodecagon inscribed in a disk. If we call these regular polygons a 3-gon, 6-gon, and 12-gon, respectively, then the next regular polygon in our approximative scheme would be a 24-gon. In general, the area of the disk can be approximated by the area of a 3×2^n -gon, for any nonnegative integer n.

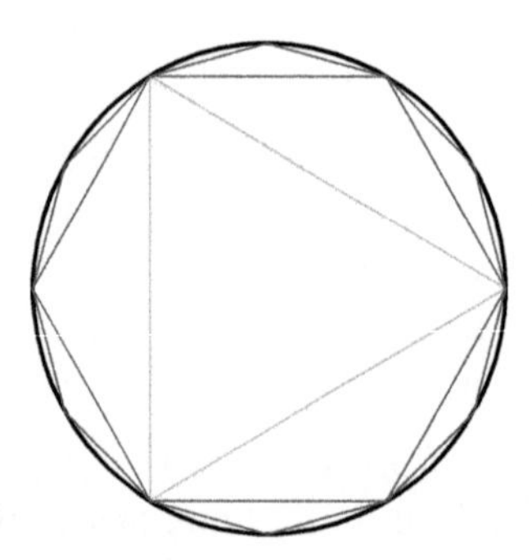

FIGURE 7.1. Polygons inscribed in a disk.

Figure 7.2 shows how we can compute the area of the triangle and the hexagon. We will assume that the radius of the disk is 1. In the diagram on the left, the area of the big triangle is six times the area of the shaded right triangle. The area of the latter is

$$\frac{1}{2}xh=\frac{1}{2}\sin\left(\frac{\pi}{3}\right)\cos\left(\frac{\pi}{3}\right)=\sin\frac{\left(\frac{2\pi}{3}\right)}{4}.$$

Therefore, the area of the big triangle is $\left(\frac{3}{2}\right)\sin\left(\frac{2\pi}{3}\right)=\frac{3\sqrt{3}}{4}\approx 1.299.$

Similarly, in the diagram on the right, the area of the hexagon is twelve times the area of the shaded right triangle. The area of the latter is

$$\frac{1}{2}xh=\frac{1}{2}\sin\left(\frac{\pi}{6}\right)\cos\left(\frac{\pi}{6}\right)=\frac{\sin\left(\frac{\pi}{3}\right)}{4}$$

so the area of the hexagon is $3\sin\left(\frac{\pi}{3}\right)=\frac{3\sqrt{3}}{2}\approx 2.598$. By similar reasoning, the area of the dodecagon in figure 7.1 is $24\left(\frac{1}{4}\sin\left(\frac{\pi}{6}\right)\right)=6\sin\left(\frac{\pi}{6}\right)=3$. The area of a 3×2^n-gon in this progression, beginning with the traingle ($n=0$), is

$$3\cdot 2^{n-1}\sin\left(\frac{\pi}{3\cdot 2^{n-1}}\right),\qquad \text{for}\quad n=0,1,2,3,\ldots$$

(verify this!). We can think of this as a formula that generates an infinite sequence of numbers

$$1.299, 2.598, 3, 3.106, 3.133, \ldots$$

(check these calculations) that approximate the area of a disk with radius 1, and the larger the value of n, the better the approximation. On the other hand, the Archimedean formula for the area of the disk with radius 1 is $\pi(1)^2 = \pi$. Thus, plugging in any value for n in the formula above gives an approximate value for π. In his treatise measurement of a circle, Archimedes did the calculation for $n = 5$ (i.e., by inscribing a 96-gon inside a disk) and found the value of π to be larger than $3\frac{10}{71}$ but less than $3\frac{1}{7}$. ◈

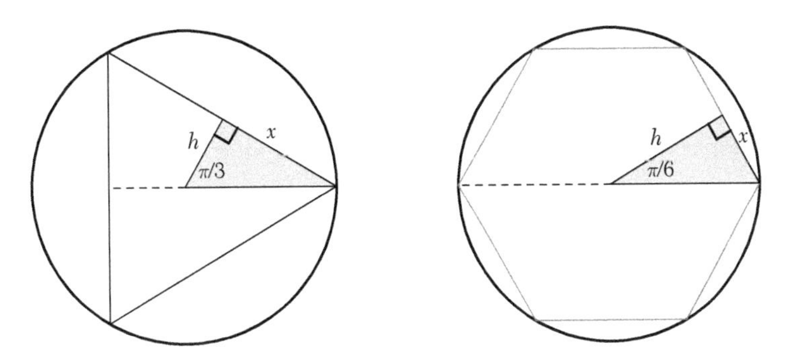

FIGURE 7.2. A triangle and a hexagon inscribed in a disk.

7.3 SEQUENCES

Infinite sequences of numbers, usually called *sequences*, are interesting to mathematicians. The numbers that form a sequence might appear to be random, or they might follow some sort of pattern, in which case the numbers can be given by some kind of formula. We will look at a few examples, in order to develop the concept of a limit.

EXAMPLE 7.3.1. We consider the sequence of numbers given by the formula $a_n = \frac{2^n - 1}{2^n}$, where a_n stands for the nth term of the sequence. Any natural number n may be substituted into the formula to obtain the corresponding term of the sequence. For example, if $n = 3$, then $a_3 = \frac{2^3 - 1}{2^3} = \frac{7}{8}$. Here is the sequence with the first five terms computed:

$$\frac{1}{2}, \frac{3}{4}, \frac{7}{8}, \frac{15}{16}, \frac{31}{32}, \ldots$$

The ellipsis (three dots) at the end indicates that the sequence is infinite (we cannot list all of the terms). Note that each term of the sequence is smaller than the term succeeding it because

$$\frac{2^n - 1}{2^n} = 1 - \frac{1}{2^n} < 1 - \frac{1}{2^{n+1}} = \frac{2^{n+1} - 1}{2^{n+1}}.$$

We want to know whether the sequence has a limit as n gets larger and larger (mathematicians use the expression as n tends to infinity). In other words, we want to know whether the formula $\frac{2^n - 1}{2^n}$ approaches a particular value as n gets larger and larger. It is obvious that $\frac{2^n - 1}{2^n}$ can never be larger than 1, because the numerator is smaller than the denominator.

However, can we conclude that the sequence approaches 1 as n tends to infinity? A mathematical answer to this question requires a mathematical interpretation of the question, which is to phrase the question this way: if we choose any number x smaller than 1, is there an element of the sequence such that it and all elements of the sequence beyond this element are larger than x? For example, if we choose the number $x = 0.9999999999$, is there a natural number N such that for any $n \geq N$ it is the case that $0.9999999999 < a_n \leq 1$? The answer is YES: if $N = 34$, then $a_{34} = \frac{2^{34}-1}{2^{34}} = 0.99999999994172...$, which is clearly larger than 0.9999999999, and all terms of the sequence beyond a_{34} will be larger than a_{34}.

We see that answering the question amounts to finding a real number t that solves the equation $\frac{2^t-1}{2^t} = 0.9999999999$ and then taking N to be any natural number bigger than t. In general, we can solve for t in the equation $\frac{2^t-1}{2^t} = x$, where x is any positive real number less than 1 (and as close to 1 as we would like it to be), and then we take N to be any natural number larger than t, so that $x < a_N < 1$. (The equation $\frac{2^t-1}{2^t} = x$ can, in fact, be solved using basic algebra and taking a logarithm. Do it yourself!)

Our conclusion is that the limit of the sequence is 1. The mathematical notation we use for this statement is

$$\lim_{n\to\infty} a_n = \lim_{n\to\infty} \frac{2^n-1}{2^n} = 1$$

This should be read as: "The limit as n tends to infinity of a_n is 1." This notation will be used again. ◈

EXAMPLE 7.3.2. As a variation of example 7.3.1, consider the sequence

$$\frac{1}{2}, 1, \frac{3}{4}, 1, \frac{7}{8}, 1, \frac{15}{16}, 1, \frac{31}{32}, \ldots$$

which consists of the terms of the sequence in example 7.3.1 alternating with "1." As in example 7.3.1, for any given value $x < 1$, we can find an odd natural number N such that for any $n \geq N$ it is the case that $x < a_n \leq 1$. Therefore, we conclude again that the limit of the sequence is 1. (It does not matter that infinitely many terms of the sequence are also equal to 1.) ◈

EXAMPLE 7.3.3. As a variation of example 7.3.2, consider the sequence

$$\frac{1}{2}, 1, \frac{3}{4}, 1, \frac{7}{8}, 1, 1, 1, \ldots \text{ (all terms equal to 1).}$$

in which all terms of the sequence beyond the fifth term are equal to 1. In this case, it is trivially true that $x < a_n \leq 1$ if $n \geq 6$, for any value $x < 1$, so here we conclude again that the limit of the sequence is 1. ◈

EXAMPLE 7.3.4 As another variation of example 7.3.1, consider the sequence $a_n = \frac{2^n + (-1)^n}{2^n}$, for which the first seven terms are as follows:

$$\frac{1}{2}, \frac{5}{4}, \frac{7}{8}, \frac{17}{16}, \frac{31}{32}, \frac{65}{64}, \ldots$$

The terms of this sequence alternately rise above and fall below the number 1. The limit of this sequence is also 1, but it is slightly more complicated to prove this using the method of example 7.3.1 ◈

There are sequences that do not have a limit. Here is an example:

EXAMPLE 7.3.5 Let $a_n = (-1)^n \left(\frac{2^n + (-1)^n}{2^n} \right)$, for which the first seven terms are

$$-\frac{1}{2}, \frac{5}{4}, -\frac{7}{8}, \frac{17}{16}, -\frac{31}{32}, \frac{65}{64}, \ldots$$

The negative terms approach –1, whereas the positive terms approach 1, so the sequence has no limit. ◈

A great deal more can be said about sequences and limits of sequences, but now we move on to limits of functions.

7.4. LIMITS OF A FUNCTION

The following three examples introduce what is meant, graphically, by the limiting value of a function $f(x)$ if x approaches a particular value.

EXAMPLE 7.4.1. The graph of a piecewise-defined function $f(x)$ with different definitions to the left and right of $x = 1$ is shown in figure 7.3. According to the graph, we see that $f(x)$ approaches the value 1 as x approaches 1 from the left, and $f(x)$ approaches the value –2 as x approaches 1 from the right. We write this mathematically as follows:

$$\lim_{x \to 1^-} f(x) = 1, \quad \lim_{x \to 1^+} f(x) = -2.$$

The minus sign in the first case means that the approach is from the left (i.e., x approaches 1 from numbers smaller than 1), and the plus sign in the second case means that the approach is from the right (i.e., x approaches 1 from numbers larger than 1). If we omit the direction indicator in the limit, that is, if we do not specify the direction of approach (from the left or the right), then we write

$$\lim_{x \to 1} f(x) \text{ does not exist.}$$

(The abbreviation d.n.e. can be used instead.) The reason the limit does not exist is that the left and right limits are different (1 and –2, respectively). ◈

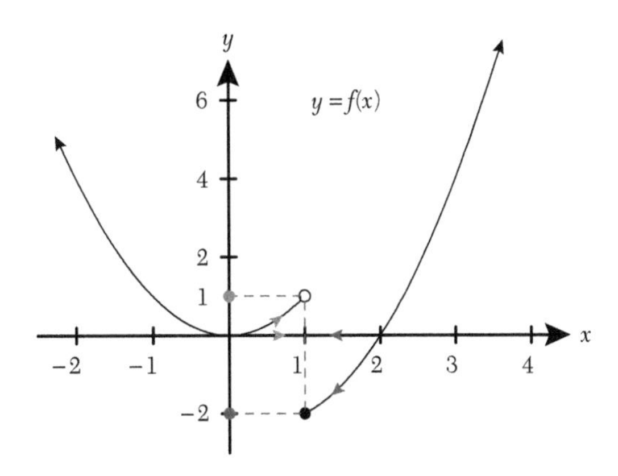

FIGURE 7.3. Limits from the left and the right.

EXAMPLE 7.4.2. The graph of a piecewise-defined function $f(x)$ with different definitions to the left and right of $x = 2$ is shown in figure 7.4. According to the graph, $f(x)$ approaches the value 1.4 as

x approaches 2 from the left, and $f(x)$ approaches the value 1.4 as x approaches 2 from the right. We write this mathematically as follows:

$$\lim_{x\to 2^-} f(x) = 1.4, \quad \lim_{x\to 2^+} f(x) = 1.4.$$

Therefore, as $f(x)$ approaches the same value 1.4 as x approaches 2 from both directions, we can omit the direction indicator in the limit. That is, we can write

$$\lim_{x\to 2} f(x) = 1.4.$$

(This means the limit above exists and is equal to 1.4, in contrast to example 7.4.1, where the limit did not exist.) ◈

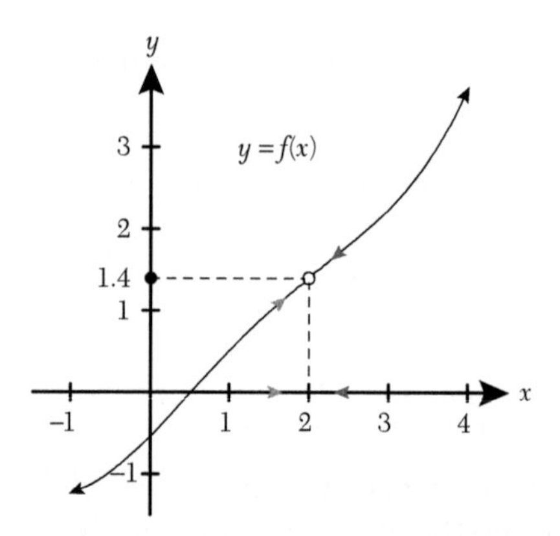

FIGURE 7.4. The limit from the left equals the limit from the right.

EXAMPLE 7.4.3. The graph of a function $f(x)$ with a vertical asymptote at $x = -1$ is shown in figure 7.5. Here $f(x)$ approaches negative infinity as x approaches –1 from the left, and $f(x)$ approaches positive infinity as x approaches –1 from the right. We write this mathematically as follows:

$$\lim_{x\to(-1)^-} f(x) = -\infty \text{ and } \lim_{x\to(-1)^+} f(x) = \infty.$$

If the direction indicator is omitted, then the limit does not exist, that is,

$$\lim_{x\to -1} f(x) \text{ d.n.e.}$$ ◈

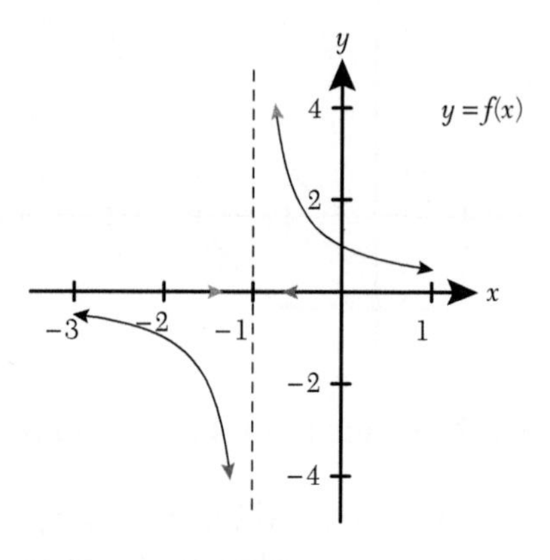

FIGURE 7.5. Diagram for Example 7.4.3.

The following concrete examples will show how these different situations can arise:

EXAMPLE 7.4.4. Given $f(x)=\frac{x-1}{x^2-1}$, we determine a piecewise expression for f and sketch the graph of f in figure 7.6:

$$f(x)=\begin{cases} \dfrac{1}{x+1} & \text{if } -\infty<x<-1 \\ \dfrac{1}{x+1} & \text{if } -1<x<1 \\ \dfrac{1}{x+1} & \text{if } 1<x<\infty \end{cases}.$$

We have the following limits:

$$\lim_{x\to(-1)^-} f(x)=-\infty,\quad \lim_{x\to(-1)^+} f(x)=\infty,\quad \lim_{x\to 1} f(x)=\frac{1}{2}.$$

◈

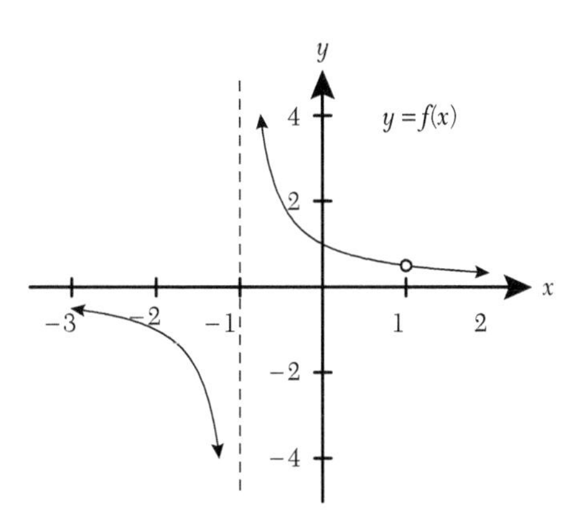

FIGURE 7.6.

EXAMPLE 7.4.5. A variation of example 7.4.4 is shown in figure 7.7. If

$$f(x)=\begin{cases} \dfrac{x-1}{x^2-1} & \text{if } x\neq\pm1 \\ 1 & \text{if } x=\pm1 \end{cases}$$

then $f(x)$ can be expressed as a piecewise-defined function as follows:

$$f(x)=\begin{cases} \dfrac{1}{x+1} & \text{if } -\infty<x<-1 \\ 1 & \text{if } x=-1 \\ \dfrac{1}{x+1} & \text{if } -1<x<1 \\ 1 & \text{if } x=1 \\ \dfrac{1}{x+1} & \text{if } 1<x<\infty \end{cases}.$$

Note that the domain of f is $\{x\in\mathbb{R}\}$. As provided in example 7.4.4, we have the same limits at $x=-1$ and $x=1$.

◈

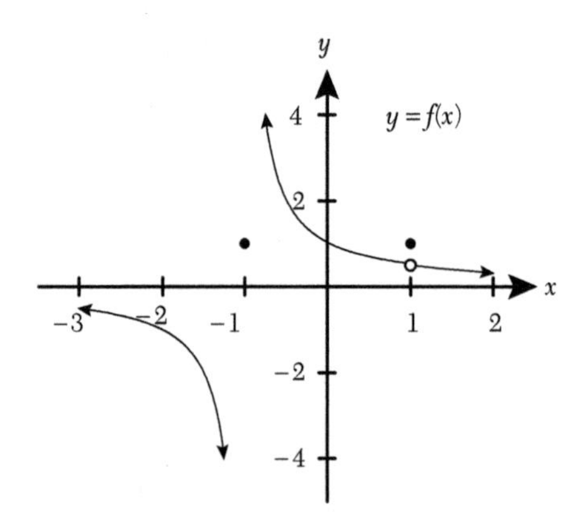

FIGURE 7.7. Diagram for Example 7.4.5.

EXAMPLE 7.4.6 Let $f(x)=\frac{1}{x-1}$ and $g(x)=\frac{x-1}{x+1}$. Then, a composition function h is defined as follows: $h(x)=f(g(x))=-\frac{1+x}{2}$, with domain $\{x \in \mathbb{R} \mid x \neq -1\}$. Here, $h(x)$ has the following piecewise definition:

$$h(x)=\begin{cases} -\dfrac{x+1}{2} & \text{if } x<-1 \\ -\dfrac{x+1}{2} & \text{if } x>-1 \end{cases}.$$

In figure 7.8, we see that $\lim_{x\to(-1)^-} h(x)=0$ and $\lim_{x\to(-1)^+} h(x)=0$. As the left and right limits are the same, we write $\lim_{x\to(-1)} h(x)=0$. ◈

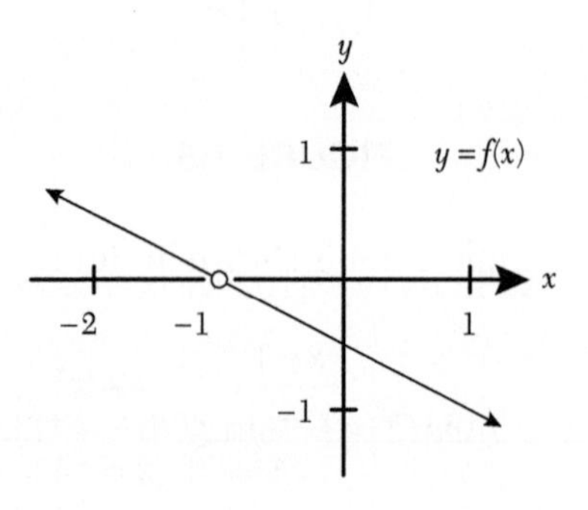

FIGURE 7.8. Diagram for Example 7.4.6.

EXAMPLE 7.4.7. If

$$f(x)=\begin{cases} \dfrac{x}{|x|} & \text{if } x\neq 0 \\ 0 & \text{if } x=0 \end{cases}$$

then we can rewrite f as the following piecewise-defined function:

$$f(x)=\begin{cases} -1 & \text{if } x<0 \\ 0 & \text{if } x=0 \\ 1 & \text{if } x>0 \end{cases}.$$

The graph of $y = f(x)$ is shown in figure 7.9. Note that the domain of f is $\{x \in \mathbb{R}\}$ and $f(0) = 0$. The left and right limits at $x = 0$ are, respectively, $\lim_{x \to 0^-} f(x) = -1$ and $\lim_{x \to 0^+} f(x) = 1$. Because the left and right limits are different, $\lim_{x \to 0} f(x)$ does not exist. ◈

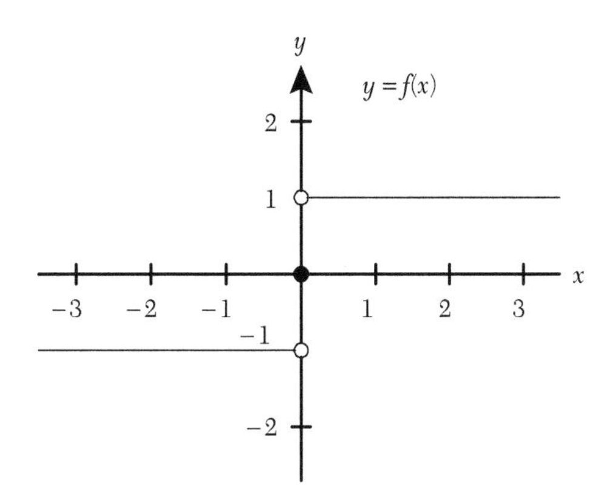

FIGURE 7.9. Diagram for Example 7.4.7.

EXAMPLE 7.4.8. We define functions f and g as follows:

$$f(x) = \begin{cases} x^2 - x + 2 & \text{if } x < 1 \\ 2 & \text{if } x = 1 \\ x^2 - x + 2 & \text{if } x > 1 \end{cases}$$

$$g(x) = \begin{cases} x^2 - x + 2 & \text{if } x < 1 \\ 3 & \text{if } x = 1 \\ x^2 - x + 2 & \text{if } x > 1 \end{cases}.$$

The graph of $y = g(x)$ is shown in figure 7.10. We compute the following limits for the functions f and g:

$$\lim_{x \to 0^-} f(x) = \lim_{x \to 0^-} g(x) = 2$$

$$\lim_{x \to 0^+} f(x) = \lim_{x \to 0^+} g(x) = 2.$$

However, the values of the functions f and g at $x = 1$ differ because $f(1) = 2$, whereas $g(1) = 3$. ◈

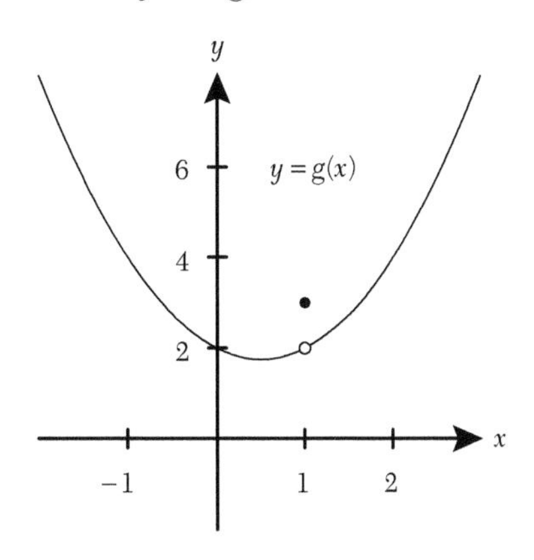

FIGURE 7.10. Diagram for Example 7.4.8.

7.5 CONTINUITY

The assumption of *continuity* of a function is a prerequisite for many of the theorems of calculus. The property of continuity of a function relates to *smoothness* of the function. Just as we might describe an object as smooth in everyday life, we might think of a function as being smooth if the function is "nice," or "well behaved," in some mathematical sense. For example, we will see in this section that if a function is continuous (smooth), then limits that involve the function can be evaluated by substitution into the function.

7.5.1 Definition of Continuity at a Point

DEFINITION 7.5.1. *If a point* a *is in the domain of a function* f(x), *and, if it is the case that*

$$\lim_{x \to a^-} f(x) = \lim_{x \to a^+} f(x) = f(a)$$

that is, both left and right limits at $x = a$ *are equal to the function value at* $x = a$, *then we say that the function* f(x) *is continuous at* $x = a$. *This statement can be abbreviated as*

$$\boxed{\lim_{x \to a} f(x) = f(a)} \tag{7.1}$$

because the left and right limits are the same. Formula (7.1) is called the equation of continuity for f(x) *at the point* $x = a$.

DEFINITION 5.2. *If it is the case that*

$$\boxed{\lim_{x \to a^-} f(x) = f(a)} \tag{7.2}$$

then f(x) *is said to be continuous from the left* at $x = a$, and if it is the case that

$$\boxed{\lim_{x \to a^+} f(x) = f(a)} \tag{7.3}$$

then f(x) *is said to be continuous from the right at* $x = a$. *Formulas (7.2) and (7.3) are called the equations of left and right continuity, respectively, for* f(x) *at* $x = a$.

EXAMPLE 7.5.1. In example 7.4.8, *f* is continuous at $x = 1$, according to definition 7.5.1, but g is not continuous at $x = 1$. ◈

7.5.2 Discontinuity at a Point

DEFINITION 7.5.3. *If a point* $x = a$ *is in the domain of a function* f(x) *and the equation of continuity fails at* $x = a$, *then we say that the function* f(x) *has a discontinuity at the point* $x = a$. *The discontinuity can be one of three types: removable, jump, or infinite discontinuity.*

DEFINITION 7.5.4. *A discontinuity is removable at* $x = a$ *in the domain of a function* f(x) *if the left and right limits at* $x = a$ *exist and are equal but not equal to* f(a). *A function* f(x) *has a jump discontinuity at* $x = a$ *in its domain if the left and right limits at* $x = a$ *exist but are not equal. A function* f(x) *has an infinite discontinuity at* $x = a$ *in its domain if either the left or right limit at* $x = a$ *is plus or minus infinity or both limits are plus or minus infinity.*

EXAMPLE 7.5.2. In example 7.4.4 (and figure 7.6), the values $x = -1$ and $x = 1$ are not in the domain of $f(x)$, so $f(x)$ has no discontinuities at these values (there is nothing to say regarding continuity at these values because the function is not defined at these values); however, in example 7.4.5 (and figure 7.7), $f(x)$ has an infinite discontinuity at $x=-1$, because

$$\lim_{x\to(-1)^-} f(x) = \lim_{x\to(-1)^+} f(x) = -\infty.$$

and a removable discontinuity at $x=1$, because

$$\lim_{x\to1^-} f(x) = \lim_{x\to1^+} f(x) = \frac{1}{2} \neq f(1) = 1.$$

In example 7.4.7 (and figure 7.9), $f(x)$ has a jump discontinuity at $x=0$, because

$$\lim_{x\to0^-} f(x) = -1 \neq \lim_{x\to0^+} f(x) = 1.$$

In example 7.4.8 (and figure 7.10), $g(x)$ has a removable discontinuity at $x=1$, because

$$\lim_{x\to1^-} f(x) = \lim_{x\to1^+} f(x) = 2 \neq f(1) = 3.$$ ◈

EXAMPLE 7.5.3. An example of a function with jump discontinuities is the *floor function*. It maps every integer to itself and every noninteger real number x to the largest integer preceding it. The notations that are used for this function are $\text{floor}(x)$ and $\lfloor x \rfloor$. The graph of $y = f(x) = \lfloor x \rfloor$ in figure 7.11 shows a jump discontinuity at every integer. A function that is related to the floor function is the *sawtooth function*. Its definition is $f(x) = x - \lfloor x \rfloor$. You can see from the second graph in figure 7.11 how this function gets its name. ◈

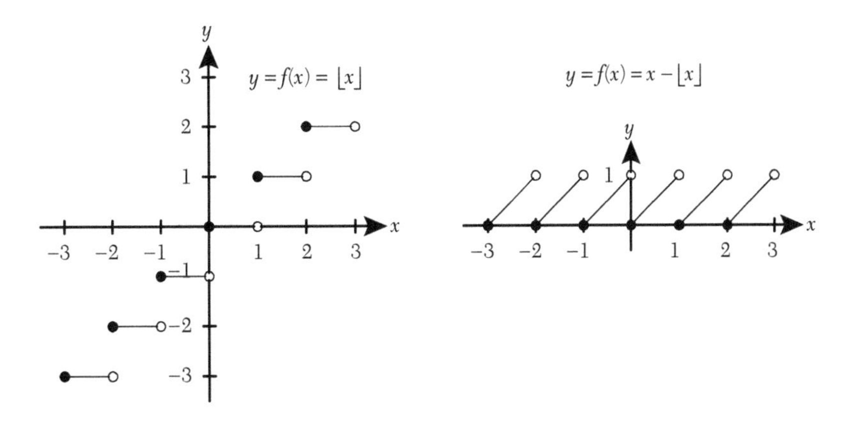

FIGURE 7.11. The floor function and the sawtooth function.

7.5.3 Continuity on an Interval

DEFINITION 7.5.5. *A function* $f(x)$ *is continuous on an interval* I *if it is continuous at each point in the interval (in terms of the definition of continuity at a point).*

REMARK 7.5.1. It is tempting to think that continuity on an interval I means "you can draw the graph of the function on the interval I without picking up your pen." However, this notion is too simplistic, in general. For example, the function

$$f(x) = \begin{cases} x\sin\left(\frac{1}{x}\right) & \text{if } x \neq 0 \\ 0 & \text{if } x = 0 \end{cases}$$

is continuous on the interval $I = (-1,1)$ according to the definition above (why?), but the graph has infinitely many "wiggles" near $x = 0$ so you cannot draw the graph on I. The graph is shown in figure 7.12.

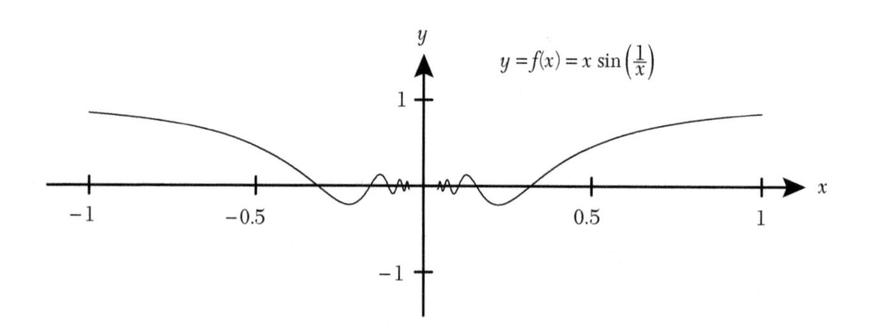

FIGURE 7.12. Diagram for Remark 7.5.1.

7.5.4 Continuous Functions

DEFINITION 7.5.6. *A continuous function is a function that is continuous at every point in its domain.*

All the standard functions that are studied in calculus are continuous. (This can be proved using methods of *real analysis* that is too advanced to explain here.) The standard functions are trigonometric, polynomials, rational, the absolute value, root, exponential, and logarithmic. The significance of this is that the equation of continuity or the equations of left and right continuity, that is, formulas (7.1)–(7.3), can be used to evaluate limits for any of these functions.

7.5.5 More Continuous Functions

Any finite number of algebraic operations (adding, subtracting, multiplying, and dividing) performed on continuous functions produces new continuous functions. For example, if $f(x) = x$ and $g(x) = \cos(x)$, then $f(x) \cdot g(x) = x\cos(x)$ is continuous and $\frac{f(x)}{g(x)} = \frac{x}{\cos(x)}$ is continuous.

Another way to produce new continuous functions is to form compositions of continuous functions. For example, the functions $f(x) = \sqrt[3]{x}$ and $g(x) = \cos(x)$ are continuous for all real values of x, so the composition $h(x) = f(g(x)) = \sqrt[3]{\cos(x)}$ is also continuous for all real values of x.

7.6 COMPUTING LIMITS

The conceptual and practical difficulties relating to the computation of limits form an important topic in the branch of mathematics called *analysis*. We are now going to look at the evaluation of limits in a few situations.

7.6.1 Limits in the Domain of a Continuous Function

In each of the following examples, the limit involves a continuous function and a value in the domain of the function, so the limit is computed by substitution into the function. In other words, the evaluation of the limit is an application of the equation of continuity (formula (7.1)) that defines continuity at a point.

EXAMPLE 7.6.1.

(i) $\lim_{x\to 3} 4x^2 + 3x - 1 = 4(3)^2 + 3(3) - 1 = 44$

Here, $f(x) = 4x^2 + 3x - 1$ is a continuous function because it is a polynomial.

(ii) $\lim_{t\to -1} (t^3 + 6t - 8)^{\frac{2}{3}} = ((-1)^3 + 6(-1) - 8)^{\frac{2}{3}} = (-15)^{\frac{2}{3}} = \sqrt[3]{225}$

Here, $f(t) = (t^3 + 6t - 8)^{\frac{2}{3}}$ is a continuous function because it is the composition of a root function with a polynomial.

(iii) $\lim_{s\to 2} \sqrt[7]{\frac{3s}{2} - s^2} + (1 + s^3)^2 = \lim_{s\to 2} \sqrt[7]{\frac{3(2)}{2} - 2^2} + (1 + 2^3)^2 = -1 + 81 = 80$

Here, $f(s) = \sqrt[7]{\frac{3s}{2} - s^2} + (1 + s^3)^2$ is a continuous function because it is the sum of compositions of continuous functions. ◈

7.6.2 Limits Involving Piecewise-Defined Functions

In the examples below, we compute the limits (where they exist) of piecewise-defined functions at the points where the definitions of the functions change. Left and right limits can be computed by substitution into the appropriate definition of the function at these points.

EXAMPLE 7.6.2.

$$f(x) = \begin{cases} x^2 + 6x - 72 & \text{if } x < 6 \\ \sqrt{x-6} & \text{if } x > 6 \end{cases}$$

$$\lim_{x\to 6^-} f(x) = \lim_{x\to 6^-} x^2 + 6x - 72 = 0, \quad \text{and} \qquad \lim_{x\to 6^+} f(x) = \lim_{x\to 6^+} \sqrt{6-6} = 0.$$

We conclude that $\lim_{x\to 6} f(x) = 0$ (because left and right limits are both equal to 0). ◈

EXAMPLE 7.6.3.

$$f(x) = \begin{cases} \sqrt{\csc\left(x + \frac{\pi}{2}\right) + (1+x)^2} & \text{if } x \le 0 \\ e^{\frac{1}{2}\ln(2+x)} & \text{if } x > 0 \end{cases}$$

$$\lim_{x\to 0^-} f(x) = \lim_{x\to 0^-} \sqrt{\csc\left(x + \frac{\pi}{2}\right) + (1+x)^2} = \sqrt{\csc\left(0 + \frac{\pi}{2}\right) + (1+0)^2} = \sqrt{2}$$

$$\lim_{x\to 0^+} f(x) = \lim_{x\to 0^+} e^{\frac{1}{2}\ln(2+x)} = e^{\frac{1}{2}\ln(2+0)} = e^{\ln\sqrt{2}} = \sqrt{2}.$$

We conclude that $\lim_{x\to 0} f(x) = \sqrt{2}$ (because left and right limits are both equal to $\sqrt{2}$). ◈

EXAMPLE 7.6.4.

$$f(x)=\begin{cases} \frac{1}{2\pi}x+\frac{1}{2} & \text{if } x\leq-\pi \\ \cos\left(\frac{x}{2}\right) & \text{if } -\pi<x\leq\pi \\ \frac{1}{2\pi}x+\frac{1}{2} & \text{if } x>\pi \end{cases}$$

$$\lim_{x\to-\pi^-} f(x)=\lim_{x\to-\pi^-}\frac{1}{2\pi}x+\frac{1}{2}=\frac{1}{2\pi}(-\pi)+\frac{1}{2}=0$$

$$\lim_{x\to-\pi^+} f(x)=\lim_{x\to-\pi^+}\cos\left(\frac{x}{2}\right)=\cos\left(\frac{-\pi}{2}\right)=0$$

$$\lim_{x\to\pi^-} f(x)=\lim_{x\to\pi^-}\cos\left(\frac{x}{2}\right)=\cos\left(\frac{\pi}{2}\right)=0$$

$$\lim_{x\to\pi^+} f(x)=\lim_{x\to\pi^+}\frac{1}{2\pi}x+\frac{1}{2}=\frac{1}{2\pi}(\pi)+\frac{1}{2}=1.$$

We conclude that $\lim_{x\to-\pi} f(x)=0$ but $\lim_{x\to\pi} f(x)$ does not exist. ◈

7.6.3 Computing Limits by Simplification

If a point $x=a$ is not in the domain of a function $f(x)$, then $\lim_{x\to a} f(x)$ might or might not exist. We will look at some examples where it is possible to simplify the limit and, consequently, the limit can be evaluated by means of a substitution, as in section 7.6.1.

EXAMPLE 7.6.5. The domain of the rational function

$$f(x)=\frac{x^2+5x+4}{x^2+x-12}$$

excludes $x=-4$. If we try to evaluate $\lim_{x\to-4} f(x)$ by substitution, the result is

$$\frac{(-4)^2+5(-4)+4}{(-4)^2+(-4)-12}=\text{"}\left(\frac{0}{0}\right)\text{"}.$$

We have placed this result in quotes because division by zero is not allowed. However, we refer to "$\frac{0}{0}$" as the type of the limit and it does give us a clue about how to evaluate the limit because the Factor Theorem (theorem 3.4.2) tells us that $(x+4)$ is a factor of both of the polynomials in the numerator and denominator of $f(x)$. If we proceed by factorizing the numerator and denominator of $f(x)$, then we obtain

$$f(x)=\frac{(x+4)(x+1)}{(x+4)(x-3)}.$$

Clearly, a function that is related to $f(x)$ is the function

$$g(x)=\frac{x+1}{x-3}.$$

Note that $f(x)$ and $g(x)$ are not the same function because the domain of f is $\{x\in\mathbb{R} \mid x\neq-4,\ x\neq3\}$, while the domain of g is $\{x\in \boldsymbol{R} \mid x\neq3\}$. As we are interested in computing $\lim_{x\to-4} f(x)$, it does not matter that $f(x)$ is not defined at $x=-4$. What does matter is finding the value that $f(x)$

approaches as x approaches –4. Therefore, it is helpful to write the following piecewise definition for $f(x)$:

$$f(x)=\begin{cases} g(x) & \text{if } x<-4 \\ g(x) & \text{if } x>-4 \end{cases}.$$

We can do this because we are excluding the value $x=-4$ at which $f(x)$ and $g(x)$ differ. We now see that

$$\lim_{x\to-4^-} f(x)=\lim_{x\to-4^-} g(x)=g(-4)=\frac{(-4)+1}{(-4)-3}=\frac{3}{7},$$

$$\lim_{x\to-4^+} f(x)=\lim_{x\to-4^+} g(x)=g(-4)=\frac{(-4)+1}{(-4)-3}=\frac{3}{7},$$

and so

$$\lim_{x\to-4} f(x)=\lim_{x\to-4} g(x)=g(-4)=\frac{3}{7}.$$

It is easier (although the underlying reasoning should be remembered) to write the evaluation of this limit by means of the following steps:

$$\begin{aligned}\lim_{x\to-4} f(x) &= \lim_{x\to-4}\frac{x^2+5x+4}{x^2+x-12} \\ &= \lim_{x\to-4}\frac{(x+4)(x+1)}{(x+4)(x-3)} \\ &= \lim_{x\to-4}\frac{(x+1)}{(x-3)} \\ &= \frac{(-4)+1}{(-4)-3} \\ &= \frac{3}{7}.\end{aligned}$$

◈

EXAMPLE 7.6.6.

$$\begin{aligned}\lim_{x\to-1}\frac{x^2-x-2}{x+1} &= \lim_{x\to-1}\frac{(x-2)(x+1)}{(x+1)} \\ &= \lim_{x\to-1}(x-2) \\ &= (-1)-2 \\ &= -3.\end{aligned}$$

◈

EXAMPLE 7.6.7.

$$\begin{aligned}\lim_{t\to1}\frac{t^3-1}{t^2-1} &= \lim_{t\to1}\frac{(t-1)(t^2+t+1)}{(t-1)(t+1)} \\ &= \lim_{t\to1}\frac{t^2+t+1}{t+1} \\ &= \frac{(1)^2+(1)+1}{(1)+1} \\ &= \frac{3}{2}.\end{aligned}$$

◈

The next example is typical of the expression of a limit for computing the derivative of a function (as we will see later).

EXAMPLE 7.6.8.

$$\begin{aligned}
\lim_{h\to 0}\frac{(3+h)^2-9}{h} &= \lim_{h\to 0}\frac{9+6h+h^2-9}{h} \\
&= \lim_{h\to 0}\frac{h(6+h)}{h} \\
&= \lim_{h\to 0} 6+h \\
&= 6+(0) \\
&= 6.
\end{aligned}$$

◈

The same method of canceling common factors across the numerator and denominator also works with nonrational expressions, as we demonstrate in the following examples.

EXAMPLE 7.6.9.

$$\begin{aligned}
\lim_{u\to 9}\frac{9-u}{3-\sqrt{u}} &= \lim_{u\to 9}\frac{(3-\sqrt{u})(3+\sqrt{u})}{3-\sqrt{u}} \\
&= \lim_{u\to 9} 3+\sqrt{u} \\
&= 3+\sqrt{9} \\
&= 6.
\end{aligned}$$

◈

EXAMPLE 7.6.10.

$$\begin{aligned}
\lim_{x\to 0}\frac{\sqrt{2-x}-\sqrt{2}}{x} &= \lim_{x\to 0}\frac{\sqrt{2-x}-\sqrt{2}}{x}\left(\frac{\sqrt{2-x}+\sqrt{2}}{\sqrt{2-x}+\sqrt{2}}\right) \\
&= \lim_{x\to 0}\frac{(\sqrt{2-x})^2-(\sqrt{2})^2}{x(\sqrt{2-x}+\sqrt{2})} \\
&= \lim_{x\to 0}\frac{2-x-2}{x(\sqrt{2-x}+\sqrt{2})} \\
&= \lim_{x\to 0}\frac{-x}{x(\sqrt{2-x}+\sqrt{2})} \\
&= \lim_{x\to 0}\frac{-1}{\sqrt{2-x}+\sqrt{2}} \\
&= \frac{-1}{\sqrt{2-(0)}+\sqrt{2}} \\
&= \frac{-1}{2\sqrt{2}}.
\end{aligned}$$

In the first step above, a technique for rationalizing the numerator was used. This technique involves multiplying and dividing by a conjugate radical. (In this case, the conjugate radical is "$\sqrt{2-x}+\sqrt{2}$.") A "conjugate radical" is not a bad-tasting vegetable: the word conjugate refers to the sign between the terms $\sqrt{2-x}$ and $\sqrt{2}$ in the conjugate radical being the opposite ("+," in this case) from the sign that appears in the numerator of the limit being computed, and the word radical refers to the presence of root terms "$\sqrt{2-x}$" and "$\sqrt{2}$." ◈

7.7 APPLICATIONS OF CONTINUITY

Continuous functions are important because of their mathematical properties, which we will begin to explore in this chapter. We have already seen a connection between continuity and the evaluation of limits.

Many processes in the real world happen in a continuous way (without breaks and jumps), and so these processes can be represented by continuous functions. For example, if a hiker walks from the bottom to the top of a mountain, then his altitude is a continuous function of the time elapsed. His altitude as a function of the horizontal distance he has walked could also be a continuous function (this function would not be continuous if the hiker has to climb up a vertical rope at a cliff face!).

7.7.1 The Intermediate Value Theorem

We now state an important theorem relating to continuous real-valued functions:

Theorem 7.7.1. The Intermediate Value Theorem: *if a real-valued function* $f(x)$ *is continuous on a closed interval* $[a,b]$ *and* v *is any real number between* $f(a)$ *and* $f(b)$ *(assuming* $f(a) \neq f(b)$*), then there exists a real number* c *in the interval* (a,b) *such that* $f(c) = v$.

The proof of the Intermediate Value Theorem (IVT) uses methods of real analysis that are too advanced to be presented here, so the proof is omitted.

REMARK 7.7.1. Loosely speaking, the IVT states that, if the domain of a continuous real-valued function is a closed interval I, then the range of the function always contains the closed interval determined by the values of the function at the end points of the interval I. For example, if the domain of a continuous real-valued function is the closed interval [2,3], then its range contains the interval $[f(2), f(3)]$ or $[f(3), f(2)]$, depending on whether $f(2) < f(3)$ or $f(3) < f(2)$. (In the case that $f(2) = f(3)$, the IVT does not say anything.)

The assumption of continuity of a real-valued function on a closed interval is essential for the conclusion of the IVT to be valid. Here are a few examples that demonstrate this:

EXAMPLE 7.7.1. Consider the floor function $f(x) = \lfloor x \rfloor$ on the closed interval [0.5, 1.5]. We know that $f(x)$ has a jump discontinuity at $x = 1$, where the value of $f(x)$ changes from 0 to 1, and so, for every real number v between 0 and 1, there is no real number c between 0.5 and 1.5 such that $f(c) = v$. ◈

EXAMPLE 7.7.2. Recall from example 7.4.8 (and figure 7.10) that the function

$$f(x) = \begin{cases} x^2 - x + 2 & \text{if } x < 1 \\ 3 & \text{if } x = 1 \\ x^2 - x + 2 & \text{if } x > 1 \end{cases}$$

has a removable discontinuity at $x = 1$. The conclusion of the IVT does not apply on the interval [0.5,3], where $f(0.5) = 1.75$ and $f(3) = 8$ because there is no number c in the interval (0.5,3) for which $f(c) = 2$. ◈

REMARK 7.7.2. The IVT is an existence theorem because it guarantees the existence of at least one real number c in the interval $[a,b]$ with the property $f(c) = v$. This does not exclude the possibility of there being more than one value between a and b with this property. For example, we can consider the function $f(x) = \sin(x)$ on the interval $\left[\frac{-\pi}{2}, \frac{5\pi}{2}\right]$. Because $f\left(\frac{-\pi}{2}\right) = -1$ and $f\left(\frac{5\pi}{2}\right) = 1$, the IVT guarantees the existence of at least one number $x=c$ in the interval $\left(\frac{-\pi}{2},\frac{5\pi}{2}\right)$ such that $f(c)=0$; however, we know there are three such numbers: $c=0$, $c=\pi$, and $c=2\pi$.

REMARK 7.7.3. The converse of the IVT is not true, meaning that a function that satisfies the conclusion of the IVT on a closed interval need not be a continuous function. We can illustrate this with the function

$$f(x)=\begin{cases}\sin\frac{1}{x} & \text{if } x\neq 0\\ 0 & \text{if } x=0\end{cases}$$

on the interval $\left[\frac{-2}{\pi},\frac{2}{\pi}\right]$. (We can, in fact, consider any interval containing $x=0$.) As x tends to zero, from either the right or the left, the function $f(x)$ will oscillate between –1 and 1 infinitely many times (the oscillations become more and more compressed) and so any value v between –1 and 1 will be the image of infinitely many points in $\left(\frac{-2}{\pi},\frac{2}{\pi}\right)$. Yet, $f(x)$ is not a continuous function on the interval $\left[\frac{-2}{\pi},\frac{2}{\pi}\right]$. In particular, $f(x)$ is not continuous at $x=0$ because $\lim_{x\to 0^-}\sin\frac{1}{x}$ and $\lim_{x\to 0^+}\sin\frac{1}{x}$ do not exist (why?).

We demonstrate by means of the following examples that the IVT can be used to solve certain existence problems, such as proving that an equation has a solution.

EXAMPLE 7.7.3. Show that the equation $3x+\cos(\pi x)=6$ has a solution in the interval $(0,2)$.

Answer: Define the function $f(x)=3x+\cos(\pi x)$. Because $f(x)$ is a continuous function of the interval $[0,2]$, and $f(0)=1$ and $f(2)=7$, there must exist, by the IVT, a number c such that $0<c<2$ and $f(c)=6$ (because $v=6$ is between 1 and 7). ◈

EXAMPLE 7.7.4. Show that the equation $\ln(x)=\frac{1}{\sqrt{x}}$ has a solution on the interval $[1,4]$.

Answer: Define the function $f(x)=\ln(x)-\frac{1}{\sqrt{x}}$. Because $f(x)$ is a continuous function of the interval $[1,4]$, and $f(1)=\ln(1)-\frac{1}{\sqrt{1}}=-1$ and $f(4)=\ln(4)-\frac{1}{\sqrt{4}}\approx 0.886$ there must exist, by the IVT, a number c such that $1<c<4$ and $f(c)=0$ (because $v=0$ is between –1 and 0.866). ◈

The IVT can be used repeatedly to approach closer and closer to the solution of an equation:

EXAMPLE 7.7.5. Show that there is a solution of the equation $3x^3-2x^2+4x-3=0$ in the interval $[0,1]$. Find an interval of width less than 0.01 that contains this solution.

Answer: Define the function $f(x)=3x^3-2x^2+4x-3$. Because $f(0)=-3$ and $f(1)=2$, the equation $f(x)=0$ has a solution in the interval $(0,1)$ (because $-3<0<2$). If you plot the graph of $y=f(x)$, you will find that this is the only real solution of the equation (the other two solutions are complex). We can narrow the search for this solution by subdividing the interval $[0,1]$ into the intervals $[0,0.5]$ and $[0.5,1]$. Because $f(0.5)=-1.125$, the solution is in the interval $(0.5,1)$. Continuing in this way, we evaluate the function at the midpoint of each interval we obtain and, depending of the sign, we take the subinterval to the left or the right of the midpoint. Thus, we find that the solution is in each of the following intervals that are successively smaller: (0.5, 0.75), (0.625, 0.75), (0.6875, 0.75), (0.71875, 0.75), (0.71875, 0.734375), (0.71875, 0.7265625). The width of the last interval is 0.0078125, so we can stop. The actual value of the solution is $c\approx 0.726373$. ◈

EXAMPLE 7.7.6. We began this section with the example of a hiker climbing a mountain. Suppose that the hiker starts walking at a steady pace on a path from the base of the mountain at 8 a.m. on a particular day, takes no breaks, and reaches the summit of the mountain at 6 p.m. on the same day. He camps overnight and starts his descent the next day at 8 a.m. on the same path, at the same steady pace and also takes no breaks, and reaches the base of the mountain again at 6 p.m. Prove that the hiker was at a particular point on the path at exactly the same time each day.

Answer: It is easy to understand why the statement should be true: if a second hiker begins the descent of the mountain at exactly the same time that the first hiker begins his ascent (along the same path and in similar fashion), then they will meet somewhere along the path.

Now we can think of the second hiker as being the first hiker on his second day, and this solves the problem. However, we can also turn the problem into a mathematical problem and apply the IVT: Lets suppose that the distance the hiker needs to walk along the path from the base to the summit of the mountain is 3 miles (this number does not matter) and let $f(t)$ be the distance the hiker has walked from the base of the mountain (on the first day) at time t. We may suppose that $t=0$ corresponds to 8 a.m. and $t=10$ corresponds to 6 p.m. Thus, $f(0)=0$ and $f(10)=3$. Furthermore, we let $g(t)$ be the hiker's distance from the base of the mountain (on the second day) at time t. So, $g(0)=3$ and $g(10)=0$.

We now define the function $h(t)=f(t)-g(t)$ and suppose that $f(t)$, $g(t)$, and $h(t)$ are continuous functions. Because $h(0)=f(0)-g(0)=-3$ and $h(10)=f(10)-g(10)=3$ there must, by the IVT, be a time t_0 at which $h(t_0)=0$, that is, $f(t_0)=g(t_0)$. At this time t_0, the hiker will be at the same point on the path on each day. ◈

7.8 HORIZONTAL ASYMPTOTES

Consider the functions $f(x)=\frac{1}{x}$ and $g(x)=\frac{1}{x}+1=\frac{1+x}{x}$. Their graphs are shown in figure 7.13. In both graphs, the y-axis is a *vertical asymptote* because the graph of $y=f(x)=\frac{1}{x}$ grows closer and closer to the y-axis, that is, as the value of x grows closer and closer to zero, the corresponding y value increases toward infinity or minus infinity depending on whether x approaches zero through positive or negative values, respectively.

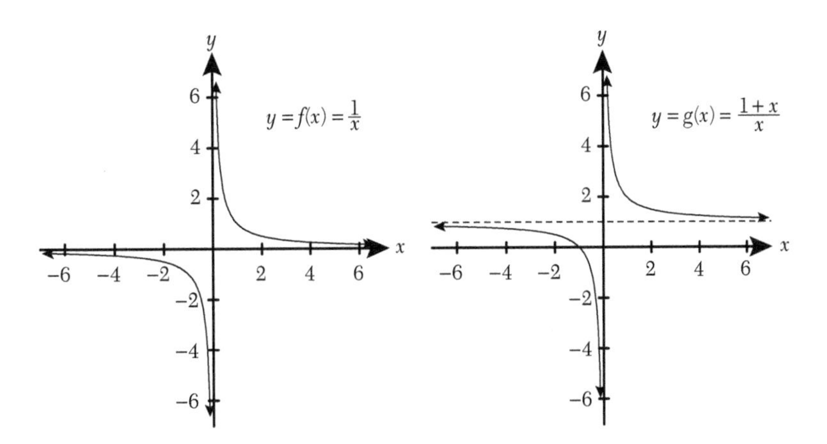

FIGURE 7.13. Functions with horizontal asymptotes.

What's more, in the first graph, the x-axis is a *horizontal asymptote* because the graph of $y = f(x) = \frac{1}{x}$ grows closer and closer to the x-axis: as the value of x grows larger and larger in both the positive and the negative directions, the corresponding y value approaches 0; while, in the second graph, the line $y = 1$ (shown as a horizontal dotted line) is a horizontal asymptote because the graph of $y = g(x) = \frac{1+x}{x}$ grows closer and closer to the line $y = 1$.

Here are a few remarks regarding horizontal asymptotes.

REMARK 7.8.1. Figure 7.14 contains a graph with oscillations that grow narrower and narrower, causing the graph to get closer and closer to the dotted line, as x tends to infinity; therefore, the dotted line is a horizontal asymptote (i.e., the graph does not have to stay above or below the asymptote).

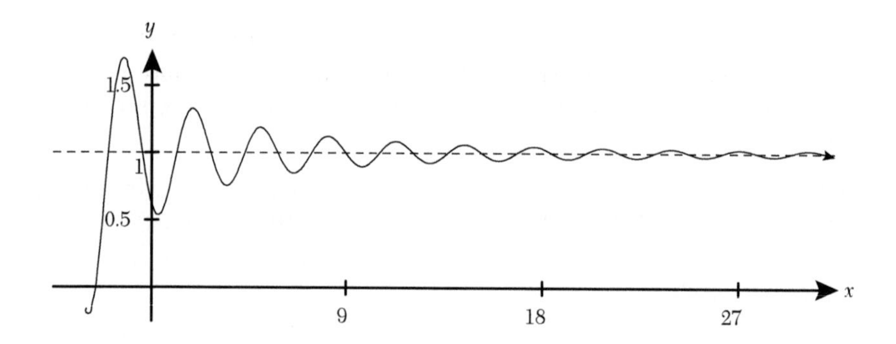

FIGURE 7.14. Oscillations approaching the horizontal asymptote.

REMARK 7.8.2. Although it might appear that the graphs of some functions "flatten out," they do not necessarily have horizontal asymptotes. The examples shown in figure 7.15 are the graphs of $y = f(x) = \sqrt[4]{x}$ and $y = g(x) = \ln(\ln(x))$. These functions do not have horizontal asymptotes because y tends to infinity as x tends to infinity.

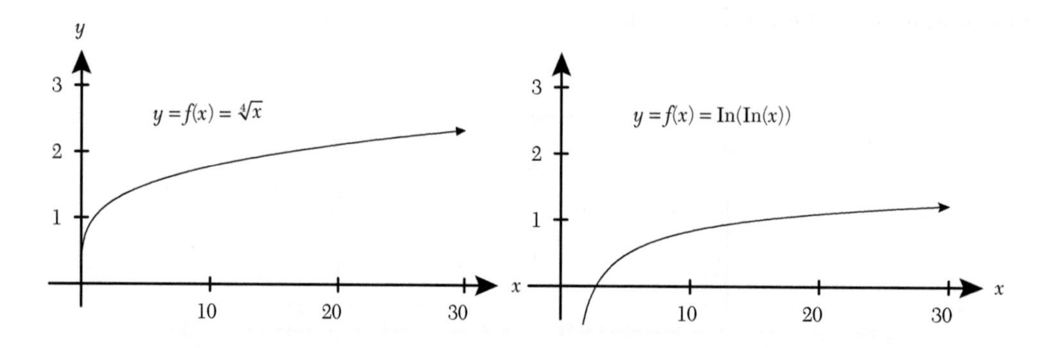

FIGURE 7.15. Functions that tend to infinity as x tends to infinity.

REMARK 7.8.3. The x-axis is not a horizontal asymptote for $f(x) = \sin(x)$ because there are y values as much as a unit distance from the x-axis no matter how large x becomes, that is (as shown in figure 7.16), the graph does not grow closer and closer to the x-axis.

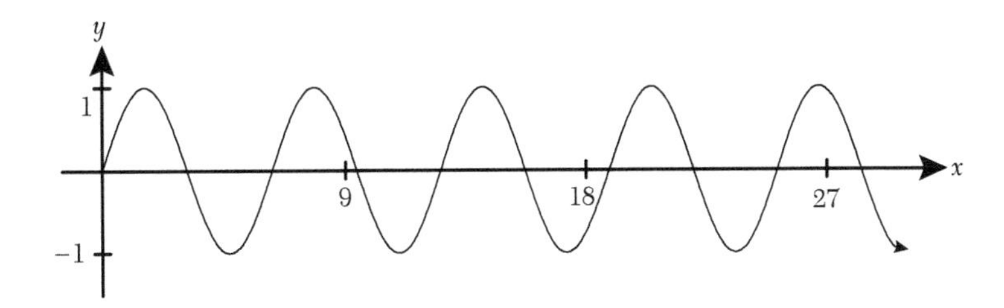

FIGURE 7.16. A function with no horizontal asymptotes.

Remark 7.8.4. A function can have the same horizontal asymptote as x approaches infinity in both positive and negative directions. An example is $f(x)=\frac{x^2-1}{x^2+1}$. Its graph is shown in figure 7.17.

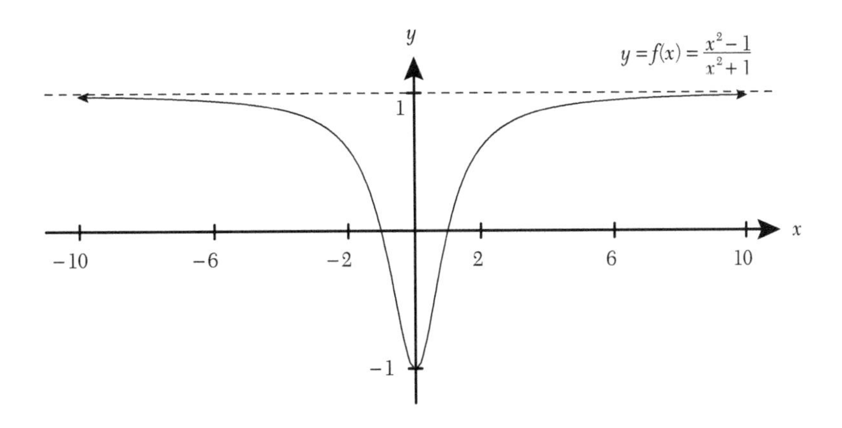

FIGURE 7.17. Diagram for remark 7.8.4.

Remark 7.8.5. A function might have two (different) horizontal asymptotes depending on whether x approaches infinity through positive or negative values. An example is $f(x)=\frac{\sqrt{2x^2+1}}{3x-5}$. Its graph is shown in figure 7.18. (It also has a vertical asymptote at $x=1.6$.)

Think about this: is it possible for a real-valued function to have three horizontal asymptotes?

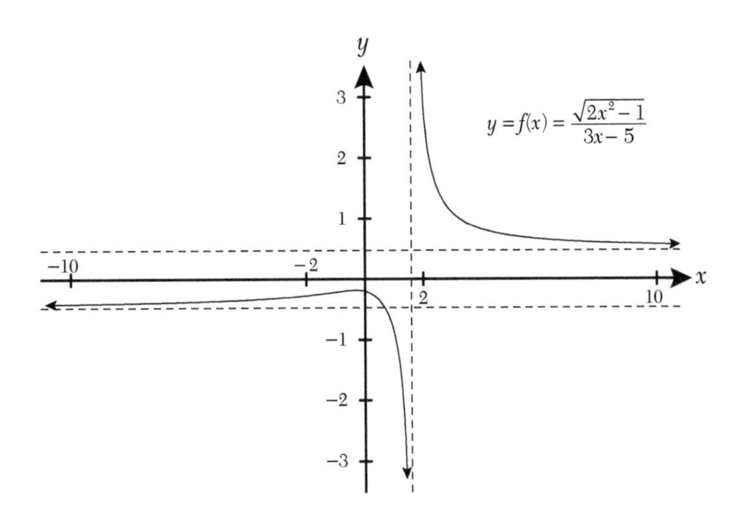

FIGURE 7.18. A function with two horizontal asymptotes.

We will state the formal definition of a horizontal asymptote in terms of limits. The interpretation of the statement $\lim_{x\to\infty} f(x)=L$ is that, in *any* arbitrarily small interval I containing the value L, there will be *some* point on the x-axis beyond which (in the positive or negative direction) the value of the function will always be in the interval I. (This is a mathematically precise way of saying that the graph of $y=f(x)$ approaches the line $y=L$.)

DEFINITION 7.8.1. *A line* $y=L$ *is a horizontal asymptote of a function* $f(x)$ *if and only if*

$$\lim_{x\to\infty} f(x)=L \text{ or } \lim_{x\to-\infty} f(x)=L \text{ (or both).}$$

We demonstrate, by means of the examples below, how to compute a limit as x tends to infinity. The basic idea is that any fraction of the form $\frac{c}{x^k}$, where c is a constant and k is any natural number, tends to 0 as x tends to infinity (through positive or negative numbers).

Recall that a rational function is a function of the form $f(x)=\frac{p(x)}{q(x)}$, where $p(x)$ and $q(x)$ are polynomials. The function $f(x)=\frac{1}{x}$ is a basic example of a rational function. We will see that a rational function has a horizontal asymptote whenever the degree of the numerator (the highest power of x in the numerator) is the same as, or less than, the degree of the denominator (the highest power of x in the denominator). In the following examples, we will let m be the degree of $p(x)$ and n be the degree of $q(x)$. The method of computing limits at infinity will require us to know the larger of the values of m and n. This is denoted "$\max\{m,n\}$" (the *maximum* of m and n).

EXAMPLE 7.8.1. Find the horizontal asymptotes of the rational function $f(x)=\frac{x^2}{x^2+1}$.

Answer: The degree of $p(x):=x^2$ is $m=2$ and $q(x):=x^2+1$ is $n=2$. Because $\max\{m,n\}=2$, we will factor x^2 from the numerator and the denominator of $f(x)$ so that all the terms that remain (after cancellation of the common factor) in the numerator and the denominator are constant terms or terms of the form $\frac{c}{x^n}$, where c is a constant and n is a natural number:

$$\frac{x^2}{x^2+1}=\frac{x^2(1)}{x^2\left(1+\frac{1}{x^2}\right)}=\frac{1}{1+\frac{1}{x^2}}.$$

Now, when we compute the limit, the term $\frac{1}{x^2}$ tends to 0. Thus,

$$\lim_{x\to\infty} f(x)=\lim_{x\to\infty}\frac{1}{1+\frac{1}{x^2}}=\frac{1}{1+(0)}=\frac{1}{1}=1.$$

By the same reasoning, the limit as x tends to $-\infty$ is also equal to 1:

$$\lim_{x\to-\infty} f(x)=\lim_{x\to-\infty}\frac{1}{1+\frac{1}{x^2}}=\frac{1}{1+(0)}=\frac{1}{1}=1.$$

Thus, the horizontal asymptote is the line $y=1$ (in both directions). ◈

EXAMPLE 7.8.2. Find the horizontal asymptotes of the rational function $f(x)=\frac{2x^2-x-2}{6x^2+4x+1}$.

Answer: As in the previous example, we find that $\max\{m,n\}=2$, so we will factor x^2 from the numerator and the denominator of $f(x)$ in order to take the limit. Because the result will be the same whether x tends to ∞ through positive or negative numbers, we will take the limit as $x\to\pm\infty$:

$$\lim_{x\to\pm\infty} f(x) = \lim_{x\to\pm\infty} \frac{x^2\left(2-\frac{1}{x}-\frac{2}{x^2}\right)}{x^2\left(6+\frac{4}{x}+\frac{1}{x^2}\right)} = \frac{2-(0)-(0)}{6+(0)+(0)} = \frac{2}{6} = \frac{1}{3}.$$

Thus, the horizontal asymptote is the line $y=\frac{1}{3}$ (in both directions). ◈

EXAMPLE 7.8.3. Find the horizontal asymptotes of the rational function $f(x)=\frac{2x^2-x-2}{6x^3+4x+1}$.

Answer: Because $\max\{m,n\}=3$, we will factor x^3 from the numerator and the denominator of $f(x)$ in order to take the limit:

$$\lim_{x\to\pm\infty} f(x) = \lim_{x\to\pm\infty} \frac{x^3\left(\frac{2}{x}-\frac{1}{x^2}-\frac{2}{x^3}\right)}{x^3\left(6+\frac{4}{x^2}+\frac{1}{x^3}\right)} = \frac{(0)-(0)-(0)}{6+(0)+(0)} = \frac{0}{6} = 0.$$

Thus, the horizontal asymptote is the line $y=0$ (in both directions). ◈

EXAMPLE 7.8.4. Find the horizontal asymptotes of the rational function $f(x)=\frac{x^2+x}{3-x}$.

Answer: Because $\max\{m,n\}=2$, we will factor x^2 from the numerator and the denominator of $f(x)$ in order to take the limit:

$$\lim_{x\to\pm\infty} f(x) = \lim_{x\to\pm\infty} \frac{x^2\left(1+\frac{1}{x}\right)}{x^2\left(\frac{3}{x^2}-\frac{1}{x}\right)} = \lim_{x\to\pm\infty} \frac{1+\frac{1}{x}}{\frac{3}{x^2}-\frac{1}{x}} = \lim_{x\to\pm\infty} \frac{1+\frac{1}{x}}{\frac{1}{x}\left(\frac{3}{x}-1\right)}.$$

The factorization of the denominator in the final step above helps us to see that

$$\lim_{x\to\infty} f(x) = \text{"}\frac{1}{(0^+)(-1)}\text{"} = -\infty \quad \text{and} \quad \lim_{x\to-\infty} f(x) = \text{"}\frac{1}{(0^-)(-1)}\text{"} = \infty$$

where the notation 0^+ and 0^- are used to mean that 0 is approached through positive and negative numbers, respectively. In either case, there is no horizontal asymptote. ◈

This method for determining the horizontal asymptotes of rational functions also works for nonrational functions.

EXAMPLE 7.8.5. Find the horizontal asymptote of the function $f(x)=\frac{1-\sqrt{x}}{1+\sqrt{x}}$.

Answer: We factor $\sqrt{x}$ from the numerator and the denominator of $f(x)$ in order to take the limit: because $f(x)$ is only defined for positive values of x, we take the limit as $x\to\infty$.

$$\lim_{x\to\infty} f(x) = \lim_{x\to\infty} \frac{\sqrt{x}\left(\frac{1}{\sqrt{x}}-1\right)}{\sqrt{x}\left(\frac{1}{\sqrt{x}}+1\right)} = \lim_{x\to\infty} \frac{\frac{1}{\sqrt{x}}-1}{\frac{1}{\sqrt{x}}+1} = \frac{(0)-1}{(0)+1} = \frac{-1}{1} = -1.$$

Thus, the horizontal asymptote is the line $y=-1$. ◈

EXAMPLE 7.8.6. Find the horizontal asymptotes of the function $f(x)=\dfrac{\sqrt{2x^2+1}}{3x-5}$.

Answer: We factor x^2 inside the square root in the numerator and factor x in the denominator:

$$\lim_{x\to\pm\infty} f(x)=\lim_{x\to\pm\infty}\frac{\sqrt{x^2\left(2+\frac{1}{x^2}\right)}}{x\left(3-\frac{5}{x}\right)}=\lim_{x\to\pm\infty}\frac{\sqrt{x^2}\sqrt{\left(2+\frac{1}{x^2}\right)}}{x\left(3-\frac{5}{x}\right)}=\lim_{x\to\pm\infty}\left(\frac{|x|}{x}\right)\frac{\sqrt{\left(2+\frac{1}{x^2}\right)}}{x\left(3-\frac{5}{x}\right)}.$$

Because the factor $\left(\dfrac{|x|}{x}\right)$ is $+1$ or -1, depending on whether x is positive or negative, respectively, we have to consider the limits $x\to\infty$ and $x\to-\infty$ separately:

$$\lim_{x\to\infty}\left(\frac{|x|}{x}\right)\frac{\sqrt{2+\frac{1}{x^2}}}{\left(3-\frac{5}{x}\right)}=\frac{\sqrt{2+(0)}}{3-(0)}=\frac{\sqrt{2}}{3},$$

$$\lim_{x\to-\infty}\left(\frac{|x|}{x}\right)\frac{\sqrt{2+\frac{1}{x^2}}}{\left(3-\frac{5}{x}\right)}=(-1)\frac{\sqrt{2+(0)}}{3-(0)}=-\frac{\sqrt{2}}{3}.$$

This explains why the function $f(x)=\dfrac{\sqrt{2x^2+1}}{3x-5}$ has two horizontal asymptotes: $y=\dfrac{\sqrt{2}}{3}$ as $x\to\infty$ and $y=-\dfrac{\sqrt{2}}{3}$ as $x\to-\infty$. ◈

Certain limits to infinity can be computed after the given expression is converted into fractional form by some means. In the next example, the method of multiplying and dividing by a conjugate radical will be used to do this.

EXAMPLE 7.8.7.

$$\lim_{x\to\infty}\left(\sqrt{x^2+1}-x\right)=\lim_{x\to\infty}\frac{\left(\sqrt{x^2+1}-x\right)}{1}\frac{\left(\sqrt{x^2+1}+x\right)}{\left(\sqrt{x^2+1}+x\right)}$$

$$=\lim_{x\to\infty}\frac{(x^2+1)-x^2}{\sqrt{x^2+1}+x}$$

$$=\lim_{x\to\infty}\frac{1}{\sqrt{x^2\left(1+\frac{1}{x^2}\right)}+x}$$

$$=\lim_{x\to\infty}\frac{1}{|x|\sqrt{1+\frac{1}{x^2}}+x}$$

$$=\lim_{x\to\infty}\frac{1}{x\left(\sqrt{1+\frac{1}{x^2}}+1\right)}$$

$$=0.$$

7.9 VERTICAL ASYMPTOTES OF RATIONAL FUNCTIONS

We now do a precise analysis of the behavior of the graphs of rational functions near their vertical asymptotes. We already know that the positions of the vertical asymptotes of the graph of a rational function coincide with the real roots of the polynomial that remains in the denominator after all factors that are common to the numerator and denominator of the rational function have been canceled. Suppose now that $f(x)=\frac{p(x)}{q(x)}$ is a rational function and that $\frac{p(x)}{q(x)}$ is a rational expression in simplest form. If $x=a$ is a real root of $q(x)$, then $f(x)$ can be expressed in the form $\frac{1}{(x-a)^k}g(x)$ where k is a natural number, $g(x)$ is another rational function, $g(a)\neq 0$, and the denominator of $g(x)$ contains no factors of the form $(x-a)$. The behavior of the graph of $y=f(x)$ near $x=a$ depends on whether k is an odd or even number and also depends on whether $g(a)$ is positive or negative. This allows four possible behaviors, which are shown in figure 7.19. The behavior of a graph near a vertical asymptote can be referred to as its *asymptotic behavior*. In the first diagram, $y=f(x)$ approaches $+\infty$ as x tends to a from the left and right, while in the second, $y=f(x)$ approaches $-\infty$ as x tends to a from the left and approaches $+\infty$ as x tends to a from the right. In the third diagram, $y=f(x)$ approaches $-\infty$ as x tends to a from the left and right, while in the fourth, $y=f(x)$ approaches $+\infty$ as x tends to a from the left and approaches $-\infty$ as x tends to a from the right.

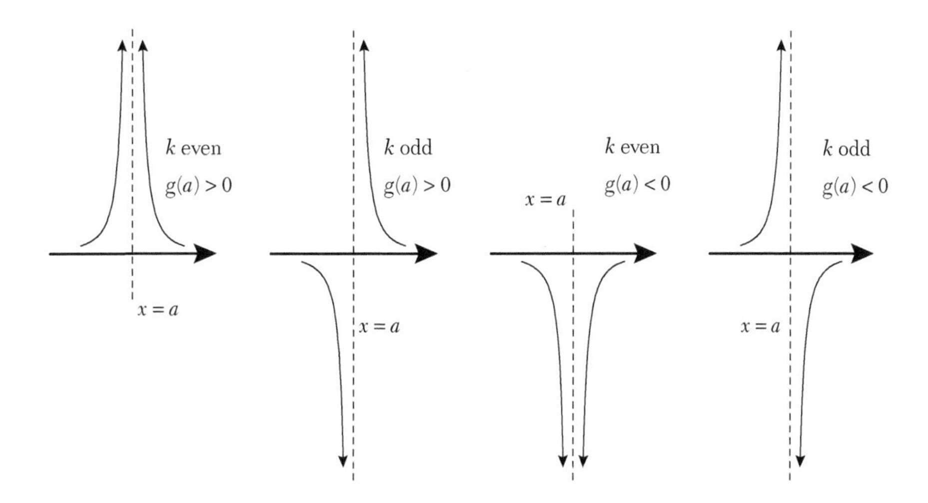

FIGURE 7.19. Four cases of asymptotic behavior.

EXAMPLE 7.9.1. The graph of the rational function $f(x)=\frac{x^4-2x^2+10}{10(x^2-4)}$ has vertical asymptotes at $x=-2$ and $x=2$. We will investigate the behavior of the graph of f near these vertical asymptotes and explore some other properties of the graph of f.

- To determine the behavior of the graph near $x=-2$, we write $f(x)=\frac{1}{x+2}g(x)$, where $g(x)=\frac{x^4-2x^2+10}{10(x-2)}$. Because $g(-2)=\frac{(-2)^4-2(-2)^2+10}{10(-2-2)}=-\frac{9}{20}$, the behavior of the graph near $x=-2$ should be as shown in the fourth diagram in figure 7.19.

- Similarly, in order to determine the behavior of the graph near $x=2$, we write $f(x)=\frac{1}{x-2}g(x)$, where $g(x)=\frac{x^4-2x^2+10}{10(x+2)}$. Because $g(2)=\frac{(2)^4-2(2)^2+10}{10(2+2)}=\frac{9}{20}$, the behavior of the graph near $x=2$ should be as shown in the second diagram in figure 7.19.

- We can determine some additional facts regarding the graph of $y=f(x)$. First, because the numerator of $f(x)$ in completed square form is $(x^2-1)^2+9$ (which is never equal to zero), the graph of $y = f(x)$ never cuts the x-axis; second, because the degree of the numerator is larger than the degree of the denominator, the graph does not have any horizontal asymptotes (in fact, $\lim_{x\to\pm\infty} f(x) = \infty$).

The graph of $y = f(x)$ is shown in figure 7.20. ◈

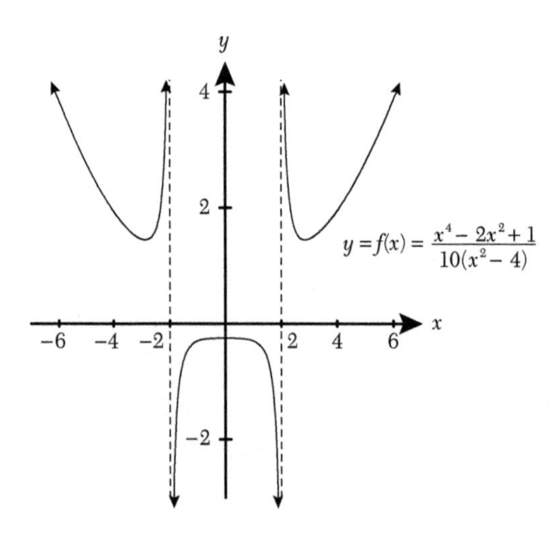

FIGURE 7.20. Diagram for example 7.9.1.

EXAMPLE 7.9.2. Sketch the graph of the rational function $f(x) = \dfrac{2x^2 + 5x + 2}{x^2 - x - 6}$.

Answer: In factored form, the expression for $f(x)$ is $f(x) = \frac{(2x+1)(x+2)}{(x-3)(x+2)}$. This means that the domain of f excludes $x = 3$ and $x = -2$. The corresponding rational expression in simplest form is $\frac{(2x+1)}{(x-3)}$. The graph of $y = \frac{(2x+1)}{(x-3)}$ has a horizontal asymptote at $y = 2$ and a vertical asymptote at $x = 3$. Because y can be expressed in the form $y = \frac{1}{(x-3)} g(x)$, where $g(x) = (2x+1)$ and $g(3) = 7 > 0$, the behavior of the graph of y near $x = 3$ should be as in the second diagram in figure 7.19. The x intercept of the graph is at $x = -\frac{1}{2}$ and its y intercept is at $y = -\frac{1}{3}$. Without additional information about what the graph should look like, it is best to sketch the simplest graph that satisfies all of the stated properties. The graph of $y = f(x)$ is shown in figure 7.21. ◈

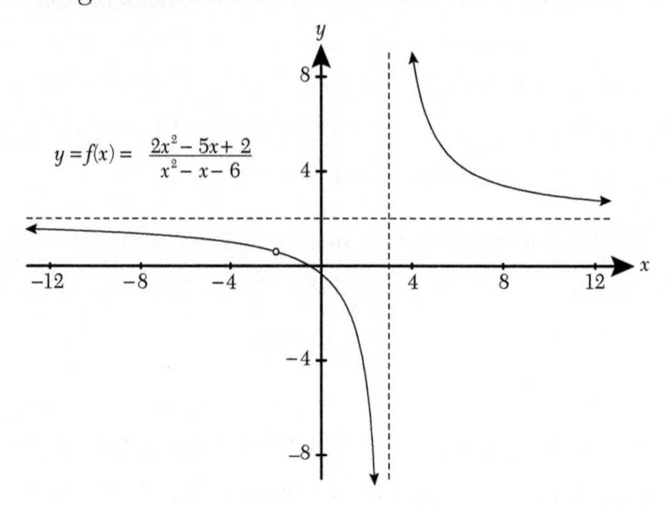

FIGURE 7.21. Diagram for example 7.9.2.

7.10 THE SQUEEZE THEOREM AND RULES FOR LIMITS

Limits involving trigonometric ratios of the form $\lim_{x\to 0}\frac{\sin x}{x}$, $\lim_{x\to 0}\frac{\sin 4x}{7x}$, or $\lim_{x\to 0}\frac{x}{\tan x}$, for example, cannot be evaluated using any of the methods learned so far. We are going to introduce the *Squeeze Theorem* (also called the *Pinching Theorem* or *Sandwich Theorem*) and demonstrate how, together with some techniques of estimation, it can be used to evaluate $\lim_{x\to 0}\frac{\sin x}{x}$. We will then introduce the *rules for limits* and apply them in order to evaluate the limits mentioned above and some other limits involving trigonometric ratios.

7.10.1 The Squeeze Theorem

Theorem 7.10.1. The Squeeze Theorem: *if a point* $x = a$ *belongs to an open or closed interval* I, *and, if* $f(x)$, $g(x)$, *and* $h(x)$ *are functions defined on the interval* I *(except, perhaps, the point* $x = a$*) such that, for every* x *in* I *not equal to* a, $g(x) \le f(x) \le h(x)$ *(i.e.,* $g(x)$ *is a lower bound, and* $h(x)$ *is an upper bound for* $f(x)$*) and* $\lim_{x\to a} g(x) = \lim_{x\to a} h(x) = L$, *then* $\lim_{x\to a} f(x) = L$.

Remark 7.10.1. If I is a closed interval and $x = a$ is a left or right end point of the interval, then the statement of the theorem applies with limits from the right or left, respectively. Also, if I is an infinite interval, then the statement could hold with $a = \pm\infty$.

The Squeeze Theorem is proved in real analysis. We will not provide the proof here. The Squeeze Theorem is, amusingly, also known as the *two policemen* and *a drunk theorem* because, if two policemen escort a drunken prisoner between them to his jail cell, then the prisoner will end up in the cell if the policemen end up in the cell, no matter how much he wobbles about!

We did (without mentioning it) previously appeal to the Squeeze Theorem, where we stated in remark 7.5.1 that the function

$$f(x) = \begin{cases} x\sin\left(\dfrac{1}{x}\right) & \text{if } x \neq 0 \\ 0 & \text{if } x = 0 \end{cases}$$

is continuous on the interval $I = (-1, 1)$. The reason is that if we define $f(x) = x\sin\left(\frac{1}{x}\right)$, $g(x) = -x$ and $h(x) = x$, then f, g, and h are defined on the interval $(-1, 1)$ except at the point $x = 0$. Furthermore, because $-1 \le \sin\left(\frac{1}{x}\right) \le 1$ for all values of x except $x = 0$, it is also the case that $-x \le x\sin\left(\frac{1}{x}\right) \le x$ for all values of x except $x = 0$.

Now, because $\lim_{x\to 0} g(x) = \lim_{x\to 0}(-x) = 0 = \lim_{x\to 0}(x) = \lim_{x\to 0} h(x)$, we can conclude, by the Squeeze Theorem, that $\lim_{x\to 0} f(x) = \lim_{x\to 0} x\sin\left(\frac{1}{x}\right) = 0 = f(0)$. This means that $f(x)$ is continuous at $x = 0$.

We turn now to the problem of computing $\lim_{\theta\to 0}\frac{\sin\theta}{\theta}$. (We will see in section 8.6 that this limit computes the derivative of $\sin x$ at $x = 0$.) We will compute this limit by applying the Squeeze Theorem to the following inequalities:

$$\cos(\theta) \le \frac{\sin\theta}{\theta} \le 1, \quad \text{for } -\frac{\pi}{2} < \theta < 0 \text{ and } 0 < \theta < \frac{\pi}{2}. \tag{7.4}$$

However, we first need to prove these inequalities. To this end, we begin by proving the first inequality, that is, $\frac{\sin\theta}{\theta} > \cos(\theta)$ for $0 < \theta < \frac{\pi}{2}$, with the help of the first diagram in figure 7.22.

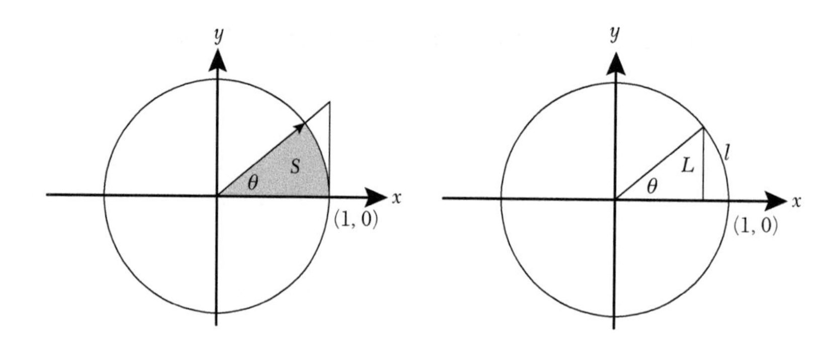

FIGURE 7.22. A proof of formula (7.4).

By symmetry of the circle, the area of the shaded sector S of the unit disk is in the same proportion to the area ($\pi(1)^2 = \pi$) of the full disk, as the angle θ is in proportion to the full angle (2π). This statement is expressed as the equation

$$\frac{\text{area }(S)}{\pi} = \frac{\theta}{2\pi},$$

which simplifies to $\text{area}(S) = \frac{\theta}{2}$. What's more, the height of the triangle in the first diagram is $\tan(\theta)$ (why?), and because the area of the triangle is greater than the area of the shaded sector S (which it contains), we have the inequality

$$\frac{\tan(\theta)}{2} > \frac{\theta}{2}, \quad \text{for } 0 < \theta < \frac{\pi}{2}.$$

By means of a trigonometric identity, this is the same as

$$\frac{\sin(\theta)}{\cos(\theta)} > \theta, \quad \text{for } 0 < \theta < \frac{\pi}{2}$$

and this is equivalent to the inequality we want to prove. The inequality $\frac{\sin\theta}{\theta} > \cos(\theta)$ is also true for $-\frac{\pi}{2} < \theta < 0$ because the functions on both sides of the inequality are even functions.

Next, we prove the second inequality in formula (7.4), that is, $\frac{\sin\theta}{\theta} < 1$ for $0 < \theta < \frac{\pi}{2}$, with the help of the second diagram in figure 7.22. It is clear that the height of the triangle (labeled L) is less than the length of the corresponding arc of the circle (labeled l).

Because $L = \sin(\theta)$ (why?) and l is the radian measure of θ, we have the inequality

$$\sin(\theta) < \theta, \quad \text{for } 0 < \theta < \frac{\pi}{2}.$$

If we divide both sides by θ, this gives us the inequality we want to prove. For the same reason as above, the inequality is also true if $-\frac{\pi}{2} < \theta < 0$.

We have now established formula (7.4) and so we can apply the Squeeze Theorem to conclude that

$$\lim_{\theta \to 0}\left(\frac{\sin\theta}{\theta}\right) = \lim_{\theta \to 0} \cos(\theta) = \lim_{\theta \to 0}(1) = 1.$$

The inequality in formula (7.4) is demonstrated in figure 7.23, which shows the graphs of $y = 1$, $y = \frac{\sin(x)}{x}$, and $y = \cos(x)$ on an interval containing $[-2\pi, 2\pi]$.

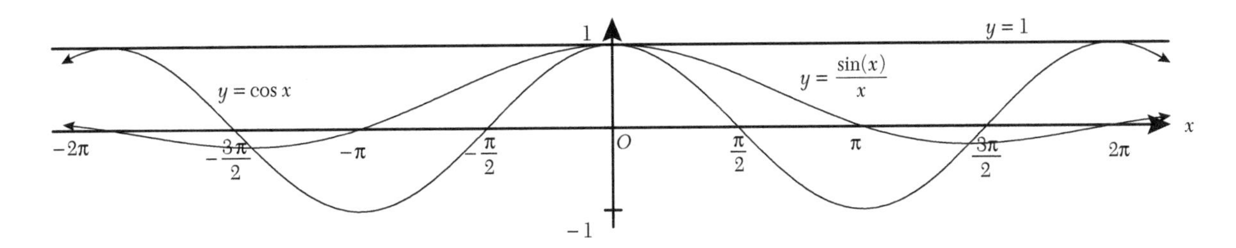

FIGURE 7.23. A graph of formula (7.4).

7.10.2 The Rules for Limits

The *rules for limits* allow us to compute the limit of an expression when limits of separate terms or factors of the expression are known. These rules can be proved using methods of real analysis. It is enough for us, at this stage, to know how to apply these rules.

Theorem 7.10.2. *The rules for limits: suppose that* c *is a constant and that the limits* $\lim_{x\to a} f(x)$ *and* $\lim_{x\to a} g(x)$ *exist. Then*

(I) $\lim_{x\to a}(f(x) \pm g(x)) = \lim_{x\to a} f(x) \pm \lim_{x\to a} g(x)$

(II) $\lim_{x\to a} cf(x) = c\lim_{x\to a} f(x)$

(III) $\lim_{x\to a} f(x)g(x) = \lim_{x\to a} f(x)\lim_{x\to a} g(x)$

(IV) $\lim_{x\to a}\frac{f(x)}{g(x)} = \frac{\lim_{x\to a} f(x)}{\lim_{x\to a} g(x)}$, provided $\lim_{x\to a} g(x) \neq 0$

(V) $\lim_{x\to 0} f(x) = \lim_{x\to 0} f(kx)$ for any $k \neq 0$,

Remark 7.10.2. It is to be understood from the statement of theorem 7.10.2 that the functions $f(x)$ and $g(x)$ in the theorem are defined on an interval containing the point $x = a$. Whenever $x = a$ is the end point of the interval, then left or right limits would be used, as appropriate. Also, if the interval is infinite, then rules (I)–(IV) could hold with $[a = \pm\infty]$.

The difficulty with using the rules for limits is that it may not be obvious how to rewrite a given expression in such a way that the rules for limits can be applied. Some practice with this is given in the next example.

Example 7.10.1.

(i) $$\lim_{x\to 0}\left(\frac{x}{\sin(x)}\right) = \lim_{x\to 0}\left(\frac{1}{\frac{\sin(x)}{x}}\right) = \frac{1}{\lim_{x\to 0}\left(\frac{\sin(x)}{x}\right)} = \frac{1}{1} = 1$$

(This is an application of rule (IV).)

(ii) $$\lim_{x\to 0}\left(\frac{\sin(7x)}{4x}\right) = \lim_{x\to 0}\left(\frac{7}{4}\right)\left(\frac{\sin(7x)}{7x}\right) = \left(\frac{7}{4}\right)\lim_{x\to 0}\left(\frac{\sin(7x)}{7x}\right) = \left(\frac{7}{4}\right)(1) = \frac{7}{4}$$

(Rule (II) was applied in the second step, and then rule (V) was applied with $k = 7$.)

(iii) $\lim_{x\to 0} x \cot x = \lim_{x\to 0}\left(\frac{x}{\sin x}\right)\cos x = \lim_{x\to 0}\left(\frac{x}{\sin x}\right)\lim_{x\to 0}\cos x = (1)(1) = 1$

(This is an application of rule (III).)

(iv)
$$\begin{aligned}\lim_{x\to 0}\frac{\cos(x)-1}{x} &= \lim_{x\to 0}\left(\frac{\cos(x)-1}{x}\right)\left(\frac{\cos(x)+1}{\cos(x)+1}\right)\\ &= \lim_{x\to 0}\frac{-\sin^2(x)}{x(\cos(x)+1)}\\ &= \lim_{x\to 0}\left(\frac{\sin(x)}{x}\right)\left(\frac{-\sin(x)}{\cos(x)+1}\right)\\ &= (1)(0) = 0\end{aligned}$$

(This is another application of rule (III).) ◈

For future reference, we record two of the limit formulas we have derived in this section:

$$\boxed{\lim_{\theta\to 0}\left(\frac{\sin\theta}{\theta}\right) = 1 \text{ and } \lim_{\theta\to 0}\frac{\cos(\theta)-1}{\theta} = 0}. \tag{7.5}$$

EXERCISES

7.1. Verify the calculation done by Archimedes to prove that $\pi > 3^{10}/_{71}$ by inscribing a 96-gon inside a disk, as explained in example 7.2.1.

7.2. How would you phrase the argument to prove that the limit of the sequence in example 7.3.4 of section 7.3 is equal to 1?

7.3. Which of the sequences below, if any, has a limit? If any does, what is its limit?

(a) $\frac{1}{2}, 0, \frac{3}{4}, 0, \frac{7}{8}, 0, \frac{15}{16}, 0, \frac{31}{32}, \ldots$ (sequence in example 7.3.1, terms alternating with 0)

(b) $1, 2, 3, 1, 2, 3, 1, 2, 3, \ldots$ (1,2,3 repeating)

(c) $2, 2, 2, 2, 2, 2, 2, 2, 2, \ldots$ (2 repeating)

(d) $\pi+12,\ \pi-\frac{1}{4},\ \pi+\frac{1}{8},\ \pi-\frac{1}{16},\ \pi+\frac{1}{32}, \ldots$

(e) $3, 1, 4, 1, 5, 9, \ldots$ (What do these digits remind you of ?)

7.4. Relating to the first graph in figure 7.24, compute the limits below. What kinds of discontinuity does $f(x)$ have at $x = 0$ and $x = 2$?

(a) $\lim_{x\to(-1)^-} f(x) =$

(b) $\lim_{x\to 0} f(x) =$

(c) $\lim_{x\to 2^-} f(x) =$

(d) $\lim_{x\to 2^+} f(x) =$

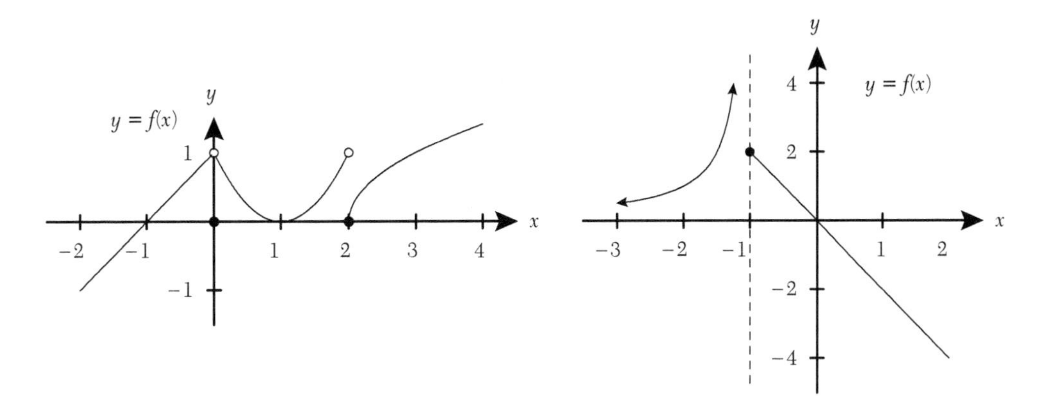

FIGURE 7.24. Diagram for Exercise 7.4.

7.5. Relating to the second graph in figure 7.24, compute the limits below. What kind of discontinuity does $f(x)$ have at $x = -1$? (The dotted line is an asymptote.)

(a) $\lim_{x \to (-1)^-} f(x) =$ **(b)** $\lim_{x \to (-1)^+} f(x) =$

7.6. Given $f(x) = \frac{x+1}{x^3+1}$, determine a piecewise expression for *f*, then sketch the graph of $y = f(x)$. Determine the limiting value (if it exists) of *f* at the point $x = -1$.

7.7. Decide whether or not the following piecewise-defined function is continuous at $x = -1$.

$$f(x) = \begin{cases} \dfrac{x+1}{x^3+1} & \text{if } x \neq -1 \\ \dfrac{1}{3} & \text{if } x = -1 \end{cases}$$

7.8. If $f(x) = \frac{2}{x}$, consider the function

$$g(x) = \begin{cases} f \circ f(x) & \text{if } x \neq 0 \\ 0 & \text{if } x = 0 \end{cases}$$

What kind of discontinuity does $g(x)$ have at $x = 0$?

7.9. Consider the function

$$f(x) = \begin{cases} \dfrac{x^2}{|x|} & \text{if } x \neq 0 \\ 1 & \text{if } x = 0 \end{cases}$$

Rewrite the piecewise definition for $f(x)$ in the form

$$f(x) = \begin{cases} (\quad) & \text{if } x < 0 \\ (\quad) & \text{if } x > 0 \\ 1 & \text{if } x = 0 \end{cases}$$

(fill in the parentheses) in order to answer the following question: What kind of discontinuity does $f(x)$ have at $x = 0$?

7.10. The diagram in figure 7.25 shows part of the graph of $y = f(x) = -\lfloor x \rfloor + x|x| - x$. By inspection of the graph, find the values of the limits below (all of your answers should be integer values).

(a) $\lim_{x\to(-1)^+} f(x) =$

(b) $\lim_{x\to 0^-} f(x) =$

(c) $\lim_{x\to 0^+} f(x) =$

(d) $\lim_{x\to 1^-} f(x) =$

(e) $\lim_{x\to 1^+} f(x) =$

(f) $\lim_{x\to 2^-} f(x) =$

(g) $\lim_{x\to 2^+} f(x) =$

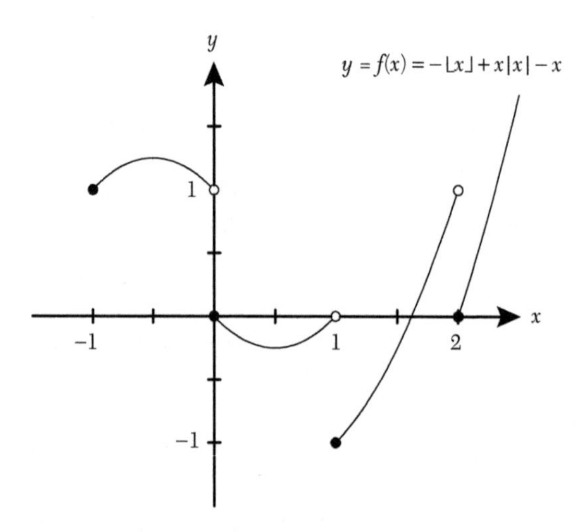

FIGURE 7.25. Diagram for exercise 7.10.

7.11. Rewrite the definition for $f(x) = -\lfloor x \rfloor + x|x| - x$ as a piecewise-defined function on the intervals $[-1,0)$, $[0,1)$, and $[1,2)$, so that the expression on each interval is a quadratic polynomial. (Hint: rewrite $\lfloor x \rfloor$ as an integer, and rewrite $|x|$ as (x) or $(-x)$, depending on the interval to which x belongs, and simplify the resulting expression.)

7.12. Evaluate the following limits, for each of the functions below.
(i) $\lim_{x\to 0^-}$, (ii) $\lim_{x\to 0^+}$, (iii) $\lim_{x\to 1^-}$, (iv) $\lim_{x\to 1^+}$, (v) $\lim_{x\to 2^-}$, and (vi) $\lim_{x\to 2^+}$

(a) $f(x) = \text{floor}(x)$

(b) $f(x) = x^2 - \text{floor}(x)$

(c) $f(x) = x^2 - \text{floor}(2x)$

(d) $f(x) = x - \text{floor}\left(\frac{x^2}{4}\right)$

7.13. Evaluate the following limits for each of the functions below. (i) $\lim_{x\to 0^-}$ and (ii) $\lim_{x\to 0^+}$

(a) $f(x) = \frac{|x|}{x} + 0.5$

(b) $f(x) = 0.5\frac{|x|}{x} + 0.5x^2$

7.14. If $f(x) = \frac{1}{x}$ and $g(x) = \frac{x}{x-1}$, what is the domain of $f \circ g(x)$? Draw the graph of $y = f \circ g(x)$. How should the graph of $y = f \circ g(x)$ be modified to create a function that is continuous on the set $\{x \in \mathbb{R} | x \neq 0\}$?

7.15. Sketch the graph of the function $f(x) = |x| + \lfloor x \rfloor$ (i.e., the absolute value function plus the floor function) for $-2 \le x \le 2$. What kind of discontinuity does $f(x)$ have at $x = 0$?

7.16. Write a piecewise definition for the function $f(x) = \frac{2|x| + x}{2|x| - x}$ (simplify the expressions for $x < 0$ and $x > 0$ as far as possible). Does $\lim_{x\to 0} f(x)$ exist?

7.17. Find the value of the constant c for which the function $f(x)$ below will be continuous for all real numbers (in particular, at $x=3$). (Hint: you can derive an equation that involves c by evaluating left and right limits at $x=3$. Then solve for c.)

$$f(x)=\begin{cases}(cx+1) & \text{if } x\leq 3\\ (cx^2-1) & \text{if } x>3\end{cases}$$

7.18. Find the values of the constants c and d for which the function $f(x)$ below will be continuous for all real numbers (in particular, at $x=1$ and $x=2$). (Hint: derive a pair of equations that involves c and d by evaluating left and right limits at $x=1$ and $x=2$, then solve for c and d.)

$$f(x)=\begin{cases}2x & \text{if } x<1\\ cx^2+d & \text{if } 1\leq x\leq 2\\ 4x & \text{if } x>2\end{cases}$$

7.19. Evaluate the following limits.

(a) $\lim_{x\to\frac{\pi}{2}}(\cos(\pi\sin(x))+\sin(\pi\cos(x)))$

(b) $\lim_{x\to\frac{1}{4}}\frac{\sqrt{\sin(\pi x)+x^2}}{x+4}$

(c) $\lim_{x\to-2}\frac{e^{|x-x^2|}}{\sqrt{x+x^2}}$

(d) $\lim_{x\to1}\frac{x^3+1}{x^2+3x+2}$

(e) $\lim_{x\to-2}\frac{1+2|t|}{|t|+2}$

7.20. Compute $\lim_{x\to0}f(x)$ (if the limit exists) for $f(x)=\frac{x^2}{|x|}+\frac{|x|+1}{|x|-1}$.

(Hint: write a piecewise definition for $f(x)$.)

7.21. Compute $\lim_{x\to0}f(x)$ (if the limit exists) for $f(x)=\frac{x}{|x|}+\frac{|x|+1}{|x|-1}$.

(Hint: write a piecewise definition for $f(x)$.)

7.22. Sketch the graph of $y=f(x)$ for $f(x)$ defined below, and compute the indicated limits.

(a) $f(x)=\begin{cases}1-|x-1| & \text{if } 0\leq x\leq 2\\ 2-|x-3| & \text{if } 2<x\leq 4\\ 1-|x-5| & \text{if } 4<x\leq 6\end{cases}$

(b) $\lim_{x\to2^-}f(x),\ \lim_{x\to2^+}f(x),\ \lim_{x\to2}f(x)$

(c) $\lim_{x\to4^-}f(x),\ \lim_{x\to4^+}f(x),\ \lim_{x\to4}f(x)$

7.23. Compute $\lim_{x\to1}f(x)$ (if the limit exists) for $f(x)$ defined below.

$$f(x)=\begin{cases}\dfrac{1-x+x^2}{x^2+2x+1} & \text{if } x<1\\ \dfrac{3-x}{7+x} & \text{if } x>1\end{cases}$$

7.24. Compute $\lim_{x\to-2}f(x)$ (if the limit exists) for $f(x)$ defined below.

$$f(x)=\begin{cases}\lfloor x\rfloor+x|x|+6 & \text{if } x<-2\\ x+|x|-\lfloor 2x\rfloor-6 & \text{if } x>-2\end{cases}$$

7.25. Evaluate the following limits.

(a) $\lim_{x\to -3}\dfrac{x+3}{2x^2+5x-3}$

(b) $\lim_{h\to -3}\dfrac{h^2-h-12}{h+3}$

(c) $\lim_{v\to 2}\dfrac{v^2+2v-8}{v^4-16}$

(d) $\lim_{t\to 2}\dfrac{\sqrt{t+2}-\sqrt{2t}}{t^2-2t}$

(e) $\lim_{y\to 9}\dfrac{y^2-81}{\sqrt{y}-3}$

(f) $\lim_{x\to -8}\dfrac{\sqrt[3]{x}}{(x+6)^4}$

(g) $\lim_{u\to 3\sqrt{3}}\dfrac{u^6-6u^3+9}{u^3+u^4-3-3u}$

(h) $\lim_{h\to 0}\dfrac{(x+h)^3-x^3}{h}$

(i) $\lim_{h\to 0}\dfrac{(1+h)^{-2}-1}{h}$

(j) $\lim_{x\to 0}\dfrac{x}{\sqrt{1+3x}-1}$

7.26. Explain why the floor function is an increasing function but not a strictly increasing function.

7.27. Explain why the IVT cannot be applied to $f(x)=\frac{1}{x}$ to conclude that the equation $f(x)=0$ has a solution in the interval $[-1,1]$.

7.28. Prove that there is a v with $0<v<2$ such that $v^2+\cos(\pi v)=4$.

7.29. Prove that $f(x)=\log\frac{1}{x}-10^x$ has a root between $\frac{1}{100}$ and $\frac{1}{10}$.

7.30. Prove that the function $1-4x\sin(\pi x)+x^2$ has two roots in the interval $[2,3]$.

7.31. Prove that there is a solution of the equation $-3x^3+2x^2+4x-5=0$ in the interval $[-2,-1]$. Find an interval of width less than 0.1 that contains this solution.

7.32. Answer True or False for each of the following statements. You may assume that all functions are real-valued functions.

(a) If f is continuous on $[-3,3]$, then $\lim_{x\to 0} x\,f(x)=0$.

(b) If f is continuous on $[-3,3]$, then it must be the case that $f(c)=1$ for some number c in $(-3,3)$.

(c) If f is continuous on $[-3,3]$ and $f(-3)=f(3)=1$, then $f(x)=0$ must have a solution in the interval $(-3,3)$.

(d) If f is continuous on $[-3,3]$ and $f(-3)=f(3)=1$, then $f(x)+x=0$ must have a solution in the interval $(-3,3)$.

(e) If f is continuous at $x=0$, then $g(x)=x^2f(x)$ is continuous at $x=0$.

(f) If f is continuous on the intervals $[0,1]$ and $[2,3]$, and $f(1)=-6$ and $f(2)=8$, then it has to be the case that $f(c)=7$ for some value c in the interval $(1,2)$.

(g) If $f(0)=2$ and $f(1)=-2$, then there must be a number c between 0 and 1 such that $f(c)=0$.

(h) If $f(2)=2$ and $f(5)=5$, and f is continuous on the interval $[2,5]$, then it must be the case that $f(3)=3$.

(i) If $f(-2)=-3$ and $f(5)=5$, and f is continuous on the interval $[-2,5]$, then it must be true that $f(c)=0$ for some number c with $-3<c<5$.

(j) If $f(-2)=1$ and $f(2)=-1$, and f is continuous on the interval $[-2,2]$, then it must be true that $f(c)=0$ for some number c with $-1<c<1$.

(k) If $f\left(\frac{\pi}{2}\right)=1$ and $f\left(\frac{3\pi}{2}\right)=-1$, then it must be the case that $f(x)=\sin(x)$.

(l) If $f(-3)=3$ and $f(-1)=-3$, and if $f(x)$ is decreasing on $[-3,-1]$, then it must be the case that $f(c)=0$ for some value c in the interval $(-3,-1)$ (Hint: increasing and decreasing functions need not be continuous functions [think of the floor function].)

(m) If f is continuous on the interval $\left[-\frac{\pi}{2}, \frac{\pi}{2}\right]$, then $g(x):=\frac{f(x)}{\cos(x)}$ is continuous on the interval $\left[-\frac{\pi}{2}, \frac{\pi}{2}\right]$.

(n) Suppose that f is continuous on the interval $(0,1)$. If $\lim_{x\to 0+} f(x)=6$ and $\lim_{x\to 1-} f(x)=1$, then there must be a number c between 0 and 1 such that $f(c)=3$.

(o) If $f(x)$ is continuous on $(0,3)$, then $f(3x)$ is continuous on $(0,1)$.

7.33. Determine the following limits.

(a) $\lim_{t\to\infty} \dfrac{4t^3+7t}{-t^3+2t^2+1}$

(b) $\lim_{x\to\infty} \dfrac{6x^2+27x-28}{x^2+x-12}$

(c) $\lim_{x\to-\infty} \dfrac{\sqrt{x^2+4x}}{4x+1}$

(d) $\lim_{t\to-\infty} \dfrac{3t^2+5t}{(t-2)(3-2t)}$

(e) $\lim_{x\to-\infty} \left(\sqrt{3x^2+3x-1}-x\right)$

(f) $\lim_{x\to-\infty} \left(\sqrt{3x^2+3x-1}+x\right)$

(g) $\lim_{x\to\infty} \dfrac{\sqrt{x^2-9}}{2x-6}$

(h) $\lim_{u\to\infty} \dfrac{(2-u)(1+u)}{2u(1-3u)}$

(i) $\lim_{x\to\infty} \dfrac{3t+5\sqrt{t}}{(\sqrt{t}-2)(3-2\sqrt{t})}$

7.34. Sketch the graphs of the following rational functions.

(a) $f(x)=\dfrac{-1}{x+1}$

(b) $f(x)=\dfrac{x-1}{x+1}$

(c) $f(x)=\dfrac{(x-1)^2}{(x+1)^2}$

(d) $f(x)=\dfrac{x^2-1}{x^2+1}$

(e) $f(x)=\dfrac{x^2-2x+1}{x^2-1}$

(f) $f(x)=\dfrac{x^2-2x+1}{x^3-1}$

(g) $f(x)=\dfrac{x^2+3x-4}{x^2+x-6}$

(h) $f(x)=\dfrac{x^4-2x^2+1}{10(x^2-4)}$

(i) $f(x)=\dfrac{10(x^2-4)}{x^4-2x^2+10}$

7.35. Use the rules for limits to compute the following limits (if the limits exist).

(a) $\lim_{x\to 0} \dfrac{\tan(5x)}{4\tan(2x)}$

(b) $\lim_{\theta\to 0} \dfrac{\cos(\theta)-1}{\sin(\theta)}$

(c) $\lim_{x\to 0} \dfrac{x}{\tan x}$

(d) $\lim_{h\to 0} \dfrac{\sin(5h)}{\tan(6h)}$

(e) $\lim_{t\to 0} \dfrac{t^2}{\sin^2(7t)}$

(f) $\lim_{\theta\to 0} \left(\dfrac{2}{\sin(3\theta)}\right)^2$

(g) $\lim_{x\to 0} \dfrac{x\cos x+\sin x}{x}$

CHAPTER 8

DIFFERENTIAL CALCULUS

8.1 INTRODUCTION

Differential calculus, the topic of this chapter, is, in large part, the legacy of Sir Isaac Newton. In his gigantic publication, *The Principia Mathematica*, Newton introduced the mathematics of calculus and applied it to the scientific study of the orbits of the planets around the sun. Thus, he not only revolutionized science but also created the mathematics he needed in order to do so.

This chapter deals with the derivatives of functions. While we study a function abstractly as mathematical object, in real-world applications it is a representation of motion (in a very general sense). In differential calculus, the notion of the derivative of a function, and the corresponding geometric notion of the slope of a tangent line to the graph of the function, relates to the idea of "instantaneous motion." As we will begin to explain, in this chapter, the knowledge gained about the "instantaneous motion" of a function enables us to investigate and discover important properties of the function; for example, where the peaks of a function occur, or the approximate behavior of a function near any particular point in the domain of the function.

In this chapter we will give the definition of the derivative, introduce the notion of a derivative function and present the rules (the power rule, sum rule, product rule, quotient rule and the chain rule) for computing it, and apply the definition and rules to compute the derivatives of the standard functions (such as polynomial, rational, root, trigonometric, exponential and logarithmic, inverse trig), and vector functions.

In section 8.2, the definition of the derivative is introduced in four parts: in section 8.2.1, the notion of a graph "leveling out" at the origin and its mathematical statement in terms of a limit is an intuitive and mathematically simplest starting point. In section 8.2.2, the tangent line to a graph at the origin is introduced as the *best approximating line* by means of a graphical example. This is interpreted in section 8.2.3, as it is stated that the *difference function* (graph minus tangent line) "levels out" at the origin. Consequently, the limit formula from section 8.2.1 can be applied to derive the slope of the tangent line at the origin. Finally, in section 8.2.4, this limit formula is generalized to compute the slope of the tangent line at any point (wherever a tangent line exists) by means of the simple trick of shifting the graph, so that the point in question is located at the origin.

Derivative functions are introduced in section 8.3 by means of a graphical demonstration of $f(x) = x^2$ and $f(x) = x^3$. This is followed by the introduction of the power, sum, product, and quotient rules and their applications to computing the derivatives of polynomial and rational functions. Tangent line problems and applications (including Newton's method for finding the roots of a function) are the topic of section 8.4. The power rule is generalized to include rational exponents in section 8.5, and the derivatives of trigonometric function are presented in section 8.6.

Students can gain a glimpse of the power of calculus for solving practical problems in the examples that are presented section 8.7. The first example deals with the derivative as a calculation of instantaneous speed (of a motorcar). This is followed by a discussion of the concept of linear approximation and the application of calculating approximate values of a function. The third and fourth examples deal with maximization and minimization problems, which are typical applications of calculus.

The chain rule, for computing the derivatives of compositions of functions, is explained in section 8.8. The calculus and the notion of a tangent vector to the trajectory of a vector-valued function are discussed in section 8.9. This is followed by the calculus of exponential and logarithmic functions (including a limit formula for Euler's number e) in section 8.10 and the derivative formulas for inverse trigonometric functions in section 8.11. We end with the application of finding the maximum viewing angle in a movie theater (which involves taking the derivative of the arctan function).

8.2 DEFINITION OF THE DERIVATIVE

The notion of a graph "flattening" or "leveling" out at the origin, as explained in section 8.2.1, will plays a crucial role in our derivation of the limit formula in section 8.2.3.

8.2.1 Graphs Tangential to the x-axis at the Origin

Compare the graphs of $f(x)=x,\ x^2,\ x^3$ in figure 8.1 and $f(x)=\sin(x),\ \cos(x)-1$ in figure 8.2.

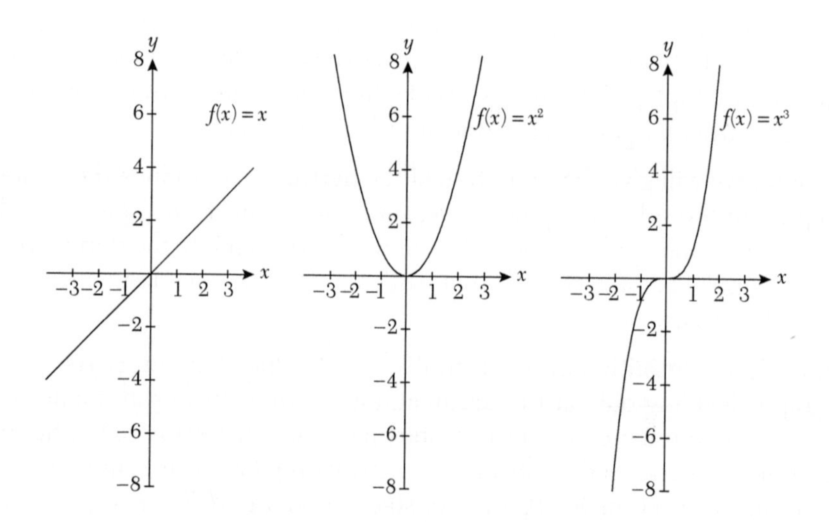

FIGURE 8.1. Graphs of polynomial functions.

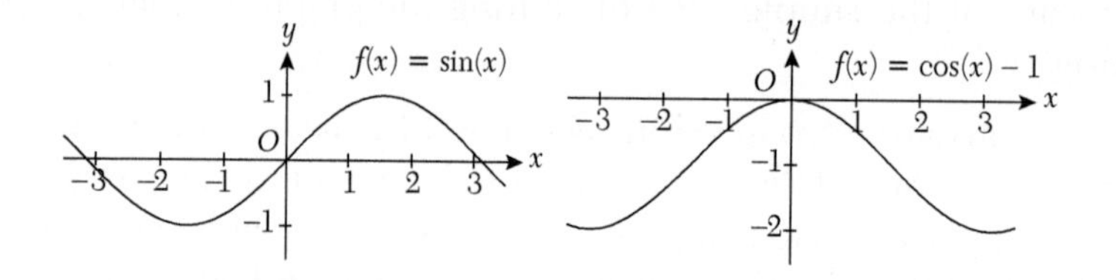

FIGURE 8.2. Graphs of trigonometric functions.

Observe that all the graphs pass through the origin and, furthermore, the graphs of $f(x)=x^2$, x^3, $\cos(x)-1$ "level out" at the origin. Mathematically speaking, we say that the graphs of the latter are *tangential* to the x-axis at the origin. In the case of the function $f(x)=x^2$, imagine an airplane coming in to land but at the moment of landing starts to takeoff again because the landing gear has not come out. At the moment of landing and takeoff, the motion of the airplane is tangential to the landing strip.

The following equation is a mathematical statement about the behavior of the function $f(x)=x^2$ near the origin:

$$\lim_{x\to 0}\frac{f(x)}{x}=\lim_{x\to 0}\frac{x^2}{x}=\lim_{x\to 0}x=0.$$

This leads to the following general statement:

DEFINITION 8.2.1. *A function* $g(x)$ *is tangential to the x-axis at the origin if and only if it is the case that* $g(0)=0$ *and* $\lim_{x\to 0}\frac{g(x)}{x}=0.$

Note that the functions $g(x)=x^3$ and $g(x)=\cos(x)-1$ satisfy this definition, and it is not hard to fabricate some other examples; for instance, $g(x)=x^3+x^2$, $g(x)=x\sin(x)$, and $g(x)=x\,|\,x\,|$ all satisfy this definition.

EXAMPLE 8.2.1. The absolute value function $g(x)=|\,x\,|$ is not tangential to the x-axis at the origin, because

$$\lim_{x\to 0^-}\frac{|x|}{x}=\lim_{x\to 0^-}\frac{-x}{x}=-1 \quad\text{and}\quad \lim_{x\to 0^+}\frac{|x|}{x}=\lim_{x\to 0^+}\frac{x}{x}=1$$

and so

$$\lim_{x\to 0}\frac{|x|}{x}\ \text{does not exist.}\quad \left(\text{In particular, } \lim_{x\to 0}\frac{|x|}{x}\neq 0.\right)$$

◈

8.2.2 The Tangent Line to a Graph at the Origin

Figure 8.3 shows the graph of a function $f(x)$ that passes through the origin, that is, $f(0)=0$ along with the graphs of four lines. Now, the question of interest is: *Which of the straight lines through the origin "best" approximates the behavior of $f(x)$ close to the origin?*

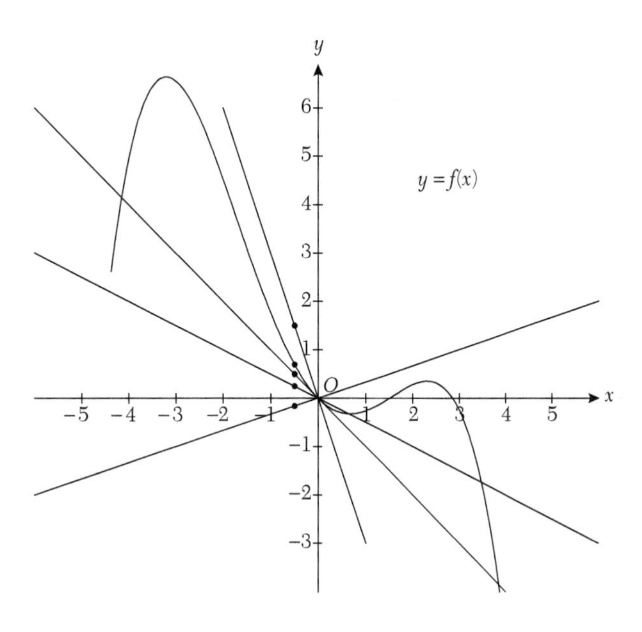

FIGURE 8.3. Approximating lines near the origin.

We can consider infinitely many different lines passing through the origin, but for simplicity, we are just comparing the four lines and asking which line best matches or best follows the graph of the function $f(x)$, as x approaches the origin. A method for deciding this is to examine the dots that are placed on each of the lines and the graph of $f(x)$, corresponding to a particular value of x close to zero, as shown in the diagram. If we visualize the movement of the dots as x approaches zero, then one of the lines will contain the dot that remains nearest to the dot on the graph of the function.

The line that we select in this way will be a candidate for the best approximating line to the graph of $f(x)$ at the origin. This method is not mathematically precise, but it is a helpful way to think about it. In section 8.2.3, we determine a formula that enables us to precisely find the best approximating line called the *tangent line*. The term *derivative* is used for the slope of the tangent line. Depending on the function, a tangent line might or might not exist (This will be explained in more detail below, with examples).

8.2.3 A Formula for the Derivative

We can write a decomposition of any given real-valued function $f(x)$ as a sum of a *linear function* mx and a *difference function* $g(x)$ as follows:

$$f(x) = mx + \underbrace{\left(f(x) - mx\right)}_{g(x)}. \tag{8.1}$$

The difference function $g(x)$ calculates the vertical distance between the function $f(x)$ and the line $y = mx$ for any value of x.

As a demonstration, we can subtract the line that we selected as the best approximating line (i.e., the tangent line) from the function $f(x)$ in figure 8.3. The graph of the resulting difference function $g(x)$ is shown in figure 8.4.

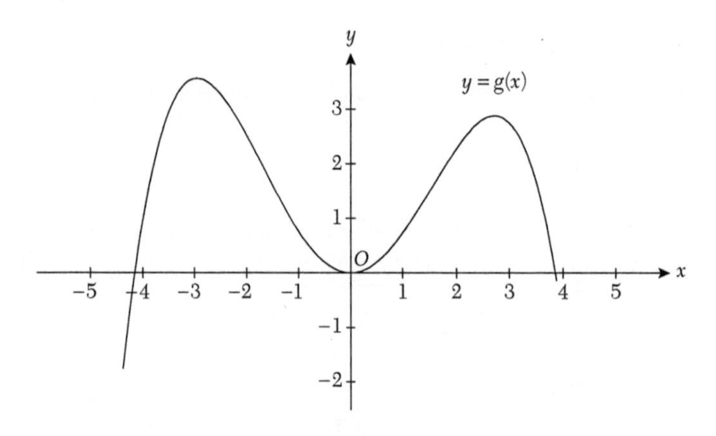

FIGURE 8.4. The graph of the difference function $g(x)$.

Suppose that the slope of the tangent line in figure 8.3 is some undetermined value $m = \alpha$, that is, the line $y = \alpha x$ is a tangent line to $f(x)$ at $x = 0$ for some value of α. We observe in figure 8.4 that the difference function $g(x) = f(x) - \alpha x$ is tangential to the x-axis at the origin. Therefore, according to definition 8.2.1

$$\lim_{x \to 0} \frac{g(x)}{x} = \lim_{x \to 0} \frac{f(x) - \alpha x}{x} = 0. \tag{8.2}$$

In order to calulate the value of α, we substitute $m=\alpha$ in formula (8.1) and divide both sides of the equation by x to obtain

$$\frac{f(x)}{x}=\alpha+\frac{f(x)-\alpha x}{x}.$$

We now take the limit on both sides, as x tends to 0. As a result of formula (8.2), we have

$$\begin{aligned}\lim_{x\to 0}\frac{f(x)}{x}&=\lim_{x\to 0}\left(\alpha+\frac{f(x)-\alpha x}{x}\right)\\&=\alpha+\lim_{x\to 0}\left(\frac{f(x)-\alpha x}{x}\right)\\&=\alpha+0\\&=\alpha.\end{aligned}$$

Reading the equation from right to left gives us a formula for α:

$$\alpha=\lim_{x\to 0}\frac{f(x)}{x}.$$

REMARK 8.2.1. This limit formula guarantees that any α that satisfies formula (8.2) is unique (and it can be calculated by means of this formula). In other words, if there is a tangent line to $f(x)$ at the origin, then there is only one tangent line.

REMARK 8.2.2. It is a good idea to use a letter different from x in the limit above because the evaluation of the limit is a separate calculation from evaluating the function. We use h in the continuation because this is the letter that is traditionally used in the limit formulas for computing derivatives.

We now introduce some important notation and summarize the results of this section:

DEFINITION 8.2.2. *If a function* $f(x)$ *is defined on an interval containing* $x=0$, *with* $f(0)=0$, *then the derivative* $f'(0)$ *is the slope of the tangent line to* $f(x)$ *at* $x=0$, *if a tangent line exists.*

REMARK 8.2.3. The symbol above the f is called a prime and we say "f'."

DEFINITION 8.2.3. *If a function* $f(x)$ *is defined on an interval containing* $x=0$ *with* $f(0)=0$, *then* $f(x)$ *is said to be differentiable at* $x=0$ *with derivative* $f'(0)$ *if and only if the function*

$$g(x)=f(x)-f'(0)x$$

satisfies

$$\lim_{h\to 0}\frac{g(h)}{h}=0.$$

Furthermore, if $f(x)$ *is differentiable at* $x=0$, *then*

$$\boxed{f'(0)=\lim_{h\to 0}\frac{f(h)}{h}}. \qquad (8.3)$$

REMARK 8.2.4. If, for some function f, $f(0)=0$, but the limit $\lim_{h\to 0}\frac{f(h)}{h}$ does not exist (i.e., f is not differentiable at the origin), then it might be the case that the graph of $f(x)$ has a vertical tangent line at the origin, as shown in example 8.2.2.

EXAMPLE 8.2.2. If $f(x)=\sqrt[3]{x}$, then

$$\lim_{h\to 0}\frac{f(h)}{h}=\lim_{h\to 0}\frac{\sqrt[3]{h}}{h}=\lim_{h\to 0}\frac{1}{\sqrt[3]{h^2}}=\infty.$$

Because the limit tends to infinity, we conclude that the y-axis, that is, the line $x=0$, is a tangent line to the graph of $f(x)=\sqrt[3]{x}$ at the origin (see figure 8.5). ◈

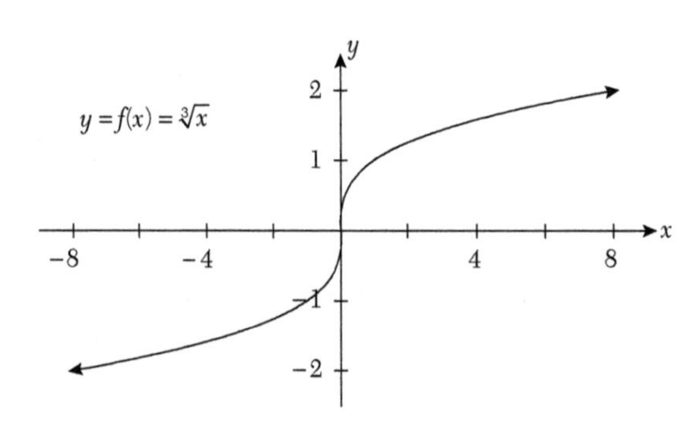

FIGURE 8.5. Graph of the cube root function.

EXAMPLE 8.2.3.

(i) If $f(x)=2x+x^3$, then

$$f'(0)=\lim_{h\to 0}\frac{f(h)}{h}=\lim_{h\to 0}\frac{2h+h^3}{h}=\lim_{h\to 0}\frac{h(2+h^2)}{h}=\lim_{h\to 0}(2+h^2)=2.$$

(ii) If $f(x)=\sqrt{x+1}-1$, then

$$\begin{aligned}f'(0)&=\lim_{h\to 0}\frac{f(h)}{h}\\&=\lim_{h\to 0}\frac{\sqrt{h+1}-1}{h}\\&=\lim_{h\to 0}\frac{\sqrt{h+1}-1}{h}\left(\frac{\sqrt{h+1}+1}{\sqrt{h+1}+1}\right)\\&=\lim_{h\to 0}\frac{h}{h\left(\sqrt{h+1}+1\right)}\\&=\lim_{h\to 0}\frac{1}{\sqrt{h+1}+1}=\frac{1}{2}.\end{aligned}$$

◈

8.2.4 Definition of the Derivative (General Case)

Our goal is to determine a formula for the tangent line (if a tangent line exists) at any point $x=a$ in the domain of a function $f(x)$. We use the notation $f'(a)$ for the slope of the tangent line at $x=a$. In order to relate the general case to the special case that we considered in section 8.2.3, we define

$$F(x)=f(x+a)-f(a). \tag{8.4}$$

Note that $F(0)=0$ and the graph of F can be obtained by shifting the graph of f to the left or right depending on whether a is positive or negative and up or down depending on whether $f(a)$

is negative or positive. Note that the tangent line to F at the origin will have the same slope as the tangent line to the graph of f at $x = a$, that is, $f'(a) = F'(0)$, as shown in figure 8.6.

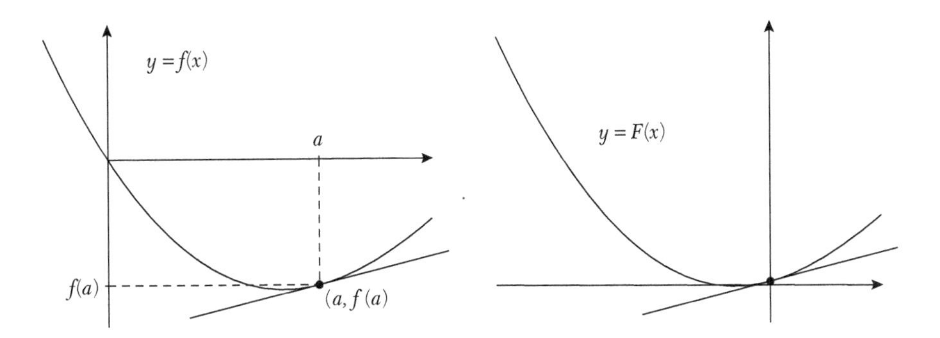

FIGURE 8.6. Shifted tangent lines that have the same slope.

According to Definition 8.2.1, if $F(x)$ is differentiable at the origin, then the derivative will be

$$\boxed{F'(0) = \lim_{h \to 0} \frac{F(h)}{h} = \lim_{h \to 0} \frac{f(a+h) - f(a)}{h}}. \tag{8.5}$$

Because we are equating $f'(a)$ with $F'(0)$, we have the following definition of the derivative.

Definition 8.2.4. *If a function* $\mathrm{f(x)}$ *is defined on an interval containing* $\mathrm{x = a}$, *then* $\mathrm{f(x)}$ *is said to be differentiable at* $\mathrm{x = a}$ *with derivative* $\mathrm{f'}(a)$ *if and only if the limit exists in the following formula.*

$$\boxed{f'(a) = \lim_{h \to 0} \frac{f(a+h) - f(a)}{h}} \tag{8.6}$$

Remark 8.2.5. If the limit in formula (8.6) does not exist, then $f(x)$ is not differentiable at $x = a$ and $f'(a)$ is not defined.

By means of an algebraic reformulation of the right-hand side of formula (8.6), we also have the alternative formula

$$\boxed{f'(a) = \lim_{x \to a} \frac{f(x) - f(a)}{x - a}}. \tag{8.7}$$

Note that, for a value of x close to a, the quantity

$$\frac{f(x) - f(a)}{x - a}$$

is the slope of a *secant line*, that is a line intersecting the graph of $y = f(x)$ at points $(a, f(a))$ and $(x, f(x))$. This observation allows an interesting geometric interpretation of a derivative:

Remark 8.2.6. The number $f'(a)$ is *the limit of slopes of secant lines as x tends to a.* Figure 8.7 shows secant and tangent lines passing through $(a, f(a))$.

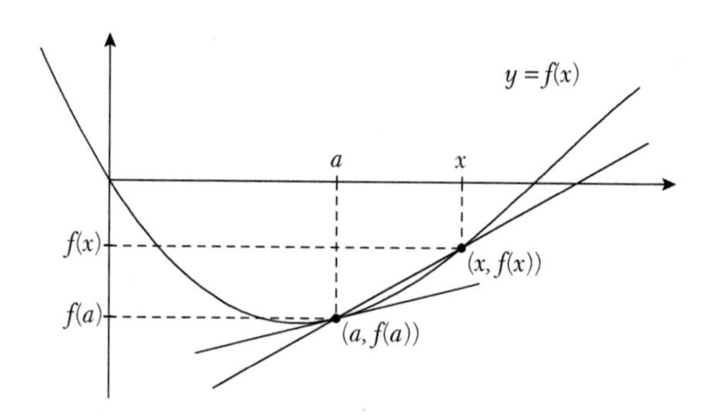

FIGURE 8.7. A secant line and a tangent line.

Because $f'(a)$ is the slope of the tangent line passing through $(a, f(a))$, we have the following important formula for the point-slope of the equation for the tangent line to the graph of $y = f(x)$ at $(a, f(a))$.

$$\boxed{y - f(a) = f'(a)(x - a)} \tag{8.8}$$

EXAMPLE 8.2.4.

(i) If $f(x) = 2x + x^3$, then

$$\begin{aligned}
f'(1) &= \lim_{h\to 0}\frac{f(1+h)-f(1)}{h} \\
&= \lim_{h\to 0}\frac{\left(2(1+h)+(1+h)^3\right)-\left(2(1)+(1)^3\right)}{h} \\
&= \lim_{h\to 0}\frac{(2+2h+1+3h+3h^2+h^3)-3}{h} \\
&= \lim_{h\to 0}\frac{5h+3h^2+h^3}{h} \\
&= \lim_{h\to 0}(5+3h+h^2) \\
&= 5
\end{aligned}$$

The equation of the tangent line passing through $(1, 3)$ is $y - 3 = 5(x - 1)$. ◈

(ii) If $f(x) = \sqrt{x+1}$, then

$$\begin{aligned}
f'(1) &= \lim_{h\to 0}\frac{f(1+h)-f(1)}{h} \\
&= \lim_{h\to 0}\frac{\sqrt{(h+1)+1}-\sqrt{(1)+1}}{h} \\
&= \lim_{h\to 0}\frac{\sqrt{h+2}-\sqrt{2}}{h}\left(\frac{\sqrt{h+2}+\sqrt{2}}{\sqrt{h+2}+\sqrt{2}}\right) \\
&= \lim_{h\to 0}\frac{h+2-2}{h\left(\sqrt{h+2}+\sqrt{2}\right)} \\
&= \lim_{h\to 0}\frac{1}{\sqrt{h+2}+\sqrt{2}} \\
&= \frac{1}{2\sqrt{2}}
\end{aligned}$$

The equation of the tangent line passing through $(1, \sqrt{2})$ is $y - \sqrt{2} = \frac{1}{\sqrt{2}}(x - 1)$. ◈

8.3 DERIVATIVE FUNCTIONS

In section 8.2, we introduced the notion of a function being differentiable at a point in its domain. It is useful to know when a function is differentiable at *every* point in its domain.

DEFINITION 8.3.1. *A function* $f(x)$ *is differentiable if and only if its derivative is defined at each point in its domain; that is, at any value of x in the domain of* $f(x)$, *there is a nonvertical tangent line passing through* $(x, f(x))$.

It is a fact that all of the following standard functions are differentiable: polynomial, trigonometric, rational, exponential and logarithmic functions.

Imagine an experiment done on graph paper, as shown in figure 8.8, where tangent lines are drawn to the graphs of the parabola $y = x^2$ and cubic $y = x^3$, and their slopes are measured accurately by counting blocks on the graph paper. If the slope values are plotted against the corresponding x values in a new coordinate system, then they create the graph of a new function, called the *derivative function*. In the case of the parabola, the graph of the derivative function is the line $y = 2x$, and in the case of the cubic it is the parabola $y = 3x^2$.

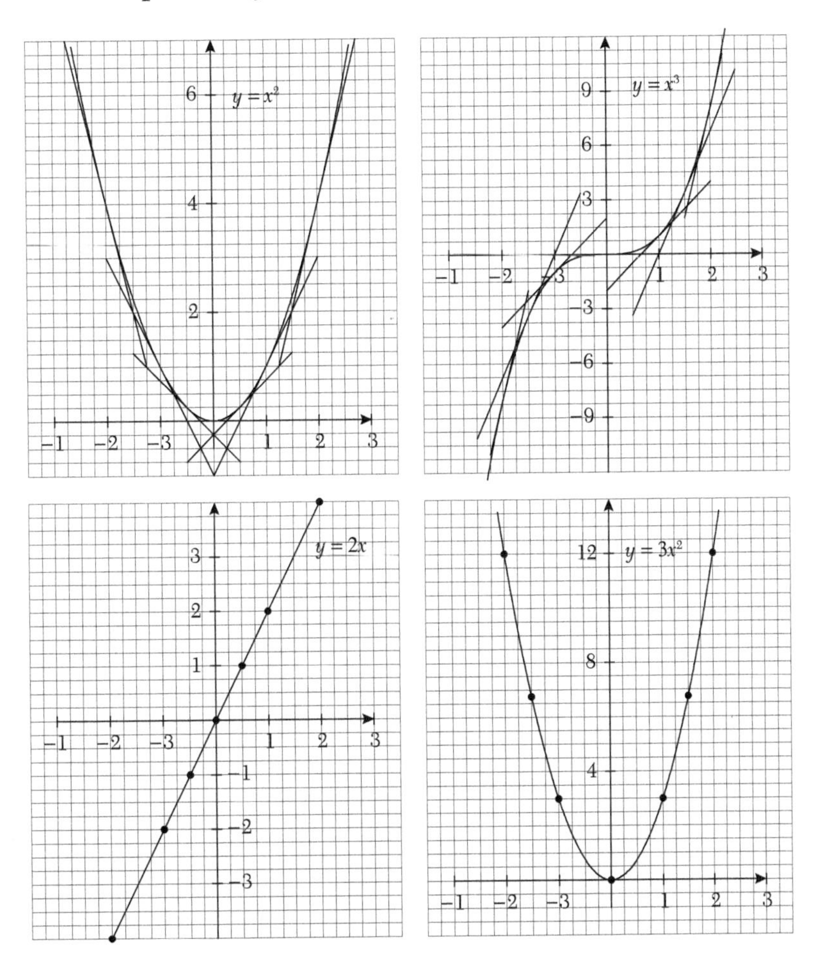

FIGURE 8.8. The derivative of the parabolic and cubic graphs.

In general, we have the following definition.

DEFINITION 8.3.2. *If a function* $f(x)$ *is differentiable, then the function* $f'(x)$ *that specifies the derivative of* $f(x)$ *at each value of x is called the derivative function.*

Remark 8.3.1. For any given differentiable function, the domain of the corresponding derivative function will be the same as the domain of the function.

Remark 8.3.2. If a function is not differentiable, that is, if there are points in the domain where the derivative does not exist, then a derivative function can be defined on the restricted domain that excludes those points.

Example 8.3.1. Figures 8.9 and 8.10 show the graphs of four functions and their corresponding derivative functions directly below them. Note that for graph (a), the value $x=0$ is excluded from the domain of the derivative function, and for graph (b), the values $x=-\frac{1}{2}, x=0, x=\frac{1}{2}$ are excluded from the domain of the derivative function. ◈

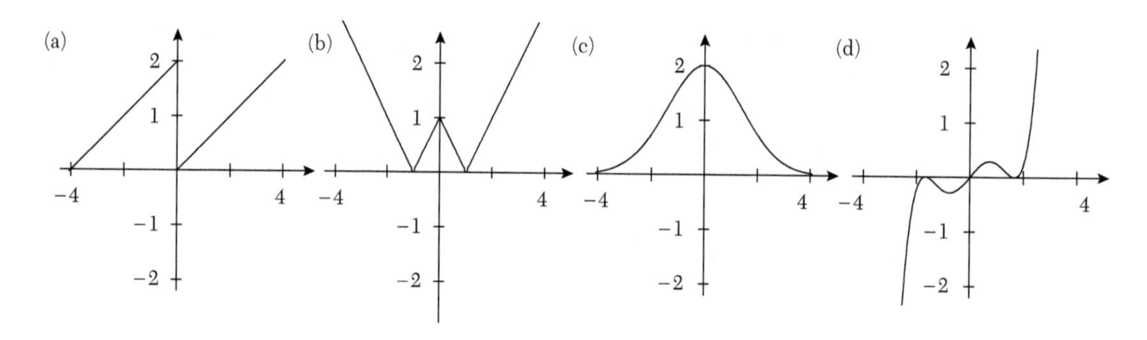

FIGURE 8.9. Diagram for Example 8.3.1.

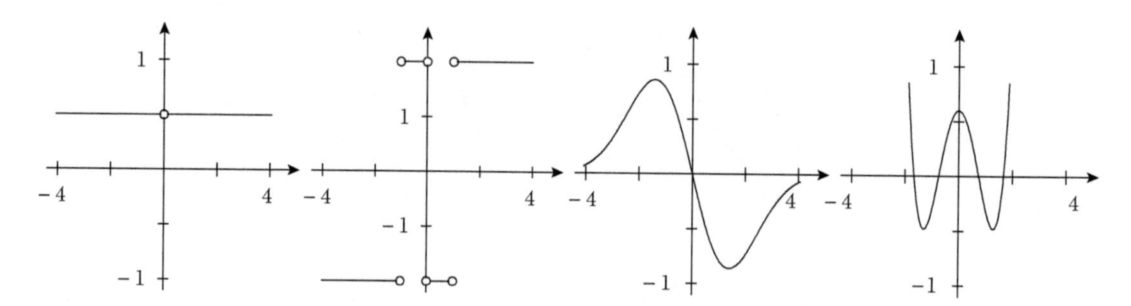

FIGURE 8.10. Diagram for Example 8.3.1.

An important observation regarding derivative functions is that the graph of a derivative function passes through the x-axis at every value for x where the function is differentiable and the graph of the function has an upward or downward peak. The reason for this is that a tangent line at any peak of a graph is horizontal (that is, having slope equal to zero). For example, in graph (c), in figure 8.9, the graph of the function has a peak at $x=0$, and the graph of the corresponding derivative function in figure 8.10 passes through the origin. Similarly, in graph (d), the function has four peaks, and the graph of the derivative function passes through the x-axis four times at the values of x where the peaks occur. Look closely at the graphs to verify this.

8.3.1 The Power Rule for Natural Numbers

We now use formula (8.6) to determine the derivative functions that are powers of x, that is, $f(x)=x^n$, where n is any natural number. In the case that $n=1$, the tangent line at any point on the graph coincides with the graph itself (a line through the origin with slope equal to 1); therefore, if

$f(x)=x$, then $f'(x)=1$ (the identity function). If $n=2$, that is, $f(x)=x^2$, then for any real number a we have

$$\begin{aligned}f'(a)&=\lim_{h\to0}\frac{f(a+h)-f(a)}{h}\\&=\lim_{h\to0}\frac{(a+h)^2-a^2}{h}\\&=\lim_{h\to0}\frac{a^2+2ah+h^2-a^2}{h}\\&=\lim_{h\to0}\frac{h(2a+h)}{h}\\&=2a.\end{aligned}$$

We can replace a with x, that is, if $f(x)=x^2$, then $f'(x)=2x$. Similarly, we can do the following calculation if $n=3$:

$$\begin{aligned}f'(a)&=\lim_{h\to0}\frac{f(a+h)-f(a)}{h}\\&=\lim_{h\to0}\frac{(a+h)^3-a^3}{h}\\&=\lim_{h\to0}\frac{a^3+3a^2h+3ah^2+h^3-a^3}{h}\\&=\lim_{h\to0}\frac{h(3a^2+3ah+h^2)}{h}\\&=3a^2.\end{aligned}$$

Again, we can replace a with x, that is, if $f(x)=x^3$, then $f'(x)=3x^2$. We now have the following formula, which will be proved in exercise 8.11. at the end of this chapter.

$$\boxed{\text{If } f(x)=x^n, \text{then } f'(x)=nx^{n-1}, \quad \text{for any natural number } n} \tag{8.9}$$

This is a special case of a more general formula, called the *power rule*, which is stated and proved in section 8.10.4.

REMARK 8.3.3. As the slope of a horizontal line is zero, the formula above is, in fact, also true if $n=0$, that is, if $f(x)=1=x^0$ (the identity function), then $f'(x)=0=0\cdot x^{-1}$ (the zero function).

EXAMPLE 8.3.2. If $f(x)=x^{101}$, then $f'(x)=101x^{100}$. ◈

The special case of the power rule that we have given above is the first of many rules and formulas that will be derived in this chapter for the computation of derivative functions. Before continuing with this, we now present another useful notation for the expression of derivatives and derivative functions.

8.3.2 Leibniz Notation

The *prime* notation being used up to now to express derivatives was introduced by the French Mathematician Joseph-Louise Lagrange (1736–1813), one of the great successors of Isaac Newton in the area of *classical mechanics* (the study of the motion of projectiles). (He also proved, among other things, that every natural number is a sum of four squares of whole numbers.)

Another major figure in the development of calculus was the French Mathematician Gottfried Wilhelm Leibniz (1646–1716), who developed much of calculus independently of, but slightly later than, Isaac Newton. Unfortunately, the relationship between Newton and Leibniz turned very sour after Leibniz published his results ahead of Newton in 1684 (Newton published the first edition of his *Principia Mathematica* in 1687), and Newton accused Leibniz of plagiarism. It can be said, however, that Leibniz's calculus had a much better grounding in terms of the notation that he invented; for example, the notation that is now called Leibniz notation for the expression of a derivative function is suggestive of the derivative as the limit of fractions (the limit of slopes of secant lines). In Leibniz notation, instead of $f'(x)$, we write

$$\boxed{\frac{d}{dx}f(x) \quad \text{or} \quad \frac{df}{dx} \quad \text{or} \quad \frac{dy}{dx}}. \tag{8.10}$$

If we want to express the derivative at a particular value of x, at $x = a$, for example, then instead of $f'(a)$, we can write

$$\boxed{\left.\frac{df}{dx}\right|_{x=a} \quad \text{or} \quad \left.\frac{dy}{dx}\right|_{x=a}}. \tag{8.11}$$

Formula (8.9) can, using Leibniz notation, be expressed as

$$\boxed{\frac{d}{dx}x^n = nx^{n-1}, \quad \text{for any natural number } n}. \tag{8.12}$$

EXAMPLE 8.3.3. $\left.\dfrac{dx^8}{dx}\right|_{x=2} = 8x^7\Big|_{x=2} = 8(2)^7 = 1{,}024.$ ◈

8.3.3 The Sum, Product, and Quotient Rules

There are rules for computing the derivative of algebraic combinations of functions; in particular, the formula for computing the derivative of a sum or difference of functions or a function multiplied by a real number is called the *sum rule*, the formula for computing the derivative of a product of two functions is called the *product rule*, and the formula for computing the derivative of a quotient of two functions is called the *quotient rule*.

THEOREM 8.3.1. *If* f(x) *and* g(x) *are continuous, differentiable functions, then the derivative of the sum or difference of* f *and* g *can be calculated using*

$$\boxed{\text{the sum rule}: \quad (f \pm g)'(x) = f'(x) \pm g'(x), \quad (cf)'(x) = cf'(x)}. \tag{8.13}$$

THEOREM 8.3.2. *If* f *and* g *are continuous, differentiable functions, then the derivative of the product of* f *and* g *can be calculated using*

$$\boxed{\text{the product rule}: \quad (f \cdot g)'(x) = f'(x)g(x) + f(x)g'(x)}. \tag{8.14}$$

THEOREM 8.3.3. *If* f *and* g *are continuous differentiable functions, then the derivative of the quotient of* f *and* g *can be calculated using*

$$\boxed{\text{the quotient rule}: \quad \left(\frac{f}{g}\right)'(x) = \frac{f'(x)g(x) - f(x)g'(x)}{g(x)^2}}. \tag{8.15}$$

REMARK 8.3.4. The sum rule states that a derivative distributes through a sum or difference of functions. The product and quotient rules, however, are not so simple: a derivative does not distribute through a product or quotient of functions as you might expect.

REMARK 8.3.5. It's important to write the correct order of terms in the numerator of the quotient rule because reversing the terms causes an incorrect sign:

$$g'(x)f(x)-g(x)f'(x)=-(f'(x)g(x)-f(x)g'(x)).$$

The sum, product and quotient rules are often expressed in the following abbreviated formats:

$$\text{the sum rule: } (f\pm g)' = f'\pm g', \quad (cf)' = cf' \tag{8.16}$$

$$\text{the product rule: } (fg)' = f'g+fg' \tag{8.17}$$

$$\text{the quotient rule: } \left(\frac{f}{g}\right)' = \frac{f'g+fg'}{g^2} \tag{8.18}$$

They can be proved using formula (8.6), and the rules for limits stated in section 7.10.2:

Proof of the sum rule: If f and g are continuous, differentiable functions, then

$$\begin{aligned}(f+g)'(a) &= \lim_{h\to 0}\frac{(f+g)(a+h)-(f+g)(a)}{h}\\ &= \lim_{h\to 0}\frac{f(a+h)+g(a+h)-f(a)-g(a)}{h}\\ &= \lim_{h\to 0}\frac{f(a+h)-f(a)}{h}+\lim_{h\to 0}\frac{g(a+h)-g(a)}{h}\\ &= f'(a)+g'(a).\end{aligned}$$

□

The proof of the product rule is more tricky, as it requires adding and subtracting the same term in the numerator in the second line of the proof and grouping and factoring terms in the third line of the proof.

Proof of the product rule: If f and g are continuous differentiable functions, then

$$\begin{aligned}(f\cdot g)'(a) &= \lim_{h\to 0}\frac{(f\cdot g)(a+h)-(f\cdot g)(a)}{h}\\ &= \lim_{h\to 0}\frac{f(a+h)g(a+h)-f(a+h)g(a)+f(a+h)g(a)-f(a)g(a)}{h}\\ &= \lim_{h\to 0}f(a+h)\left(\frac{g(a+h)-g(a)}{h}\right)+g(a)\left(\frac{f(a+h)-f(a)}{h}\right)\\ &= \lim_{h\to 0}f(a+h)\lim_{h\to 0}\left(\frac{g(a+h)-g(a)}{h}\right)+g(a)\lim_{h\to 0}\left(\frac{f(a+h)-f(a)}{h}\right)\\ &= f(a)g'(a)+g(a)f'(a).\end{aligned}$$

In the last line of this proof, we are allowed to replace $\lim_{h\to 0}f(a+h)$ with $f(a)$ because we are assuming that f is a continuous function (in particular, continuous at $x=a$). □

The proof of the quotient rule is left as an exercise at the end of this chapter.

We will now look at some examples and applications of the rules above. First, the sum rule can be combined with the power rule to compute derivatives of polynomials.

EXAMPLE 8.3.4. If $f(x)=1+2x+3x^2+4x^3$, then

$$\begin{aligned}f'(x) &= 0+2(1)+3(2x)+4(3x^2)\\ &= 2+6x+12x^2.\end{aligned}$$

◈

As an application of the product rule consider the following example:

EXAMPLE 8.3.5. If $f(x)=x^6$ and $g(x)=x^7$, then by the product and power rules,

$$(f\cdot g)'(x)=f'(x)g(x)+f(x)g'(x)=(6x^5)(x^7)+(x^6)(7x^6)=13x^{12}.$$

Of course, we could have first multiplied the functions and then applied the power rule to obtain the same answer, that is,

$$(f\cdot g)'(x)=\frac{d}{dx}x^{13}=13x^{12}.$$ ◈

Similarly, the product rule can be applied to a product of polynomials:

EXAMPLE 8.3.6.

$$\frac{d}{dx}(11+7x+2x^3)(2x+x^2+5x^4)=(7+6x^2)(2x+x^2+5x^4)+(11+7x+2x^3)(2+2x+20x^3)$$

There is no need to expand the answer. ◈

The product rule will become more useful as we learn derivative formulas for different types of function in this chapter. On the other hand, we can use the quotient rule immediately to obtain a generalization of the power rule: if m is a positive integer, then by the quotient rule,

$$\begin{aligned}\frac{d}{dx}x^{-m}&=\frac{d}{dx}\left(\frac{1}{x^m}\right)=\frac{(0)(x^m)-(1)(mx^{m-1})}{(x^m)^2}\\&=-mx^{m-1}x^{-2m}\\&=-mx^{-m-1}\end{aligned}$$

(The quotient rule was applied with $f(x)=1$ and $g(x)=x^m$.) This verifies the following statement of the power rule.

$$\boxed{\text{If } f(x)=x^n, \text{then } f'(x)=nx^{n-1}, \quad \text{for any integer } n} \qquad (8.19)$$

The quotient rule can be combined with the power rule and sum rule to compute derivatives of rational expressions.

EXAMPLE 8.3.7.

$$\begin{aligned}\frac{d}{dx}\left(\frac{x^2}{1+5x^3}\right)&=\frac{(2x)(1+5x^3)-(x^2)(15x^2)}{(1+5x^3)^2}\\&=\frac{2x+10x^4-15x^4}{(1+5x^3)^2}\\&=\frac{2x-5x^4}{(1+5x^3)^2}.\end{aligned}$$ ◈

REMARK 8.3.5. It frequently happens, when the quotient rule is applied (as in the example above), that the expression in the numerator can be simplified. However, there is usually no need to expand the expression in the denominator.

It might be the case that a function is expressed in terms of some unknown constants. When taking the derivative, the sum rule applies to these constants in the same way that it applies to real numbers (as constants). In the following examples, the constants are the numbers a, b, c, and d.

EXAMPLE 8.3.8.

(i) $\frac{d}{dx}(a+bx+cx^2+dx^3)=b+2cx+3dx^2$

(ii) $\frac{d}{dx}(a+bx)(c+dx)=(b)(c+dx)+(a+bx)(d)=ad+bc+2bdx$

(iii) $$\frac{d}{dx}\left(\frac{a+bx}{c+dx}\right)=\frac{(b)(c+dx)-(a+bx)(d)}{(c+dx)^2}$$
$$=\frac{bc-ad}{(c+dx)^2}$$ ◈

8.4 TANGENT LINE PROBLEMS

At this point, we take a break from the theoretical development of calculus to look at some applications that involve tangent lines. A calculus problem can typically be expressed as a problem that involves finding a tangent line with a particular slope or constructing a particular tangent line. In this section, we are going to begin with simple examples and then move on to an interesting application, called *Newton's method*.

The simplest application involving a tangent line is to apply formula (8.8) to find the equation at a given point on a graph (assuming a tangent line exists at the given point). Here is an example:

EXAMPLE 8.4.1. Find the equation of the line that is tangent to the graph of $f(x)=\frac{1}{x+2}$ at $x=1$.

Answer: The corresponding derivative function is

$$f'(x)=\frac{(0)(x+2)-(1)(1)}{(x+2)^2}=\frac{-1}{(x+2)^2},$$

so the slope of the tangent line at $x=1$ is $f'(1)=\frac{-1}{(1+2)^2}=\frac{-1}{9}$, and the tangent line passes through coordinates $\left(1,\frac{1}{3}\right)$. Therefore, the equation of the tangent line is

$$y-\frac{1}{3}=-\frac{1}{9}(x-1).$$ ◈

A slightly more complicated problem is finding a tangent line with specified slope.

EXAMPLE 8.4.2. At what point(s) on the graph of $y=\frac{x}{x^2+1}$ is (are) the tangent line(s) parallel to the line $x=2y$?

Answer: The slope at any point on the graph of $y=\frac{x}{x^2+1}$ is given by

$$\frac{dy}{dx}=\frac{(1)(x^2+1)-(x)(2x)}{(x^2+1)^2}=\frac{1-x^2}{(x^2+1)^2}.$$

In order for a tangent line to the graph to be parallel to the line $x=2y$, it must have the same slope as this line, which is $\frac{1}{2}$ (write the equation for the line as $y=\frac{1}{2}x$). Thus, we set

$$\frac{dy}{dx}=\frac{1}{2}.$$

Now, using the above expression for $\frac{dy}{dx}$, we need to solve for x in the equation

$$\frac{1-x^2}{(x^2+1)^2}=\frac{1}{2}.$$

This reduces to solving the quartic equation

$$1-4x^2-x^4=0,$$

which can be solved as a quadratic equation by means of the substitution $X=x^2$. Thus, the roots of the equation are $x=\pm\sqrt{\sqrt{5}-2}$, and, consequently, the required points on the graph are

$$\left(\sqrt{\sqrt{5}-2},\frac{\sqrt{\sqrt{5}-2}}{\sqrt{5}-1}\right) \text{ and } \left(-\sqrt{\sqrt{5}-2},\frac{-\sqrt{\sqrt{5}-2}}{\sqrt{5}-1}\right).$$

These coordinates are approximately $(0.4859, 0.393)$ and $(-0.4859, -0.393)$, respectively. ◈

The following problem involves finding the tangent lines (to a specified graph) that pass through a given point in the plane (which need not be a point on the graph). The solution to this problem involves introducing an additional variable a to represent an arbitrary point on the graph that the tangent line passes through. The required value(s) of a are then obtained by solving an equation determined by the requirement that the line passes through the given point.

Example 8.4.3. How many tangent lines to the graph of $y=\frac{x}{x-1}$ pass through the point $\left(2,\frac{1}{2}\right)$? Find the coordinates of the points at which the tangent lines touch this graph.

Answer:

$$\frac{dy}{dx}=\frac{(1)(x-1)-(x)(1)}{(x-1)^2}=\frac{-1}{(x-1)^2}.$$

Therefore, the equation for a tangent line passing through coordinates $\left(a,\frac{a}{a-1}\right)$ with slope $\frac{-1}{(a-1)^2}$ is

$$y-\frac{a}{a-1}=-\frac{1}{(a-1)^2}(x-a).$$

Because we require the tangent line to pass through coordinates $\left(2,\frac{1}{2}\right)$, we substitute $x=2$ and $y=\frac{1}{2}$ into this equation to produce

$$\frac{1}{2}-\frac{a}{a-1}=-\frac{1}{(a-1)^2}(2-a).$$

This simplifies to the equation

$$a^2+2a-5=0.$$

The two solutions for a (corresponding to two tangent lines) are $-1\pm\sqrt{6}$, and the corresponding coordinates on the graph of $y=\frac{x}{x-1}$ are

$$\left(-1-\sqrt{6},\frac{1+\sqrt{6}}{2+\sqrt{6}}\right) \text{ and } \left(-1+\sqrt{6},\frac{1-\sqrt{6}}{2-\sqrt{6}}\right).$$

The coordinates are approximately $(-3.45, 0.775)$ and $(1.45, 3.22)$, respectively.

Figure 8.11 shows the two tangent lines passing through these points on the graph. ◈

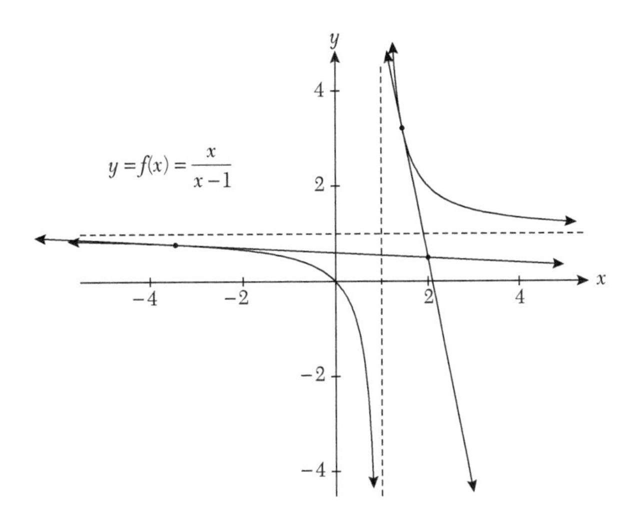

FIGURE 8.11. Diagram for Example 8.4.3.

A widely used application involving tangent lines is *Newton's method* (also called the *Newton–Raphson method*). This is a procedure for calculating accurately and efficiently the intercept of a graph of a function with the x-axis when the exact value of this intercept cannot be determined precisely (or cannot be determined easily). The method involves starting with an arbitrary value x_1 that is close to the intercept (it might be helpful to have a computer-generated graph of the function to begin with), and then, by means of an algorithmic procedure involving tangent lines, as demonstrated in the example below, finding values x_2, x_3, x_4, and so on that move closer and closer to the intercept. This method will always work if the following conditions are satisfied:

- the function is differentiable in an interval containing the x-intercept in question,
- the derivative of the function is not zero at the x-intercept in question, and
- the value x_1 is close enough to the intercept (if it is not close enough for the method to work, then a closer value should be used).

Example 8.4.4. Use Newton's method to find $\sqrt[5]{7}$ correct to three decimal positions (four significant digits).

Answer: If $f(x)=x^5-7$, then the root of this function in the interval $(1,2)$ is the value we are looking for because if we label this root $x=c$, then $f(c)=0$ means that $c=\sqrt[5]{7}$. (Note that $f(1)=-6$ and $f(2)=25$, so, by the Intermediate Value Theorem, we know there is a root in the interval $(1,2)$.) In the first step of the algorithm, we set $x_1=2$. Figure 8.12 shows a tangent line to the graph of $y=x^5-7$ through the coordinates $(2,25)$. Because $f'(x)=5x^4$, the slope of the tangent line is $f'(2)=80$, and the equation for this tangent line is $y-25=80(x-2)$ or $y=80x-135$. We will label the x-intercept of this tangent line as x_2. Then $0=80x_2-135$ determines that $x_2=\frac{13}{580}=\frac{27}{16}\approx 1.6875$. The second diagram demonstrates the next stage of the algorithm: another tangent line through the coordinates $\left(\frac{27}{16},\left(\frac{27}{16}\right)^5-7\right)$ is drawn, and its intercept $x_3=\frac{3236783}{2125764}\approx 1.5226$ is found. We can see in the diagram that the values x_2 and x_3 are closer to the intercept. The values $x_4\approx 1.4786$, $x_5\approx 1.47578$, and $x_6\approx 1.47577$ can be computed by continuing this procedure. Because the values x_5 and x_6 are stagnant in the first four decimal positions, we can state with confidence that $\sqrt[5]{7}\approx 1.475$. ◈

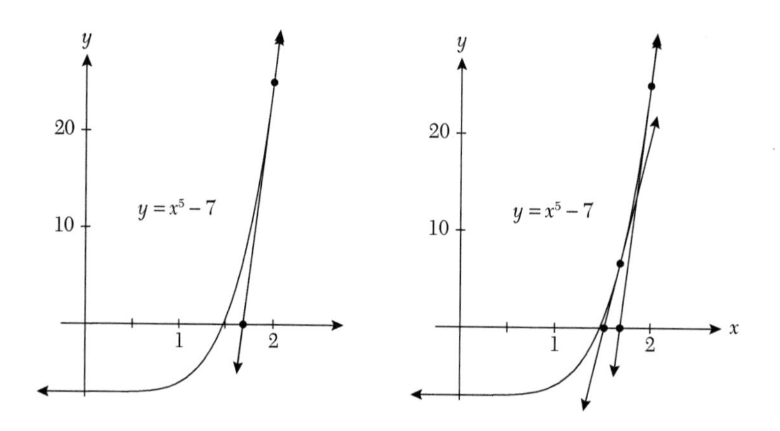

FIGURE 8.12. Two iterations of Newton's method.

8.5 THE POWER RULE FOR RATIONAL EXPONENTS

We now state the power rule for rational exponents. Again, this is a special case of the general formula for the power rule stated in section 8.10.4.

$$\boxed{\frac{dx^r}{dx} = rx^{r-1}, \quad \text{for any rational number } r} \tag{8.20}$$

Taking $r = \frac{1}{2}$ gives the formula for the derivative of a square root:

$$\frac{d\sqrt{x}}{dx} = \frac{dx^{1/2}}{dx} = \frac{1}{2}x^{-1/2} = \frac{1}{2\sqrt{x}}.$$

Because this formula will be used frequently, it is helpful to memorize it in the following form.

$$\boxed{\left(\sqrt{\cdot}\right)' = \frac{1}{2\sqrt{\cdot}}} \tag{8.21}$$

Loosely speaking, this states that "the derivative of a square root is one over twice a square root." Similar formulas can be obtained for the third and fourth roots:

$$\frac{d\sqrt[3]{x}}{dx} = \frac{dx^{1/3}}{dx} = \frac{1}{3}x^{-2/3} = \frac{1}{3\sqrt[3]{x^2}} \quad \text{and} \quad \frac{d\sqrt[4]{x}}{dx} = \frac{dx^{1/4}}{dx} = \frac{1}{4}x^{-3/4} = \frac{1}{4\sqrt[4]{x^3}}.$$

The power rule for rational exponents can be combined with the sum, product, and quotient rules to compute the derivative of any *rational algebraic expression* in one variable, that is, any expression that contains a variable and a finite number of algebraic operations (addition, subtraction, multiplication, division, and exponentiation with a rational exponent).

EXAMPLE 8.5.1. Using the sum, product, and quotient rules, respectively,

(i) $\frac{d}{dx}(2\sqrt{x}+3\sqrt{x^3})=\frac{d}{dx}(2\sqrt{x}+3x^{3/2})=\frac{1}{\sqrt{x}}+\frac{9\sqrt{x}}{2}$

(ii) $\frac{d}{dx}\sqrt{x}(x^2+1)=\left(\frac{1}{2\sqrt{x}}\right)(x^2+1)+\sqrt{x}(2x)=\frac{5x^2+1}{2\sqrt{x}}$

(iii) $\frac{d}{dx}\frac{\sqrt[3]{x}-1}{\sqrt[3]{x}+1}=\frac{\left(\frac{1}{3\sqrt[3]{x^2}}\right)(\sqrt[3]{x}+1)-(\sqrt[3]{x}-1)\left(\frac{1}{3\sqrt[3]{x^2}}\right)}{(\sqrt[3]{x}+1)^2}=\frac{2}{3\sqrt[3]{x^2}(\sqrt[3]{x}+1)^2}$ ◈

8.6 DERIVATIVES OF TRIGONOMETRIC FUNCTIONS

We will use formulas (8.3) and (8.6) to compute the derivatives of trigonometric functions.

If $f(x)=\sin(x)$ and $g(x)=\cos(x)$, then

$$f'(0)=\lim_{h\to 0}\frac{f(h)-f(0)}{h}=\lim_{h\to 0}\frac{\sin(h)}{h}=1$$

$$g'(0)=\lim_{h\to 0}\frac{g(h)-g(0)}{h}=\lim_{h\to 0}\frac{\cos(h)-1}{h}=0.$$

(These limits are given in formula (7.5).) Furthermore, we can compute $f'(x)$ and $g'(x)$ for any x by making use of the trigonometric identities for $\sin(A+B)$ and $\cos(A+B)$, the rules for limits and the two limits computed above:

$$\begin{aligned}f'(x)&=\lim_{h\to 0}\frac{f(x+h)-f(x)}{h}\\&=\lim_{h\to 0}\frac{\sin(x+h)-\sin(x)}{h}\\&=\lim_{h\to 0}\frac{\sin(x)\cos(h)+\cos(x)\sin(h)-\sin(x)}{h}\\&=\sin(x)\lim_{h\to 0}\frac{\cos(h)-1}{h}+\cos(x)\lim_{h\to 0}\frac{\sin(h)}{h}\\&=\sin(x)(0)+\cos(x)(1)\\&=\cos(x).\end{aligned}$$

$$\begin{aligned}g'(x)&=\lim_{h\to 0}\frac{g(x+h)-g(x)}{h}\\&=\lim_{h\to 0}\frac{\cos(x+h)-\cos(x)}{h}\\&=\lim_{h\to 0}\frac{\cos(x)\cos(h)-\sin(x)\sin(h)-\cos(x)}{h}\\&=\cos(x)\lim_{h\to 0}\frac{\cos(h)-1}{h}-\sin(x)\lim_{h\to 0}\sin(h)h\\&=\cos(x)(0)-\sin(x)(1)\\&=-\sin(x).\end{aligned}$$

We have thus obtained the following derivative formulas.

$$\boxed{\frac{d}{dx}\sin(x)=\cos(x), \qquad \frac{d}{dx}\cos(x)=-\sin(x)} \tag{8.22}$$

The derivatives of the remaining four trigonometric ratios can be obtained by means of the reciprocal trigonometric identities and the quotient and product rules. For example,

$$\begin{aligned}\frac{d}{dx}\tan(x)&=\frac{d}{dx}\frac{\sin(x)}{\cos(x)}\\&=\frac{\cos(x)\cdot\cos(x)-\sin(x)\cdot(-\sin(x))}{\cos^2(x)}\\&=\frac{\cos^2(x)+\sin^2(x)}{\cos^2(x)}\\&=\frac{1}{\cos^2(x)}\\&=\sec^2(x).\end{aligned}$$

The proofs of the remaining trigonometric ratios are left as exercises. It is well worthwhile memorizing their derivatives given in table 8.1.

TABLE 8.1. Derivatives of trigonometric ratios

$\frac{d}{dx}\sin(x)=\cos(x)$	$\frac{d}{dx}\cos(x)=-\sin(x)$
$\frac{d}{dx}\tan(x)=\sec^2 x$	$\frac{d}{dx}\cot(x)=-\csc^2 x$
$\frac{d}{dx}\sec(x)=\sec(x)\tan(x)$	$\frac{d}{dx}\csc(x)=-\csc(x)\cot(x)$

The following examples demonstrate the product rule combined with the derivatives of trigonometric functions.

Example 8.6.1.

(i) $\frac{d}{dx}x\sin(x)=\sin(x)+x\cos(x)=\sin(x)+x\cos(x)$

(ii) $\frac{d}{dx}x^2\sec(x)=2x\sec(x)+x^2\sec(x)\tan(x)$ ◈

8.7 SOME BASIC APPLICATIONS OF CALCULUS

By means of the examples below, we demonstrate how the techniques of calculus can be used to solve many kinds of practical problems.

In the first example, we introduce the important notion of *instantaneous rate of change* and explain how it relates to the derivative of a *time–displacement function* of an object in motion. The general study or science of motion is called *dynamics*, and, because of this relationship of the derivative to instantaneous change, calculus is an essential tool in dynamics. A more general and sophisticated expression for a moving body than a simple time–displacement function is a *differential equation*. Students often ask why they need to learn calculus. The best answer to this question is that calculus is the foundation for the mathematics of differential equations.

EXAMPLE 8.7.1. A car approaches an intersection with time–displacement function $s(t)=t^2-4$, where s is measured in meters and t is measured in seconds. (The letter s is the variable that physicists typically use to represent displacement or position, and t is the letter used to represent time, so $s(t)$ is a function that gives the position of the car at any given time.) The value $s(0)=-4$ is the initial displacement. If the car is 4 m from the intersection when $t=0$ (when an observer starts his timer), then the car passes through the intersection when $t=2$. The graph of the time–displacement function $s(t)=t^2-4$ is shown in figure 8.13 with time t on the horizontal axis and displacement s on the vertical axis.

In the first diagram in figure 8.13, the slope of the secant line determines the average speed of the car from $t=0$ to $t=2$ because average speed is calculated using the formula $\frac{\Delta s}{\Delta t}$, where the symbol "$\Delta$" denotes change, that is, Δs is the change in displacement and Δt is the change in time. For the secant line, $\frac{\Delta s}{\Delta t}=\frac{4}{2}=2$. As s is measured in meters and t is measured in seconds, the average speed of the car from $t=0$ to $t=2$ is 2 m/s.

Similarly, in the second diagram in figure 8.13, the slope of the secant line determines the average speed of the car from $t=1$ to $t=2$. Here, $\frac{\Delta s}{\Delta t}=\frac{3}{1}=3$, so the average speed of the car from $t=1$ to $t=2$ is 3 m/s.

Finally, in the third diagram in figure 8.13, there is a tangent line to the graph at $t=2$. We can imagine secant lines as in the first and second diagrams merging with the tangent line as Δt gets smaller and smaller. Therefore, we calculate the slope of the tangent line in order to determine the instantaneous speed of the car when $t=2$ (the reading on the speedometer at the moment the car passes through the intersection). Because $\left.\frac{ds}{dt}\right|_{t=2}=2t\big|_{t=2}=4$, the instantaneous speed of the car at $t=2$ is 4 m/s. ◈

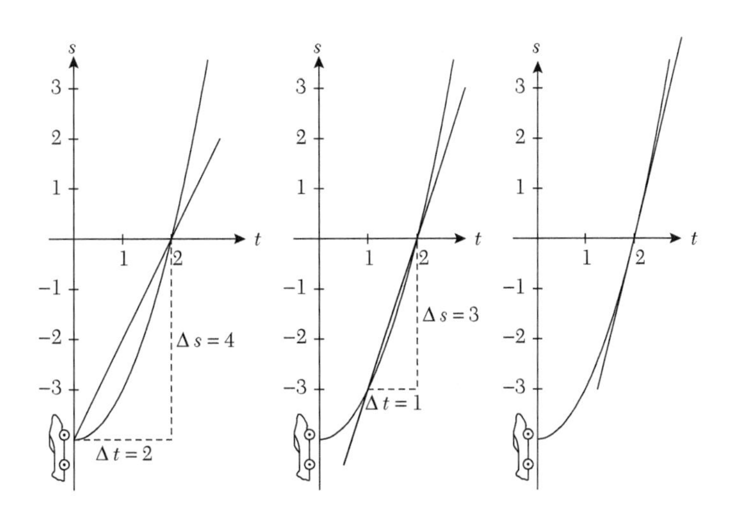

FIGURE 8.13. A car approaching an intersection.

In a certain sense, calculus is a *science of approximation*. The reason for this begins with the observation that a tangent line is a good approximation to a function close to the point of tangency. Indeed, this was the criterion used in section 8.2.2 to define the tangent line.

Now, recall formula (8.8), which is the equation for the tangent line if $f(x)$ is differentiable at $x=a$ and the point of tangency is $(a, f(a))$. If x is close to a (here, "close" just means close enough for the purpose in mind), then the factor $(x-a)$ can be abbreviated by Δx, meaning a small increment in the independent variable x, and $y-f(a)$ can be abbreviated by Δy, meaning a small increment in the dependent variable y. The equation for the tangent line can thus be expressed as $\Delta y = f'(a)\Delta x$. This states that the increment in y along the tangent line is proportional to the increment in x according to the factor $f'(a)$. The actual amount by which the function f changes corresponding to the increment Δx is $f(a+\Delta x)-f(a)$, and this is approximately the value Δy. Another way to write this approximation is

$$f(a+\Delta x) \approx f(a)+\Delta y = f(a)+f'(a)\Delta x. \tag{8.23}$$

This is shown in figure 8.14.

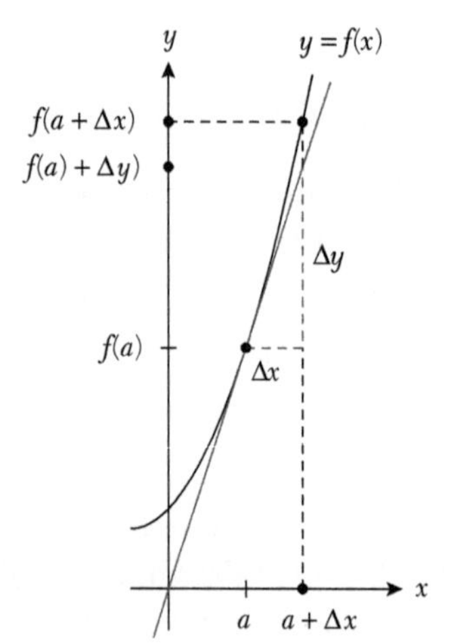

FIGURE 8.14. Linear approximation of a function.

Remark 8.7.1. The equation $\Delta y = f'(a)\Delta x$ or (with x replacing a) $\Delta y = f'(x)\Delta x$ resembles the equivalence $\frac{dy}{dx} = f'(x)$. We do not regard "$\frac{dy}{dx}$" as a fraction in the usual sense; however, in the advanced mathematics of tensor algebra, expressions such as dy, df, or dx are called differential forms, and they are given a precise meaning. We cannot explain this in detail here, but we will regard the equation $dy = f'(x)dx$ as a valid statement and call the terms differential forms. Formula (8.23) can now be expressed as the differential approximation formula

$$\boxed{f(x+dx) \approx f(x)+dy = f(x)+f'(x)dx}. \tag{8.24}$$

Example 8.7.2. Suppose we want to know an approximae value for $\cos(2.1)$, given that $\cos(2.0) \approx -0.416$ and $\sin(2.0) \approx -0.9$. We define $f(x)=\cos(x)$ and use formula (8.23) with $a=2.0$ and $\Delta x = 0.1$ to obtain

$$\cos(2.1) \approx \cos(2)-0.09(0.1) \approx -0.416-0.009 = -0.506\,.$$

(The actual value of $\cos(2.1)$ is $-0.50484\ldots$.) ◈

It has already been mentioned (at the end of section 8.3) that the slope of the tangent line is zero wherever the graph of a differentiable function has an upward or a downward peak. This is known as Fermat's Theorem, after the French Mathematician Pierre de Fermat, and it can be proved rigorously using the definition of the derivative. (We will not provide the proof here.)

Definition 8.7.1. *A local extremum* x_0 *of a real-valued function* f *is a value for* x *where the graph of* f(x) *has an upward or downward peak.*

THEOREM 8.7.1. Fermat's Theorem: *if* x_0 *is a local extremum of a real-valued function f defined on an interval* (a,b) *containing* x_0 *and, if f is differentiable at* $x = x_0$, *then* $f'(x_0)=0$.

Thus, a method of locating an upward or downward peak of graph (an extremum) involves taking the derivative of the function, setting it equal to zero and solving for x. The solutions for x are called *critical values*. Additional information about the shape of the graph can then be used to determine whether a particular critical value obtained in this way is a local extremem. The next two examples demonstrate this.

EXAMPLE 8.7.3. A rectangular open box that is twice as long as it is wide has a volume of 48 cm³. What are the dimensions of the box if it has the smallest possible surface area.

Answer: If the unknown width is labeled x, then the length should be labeled $2x$. The unknown height of the box can be labeled h (see figure 8.15). We have the following formulas for the volume (V) and surface area (SA):

$$V = 2x^2h = 48$$

$$SA = 2(2xh) + 2(xh) + 2x^2 = 6xh + 2x^2.$$

Now, if we solve for h in the first equation above and substitute the result in the second equation, then we can define the following function of x, which determines the surface area:

$$f(x) = 6x\left(\frac{48}{2x^2}\right) = \frac{144}{x} + 2x^2.$$

Note that

$$\lim_{x\to 0^+} f(x) = \infty \quad \text{and} \quad \lim_{x\to\infty} f(x) = \infty.$$

Furthermore, the following calculation determines that there is a single critical point, which must be the point where f attains its minimum value.

$$f'(x) = -\frac{144}{x^2} + 4x = 0$$

$$-144 + 4x^3 = 0$$

$$x^3 = 36$$

$$x = \sqrt[3]{36}.$$

Therefore, the box should have a width of $\sqrt[3]{36} \approx 3.3$ cm, a length of $2\sqrt[3]{36} \approx 6.6$ cm, and a height of $\frac{24}{\sqrt[3]{36^2}} \approx 2.2$ cm. ◇

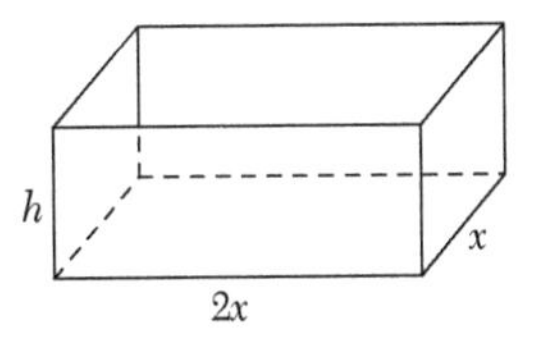

FIGURE 8.15. A rectangular, open box.

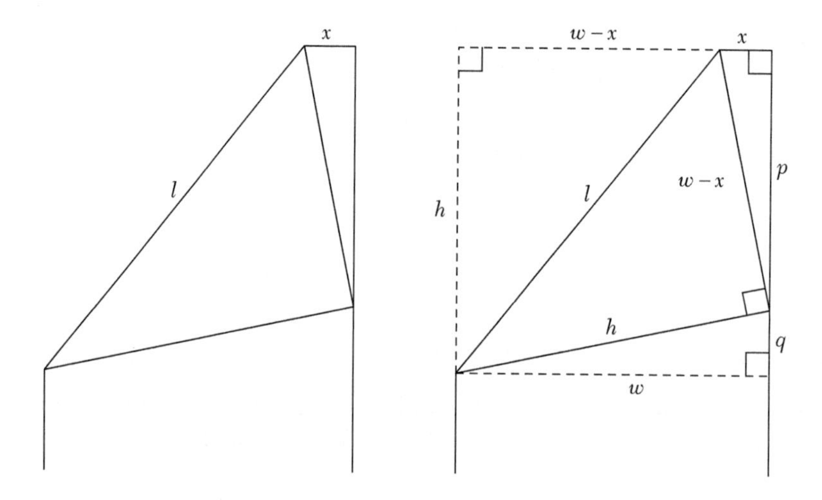

FIGURE 8.16. Folding a rectangular sheet.

EXAMPLE 8.7.4. How would you fold an arbitrarily long rectangular sheet of paper with centimeters wide to bring the upper left corner to the right-hand edge and minimize the length of the fold? That is, in figure 8.16, how do you choose x in order to minimize l?

Answer: It is generally a good problem-solving strategy to consider a few possibilities before writing down any equations: first, if $x = 0$, then folding the upper left corner to the right-hand edge creates an equilateral, right triangle, where the two short sides have length w centimeters and the length of the hypotenuse (the fold) is equal to $\sqrt{2}w$ centimeters; second, the maximum value of x for which it is possible for the upper left corner to be folded to the right-hand edge is $\frac{w}{2}$ centimeters (half of the width), and this would result in a fold of infinite length (as we are assuming the rectangular sheet is arbitrarily long). So, presumably, the value for x we are looking for must be somewhere between 0 and $\frac{w}{2}$ centimeters (the answer clearly cannot be $x=\frac{w}{2}$ centimeters, although it can be 0 centimeters).

We now proceed by setting up some equations and applying the methods of calculus. We begin by identifying and labeling all the parameters of the problem, as shown in figure 8.16. The fold forms the hypotenuse of two congruent right triangles that have one side of length $w-x$ and another side of length h. Along the right edge of the sheet, p and q add up to h. Our strategy will be to find an expression for h in terms of x and the constant w and then to use it to find an expression for l in terms of x and w. (Try to do this yourself before reading ahead.)

The first step is to solve for p in the Pythagorean formula

$$p^2 + x^2 = (w-x)^2.$$

Thus,

$$p = \sqrt{w(w-2x)}.$$

Consequently,

$$q = h - p = h - \sqrt{w(w-2x)}.$$

We now substitute this expression for q into the Pythagorean formula

$$q^2 + w^2 = h^2$$

to obtain

$$\left(h - \sqrt{w(w-2x)}\right)^2 + w^2 = h^2.$$

By expanding the left-hand side of this equation and subtracting h^2 from both sides, we are left with

$$-2h\sqrt{w(w-2x)}+w(w-2x)+w^2=0.$$

Thus, we can solve for h to obtain

$$h=\frac{w(w-x)}{\sqrt{w(w-2x)}}.$$

Finally, we have

$$\begin{aligned} l^2=(w-x)^2+h^2 &= (w-x)^2+\frac{w(w-x)^2}{w-2x} \\ &= (w-x)^2\left[1+\frac{w}{w-2x}\right] \\ &= (w-x)^2\left[\frac{2w-2x}{w-2x}\right] \\ &= \frac{2(w-x)^3}{w-2x}. \end{aligned}$$

We can take the square root on both sides so obtain an explicit expression for l in terms of x, but the calculus will be easier if we set $L=l^2$ and then find the value for x that minimizes L. (This is a good trick that is worth remembering.) So, now we apply the quotient rule to the function

$$L(x)=\frac{2(x-w)^3}{2x-w}$$

to obtain

$$\begin{aligned} L'(x) &= \frac{6(x-w)^2(2x-w)-2(x-w)^3(2)}{(2x-w)^2} \\ &= (x-w)^2\,\frac{(12x-6w)-(4x-4w)}{(2x-w)^2} \\ &= 2(x-w)^2\,\frac{(4x-w)}{(2x-w)^2}. \end{aligned}$$

Now solving for x in the equation

$$L'(x)=2(x-w)^2\,\frac{(4x-w)}{(2x-w)^2}=0$$

yields the critical values

$$x=\frac{w}{4} \quad \text{and} \quad x=w.$$

(We are not interested in the critical value $x=w$.) Now, it is a consequence of the following calculation:

$$L\left(\frac{w}{4}\right)=\frac{2\left(\frac{w}{4}-w\right)^3}{2\left(\frac{w}{4}\right)-w}=\frac{2\left(\frac{3w}{4}\right)^3}{\left(\frac{w}{2}\right)}=\frac{27w^2}{16}<2w^2=L(0)$$

and

$$\lim_{x \to \left(\frac{w}{2}\right)^-} L(x) = \infty$$

that we get a shorter fold by folding at $x=\frac{w}{4}$ than by folding at $x=0$ or $x=\frac{w}{2}$ and, because $x=\frac{w}{4}$ is the only critical value in the interval $\left[0,\frac{w}{2}\right)$, folding at one quarter of the width must produce the shortest possible fold. ◈

8.8 THE CHAIN RULE

The *chain rule* is a method for taking the derivative of a composition of functions. In the simplest case, we can consider the composition of lines passing through the origin. For example, if the lines are the graphs of equations $y=L_2(u)=-\frac{1}{2}u$ and $u=L_1(x)=3x$, then the composition $L_2 \circ L_1$ of these lines is the line

$$y = L_2 \circ L_1(x) = L_2(L_1(x)) = L_2(3x) = -\frac{1}{2}(3x) = -\frac{3}{2}x.$$

The graphs of the lines are shown in figure 8.17 in the (x,u)-coordinate plane, the (u,y) coordinate plane, and the (x,y)-coordinate plane, respectively.

It is important to note that the slope of the third line is the product of the slopes of the first two lines, that is, *taking a composition of lines results in a new line whose slope is the product of slopes of the lines*.

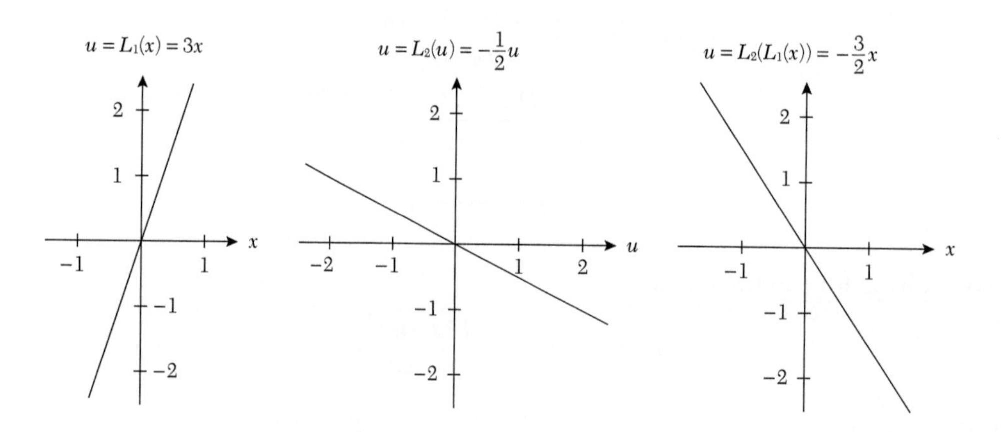

FIGURE 8.17. A composition of lines.

Suppose that, instead of lines $L_1(x)$ and $L_2(u)$ passing through the origin, we have functions $f(u)$ and $g(x)$ whose graphs pass through the origin, that is, $f(0)=0$ and $g(0)=0$. Suppose, furthermore, that g and *f* are differentiable at the origin, that is, $g'(0)$ and $f'(0)$ exist and so, by formula (8.3), can be expressed as:

$$g'(0) = \lim_{x \to 0} \frac{g(x)}{x} \quad \text{and} \quad f'(0) = \lim_{u \to 0} \frac{f(u)}{u}.$$

We can use formula (8.3) again and rules for limits to compute the derivative $(f \circ g)'(0)$ of the composition $f \circ g$.

$$
\begin{aligned}
(f \circ g)'(0) &= \lim_{x \to 0} \frac{f \circ g(x)}{x} \\
&= \lim_{x \to 0} \frac{f(g(x))}{x} \\
&= \lim_{x \to 0} \frac{f(g(x))}{g(x)} \frac{g(x)}{x} \\
&= \lim_{u \to 0} \frac{f(u)}{u} \lim_{x \to 0} \frac{g(x)}{x} \\
&= f'(0)g'(0).
\end{aligned}
$$

We have proved the following special case of the chain rule: if $f(0)=0$ and $g(0)=0$, then

$$\boxed{(f \circ g)'(0) = f'(0)g'(0)} \tag{8.25}$$

provided $g'(0)$ and $f'(0)$ exist.

From this result we can conclude, furthermore, that the tangent line (through the origin) to the composition function $f \circ g$ is the composition of the tangent lines (through the origin) to the functions f and g. This is illustrated in figure 8.18.

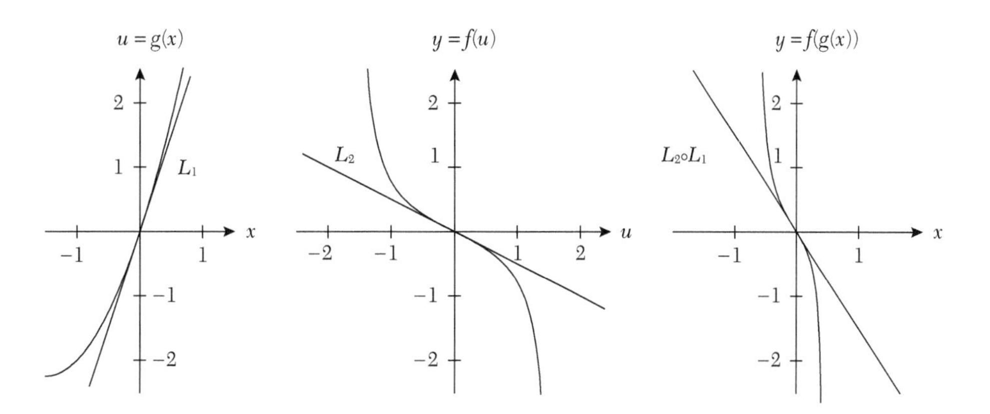

FIGURE 8.18. A composition of graphs.

EXAMPLE 8.8.1. The diagram in figure 8.18 was generated using $f(u) = -\frac{1}{2}\tan(u)$, $g(x) = x^2 + 3x$ and $f \circ g(x) = f(g(x)) = -\frac{1}{2}\tan(g(x)) = -\frac{1}{2}\tan(x^2 + 3x)$.

Because $f'(u) = -\frac{1}{2}\sec^2(u)$ and $g'(x) = 2x + 3$, we get $f'(0) = -\frac{1}{2}$ and $g'(0) = 3$.

Therefore, by formula (8.25),

$$(f \circ g)'(0) = f'(0)g'(0) = \left(-\frac{1}{2}\right)(3) = -\frac{3}{2}.$$

◈

It will be proved in exercise 8.47 that formula (8.25) can be generalized to compute the derivative of a composition of functions at *any* point in the domain of the composition.

This generalization is:

THEOREM 8.8.1. The chain rule: *If* f *and* g *are continuous differentiable functions, and if* a *is in the domain of* g *and* g(a) *is in the domain of* f, *then the derivative* $(f \circ g)'(a)$ *can be computed using*

$$\boxed{\text{the chain rule:} \quad (f \circ g)'(a) = f'(g(a))g'(a)} \tag{8.26}$$

Theorem 8.8.1 states that the derivative of $f \circ g$ at $x = a$ is computed as the derivative of f evaluated at the image of g at $x = a$, multiplied by the derivative of g at $x = a$. (Loosely speaking, we evaluate the derivative of the "outside function" at the "inside function" and multiply by the derivative of the "inside function".)

The chain rule can also be expressed using Leibniz notation.

$$\boxed{\frac{dy}{dx} = \left.\frac{dy}{du}\right|_{u=u(x)} \cdot \frac{du}{dx}} \tag{8.27}$$

$$\boxed{\left.\frac{dy}{dx}\right|_{x=a} = \left.\frac{dy}{du}\right|_{u=b} \cdot \left.\frac{du}{dx}\right|_{x=a}, \quad \text{where } b = u(a)} \tag{8.28}$$

EXAMPLE 8.8.2. If $y = u^2$ and $u = \sin(x)$, then

$$\frac{dy}{dx} = 2u\Big|_{u=u(x)} \cdot \cos(x) = 2\sin(x)\cos(x).$$ ◈

EXAMPLE 8.8.3. If $y = \sin(u)$ and $u = x^2$, find $\frac{dy}{dx}$ and $\left.\frac{dy}{dx}\right|_{x=\sqrt{\pi}}$

Answer:

$$\frac{dy}{dx} = \frac{dy}{du} \cdot \frac{du}{dx} = \cos(u) \cdot 2x = 2x\cos(x^2)$$

$$\left.\frac{dy}{dx}\right|_{x=\sqrt{\pi}} = \left.\frac{dy}{du}\right|_{u=\pi} \cdot \left.\frac{du}{dx}\right|_{x=\sqrt{\pi}} = \cos(u)\Big|_{u=\pi} \cdot 2x\Big|_{x=\sqrt{\pi}} = 2\sqrt{\pi}\cos(\pi) = -2\sqrt{\pi}$$ ◈

The functions that make up a composition are not always stated explicitly. In this situation, the functions that make up the composition have to be determined.

EXAMPLE 8.8.4. The function $h(x) = (1 - x^2)^{99}$ is the composition $f(g(x))$, where $f(u) = u^{99}$ and $g(x) = 1 - x^2$. Thus,

$$\begin{aligned} h'(x) = f'(g(x))g'(x) = f'(1-x^2)(-2x) &= 99(1-x^2)^{98}(-2x) \\ &= -198x(1-x^2)^{98}. \end{aligned}$$ ◈

REMARK 8.8.1. When the "outside function" is of the form x^α (as in the previous example, with $\alpha = 99$), then the application of the chain rule is sometimes called the chain rule combined with the power rule.

In some cases, a function can be expressed in terms of several compositions and then the chain rule has to be applied repeatedly in order to take the derivative.

EXAMPLE 8.8.5. If $f(x) = \frac{1 - \sin^2(x)}{1 + \sin^2(x)}$, then $f(x)$ is a composition of the form

$$f(x) = p(q(r(x))) = p(q \circ r(x)) \quad \text{where } p(v) = \frac{1-v}{1+v}, \ q(u) = u^2, \ r(x) = \sin(x).$$

The derivatives of $p(v)$, $q(u)$, and $r(x)$ are

$$p'(v)=\frac{-2}{(1+v)^2},\quad q'(u)=2u,\quad r'(x)=\cos(x)$$

respectively, and so we compute $f'(x)$:

$$\begin{aligned}
f'(x) &= p'(q\circ r(x))\cdot(q\circ r)'(x) && \texttt{first application of the chain rule}\\
&= p'(\sin^2(x))\cdot q'(r(x))\cdot r'(x) && \texttt{second application of the chain rule}\\
&= \frac{-2}{(1+\sin^2(x))^2}\cdot 2\sin(x)\cdot\cos(x)\\
&= \frac{-4\sin(x)\cos(x)}{(1+\sin^2(x))^2}.
\end{aligned}$$

◈

8.9 THE CALCULUS OF VECTOR-VALUED FUNCTIONS

The derivative of a vector-valued function is obtained by means of the same limit formula that is used to obtain the derivative of a real-valued function. We find the limit of a vector function by finding the limit of each component (in other words, the limit distributes through the components of the vector). Consequently, we take the derivative of a vector function by taking the derivative of each component, as shown in the calculation below. We will use the notation $\vec{r}(t)=\langle x(t),y(t)\rangle$ for a vector function and suppose that the component functions $x(t)$ and $y(t)$ are differentiable. Then,

$$\begin{aligned}
\vec{r}'(a) &= \lim_{h\to 0}\frac{\vec{r}(t+h)-\vec{r}(t)}{h}\\
&= \lim_{h\to 0}\frac{\langle x(t+h),\ y(t+h)\rangle-\langle x(t),y(t)\rangle}{h}\\
&= \lim_{h\to 0}\left\langle \frac{x(t+h)-x(t)}{h},\frac{y(t+h)-y(t)}{h}\right\rangle\\
&= \left\langle \lim_{h\to 0}\frac{x(t+h)-x(t)}{h},\lim_{h\to 0}\frac{y(t+h)-y(t)}{h}\right\rangle\\
&= \langle x'(t),y'(t)\rangle
\end{aligned}$$

Thus,

$$\boxed{\frac{d\vec{r}}{dt}=\vec{r}'(t)=\langle x'(t),y'(t)\rangle}. \tag{8.29}$$

In the same way that the derivative of a real-valued function produces the slope of a tangent line, the derivative of a vector-valued function produces a *tangent vector*. This is a vector contained in a tangent line to the trajectory of a vector-valued function. In figure 8.19, a tangent vector is shown at a point $t=a$ along the trajectory of a vector function. Furthermore, the first two diagrams in figure 8.19 demonstrate a geometric interpretation of the derivation above: evidently, the tangent vector can be regarded as the limit of secant vectors $\frac{\vec{r}(t+h)-\vec{r}(a)}{h}$ as h becomes smaller and smaller.

EXAMPLE 8.9.1. The diagrams in figure 8.19 were generated using the vector function

$$\vec{r}(t)=\langle t\cos(t),\ t\sin(t)\rangle$$

with $t=a=3.75$ (recall from example 5.12.4 that the trajectory of $\vec{r}(t)$ is a spiral). By the product rule, applied to each component,

$$\vec{r}'(t)=\langle\cos(t)-t\sin(t),\ \sin(t)+t\cos(t)\rangle.$$

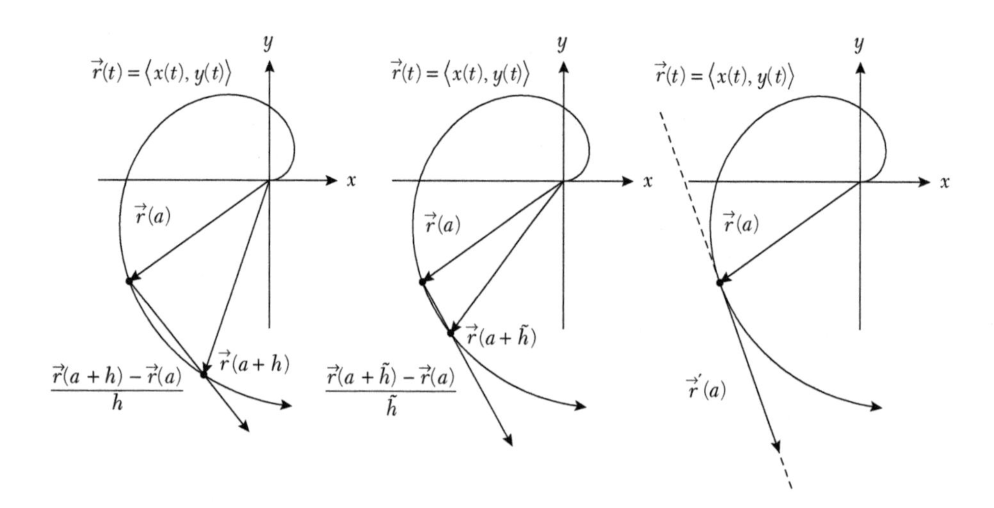

FIGURE 8.19. Secant vectors and a tangent vector.

Therefore,

$$\vec{r}'(3.75) = \langle \cos(3.75) - 3.75\sin(3.75), \sin(3.75) + 3.75\cos(3.75) \rangle \approx \langle 1.32, -3.65 \rangle.$$

This is the tangent vector shown in the third diagram. ◈

REMARK 8.9.1. In applications where t represents time, $\vec{r}'(t)$ determines a velocity vector at any point on the trajectory of $\vec{r}(t)$.

EXAMPLE 8.9.2. The trajectory of the vector-valued function $\vec{r}(t) = \langle t^2 - 2t, t+1 \rangle$ is shown in figure 5.20.The derivative of this vector-valued function is $\vec{r}'(t) = \langle 2t-2, 1 \rangle$. The tangent vectors $\vec{r}'(t)$ for $t = -2, -1, 0, 1, 2, 3$ are shown in figure 8.20. ◈

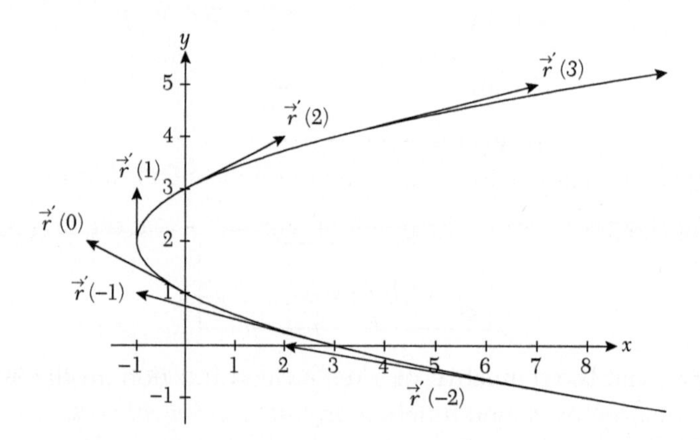

FIGURE 8.20. Tangent vectors along the trajectory of a vector function.

8.10 THE CALCULUS OF EXPONENTIAL AND LOGARITHMIC FUNCTIONS

The expression for the derivative of an exponential function $f(x) = a^x$ depends on the choice of the base a, which can be any positive real number (but not equal to 1). We will first obtain the derivative in the case that $a = e$, the number introduced in section 1.12.3 as Euler's number. There are various ways to define the number e. In order to derive the derivative formula for the exponential function with base e, we will define e below according to certain statements that should be intuitively true. There is no need, at this stage, to give a completely rigorous definition for the number e.

From an examination of graphs of $f(x)=a^x-1$ for different choices of $a>1$, as shown in figure 8.21, it can be seen that the tangent lines through the origin become steeper as the value of a increases. In particular, if a is close to 1, then the tangent line will be close to horizontal, and if a is very large, then the tangent line will be close to vertical. We can suppose, therefore, that there is a unique value for a for which the tangent line through the origin will be the line $y=x$. We will define e to be this particular value for a. Thus, the second diagram in figure 8.21 shows the graph of $f(x)=e^x-1$ together with its tangent line $y=x$ through the origin. These graphs are drawn to scale so it can be surmised, by comparison of the second diagram with the first diagram, that $2<e<4$.

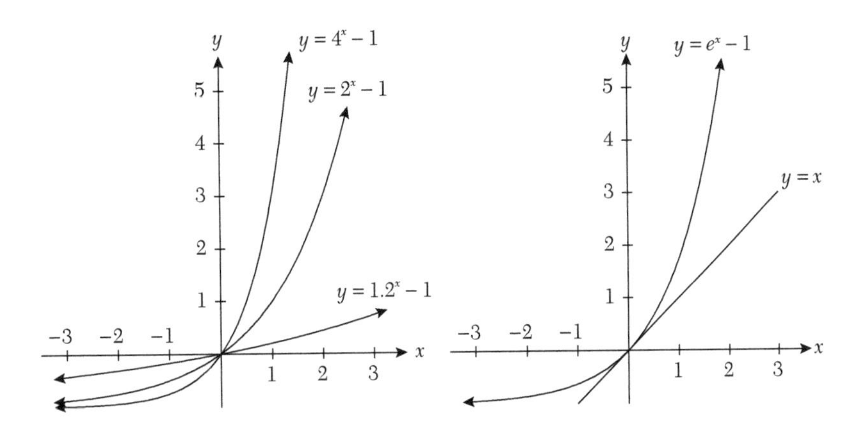

FIGURE 8.21 The definition of e.

Based on the second diagram in figure 8.21, we will make the assumption that the graph of $y=e^x-1$ lies entirely above the tangent line through the origin (this is another intuitive fact that can be proved using the property known as *convexity* of the graph of $y=e^x$). A statement of this assumption is

$$\boxed{e^x-1\geq x, \quad \text{for all real numbers } x}. \tag{8.30}$$

We will need this inequality below when we determine a limit formula e.

The tangency of the line $y=x$ to the graph of $f(x)=e^x-1$ at the origin implies

$$f'(0)=\lim_{h\to 0}\frac{f(h)}{h}=\lim_{h\to 0}\frac{e^h-1}{h}=1.$$

We can use this formula to compute the derivative of $f(x)=e^x$ at any point $x=a$

$$\begin{aligned} f'(a) &= \lim_{h\to 0}\frac{f(a+h)-f(a)}{h} \\ &= \lim_{h\to 0}\frac{e^{a+h}-e^a}{h} \\ &= \lim_{h\to 0}\frac{e^a(e^h-1)}{h} \\ &= e^a\left(\lim_{h\to 0}\frac{e^h-1}{h}\right) \\ &= e^a. \end{aligned}$$

Thus,

$$\boxed{\frac{d}{dx}e^x = e^x} \tag{8.31}$$

This fact is another characterization of the number e: *the function* $f(x)=e^x$ *is the only function that is its own derivative.*

8.10.1 A Formula for *e*

We are now going to derive a limit formula for the value of e. Refer to figure 8.22. We will start with the following interpretation of formula (8.31): for any positive integer n, the slope of the tangent line to the graph of $y=e^x$ at $x=\ln(n+1)$ is equal to $e^{\ln(n+1)}=n+1$. Consequently, the equation for the tangent line at $x=\ln(n+1)$ is

$$y-(n+1)=(n+1)(x-\ln(n+1)).$$

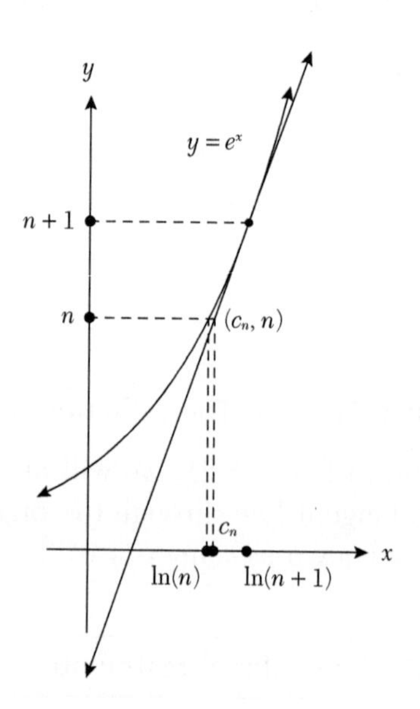

FIGURE 8.22. A tangent line to the exponential graph

By substituting $y=n$ into this equation we produce

$$1=(n+1)(\ln(n+1)-c_n),$$

or

$$\ln(n+1)-c_n=\frac{1}{n+1} \tag{8.32}$$

where (c_n, n) are the coordinates on the tangent line at $y=n$, as shown in figure 8.22.

Furthermore (as a consequence of the convexity of the exponential graph), it is the case that

$$\ln(n)<c_n<\ln(n+1).$$

Another way to express this inequality is

$$\ln(n+1)-\ln(n)>\ln(n+1)-c_n.$$

It follows now, from formula (8.32), that

$$\ln(n+1)-\ln(n) > \frac{1}{n+1}, \quad \text{for all natural numbers } n.$$

According to properties the properties of logarithms, this is equivalent to

$$\ln\left(1+\frac{1}{n}\right) > \frac{1}{n+1}, \quad \text{for all natural numbers } n.$$

Finally, using the inverse property of exponential and logarithmic functions, we obtain the inequality

$$e < \left(1+\frac{1}{n}\right)^{n+1}, \quad \text{for all natural numbers } n.$$

What's more, by substituting $x=\frac{1}{n}$ in the inequality in formula (8.30), we can derive a double inequality for e:

$$\boxed{\left(1+\frac{1}{n}\right)^{n} \le e \le \left(1+\frac{1}{n}\right)^{n+1}, \quad \text{for all natural numbers } n}. \tag{8.33}$$

For example, if $n=100$, then (in decimal form)

$$2.704 \le e \le 2.732.$$

Furthermore, if we divide the inequality in formula (8.33) by $1+\frac{1}{n}$, we obtain

$$\frac{\left(1+\frac{1}{n}\right)^{n}}{1+\frac{1}{n}} \le \frac{e}{1+\frac{1}{n}} \le \left(1+\frac{1}{n}\right)^{n} \quad \text{for all natural numbers } n. \tag{8.34}$$

As a consequence of formulas (8.33) and (8.34), we obtain

$$\frac{e}{1+\frac{1}{n}} \le \left(1+\frac{1}{n}\right)^{n} \le e \quad \text{for all natural numbers } n.$$

Now, according to the Squeeze Theorem, if we take the limit as $n \to \infty$, then

$$\boxed{e = \lim_{n\to\infty}\left(1+\frac{1}{n}\right)^{n}}.$$

The investigation of the limit on the right-hand side of this equation by the Mathematician Jacob Bernoulli led to his discovery of the number e near the end of the seventeenth century. The limit arose in some problems relating to compound interest that he was studying.

8.10.2 Derivatives of Exponential Functions

We continue with derivatives of exponential and logarithmic functions. It is helpful to know that the function $f(x)=e^x$ is sometimes expressed as $f(x)=\exp(x)$. With this notation, formula (8.31) becomes

$$\boxed{\frac{d}{dx}\exp(x)=\exp(x)}. \tag{8.35}$$

With the derivative formula in this form, it is easier to apply the chain rule to compositions that involve the exponential function, as in this example:

Example 8.10.1.

(i) If $f(x)=e^{2x}$, then

$$f'(x)=\frac{d}{dx}e^{2x}=\frac{d}{dx}\exp(2x)=\exp(2x)\cdot 2=2e^{2x}$$

(ii) If $f(x)=e^{-x}$, then

$$f'(x)=\frac{d}{dx}e^{-x}=\frac{d}{dx}\exp(-x)=\exp(-x)\cdot(-1)=-e^{-x}$$

◈

These examples have motivated the following:

$$\boxed{\frac{d}{dx}e^{\alpha x}=\alpha e^{\alpha x},\quad \text{for any real number } \alpha} \tag{8.36}$$

The following examples combine this formula with the sum rule and quotient rule.

Example 8.10.2.

(i) $\dfrac{d}{dx}\left(e^{x}+e^{-x}\right)=e^{x}-e^{-x}$

(ii) $\dfrac{d}{dx}\left(e^{x}-e^{-x}\right)=e^{x}+e^{-x}$

(iii)
$$\frac{d}{dx}\left(\frac{e^{x}+e^{-x}}{e^{x}-e^{-x}}\right)=\frac{(e^{x}-e^{-x})(e^{x}-e^{-x})-(e^{x}+e^{-x})(e^{x}+e^{-x})}{(e^{x}-e^{-x})^{2}}$$
$$=\frac{-4}{(e^{x}-e^{-x})^{2}}$$

◈

In more generality, an exponential function e^{x} can be composed with any other real-valued function $f(x)$, and the chain rule can be applied to find the derivative of the composition. Thus,

$$\frac{d}{dx}e^{f(x)}=\frac{d}{dx}\exp(f(x))=\exp(f(x))\cdot f'(x)=f'(x)e^{f(x)}.$$

This is a useful result to remember.

$$\boxed{\frac{d}{dx}e^{f(x)}=f'(x)e^{f(x)}} \tag{8.37}$$

Example 8.10.3.

(i) $\dfrac{d}{dx}e^{\sin(x)}=\cos(x)e^{\sin(x)}$

(ii) $\dfrac{d}{dx}e^{x^{2}}=2xe^{x^{2}}$

(iii) $\frac{d}{dx}(xe^{x^2}) = (1)e^{x^2} + (x)2xe^{x^2}$

$$= e^{x^2}\left(1 + 2x^2e^{x^2}\right)$$

◈

An exponential function $f(x) = a^x$, where $a > 0$ and $a \neq 1$, can be expressed as an exponential function with base e by means of a simple trick: as a consequence of the cancellation equations (formula (5.5)) and property (III) of the laws for natural logarithms in table 5.6 we can write $a^x = e^{\ln(a^x)} = e^{x\ln(a)}$; and then the application of formula (8.33) (with $\alpha = \ln(a)$) results in

$$\frac{d}{dx}a^x = \frac{d}{dx}e^{\ln(a)x} = \ln(a)e^{\ln(a)x} = \ln(a)a^x.$$

Thus, we have proved the derivative formula for exponential functions with base a:

$$\boxed{\frac{d}{dx}a^x = \ln(a)a^x, \quad \text{for any real number } a > 0 \text{ and } a \neq 1} \tag{8.38}$$

Note that this reduces to formula (8.31) if $a = e$ (because $\ln e = 1$).

EXAMPLE 8.10.4.

(i) $\frac{d}{dx}2^x = \ln(2)2^x$

(ii) $\frac{d}{dx}2^{\sin(x)} = \ln(2)2^{\sin(x)}\cos(x)$

(iii) $\frac{d}{dx}3^{x^2} = \ln(3)3^{x^2}(2x) = 2x\ln(3)3^{x^2}$

(iv) $\frac{d}{dx}\left(4^x - 4^{-x}\right) = \ln(4)\left(4^x + 4^{-x}\right)$

◈

EXAMPLE 8.10.5. A first-order differential equation is any equation in which a function is related to its own derivative. Differential equations typically arise in mathematical modeling, where mathematical equations are used to represent and study natural processes. For example, a model for population growth is the first-order differential equation

$$\boxed{y'(t) = ky(t)} \tag{8.39}$$

where $y(t)$ is the size of a population at time t, and k is a constant. If $k > 0$, then this equation states that the rate of growth of the population is proportional to the size of the population. A solution for this equation is any function $y(t)$ that solves the equation in the sense that plugging the function into the equation will make the equation a true statement. For example, if $y(t) = e^{kt}$, then $y'(t) = ke^{kt} = ky(t)$, and so $y(t)$ is a solution.

A more sophisticated equation for modeling population growth is the *logistic differential equation*

$$\boxed{y'(t) = ky(t)(M - y(t))} \tag{8.40}$$

where $y(t)$ and k are as above and M is a constant that specifies the maximum possible size of the population (that is, $y(t) < M$). If $k > 0$, then this equation models the growth of a population in an environment with limited food supply. If the population has an initial of size m, that is $y(0) = m$, then a solution for the equation is

$$y(t) = \frac{mM}{m + (M-m)e^{-kMt}}.$$

We can check that this is a solution by calculating both sides of formula (8.40):

$$\begin{aligned} y'(t) &= \frac{-mM}{(m+(M-m)e^{-kMt})^2} \cdot -kM(M-m)e^{-kMt} \\ &= ky(t)\left(\frac{M(M-m)e^{-kMt}}{m+(M-m)e^{-kMt}}\right) \end{aligned}$$

and

$$\begin{aligned} ky(t)(M-y(t)) &= ky(t)\left(\frac{Mm + M(M-m)e^{-kMt} - mM}{m+(M-m)e^{-kMt}}\right) \\ &= ky(t)\left(\frac{M(M-m)e^{-kMt}}{m+(M-m)e^{-kMt}}\right) \end{aligned}$$

We see that the left-hand side of formula (8.40) equals the right-hand side and so $y(t)$ is indeed a solution. A graph of $y(t)$ with $m = 2$, $M = 10$ and $k = 0.05$ is shown in figure 8.23. This is called a *logistic graph*.

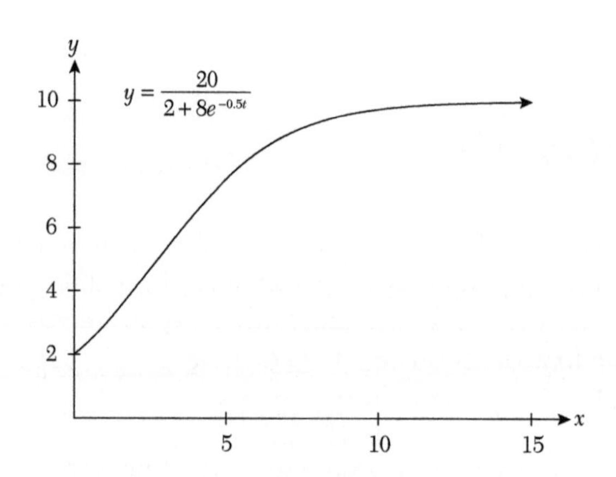

FIGURE 8.23. A logistic graph. ◈

8.10.3 Derivatives of Logarithmic Functions

If $f(x) = \ln x$, then $e^{f(x)} = x$. The derivative of the left-hand side of this equation is $f'(x)e^{f(x)}$, and the derivative of its right-hand side is 1. If we equate these derivatives, we produce

$$f'(x) = \frac{1}{e^{f(x)}} = \frac{1}{x}.$$

Thus, we have found a formula for the derivative of the natural logarithm:

$$\boxed{\frac{d}{dx}\ln(x) = \frac{1}{x}} \tag{8.41}$$

While $\ln(x)$ is defined only for $x > 0$, its derivative function $\frac{1}{x}$ is also defined for $x < 0$. However, note that $\ln|x|$ is also defined for $x < 0$ and by the chain rule and exercise 8.4,

$$\frac{d}{dx}\ln|x| = \frac{1}{|x|}\left(\frac{|x|}{x}\right) = \frac{1}{x}.$$

Thus, we have this extension of formula (8.41):

$$\boxed{\frac{d}{dx}\ln|x| = \frac{1}{x}} \tag{8.42}$$

The graph of $\ln|x|$ together with its derivative function $\frac{1}{x}$ is shown in figure 8.24.

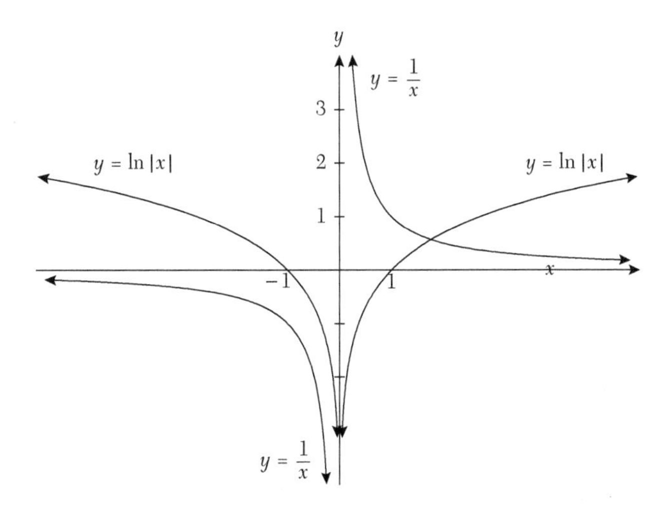

FIGURE 8.24. $\ln|x|$ and its derivative function $\frac{1}{x}$.

When taking compositions of functions involving the natural logarithm, the derivative of the natural logarithm evaluates as a reciprocal. The simplest case of this is:

$$\boxed{\frac{d}{dx}\ln(\alpha x) = \frac{1}{\alpha x}(\alpha) = \frac{1}{x} \quad \text{for any positive number } \alpha} \tag{8.43}$$

This result might seem surprising but note that, using properties of logarithms,

$$\frac{d}{dx}\ln(\alpha x) = \frac{d}{dx}(\ln\alpha + \ln x) = \frac{d}{dx}\ln\alpha + \frac{d}{dx}\ln x = \frac{1}{x} \quad \text{for any real number } \alpha,$$

(This is because $\ln(\alpha)$ is a constant.) It is also true that

$$\boxed{\frac{d}{dx}\ln|\alpha x| = \frac{1}{x} \quad \text{for any real number } \alpha}, \tag{8.44}$$

A more general formula that is a consequence of the chain rule is

$$\boxed{\frac{d}{dx}\ln|f(x)| = \frac{f'(x)}{f(x)}} \tag{8.45}$$

for any *nonvanishing* function $f(x)$ (that is, a function $f(x)$ that is never zero).

The change of base formula for logarithms (formula (5.6)) makes is easy to compute derivatives for logarithms with any base:

$$\frac{d}{dx}\log_a x = \frac{d}{dx}\left(\frac{\ln x}{\ln a}\right) = \frac{1}{(\ln a)x}$$

for $a > 0$ and $a \neq 1$. Similarly, we have the general formula

$$\boxed{\frac{d}{dx}\log_a |f(x)| = \frac{f'(x)}{(\ln a)f(x)}} \tag{8.46}$$

for $a > 0$ and $a \neq 1$, and any nonvanishing function $f(x)$.

Example 8.10.6.

(i) $\frac{d}{dx}\ln(2x) = \frac{1}{x}$

(ii) $\frac{d}{dx}\log_3(2x) = \frac{1}{(\ln 3)x}$

(iii) $\frac{d}{dx}\ln|\sin(x)| = \frac{\cos(x)}{\sin(x)}$

(iv) $\frac{d}{dx}\log_3|\sin(x)| = \frac{\cos(x)}{(\ln 3)\sin(x)}$

(v) $\frac{d}{dx}\ln\ln x = \frac{1}{x\ln x}$

(vi) $\frac{d}{dx}\log_3\log_2 x = \frac{1}{(\ln 3)\log_2 x}\cdot\frac{1}{(\ln 2)x} = \frac{1}{(\ln 3)x\ln x}$

(vii) $\frac{d}{dx}\ln\left(e^x + e^{-x}\right) = \frac{e^x - e^{-x}}{e^x + e^{-x}}$ ◈

8.10.4 The Proof of the Power Rule

We can now state and prove

$$\boxed{\text{the power rule:} \quad \frac{d}{dx}x^\alpha = \alpha x^{\alpha-1} \text{ for any real number } \alpha} \tag{8.47}$$

Proof: Because

$$x^\alpha = e^{\ln x^\alpha} = e^{\alpha\ln x}$$

we can take the derivative of x^α using the formulas for derivatives of exponential and logarithmic functions:

$$\begin{aligned}\frac{d}{dx}x^\alpha &= \frac{d}{dx}e^{\alpha\ln x}\\ &= e^{\alpha\ln x}\frac{d}{dx}\alpha\ln x\\ &= x^\alpha\left(\frac{\alpha}{x}\right)\\ &= \alpha x^{\alpha-1}.\end{aligned}$$

EXAMPLE 8.10.7.

$$\frac{d}{dx}x^{\pi} = \pi x^{\pi-1}$$

◈

8.11 DERIVATIVES OF THE INVERSE TRIGONOMETRIC FUNCTIONS

If $y = \sin^{-1} x$, then $\sin(y) = x$. Taking the derivative of both sides of this equation with respect to x (with y regarded as a function of x) results in $\cos(y)y' = 1$. Solving for y' gives

$$y' = \frac{1}{\cos(y)} = \frac{1}{\sqrt{1-\sin^2 y}} = \frac{1}{\sqrt{1-x^2}}.$$

Similarly, if $y = \cos^{-1} x$, then $y' = -\frac{1}{\sqrt{1-x^2}}$ and, if $y = \tan^{-1} x$, then $y' = \frac{1}{1+x^2}$.

We have thus proved

$$\boxed{\begin{aligned} \frac{d}{dx}\sin^{-1}(x) &= \frac{1}{\sqrt{1-x^2}} \\ \frac{d}{dx}\cos^{-1}(x) &= \frac{1}{\sqrt{1-x^2}} \\ \frac{d}{dx}\tan^{-1}(x) &= \frac{1}{1+x^2} \end{aligned}} \tag{8.48}$$

EXAMPLE 8.11.1.

(i) $\frac{d}{dx}\sin^{-1}(2x) = \frac{2}{\sqrt{1-4x^2}}$

(ii) $\frac{d}{dx}\cos^{-1} x^2 = \frac{2x}{\sqrt{1-x^4}}$

(iii) $\frac{d}{dx}\tan^{-1}(1+x) = \frac{1}{1+(1+x)^2}$

(iv) $\frac{d}{dx}\tan^{-1}(2\tan(x)) = \frac{2\sec^2(x)}{1+4\tan^2(x)}$

◈

EXAMPLE 8.11.2. A rectangular movie theater is 20 m long, with seating on a flat floor and the screen at one end. The top and bottom of the screen are 6 and 2 m from the floor, respectively. Find the position in the theater with the largest viewing angle.

Answer: In figure 8.25, the angle from a point x to the top of the screen is labeled θ and to the bottom labeled α. Thus, the viewing angle from x is $\theta - \alpha$. Because $\theta = \tan^{-1}\left(\frac{6}{x}\right)$ and $\alpha = \tan^{-1}\left(\frac{2}{x}\right)$, we can define the viewing angle as the function

$$f(x) = \tan^{-1}\left(\frac{6}{x}\right) - \tan^{-1}\left(\frac{2}{x}\right).$$

We now take the derivative:

$$f'(x) = \frac{1}{1+\left(\frac{6}{x}\right)^2}\left(-\frac{6}{x^2}\right) - \frac{1}{1+\left(\frac{2}{x}\right)^2}\left(-\frac{2}{x^2}\right) = \frac{2}{x^2+4} - \frac{6}{x^2+36}$$

$$= \frac{4(12-x^2)}{(x^2+4)(x^2+36)}$$

By setting this equal to zero, we can solve for x to produce the critical value $x = 2\sqrt{3}$. For this value of x, the viewing angle is

$$f\left(2\sqrt{3}\right) = \tan^{-1}\left(\frac{3}{\sqrt{3}}\right) - \tan^{-1}\left(\frac{1}{\sqrt{3}}\right) = \frac{\pi}{3} - \frac{\pi}{6} = \frac{\pi}{6} \text{ (or } 30^\circ\text{)}.$$

This is the maximum possible viewing angle, at a distance of $2\sqrt{3} \approx 3.464$ m from the front of the theater. Note that the viewing angle at the front of the theater is $\lim_{x\to 0^+} f(x) = 0$ and at the back $f(20) \approx 0.523$ (or 5.67°). ◈

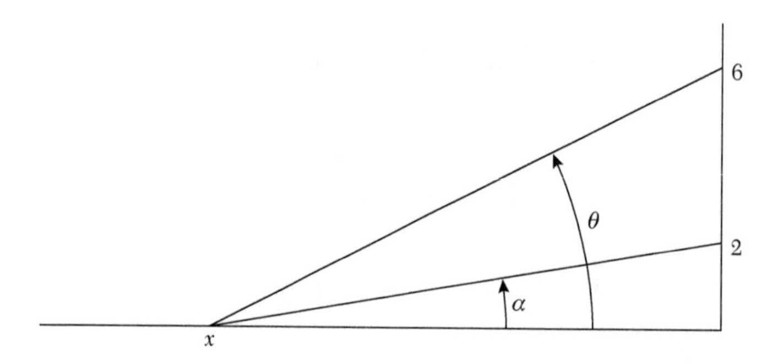

FIGURE 8.25. The viewing angle in a movie theater.

EXERCISES

8.1. In each of the diagrams in figure 8.26, decide whether the line is a tangent to the graph (at one or more points).

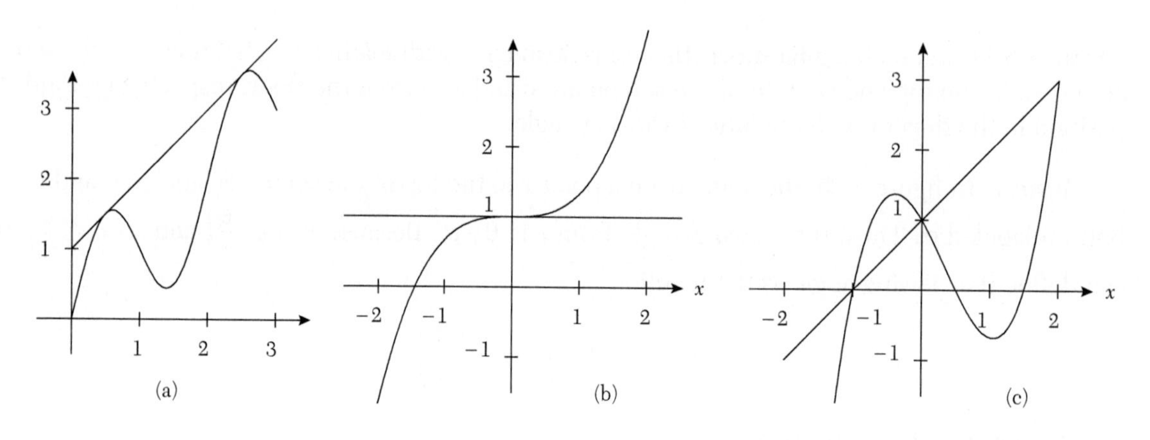

FIGURE 8.26. Diagram for exercise 8.1.

8.2. Use formula (8.3) to determine $f'(0)$ for each of the following functions.

(a) $f(x)=2x\cos(x)$

(b) $f(x)=\dfrac{x}{3+x}$

(c) $f(x)=6x+4x^2$

(d) $f(x)=x^2\log(2+x)$

(e) $f(x)=\sqrt{2+x}-\sqrt{2}$

8.3. Use formula (8.6), with $a=2$, to determine $f'(2)$ for each of the following functions, and use formula (8.8) to find the equation of the tangent line at $x=2$.

(a) $f(x)=2x+x^3$

(b) $f(x)=\dfrac{x}{3+x}$

(c) $f(x)=6x+4x^2$

(d) $f(x)=\sqrt{x}+1$

(e) $f(x)=\dfrac{1}{\sqrt{x}}$

(f) $f(x)=\dfrac{1}{2+\sqrt{x}}$

8.4. Use formula (8.6) to prove that, if $f(x)=|x|=\sqrt{x^2}$, then $f'(x)=\dfrac{x}{|x|}=\dfrac{|x|}{x}=\dfrac{\sqrt{x^2}}{x}$.

8.5. Each limit below represents the derivative of some function f at some number a. State f and a in each case. (Hint: match each limit with formula (8.6).

(a) $\lim_{h\to 0}\dfrac{\sqrt{1+h}-1}{h}$ and **(b)** $\lim_{h\to 0}\dfrac{(2+h)^3-8}{h}$.

8.6. Match each graph from (a) to (d) in figure 8.27 with the graph of its derivative from (i) to (iv) in figure 8.28.

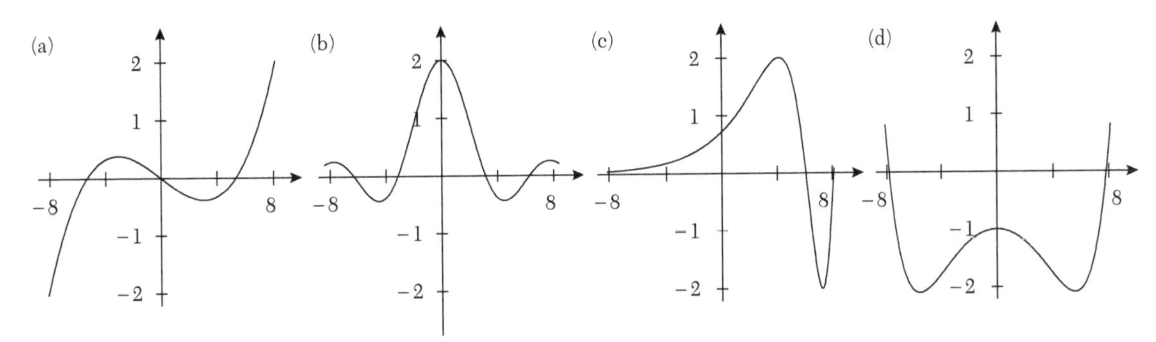

FIGURE 8.27. Diagram for Exercise 8.6

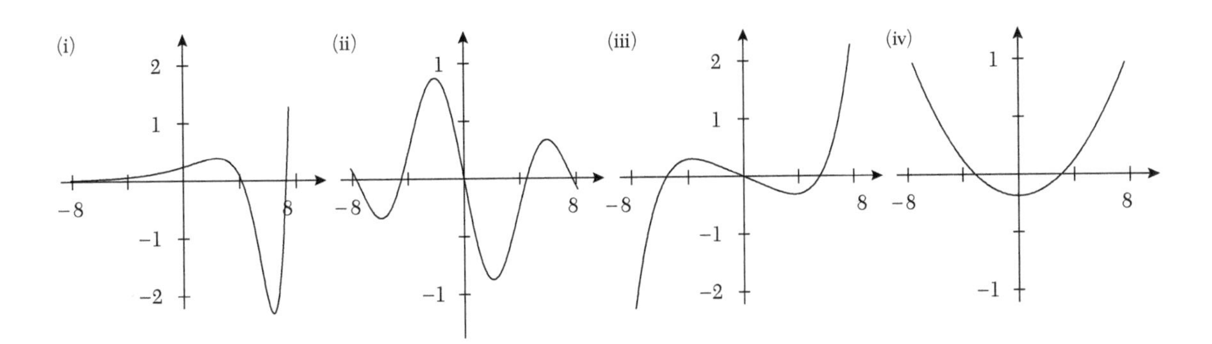

FIGURE 8.28. Diagram for Exercise 8.6

8.7. At which points on the graph in figure 8.29 is it not possible to draw a tangent line?

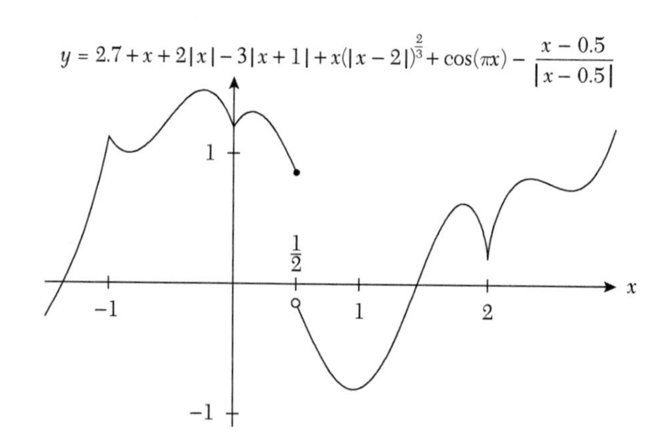

FIGURE 8.29. Diagram for Exercise 8.7

8.8. For each the graphs in figure 8.30, sketch the graph of the corresponding derivative function.

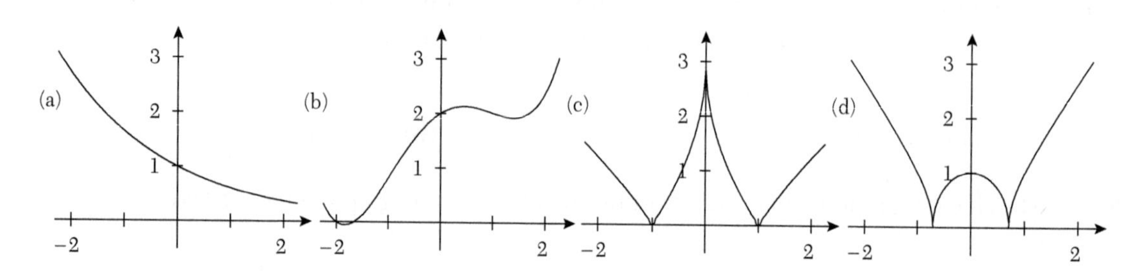

FIGURE 8.30. Diagram for Exercise 8.8

8.9. Prove the second part of the sum rule (formula (8.13)), that is, $(cf)'(x) = cf'(x)$.

8.10. If g is a differentiable function and $g(a) \neq 0$, use formula (8.6) and the rules for limits in section 7.10.2 to prove that

$$\left(\frac{1}{g}\right)'(a) = \frac{-g'(a)}{g(a)^2}.$$

Furthermore, if f is a differentiable function, use the product rule to prove that

$$\left(f \cdot \frac{1}{g}\right)'(a) = \frac{f'(a)}{g(a)} - \frac{f(a)g'(a)}{g(a)^2}.$$

This is a proof of the quotient rule.

8.11. The power rule for natural numbers (formula (8.9)) was proved in section 8.3.1 for $n = 1$, $n = 2$, and $n = 3$; that is, it was proved that $\frac{d}{dx}x = 1$, $\frac{d}{dx}x^2 = 2x$, and $\frac{d}{dx}x^3 = 3x^2$.

Now apply the product rule to $x^4 = x \cdot x^3$ to prove that $\frac{d}{dx}x^4 = 4x^3$.
Similarly, apply the product rule to $x^5 = x \cdot x^4$ to prove that $\frac{d}{dx}x^5 = 5x^4$.

In general, assume it is true that $\frac{d}{dx}x^n = nx^{n-1}$ for *any* natural number n and then apply the product rule to $x^{n+1} = x \cdot x^n$ to prove formula (8.9), that is, $\frac{d}{dx}x^{n+1} = (n+1)x^n$.

(This method of proof, called the *method of mathematical induction*, is a powerful method that can be applied to many other problems involving natural numbers.)

8.12. Use the power rule (formula (8.9)) and sum rule (formula (8.13)) to compute the derivatives of the following polynomials.

(a) $f(x) = 1 + 2x$

(b) $f(x) = \frac{x}{3} + x^2$

(c) $f(x) = 6x^6 + 7x^7 + 8x^8$

(d) $f(x) = 7 + 4x^8 + 6x^{61}$

8.13. Use the power rule (formula (8.19)) and sum rule (formula (8.13)) to find $f'(x)$ if

(a) $f(x) = x^{-2}$

(b) $f(x) = \frac{2}{x^{20}}$

(c) $f(x) = 11x^3 - \frac{6}{x^3}$

(d) $f(x) = 4x^{-3} - x^{-4}$

(e) $f(x) = 1 + \frac{1}{x} + \frac{1}{x^2} + \frac{1}{x^3}$

(f) $f(x) = 1 + 2x^{-1} + 3x^{-2} + 4x^{-3}$

8.14. Use the sum rule to compute $f'(x)$ for the following functions $f(x)$. Assume that a, b, c, and d are constants. Do this in two ways: first, by expanding the expression for $f(x)$ and then taking the derivative; second, by applying the sum rule repeatedly (without expanding). Check that your answers agree.

(a) $f(x) = a(x + b)$

(b) $f(x) = a(x + b(x + c))$

(c) $f(x) = a(x + b(x + c(x + d)))$

8.15. Use the product rule (formula (8.14)) to compute $f'(x)$ for the following functions $f(x)$. Assume that a, b, c, and d are constants.

(a) $f(x) = x(b + x)$

(b) $f(x) = (a + x)(b + x)$

(c) $f(x) = x^2(b + x)$

(d) $f(x) = (a + x)(b + x^2)$

8.16. Use the quotient rule (formula (8.15)) to compute the following derivatives of rational expressions.

(a) $\frac{d}{dx}\left(\frac{2}{1+x}\right)$

(b) $\frac{d}{dx}\left(\frac{2}{1-x}\right)$

(c) $\frac{d}{dx}\left(\frac{x}{1+x}\right)$

(d) $\frac{d}{dx}\left(\frac{x+1}{x-1}\right)$

(e) $\frac{d}{dx}\left(\frac{1-x}{1+x}\right)$

8.17. Use the quotient rule to compute the following derivatives of rational expressions.

(a) $\frac{d}{dx}\left(\frac{2}{1+x^2}\right)$

(b) $\frac{d}{dt}\left(\frac{2}{1-t^3}\right)$

(c) $\frac{d}{du}\left(\frac{u^2}{1+u^2}\right)$

(d) $\frac{d}{dv}\left(\frac{v^3+1}{v^2-1}\right)$

(e) $\frac{d}{dy}\left(\frac{1-y^8}{1+y^9}\right)$

8.18. Use the quotient rule to compute the following derivatives of rational expressions.

(a) $\frac{d}{dx}\left(\frac{2}{1+3x^2}\right)$

(b) $\frac{d}{dt}\left(\frac{2}{1-4t^3}\right)$

(c) $\frac{d}{du}\left(\frac{6u^2}{1+2u^2}\right)$

(d) $\frac{d}{dv}\left(\frac{5v^3+4}{7v^2-3}\right)$

(e) $\frac{d}{dy}\left(\frac{1-7y^8}{1+8y^9}\right)$

8.19. Compute $f'(x)$ for the following functions $f(x)$. Assume that a, b, c, and d are constants. Do this in two ways: first, by expanding the expression for $f(x)$ and then taking the derivative; second, by applying the product rule (formula (8.14)) repeatedly (without expanding). Check that your answers agree.

(a) $f(x)=x(a+x)$

(b) $f(x)=x(a+x(b+x))$

(c) $f(x)=x(a+x(b+x(c+dx)))$

8.20. Use the quotient rule to compute $f'(x)$ for the following functions $f(x)$. Assume that a, b, c, and d are constants.

(a) $f(x)=ax+b+\frac{c}{x}$

(b) $f(x)=\frac{x^2}{x+c}$

(c) $f(x)=\frac{x}{a+\frac{b}{x}}$

8.21. Find the equation of the tangent line to the graph of the given equation at $x=2$. (Hint: use your answers from exercise 8.16.)

(a) $y=\frac{2}{1+x}$

(b) $y=\frac{2}{1-x}$

(c) $y=\frac{x}{1+x}$

(d) $y=\frac{x+1}{x-1}$

(e) $y=\frac{1-x}{1+x}$

8.22. Find the intercepts of each of the tangent lines from the previous exercise with the x and y axes. Draw the graph of the equation together with the graph of the tangent line.

8.23. Find the points on the graph of the given equation where the tangent line is parallel to the line $y=-4x+1$.

(a) $y=\frac{1}{x+1}$

(b) $y=\frac{2}{x-1}$

8.24. Find the points on the graph of the given equation where the tangent line is parallel to the line $x=4y+1$.

(a) $y=\frac{-1}{1+x}$

(b) $y=\frac{2}{1-x}$

8.25. Find the points on the graph of the given equation where the tangent line is parallel to the line $\frac{1}{x}=\frac{1}{y+1}$.

(a) $y=\frac{x}{1+x^2}$

(b) $y=\frac{2}{(1-x)^2}$

8.26. How many tangent lines to the graph $y = \frac{x}{x-1}$ pass through each of the following points? Find the coordinates of the points at which the tangent lines touch this graph.

(a) $(2,0)$

(b) $(2,1)$

(c) $(2,2)$

(d) $(2,3)$

(e) $(0,2)$

8.27. Prove that the algorithm for Newton's method, as explained in section 8.4, is given by the formula

$$\boxed{x_{n+1} = x_n - \frac{f(x_n)}{f'(x_n)}} \qquad (8.49)$$

where x_1 is the first approximation to a root of $f(x)$, $x_2 = x_1 - \frac{f(x_1)}{f'(x_1)}$ is the second approximation to the root and so on. Use this formula to verify the calculations in example.

Compute the following values with accuracy to two decimal positions (three significant digits) using the method of example 8.4.4.

(a) $\sqrt[2]{2}$

(b) $\sqrt[3]{3}$

(c) $\sqrt[4]{4}$

(d) $\sqrt[5]{5}$

8.28. Use Newton's method to find the three real roots of the polynomial $f(x) = x^3 - 9x - 5$ rounded to two decimal positions. (Hint: first use the intermediate value theorem to locate three intervals containing the roots.)

8.29. Use the power rule for rational exponents (formula (8.20)) to compute the derivatives of the following rational algebraic expressions

(a) $\frac{d}{dx} 6x^{\frac{6}{5}}$

(b) $\frac{d}{dt} t\sqrt{\sqrt{t}}$

(c) $\frac{d}{du}\left(\frac{\sqrt{u}}{1+\sqrt{u}}\right)$

(d) $\frac{d}{dv}\left((3v)^{\frac{1}{2}} + (3v)^{\frac{-1}{2}}\right)$

(e) $\frac{d}{dy}\left(\sqrt[3]{y^2} - \sqrt{y^3}\right)$

8.30. Find all points on the graph of the equation $y = \sqrt{x^3} + \sqrt{x} - 2x$ at which the tangent line is horizontal.

8.31. Prove the derivative formulas for the secant, cosecant and cotangent ratios in table 8.1.

8.32. Use the derivative formulas in table 8.1 to compute the derivatives below

(a) $\frac{d}{dx}\frac{\sin(x)}{1+\cos(x)}$

(b) $\frac{d}{dt} t\cot(t)$

(c) $\frac{d}{du}\frac{\sqrt{u}}{\csc(u)}$

(d) $\frac{d}{dv}\frac{\tan(v)-1}{\sec(v)}$

(e) $\frac{d}{dy}\frac{1+2\sin(y)}{y^2}$

8.33. Find the points on the graphs of the following equations at which the tangent line is horizontal.

(a) $y = \cos(x) + \sin(x)$

(b) $y = x + \sin(x)$

(c) $y = \dfrac{\sin(x)}{2 + \cos(x)}$

(d) $y = \dfrac{2 + \cos(x)}{\sin(x)}$

(e) $y = \cot(x) + \tan(x)$

8.34. The location of a car as it approaches an intersection is given by the time–displacement equation $s = \frac{t^3 + 3t - 4}{t+1}$ where s is measured in meters and t is measured in seconds. At time $t = 0$ the car is 4 m from the intersection. What is the speed of the car as it passes through the intersection?

(Hint: it's not hard to guess the time when the car passes through the intersection.)

8.35. Use formula (8.23) to find the approximate value of each of the following numbers.

(a) $\sqrt{5}$ (take $a = 4$)

(b) $\sqrt[3]{30}$ (take $a = 27$)

(c) $\cos\left(\dfrac{5}{16}\right)$ (take $a = \dfrac{\pi}{4}$)

8.36. Find two positive numbers whose product is 121 and whose sum is a minimum. Use calculus to solve the problem.

8.37. A vacationer wants to fence off a rectangular beach-front property using 1,200 m of fencing. What are the dimensions of the property with the largest area? (Assume that the shoreline is straight and there is no fencing along the beach.)

8.38. If 2,000 m of fencing are used to enclose a rectangular area and to divide it into three equal areas, find the dimensions such that the area enclosed is the greatest. Use calculus to solve the problem.

8.39. An open-topped box is to be constructed by cutting off equal squares from each corner of a 6 m by 12 m rectangular cardboard sheet and folding up the sides. Find the maximum possible volume of such a box.

8.40. Find the point on the parabola $y = 3x^2$ that is closest to the point $(1,5)$.

8.41. A cylindrical tin can is to be manufactured to contain a volume of 8 liters. The circular end pieces will be cut from two square sheets of tin with the corners wasted and the side of the can will be a rectangular sheet of tin with two opposite ends joined together. Find the ratio of the height to the radius of the most economical can.

8.42. Find the area of the largest rectangle that can be inscribed in a semicircle of radius r.

8.43. A wire 30 in. long is cut into two pieces of possibly different lengths with one piece bent into a circle and the other piece bent into a square. How long must each piece be to minimize the total area enclosed by the circle and the square? (Assume that the two areas are separated.)

8.44. A sheet of metal is 10 m long and 4 m wide. It is bent lengthwise down the middle to form a V-shaped trough 10 m long. What should the width across the top of the trough be in order for the trough to have the maximum capacity?

8.45. Find the length of the longest straight rod that can be carried horizontally (i.e., without tilting up or down) from a rectangular corridor 6 m wide into another rectangular corridor, at right angles to it, that is 4 m wide (as in figure 8.31).

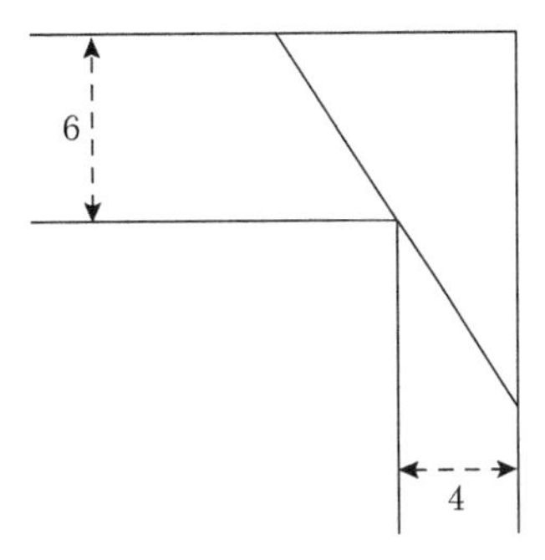

FIGURE 8.31. Carrying a rod through a rectangular corridor

8.46. Use the special case of the chain rule (formula (8.25)) to determine $(f \circ g)'(0)$ for each of the following choices of f and g.

(a) $f(u)=5u,\ g(x)=6x$

(b) $f(t)=3t,\ g(u)=11u$

(c) $f(u)=2u,\ g(x)=2x^2$

(d) $f(t)=3t^2,\ g(u)=4u$

(e) $f(u)=2\sin(u),\ g(x)=2x$

(f) $f(t)=6t,\ g(u)=\sin(u)$

(g) $f(u)=2\sin(u)+\cos(u)-1,$
$g(x)=2x^2+2x$

(h) $f(t)=6t^2+t,\ g(u)=2\tan(u)+u$

8.47. The purpose of this exercise is to prove the general formula for the chain rule by making use of the formula for the special case of the chain rule proved in section 8.8, that is, if $f(0)=0$ and $g(0)=0$, then $(f \circ g)'(0)=f'(0)g'(0)$.

Suppose that f and g are differentiable functions and $h=f \circ g$. Introduce numbers a, b, and c such that $x=a$ is in the domain of $g(x)$, $u=b=g(a)$ is in the domain of $f(u)$ and $c=f(b)=f(g(a))=h(a)$, as shown in figure 8.32.

Define the functions F, G, and H as follows:

$$F(u)=f(u+b)-c,\ G(x)=g(x+a)-b \text{ and } H(x)=h(x+a)-c.$$

Note that $F'(0)=f'(b)$, $G'(0)=g'(a)$ and $H'(0)=h'(0)$ because F, G, and H are translations of f, g, and h, as shown in figure 8.33.

Now prove the following:

(a) $(F \circ G)(x)=H(x)$

(b) $(f \circ g)'(a)=f'(g(a))g'(a)$ (Hint: apply the special case of the chain rule stated above to $H'(0)$.)

(The two diagrams in figures 8.32 and 8.33 were generated using the formulas $f(u)=\sqrt{u}$, $g(x)=1+x^2$ and $a=1$. Find the values of $f'(2)$, $g'(1)$and $(f \circ g)'(1)$.)

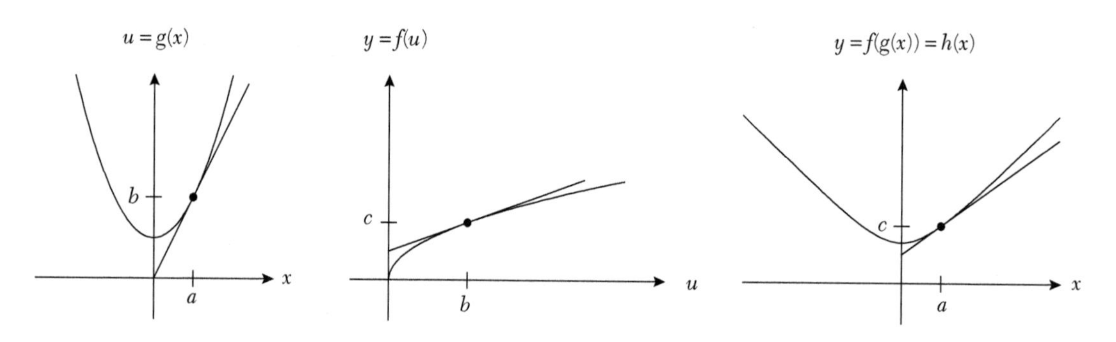

FIGURE 8.32. A proof of the chain rule

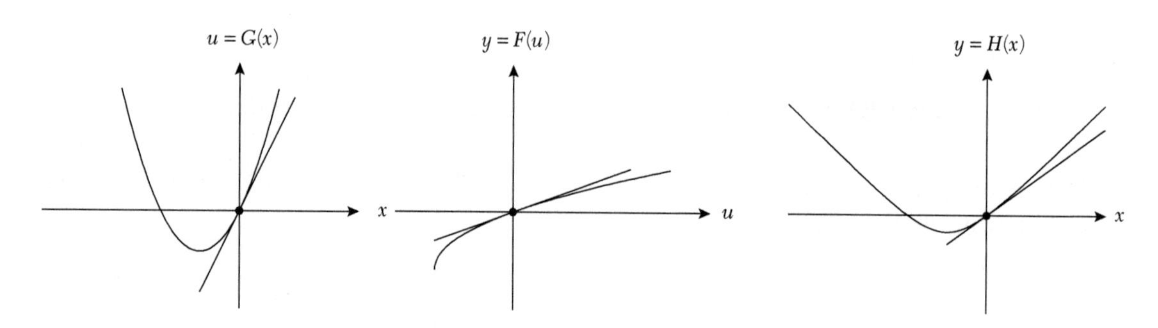

FIGURE 8.33. A proof of the chain rule

8.48. Use the chain rule (formula (8.27)) to find $\frac{dy}{dx}$ in each case below.

(a) $y = 5u,\ u = x^2$

(b) $y = u^2,\ u = 5x$

(c) $y = \cos(u),\ u = \pi x$

(d) $y = \sin(u),\ u = 3x$

(e) $y = u^2,\ u = x + \pi$

(f) $y = u^2,\ u = \cos(x)$

8.49. Use the chain rule to find $\left.\frac{dy}{dx}\right|_{x=\pi}$ in each case below.

(a) $y = 5u,\ u = \sin(x)$

(b) $y = \sin(u),\ u = 5x$

(c) $y = \cos(u),\ u = 2x$

(d) $y = \sin(u),\ u = 3x + \pi$

(e) $y = u^2,\ u = \cos(x)$

(f) $y = u^2,\ u = \tan(x)$

8.50. Use the chain rule to find $f'(x)$ for each function $f(x)$.

(a) $f(x) = (1+x)^9$

(b) $f(x) = (1-x)^9$

(c) $f(x) = (1+x^2)^8$

(d) $f(x) = (1+x+x^2)^8$

(e) $f(x) = (x^3)^4$

(f) $f(x) = (x^8)^9$

(g) $f(x) = (\sin(x))^3$

(h) $f(x) = \sin^3(x)$

(i) $f(x) = \cos^6(x)$

(j) $f(x) = (\cos^2(x))^3$

(k) $f(x) = \cos(x^2)$

(l) $f(x) = \sin(x + \pi)$

(m) $f(x) = \cos^2(x + \pi)$

(n) $f(x) = \cos(\pi x)$

(o) $f(x)=\cos(\pi^2x^2)$

(p) $f(x)=\cos(\sin(x))$

(q) $f(x)=\sin(\sin(x))$

8.51. Use the chain rule to compute $f'(1)$ for each function f.

(a) $f(x)=(1+x)^9$

(b) $f(t)=(1+t^2)^8$

(c) $f(u)=\sin(\pi u)$

(d) $f(x)=\cos^6(\pi w)$

(e) $f(x)=\dfrac{1-x^2}{1+x^2}$

(f) $f(t)=\dfrac{1+\sin(\pi t)}{1-\sin(\pi t)}$

(g) $f(t)=\dfrac{2+\cos^2 2\pi t}{2-\cos^2 2\pi t}$

(h) $f(w)=\dfrac{1}{(1+w)^3}$

(i) $f(x)=\dfrac{1}{(1+x+x^2)^3}$

(j) $f(t)=\dfrac{1}{(1+2t+t^2)^3}$

(k) $f(u)=\dfrac{1}{(1+2u-u^2)^3}$

(l) $f(w)=\dfrac{2}{(1-\cos(\pi w))^3}$

8.52. Compute the derivative of each function.

(a) $f(x)=(1+\sin(x))^9$

(b) $g(u)=\tan(\pi u^2)$

(c) $h(y)=\dfrac{1+\tan(3y)}{1-\tan(3y)}$

(d) $f(x)=\sqrt{1+\sin(2x)}$

(e) $g(u)=\tan(\pi\sqrt{u})$

(f) $h(y)=\dfrac{1+\sqrt{3y}}{1-\sqrt{3y}}$

(g) $f(x)=(1+\sin(x))^9\cos^2 x$

(h) $g(u)=\tan(\pi u^2)\cos(\pi u)$

(i) $h(y)=\dfrac{\sec(3y)}{1-\tan(3y)}$

8.53. Compute $f'(x)$ for each function $f(x)$.

(a) $f(x)=\sqrt{1+\sqrt{x}}$

(b) $f(x)=\sqrt{1-\sqrt{x+1}}$

(c) $f(x)=\sqrt{1-\sqrt{x^2+1}}$

(d) $f(x)=\sqrt{1+\sqrt[3]{x^2+1}}$

(e) $f(x)=\sin\left(\dfrac{1}{x}\right)$

(f) $f(x)=\sin\left(\dfrac{1}{x^2}\right)$

(g) $f(x)=\sin\left(\sec\left(\dfrac{1}{x}\right)\right)$

(h) $f(x)=\dfrac{1}{\sin\left(\dfrac{1}{x}\right)}$

8.54. Suppose that $h(x)=f(g(x))$ and

$$g(2)=4,\ g'(2)=4,\ f(4)=1,\ f'(4)=-1 \text{ and } f'(2)=0.$$

Use the chain rule to find $h'(2)$.
(Hint: some of the given information is dummy information.)

8.55. Suppose that $F(x) = f(g(h(x)))$ and

$$h(0)=1,\ h'(0)=2,\ g(1)=1,\ g'(1)=-3,\ f(4)=1,\ f(1)=0 \text{ and } f'(1)=2.$$

Use the chain rule to find $f'(0)$.

(Hint: some of the given information is dummy information.)

8.56. Find all points on the graphs of the following equations at which the tangent line is horizontal. (Hint: there are infinitely many points in each case.)

(a) $y=\cos(2x)+\sin(2x)$

(b) $y=x+\sin(2x)$

(c) $y=\dfrac{\sin(3x)}{2+\cos(3x)}$

(d) $y=\dfrac{2+\cos(\pi x)}{\sin(\pi x)}$

8.57. Use formula (8.29) to find $\vec{r}'(t)$ for each of the following vector-valued functions $\vec{r}(t)$.

(a) $\vec{r}(t)=\langle\cos(t),\sin(t)\rangle$

(b) $\vec{r}(t)=\langle 2+\cos(t),3+\sin(t)\rangle$

(c) $\vec{r}(t)=\langle\cos(2\pi t),\sin(3\pi t)\rangle$

(d) $\vec{r}(t)=\langle\cos(t),t\rangle$

(e) $\vec{r}(t)=\langle 1+2t,t^2\rangle$

(f) $\vec{r}(t)=\langle 1+2t^2,t^2\rangle$

(g) $\vec{r}(t)=\langle 1+2t,3t\rangle$

(h) $\vec{r}(t)=\left\langle 2t,\dfrac{1}{1+t}\right\rangle$

(i) $\vec{r}(t)=\left\langle \dfrac{1}{1+t},2t\right\rangle$

8.58. Plot the trajectory of each vector-valued function in the previous exercise and plot the tangent vector corresponding to a few values of t.

8.59. Use formula (8.36) to compute $f'(t)$ for the following functions $f(t)$.

(a) $f(t)=\exp(2t)$

(b) $f(t)=e^{5t}$

(c) $f(t)=e^{-3t}$

(d) $f(t)=e^{3t}+e^{-3t}$

(e) $f(t)=e^{3t}-e^{-3t}$

(f) $f(t)=\dfrac{e^{3t}-e^{-3t}}{e^{3t}+e^{-3t}}$

8.60. Use formula (8.37) to compute $f'(u)$ for the following functions $f(u)$.

(a) $f(u)=\exp(u^2)$

(b) $f(u)=e^{u^2}$

(c) $f(u)=e^{u\sin(u)}$

(d) $f(u)=e^{1+u}+e^{1-u}$

(e) $f(u)=e^{1u}+e^{1u^2}$

(f) $f(u)=\dfrac{e^{1+u^2}-e^{1-u^2}}{e^{u^2}+e^{-u^2}}$

8.61. Prove the following generalization of the product rule to compute the derivative of a product of three functions:

$$\boxed{(f\,g\,h)'(x)=(f'g\,h)(x)+(f\,g'h)(x)+(f\,g\,h')(x)} \tag{8.50}$$

8.62. Use formula (8.50) to compute $f'(x)$ for the following functions $f(x)$.

(a) $f(x)=(1+x)(2+x^2)(3+x^3)$

(b) $f(x)=x\sin(x)\exp(x)$

(c) $f(x)=(1+x)(2+x^2)e^x$

(d) $f(x)=(1+x)\sin(x)e^x$

(e) $f(x)=\sqrt{x}\sin(x)e^x$

(f) $f(x)=\sqrt{(1+x)(2+x^2)(3+x^3)}$

8.63. Use formula (8.38) to compute $f'(w)$ for the following functions $f(w)$.

(a) $f(w)=6^w$

(b) $f(w)=2^w+3^w$

(c) $f(w)=2^{w^3}+3^{w^2}$

(d) $f(w)=6^{\sin(w)+\cos(w)}$

(e) $f(w)=2^{\sin(w)}+3^{\cos(w)}$

(f) $f(w)=\sin(2^w)\cos(3^w)$

(g) $f(w)=\dfrac{2^w-2^{-w}}{2^w+2^{-w}}$

8.64. Compute the derivative of $f(x)=2^x3^x$ in two different ways: (i) apply the product rule and then use the laws for logarithms to simplify the answer, (ii) write the expression for $f(x)$ as a single exponent and then take the derivative. Are your answers the same?

8.65. (a) Check that $y(t)=\dfrac{1}{2}+e^{-t^2}$ is a solution of the differential equation $y'(t)+2ty(t)=t$.

(b) Check that $y(t)=\dfrac{-3}{t^3+1}$ is a solution of the differential equation $y'(t)=t^2y^2$.

(c) Check that $y(t)=1+t+2e^t$ is a solution of the differential equation $y'(t)=y-t$.

(d) Check that $y(t)=ct-1-\ln|t|$,
where c is any constant, is a solution of the differential equation $y=ty'(t)-\ln|t|$.

8.66. Use formulas (8.45) and (8.46) to compute the following derivatives.

(a) $\dfrac{d}{dx}\ln(x+1)$

(b) $\dfrac{d}{dt}\ln t^2$

(c) $\dfrac{d}{du}\ln\left|\dfrac{u+1}{u-1}\right|$

(d) $\dfrac{d}{dv}\ln|\tan(v)|$

(e) $\dfrac{d}{dy}\ln(y+e^y)$

(f) $\dfrac{d}{dw}\ln(1+\ln w)$

(g) $\dfrac{d}{dx}\log_2\sqrt{x}$

(h) $\dfrac{d}{dt}\log_2(2+2t^2)$

(i) $\dfrac{d}{du}\log_2|1+u|+\log_2|1-u|$

(j) $\dfrac{d}{dv}\log_3 5^v+\log_5 3^v$

(k) $\dfrac{d}{dy}\log_5(5^y+5^{-y})$

(l) $\dfrac{d}{dw}\log_3(\log_3 9^w)$

8.67. Use formulas (8.48) to compute the following derivatives.

(a) $\frac{d}{dx}\sin^{-1}(1-x)$

(b) $\frac{d}{dt}\sin^{-1}(e^{2t})$

(c) $\frac{d}{du}\cos^{-1}(\sin(u))$

(d) $\frac{d}{dv}\tan^{-1}\left(\frac{2}{v}\right)$

(e) $\frac{d}{dy}\tan^{-1}(\sin(y))$

8.68. One hundred meters directly above ground level of an airport tower, a helicopter and an airplane start flying due east. If the airplane travels four times faster than the helicopter, what is the greatest angle of sight between the two aircraft from the ground level of the tower?

8.69. At what point on the positive x-axis does the line-segment joining $(0,2)$ to $(2,3)$ subtend the maximum angle? Verify that you have found the maximum.

CHAPTER 9

EUCLIDEAN GEOMETRY

9.1 INTRODUCTION

In Euclidean geometry, theorems are proved by means of a process of deductive reasoning. This means that a conclusion (the theorem) is reached by stringing together observations based on certain given "facts," which might include definitions, axioms, postulates, or previously proven theorems.

The first systematic and comprehensive compilation of geometry theorems and constructions was made by the Greek Mathematician Euclid in about 300 BC. His work, called the "Elements," formed the basis for more than two thousand years for the study called "Euclidean geometry."

Euclid's geometry was based on discoveries made by the Greek geometers who lived before him, including Pythagoras and Plato. It was Plato, in particular, who emphasized the method of constructing diagrams and proving theorems using a compass and straightedge. Plato decreed that, for a construction to be acceptable, the straightedge should not be marked and the compass should not be used to transfer distances.

The Elements, in turn, were taken as a starting point by geometers who followed Euclid. Archimedes, for example, derived the formula πr^2 for the area of a disk with radius r. The Geometer Pappus, of Alexandria, who lived at the beginning of the fourth century AD (near the end of the classical period), wrote a collection of geometry books that also contained his own theorems on touching circles and many other discoveries.

The content of Euclid's Elements is outlined in section 9.2. This is followed in section 9.3 by a discussion of terminology relating to lines, angles, polygons, and circles. As preparation for understanding proofs in geometry, some basic reasoning skills in geometry are explained in section 9.4. A selection of elementary theorems from the Elements is stated and proved in sections 9.5 and 9.6, along with many examples of applications of the theorems.

Section 9.7 contains five interesting examples where the theorems of sections 9.5 and 9.6 can be applied: the first involves a construction of the *golden ratio* using a 3-4-5 right triangle; the second is a geometric illustration of the *classical means*; the third is a proof of the fact that the circum-center, centroid, and ortho-center of any triangle are collinear; the fourth is a demonstration of a compass and straightedge construction of a common tangent line to two circles; and the fifth example is a second proof of Ptolemy's theorem (theorem 9.6.8(a)) as a demonstration of the unification of algebra, trigonometry, and geometry.

9.2 EUCLID'S ELEMENTS

We have mentioned the importance of Euclid's Elements in the development of geometry. Therefore, we will begin with a discussion of the Elements as a starting point for learning geometry. The

Elements contains 13 books that mostly deal with geometry. There are a few books on number theory and the theory of ratio and proportion.

In book 1, Euclid begins with the definitions of the basic objects of geometry, such as points, straight lines, angles, triangles, quadrilaterals, and parallel lines. Following these definitions, he states five postulates. The first three postulates describe basic constructions that can always be carried out in geometry: (i) a straight line can be drawn from any point to another; (ii) a finite line can be produced continuously in a straight line (i.e., any straight line can be made longer); (iii) a circle can be drawn with any center and radius. The fourth postulate states that all right angles are equal to one another (perhaps Euclid thought this was not obvious!). The fifth is often called the "parallel postulate" because it states "that, if a straight line falling on two straight lines makes the interior angles on the same side less than two right angles, the two straight lines, if produced indefinitely, meet on that side on which the angles are less than two right angles." In terminology, this says that if cointerior angles formed by two straight lines and a transversal add up to less than 180°, then the two straight lines are not parallel; that is, they will intersect.

A postulate is a synthetic proposition, the contradiction of which, though difficult to imagine, nevertheless remains conceivable. For this reason, the parallel postulate was the subject of intense investigation until the nineteenth century because geometers did not know if it could be proved (with assumption of the first four postulates) or if Euclid was correct in stating it as a separate postulate. In the first half of the nineteenth century, the Russian Mathematician Nicolai Lobachevsky and the Hungarian Mathematician János Bolyai developed a detailed geometric theory without the assumption of the parallel postulate. (That is, the parallel postulate is not a consequence of the other postulates.) By the end of the nineteenth century, mathematicians had constructed some planar models of a "non-Euclidean" or "hyperbolic" geometry with the aid of certain three-dimensional methods of projection, in which the parallel postulate failed.

After the statements of the five postulates, Euclid states five "common notions": (i) things that are equal to the same thing are also equal to one another; (ii) if equals are added to equals, then the wholes are equal; (iii) if equals are subtracted from equals, then the remainders are equal; (iv) things that coincide with one another are equal to one another; and (v) the whole is greater than the part. These statements can be called axioms because they are the basis for reasoning in all of Euclid's proofs.

The remainder of book 1, in the form of 48 propositions, is devoted to the demonstration of elaborate constructions with a compass and a straightedge (e.g., the construction of an equilateral triangle on a line segment) and statements and proofs of basic theorems regarding parallel lines, triangles (including the Pythagorean Theorem for right triangles), and parallelograms. We list many of these theorems in section 9.5.

9.3 TERMINOLOGY

Terminology should be learned very well, so there can never be any doubt about the meaning of the statement of a problem. We will introduce terminology relating to lines, angles, polygons, and circles. The most important terminology will be introduced in numbered definitions.

9.3.1 Lines, Angles, and Polygons

When we refer to a "line" or "line segment," we will always mean a "straight line" that is extended as far as it needs to be in order for the continuing statements to make sense.

We will refer to a line segment by means of labels (usually uppercase letters) that mark the two end points.

In figure 9.1, AB and CB are line segments that meet at B, forming an *angle* with *vertex* at B. Notation for the angle is $\angle ABC$, $\angle B$, $A\hat{B}C$, or $\hat{B}$. If more than one angle is formed at some point in a diagram, then the angles can be numbered, as we will see in many diagrams below.

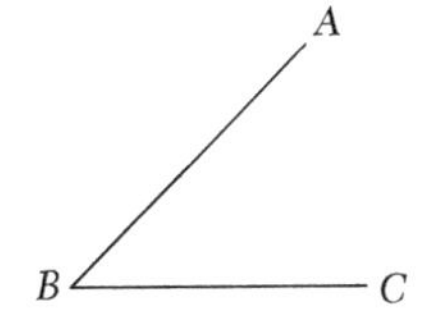

FIGURE 9.1. An angle.

DEFINITION 9.3.1. *If two straight lines intersect so that the angles formed at the point of intersection are all equal, then they are all right angles, and we say that the lines are* perpendicular *to one another.*

In figure 9.2, AB is perpendicular to CD, so we write $AB \perp CD$. A right angle can be indicated using the square symbol, as shown in figure 9.2.

Angles are commonly measured in degrees, with the measurement for a right angle being 90 degrees (notation: 90°).

DEFINITION 9.3.2. *An* acute angle *is less than* 90°. *An* obtuse angle *is greater than* 90° *but less than* 180°. *A* straight angle *is equal to* 180°. *A* reflex angle *is greater than* 180° *but less than* 360°. *A* revolution *is equal to* 360°.

DEFINITION 9.3.3. *A* triangle *is formed when three lines meet in such a way that there are exactly three distinct intersection points called* vertices.

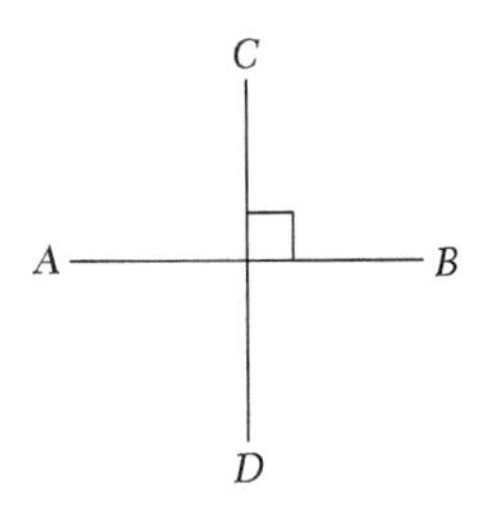

FIGURE 9.2. A right angle.

In figure 9.3, three line segments AB, AC, and BC form triangle ABC (notation: ΔABC). The angles $A\hat{B}C$, $B\hat{C}A$, and $C\hat{A}B$ are called the *interior angles* (or *angles*) of ΔABC.

DEFINITION 9.3.4. *An* exterior *angle of a triangle is formed when a side of a triangle is produced (extended). It is the angle between the produced segment and the adjacent side of the triangle.*

In figure 9.3, AC is produced to E and BC is produced to D. $\hat{C}_1$ and $\hat{C}_3$ are exterior angles of ΔABC, but $\hat{C}_2$ is *not* an exterior angle of ΔABC.

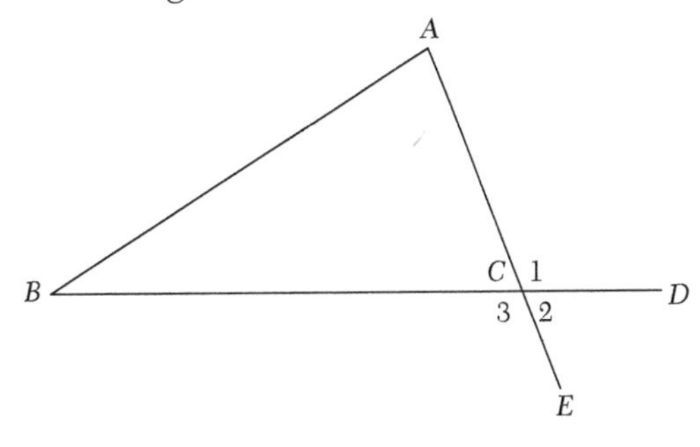

FIGURE 9.3. A triangle with exterior angles.

Definition 9.3.5. *If two angles add up to* $90°$, *they are said to be* complementary angles. *If two angles add up to* $180°$, *they are said to be* supplementary angles.

A line that intersects two other specified lines is called a *transversal.*

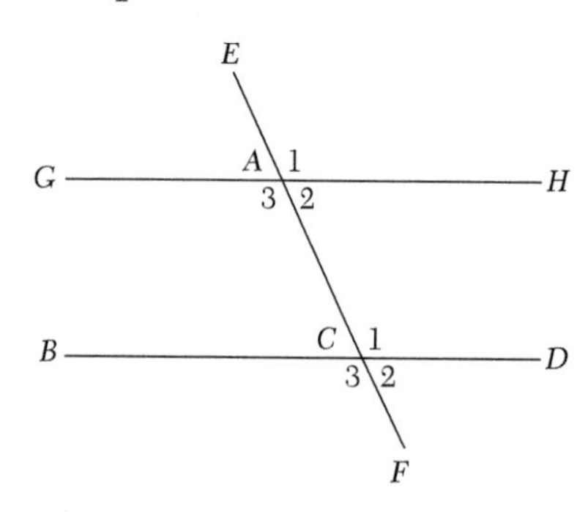

FIGURE 9.4. Parallel lines with a transversal.

In figure 9.4 the lines *GH* and *BD* (which might or might not be parallel) are intersected by the transversal *EF*. The angles $\hat{A}_3$ and $\hat{C}_1$ are a pair of *alternate angles* because they lie *either side* of the transversal and the lines *between GH* and *BD*. (There is one other pair of alternate angles in the diagram.) Angles $\hat{A}_3$ and $\hat{C}_3$ are a pair of *corresponding angles* because they are formed between the transversal and the lines *GH* and *BD* in the same way. (There are three other pairs of corresponding angles in the diagram.) The angles $\hat{A}_2$ and $\hat{C}_1$ are a pair of *cointerior angles* because they lie on the *same side* of the transversal and the lines *between GH* and *BD*. (There is one other pair of cointerior angles in the diagram.)

Definition 9.3.6. *Two angles situated at the same vertex along a common side are referred to as* adjacent angles. *When two straight lines intersect, the two nonadjacent angles are called* vertically opposite angles.

In figure 9.4 $\hat{A}_1$ and $\hat{A}_2$ are adjacent (e.g., also $\hat{C}_3$ and $\hat{C}_2$) and $\hat{C}_3$ and $\hat{C}_1$ are vertically opposite (e.g., also $\hat{A}_1$ and $\hat{A}_3$, for example). In figure 9.3, $\hat{C}_1$ and $\hat{C}_3$ are vertically opposite and $\hat{C}_2$ and $\hat{C}_3$ are adjacent.

Definition 9.3.7. Parallel lines *in the plane are lines that do not intersect no matter how far they are extended.*

We use the "||" symbol for parallel lines. For example, in figure 9.4, if *BD* and *GH* are parallel lines, then we write $BD \parallel GH$.

We now present some of the terminology relating to triangles and polygons.

If two sides of a triangle (or polygon) have the same length, then, for brevity, we will refer to them as "equal sides." For example, an *equilateral triangle* has three equal sides, an *isosceles triangle* has two equal sides, and a *scalene triangle* has no equal sides.

An *acute triangle* has three acute angles, a *right triangle* has one angle equal to $90°$, and an *obtuse triangle* has one obtuse angle. In a right triangle, the side opposite the right angle is called the *hypotenuse*.

Any side of a triangle (or another geometric figure) can be taken to be the *base* of the triangle (or geometric figure). An *altitude* of a triangle is the vertical (perpendicular) distance from the base of the triangle to its highest point. A *median* of a triangle is a line from any vertex of the triangle to the mid-point of the opposite side.

A triangle is a *polygon* with three sides (or edges). In general, a polygon has *many* (three or more) sides. In particular, a *quadrilateral* has *four* sides, a *pentagon* has *five* sides, a *hexagon* has *six* sides, and an *octagon* has *eight* sides. A *regular polygon* is a polygon with all sides equal in length and equal angles at all the vertices. Figure 9.5 shows a regular triangle (an equilateral triangle), a quadrilateral (a *square*), a pentagon, and a hexagon. A *diagonal* of a polygon is a line segment joining any two nonadjacent vertices.

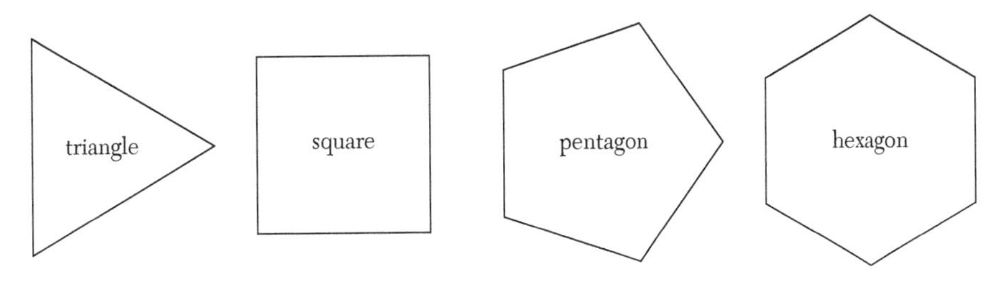

FIGURE 9.5. Regular polygons.

A *rectangle* is a quadrilateral with both pairs of opposite sides equal, and all angles equal to right angles. A *rhombus* is a quadrilateral with all sides equal, but the angles can be any size. A *kite* is a quadrilateral with two pairs of adjacent sides equal. A *trapezoid* is a quadrilateral with one pair of opposite sides parallel. A *parallelogram* is a quadrilateral with both pairs of opposite sides parallel, see figure 9.6.

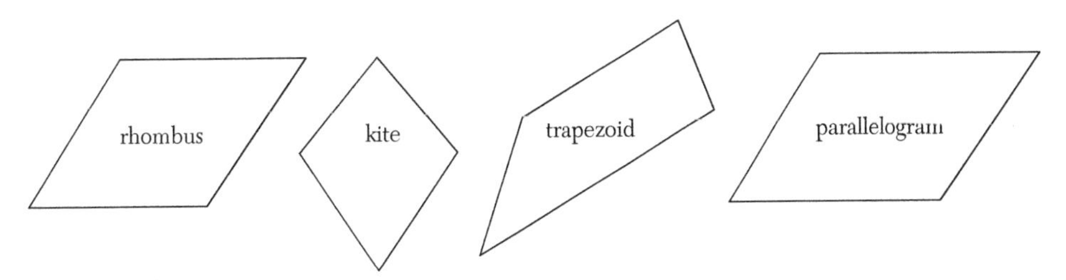

FIGURE 9.6. Special quadrilaterals.

DEFINITION 9.3.8. *A* cyclic quadrilateral *is a quadrilateral with all four vertices on the circumference of the same circle.*

See figure 9.39 for an illustration of a cyclic quadrilateral.

DEFINITION 9.3.9. *Congruent triangles (or polygons) are triangles (or polygons) that are equal in all respects.* Similar triangles *(or polygons) are scaled copies of each other; that is, they can be larger or smaller, but the angles do not change.*

We use the notation $\Delta ABC \equiv \Delta DEF$ if the triangles ΔABC and ΔDEF are congruent, and we use the notation $\Delta ABC \,|||\, \Delta DEF$ if they are similar.

9.3.2 Circles

DEFINITION 9.3.10. *The* circumference *of a circle is the distance around it. A* diameter *of a circle is a line passing through its center and joining two opposite points. A* radius *of a circle is a line joining its center with any point on the circle (plural: radii). A* chord *is a line joining any two points of the circle. A* tangent line *to a circle is a straight line that touches it at exactly one point. A* secant *is a line drawn from a point outside the circle that cuts it at two different points.*

We say that two circles are touching if they intersect at a single point (see figure 9.46).

In the first diagram in figure 9.7, *AB* is a diameter of the circle, *OT* is a radius, *MN* is a chord, *QTR* is a tangent line, and *CDEQ* is a secant line. In the second diagram, the chord *JK* divides the circle into a *major segment* and a *minor segment* of JK. The points *J* and *K* divide the circumference of the circle into a *major arc JK* and a *minor arc JK*.

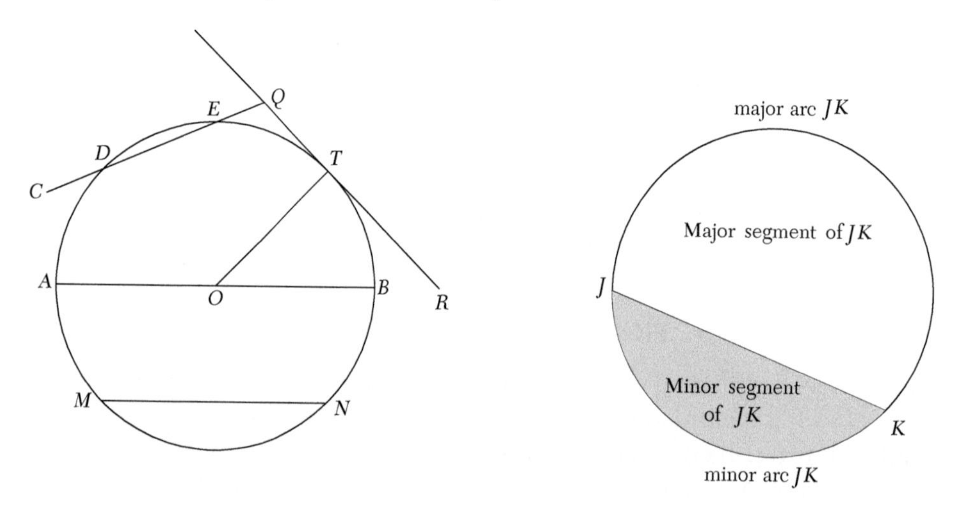

FIGURE 9.7. Terminology for a circle.

Concentric circles are circles that have the same center but different radii. Points are *concyclic* if and only if they lie on the same circle. A *semicircle* is exactly half of a circle.

DEFINITION 9.3.11. *To subtend means to stretch across or be opposite to.*

Refer to figure 9.8. In the first diagram, the minor arc *AB* subtends an acute angle $\hat{O}_1$ at the center of the circle and another acute angle $\hat{C}$ on the circle, while the major arc *AB* subtends the reflex angle $\hat{O}_2$ at the center and the obtuse angle $\hat{D}$ on the circle. In the second diagram, the boundary of *sector EQF* is composed of the two radii *QE* and *QF* and the minor arc *EF*.

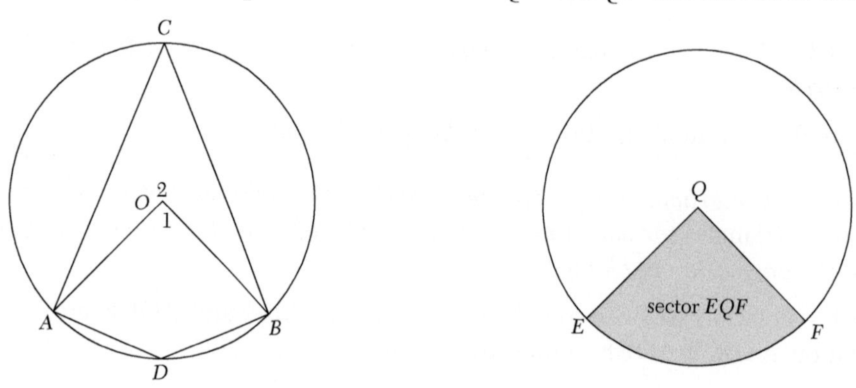

FIGURE 9.8. Subtending an angle.

9.3.3 Other Important Terms

Our statements of theorems in geometry will make use of the precise language that we introduce in the following paragraphs.

To *intersect* means to pass through or across another line or surface so as to have one or more points in common; and to *bisect* means to divide into two equal parts. Pairs of points are *equidistant* if they are equally distant from each other.

To *inscribe* means to draw one figure within another figure so that the inner touches the outer in as many points as possible. To *circumscribe* means to draw one figure around another so as to touch as many points as possible.

DEFINITION 9.3.12. *Three or more points are called* collinear *if they lie on the same straight line. If three or more different lines pass through the same point, then the lines are said to be* concurrent.

Every triangle has unique inscribed and circumscribed circles. The center of the inscribed circle is called the *in-center*, and it is the point of concurrency of the bisectors of the angles of the triangle (see theorem 9.5.12(a) in section 9.5). The center of the circumscribed circle is called the *circum-center*, and it is the point of concurrency of the perpendicular bisectors of the sides of the triangle (see theorem 9.5.12(b)). The *centroid* of a triangle is the point of concurrency of the medians of the triangle (see theorem 9.5.12(c)). The *ortho-center* of a triangle is the point of concurrency of the altitudes of the triangle (see theorem 9.5.12(e)).

A *theorem* is any statement or rule that can be proved to be true by reasoning from the definitions, axioms, and postulates. Each theorem, once proved, becomes a statement that can be accepted without proof in the proofs of subsequent theorems.

A *corollary* is something proved by inference from something else already proven, that is, a natural consequence or a result. A *converse* is a statement that is turned around; for example, that which is "given" in a theorem becomes the "required to prove" in its converse and is "required to prove" in the theorem becomes the "given" in its converse. We give the following example of a theorem and its converse:

EXAMPLE 9.3.1. If a transversal cuts two parallel lines, then the corresponding angles are equal (this is a theorem); conversely, when a transversal cuts two other lines, then these two lines are parallel if a pair of corresponding angles is equal (this is the converse of the theorem). For example, in figure 9.9, if it is given (as marked with arrows in the diagram on the left) that $AB \| CD$ then we can conclude from the theorem that $\hat{M}_1 = \hat{L}_1$. On the other hand, if it is given (as marked with arcs in the diagram on the right) that $\hat{M}_1 = \hat{L}_1$ then, by the converse of the theorem, we can conclude that $AB \| CD$. ◈

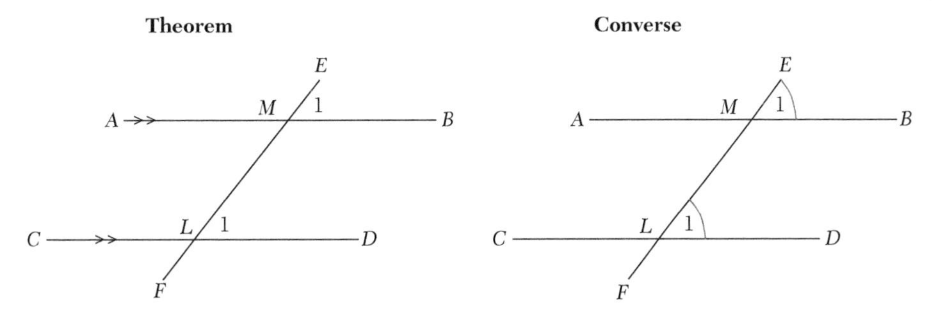

FIGURE 9.9. A theorem and its converse.

9.4 BASIC PROBLEM SOLVING IN GEOMETRY

Some basic observations and methods for proving theorems in geometry need to be emphasized. In this section, we give examples of elementary deductions that frequently form part of the longer proofs of more complicated problems.

EXAMPLE 9.4.1. As shown in the first diagram in figure 9.10, the size of an angle can be computed if it is part of a larger angle. ◈

A basic principle in mathematics is that when equals are subtracted from equals, the differences are equal. Here are the two examples in which this principle is applied to angles.

EXAMPLE 9.4.2. In the second diagram in figure 9.10, $B\hat{A}E = D\hat{A}C$, that is, $\hat{A}_1 + \hat{A}_2 = \hat{A}_2 + \hat{A}_3$. Now $\hat{A}_2$ can be subtracted from both sides to deduce that $\hat{A}_1 = \hat{A}_3$. ◈

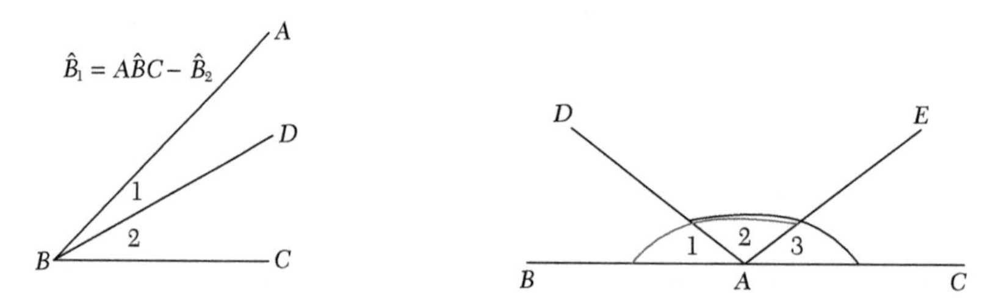

FIGURE 9.10. Parts of angles.

A good way to get used to the process of deductive reasoning in geometry is to tabulate the steps of a proof in a two-column format with each statement in the chain of reasoning given in the left column of the table, and the corresponding reason for the statement alongside it in the right column. The symbol "∴" is used as an abbreviation for "therefore."

EXAMPLE 9.4.3. In figure 9.11, $B\hat{A}C = 90°$ and $AD \perp BC$. We prove, by means of the steps in table 9.1, that $\hat{B} = \hat{A}_2$ and $\hat{A}_1 = \hat{C}$. (We need theorem 9.5.5(c), which states that the sum of the angles of a triangle is $180°$.)

TABLE 9.1.

Statement	Reason
$\hat{B} + \hat{A}_1 = 90°$	Sum $\angle$s $\Delta ABD = 180°$
$\hat{A}_2 + \hat{A}_1 = 90°$	Given
$\therefore \hat{B} + \hat{A}_1 = \hat{A}_2 + \hat{A}_1$	
$\therefore \hat{B} = \hat{A}_2$	Subtract $\hat{A}_1$ from both sides
Similarly $\hat{A}_1 = \hat{C}$	

◈

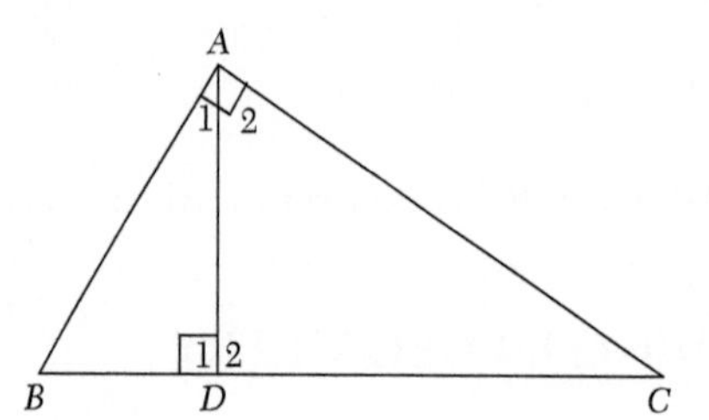

FIGURE 9.11. The duplication principle.

If angles are duplicated in a diagram, then it is useful to know the *duplication principle*, which can be illustrated as follows: if $\bullet + \bullet + * + * = 180°$, for example, then $\bullet + * = 90°$. The next two examples demonstrate this.

EXAMPLE 9.4.4. In the first diagram in figure 9.12, *AFE* is a straight line, with $\hat{F}_1 = \hat{F}_3$ and $\hat{F}_2 = \hat{F}_4$. We prove, by means of the steps in table 9.2, that $B\hat{F}D = 90°$:

TABLE 9.2.

Statement	Reason
$\hat{F}_1 + \hat{F}_2 + \hat{F}_3 + \hat{F}_4 = 180°$	AFE straight ∠
$\therefore \frac{1}{2}(\hat{F}_1 + \hat{F}_2 + \hat{F}_3 + \hat{F}_4) = 90°$	
$\therefore \hat{F}_2 + \hat{F}_3 = 90°$	$\hat{F}_1 = \hat{F}_3$ and $\hat{F}_2 = \hat{F}_4$
That is, $B\hat{F}D = 90°$	

◈

EXAMPLE 9.4.5. In the second diagram in figure 9.12, $AB \| CD$, *KF* bisects $B\hat{F}G$, and *KG* bisects $F\hat{G}D$. We prove, by means of the steps in Table 9.3, that $\hat{K} = 90°$: (in the first line of the table, we use part (iii) of theorem 9.5.3(a).)

TABLE 9.3.

Statement	Reason
$B\hat{F}G + D\hat{G}F = 180°$	Cointerior ∠s; $AB \| CD$
$\therefore 12 B\hat{F}G + 12 D\hat{G}F = 90°$	
$\therefore \hat{F}_2 + \hat{G}_1 = 90°$	$\hat{F}_1 = \hat{F}_2$ and $\hat{G}_1 = \hat{G}_2$ (given)
$\therefore \hat{K} = 90°$	Sum ∠s $\Delta FGK = 180°$

◈

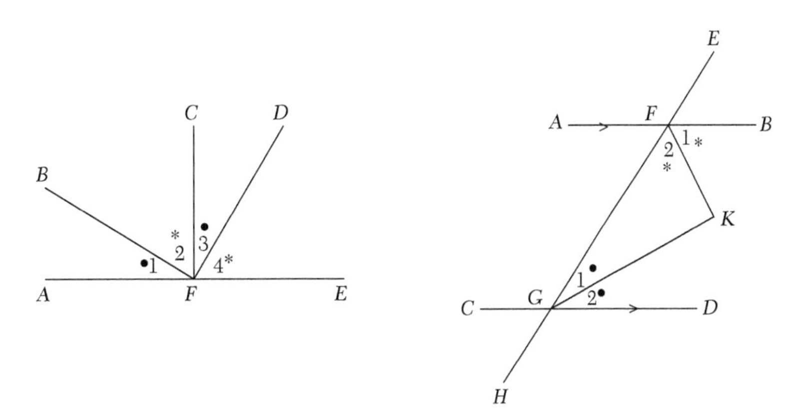

FIGURE 9.12. Diagram for examples 9.4.4 and 9.4.5.

Problems involving ratio and proportion of line segments are made easier by assigning lengths to the line segments, as we demonstrate in the next example.

EXAMPLE 9.4.6. If B is a point on the line segment AC below such that $2|AB|=3|BC|$, write $|AC|$ in terms of $|AB|$.

A B C

Answer: If we divide both sides of the equation by $2|BC|$, then $\frac{|AB|}{|BC|}=\frac{3}{2}$. Now we let $|AB|=3k$ units and $|BC|=2k$ units, as shown below.

A 3k B 2k C

Because $|AC|=|AB|+|BC|=3k+2k=5k$ units, we have $\frac{|AC|}{|AB|}=\frac{5k}{3k}$, which simplifies to $\frac{|AC|}{|AB|}=\frac{5}{3}$.

Consequently, $|AC|=\frac{5}{3}|AB|$. ◈

When two triangles overlap, it often helps to separate them. Then, one immediately sees that a pair of equal angles emerges, namely, their common angle.

EXAMPLE 9.4.7. In figure 9.13 we want to prove that $\Delta ACD \,|||\, \Delta DCB$. (It is given that $\hat{A}=\hat{D}_2$.) If the overlapping triangles are separated (as shown in figure 9.14), then it is clear that two pairs of angles are equal and, therefore, the angle at D in the triangle on the left is equal to the angle at B in the triangle on the right. Thus, the triangles are similar. ◈

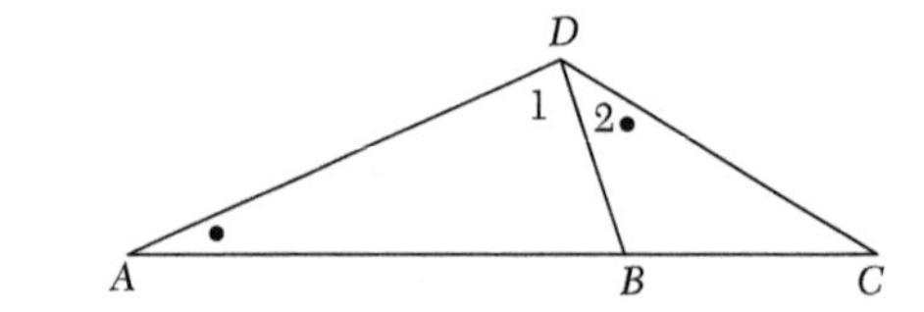

FIGURE 9.13. Similar triangles.

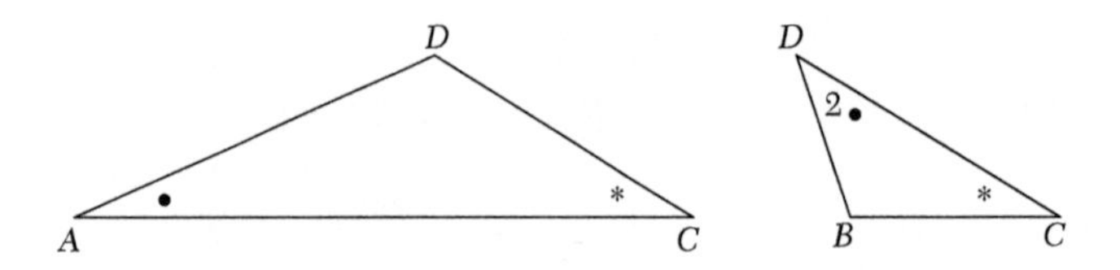

FIGURE 9.14. Separated similar triangles.

9.5 ELEMENTARY THEOREMS RELATING TO LINES AND POLYGONS

In this section, we will present sixteen theorems relating to lines and polygons (including triangles and certain types of quadrilaterals). Most of them can be found in book 1 of the Elements, but we do not follow Euclid's sequence of theorems and proofs.

The theorems can be categorized as statements about the relationships of certain angles in a geometric figure (e.g., the sum of the angles in a triangle is 180°), relationships involving the lengths of line segments in a geometric figure (e.g., the Pythagorean Theorem), or statements relating to proportions or geometric divisions (e.g., the diagonals of a parallelogram bisect each other). There are also theorems that determine when a polygon can be classified as a certain type (e.g., if the diagonals of a quadrilateral bisect each other, then the quadrilateral is a parallelogram). Lastly, there are theorems (called *incidence theorems*) that determine when lines are concurrent (e.g., the medians of a triangle are concurrent).

It is not our intention to give complete proofs of all the theorems. It will suffice, in most cases, to state briefly the reason for the truth of the theorem or to outline the proof of the theorem with the help of a diagram. Commentary and examples of applications of the theorems are included to help with the familiarization of the theorems. The theorems are restated in Appendix for ease of reference.

9.5.1 Theorems about Angles

THEOREM 9.5.1.

(a) *If two straight lines intersect, then the sum of any pair of adjacent angles is* $180°$.

(b) *(Converse) If the sum of any pair of adjacent angles is* $180°$, *then their noncommon sides lie on the same line.*

THEOREM 9.5.1(a) is clearly true if the adjacent angles are equal to each other because each angle is then a right angle (the definition of a right angle), and two right angles add up to $180°$. If the adjacent angles are not equal, then an additional line can be drawn through the common vertex of the angles so that a right angle is formed. It is then clear that the sum of the two original adjacent angles is the same as the sum of three adjacent angles, which is $180°$.

A mathematical statement can sometimes be proved by *contradiction*. This means that the *negation* of the statement is assumed to be true and, by the process of deduction, an obviously false statement is reached. Thus, the negation of the statement has to be discarded and the only possibility that remains is for the statement to be true. For example, the statement of theorem 9.5.1(b) can be proved by contradiction: if the noncommon sides do *not* lie on a line, then one of them can be extended through the common vertex forming a new (nonzero) angle with the other noncommon side. This results in the contradiction that the two original angles that sum to $180°$, together with the new angle, also sum to $180°$, by theorem 9.5.1(a). (Draw a diagram to convince yourself!)

THEOREM 9.5.2. *If two lines intersect, then the vertically opposite angles are equal.*

This can be proved using the method explained in example 9.4.2.

THEOREM 9.5.3.

(a) (i) *If a transversal intersects two parallel lines, then pairs of corresponding angles are equal to one another.*

(ii) *If a transversal intersects two parallel lines, then pairs of alternate angles are equal to one another.*

(iii) *If a transversal intersects two parallel lines, then pairs of cointerior angles are supplementary.*

(b) (i) *(Converse) If two lines are intersected by a transversal such that two corresponding angles are equal, then the two lines are parallel.*

(ii) *(Converse) If two lines are intersected by a transversal such that two alternate angles are equal, then the two lines are parallel.*

(iii) *(Converse) If two lines are intersected by a transversal such that two cointerior angles are supplementary, then the two lines are parallel.*

The statements of theorem 9.5.3(a) and (b) are a consequence of Euclid's parallel postulate and theorems 9.5.1 and 9.5.2. Indeed, Euclid's parallel postulate states that, if cointerior angles add up to *less than* $180°$, then the lines are *not* parallel. It is also clear that, if cointerior angles add up to *more than* $180°$, then the cointerior angles that are adjacent to them will add up to less than $180°$ and so, again, we conclude that the lines are not parallel. Therefore, it must be the case that, if two lines are parallel, then the pairs of cointerior angles are supplementary.

THEOREM 9.5.4. *Lines that are parallel to the same line are parallel to each other.*

This can be proved by constructing a transversal that intersects all three parallel lines and applying theorem 9.5.3.

THEOREM 9.5.5.

(a) *The exterior angle of a triangle is equal to the sum of the interior opposite angles.*

(b) *The exterior angle of a triangle is greater than either of the interior opposite angles.*

(c) *The sum of the angles of a triangle is* 180°.

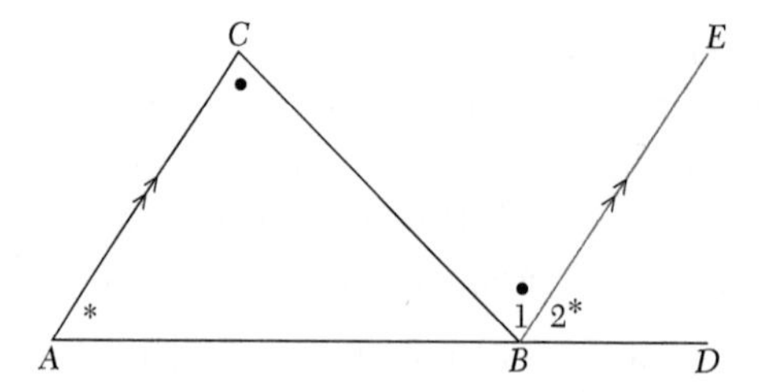

FIGURE 9.15. The exterior angle is equal to the sum of the interior opposite angles.

Theorem 9.5.5(b) is as a consequence of theorem 9.5.5(a). The reason for stating theorem 9.5.5(b) separately is that it is a theorem in non-Euclidean geometry, while theorem 9.5.5(a) is not in non-Euclidean geometry. For the proofs of theorem 9.5.5(a) and (b), refer to figure 9.15, in which the line BE has been constructed parallel to the side AC of ΔABC. The angles marked with a bullet are equal because they form a pair of alternate angles with respect to the parallel lines, and the angles marked with an asterisk are equal because they form a pair of corresponding angles with respect to the parallel lines. It follows that the exterior angle $C\hat{B}D = \hat{B}_1 + \hat{B}_2$ of ΔABC is equal to the sum of the interior opposite angles $\hat{A}$ and $\hat{C}$.

Theorem 9.5.5(c) can be proved by constructing a line parallel to any side of a triangle and through the vertex opposite that side. This creates two pairs of alternate angles with the angles at the other two vertices of the triangle. It can then be deduced that the sum of the angles of the triangle is equal to the sum of three angles along the constructed line that add up to 180°. (Draw a diagram!)

EXAMPLE 9.5.1. Refer to figure 9.16. Suppose that $\hat{A} > \hat{D}_1$. Use theorem 9.5.5(a) to prove that $\hat{D}_2 > \hat{C}$.

Answer: Theorem 9.5.5(a) asserts that, because $\hat{D}_2$ is exterior to ΔABD, $\hat{D}_2 > \hat{A}$. Similarly, $\hat{D}_1$ is exterior to ΔBCD, so $\hat{D}_1 > \hat{C}$. Therefore, $\hat{D}_2 > \hat{A} > \hat{D}_1 > \hat{C}$, and so we have proved that $\hat{D}_2 > \hat{C}$. ◈

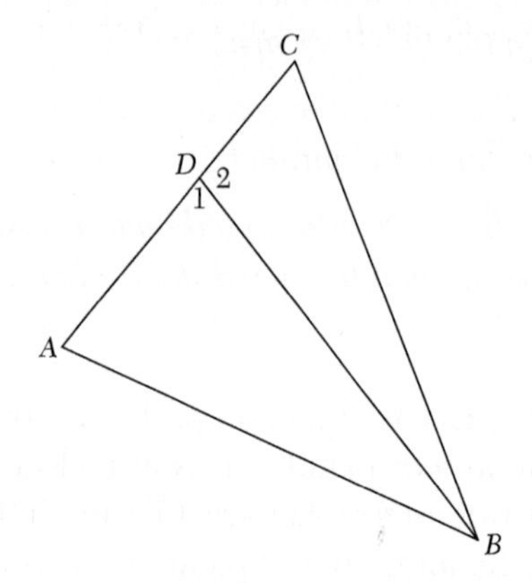

FIGURE 9.16. Diagram for example 9.5.1.

9.5.2 Theorems about Triangles

We say that two triangles are congruent if they are identical to each other. This can mean, for example, that one triangle is the *mirror image*, or *reflection*, of the other. Figure 9.17 shows a triangle T_1 with vertices A, B, and C, together with its mirror images T_2, T_3, and T_4 with respect to its sides AB, BC, and AC, respectively. Note that the triangle T_3 can be rotated in a clockwise direction around the vertex B so that it matches the triangle T_2 exactly. In the same way, any one of the triangles T_2, T_3, or T_4 can be rotated to match any one of the others. The triangle T_5 is a copy of the triangle T_4. All of these triangles are congruent to each other. (They are identical triangles.)

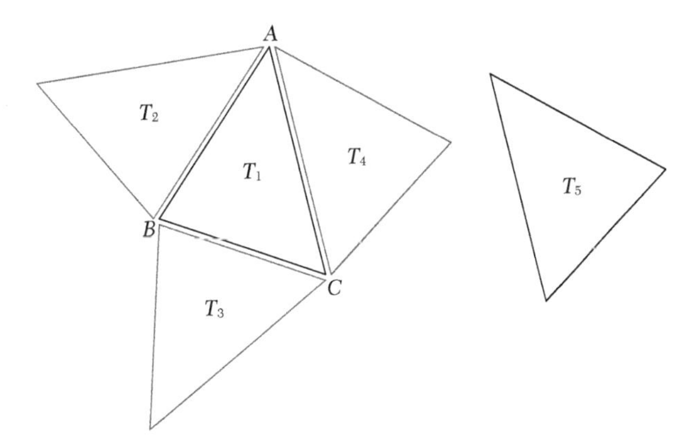

FIGURE 9.17. Congruent triangles.

If two triangles are congruent, then their vertices can be labeled A, B, C and D, E, F, for example, so that in ΔABC the angle at A is the same as the angle at D in ΔDEF, the angle at B is the same as the angle at E, and the angle at C is the same as the angle at F. Similarly, the side AB in ΔABC has the same length as side DE in ΔDEF, the side AC as side DF, and the side BC as side EF. In summary, the six *parts* of ΔABC are equal to the six *parts* of ΔDEF.

In most situations, it is enough to know that two triangles have three pairs of parts equal to each other in order to conclude that the two triangles are congruent. Theorem 9.5.6 states all possible cases in which it can be concluded that triangles are congruent if three pairs of parts are the same. These are called the *rules for congruence of triangles*. It is helpful to refer to them by their abbreviations in parentheses.

THEOREM 9.5.6.

(a) *If two sides and the included angle of one triangle are, respectively, equal to two sides and the included angle of another triangle, then the two triangles are congruent* (s∠s).

(b) *If two angles and a side of one triangle are, respectively, equal to two angles and the corresponding side of another triangle, then the triangles are congruent* (∠∠s).

(c) *If three sides of one triangle are equal to three sides of another triangle, then the triangles are congruent* (sss).

(d) *If, in two right triangles, the hypotenuse and one side of the one are, respectively, equal to the hypotenuse and one side of the other, then the triangles are congruent* (⊥hs).

Two triangles need *not* be congruent if two pairs of sides are the same, but the angles that are the same are not the *included angles* (unless the angles that are the same are right angles (which is the case *⊥hs*). A detailed explanation of this was given in section 4.13.2. It will be worthwhile to examine figure 4.33.

EXAMPLE 9.5.2. In figure 9.18, we can conclude, in the first diagram, that $\Delta ABC \equiv \Delta DEF$ by the $s\angle s$ criterion because the sides AB and DE are marked as parallel and equal in length, and so the alternate angles at B and E are equal, and the corresponding sides BC and EF are also shown to be equal in length. In the second diagram, we can conclude that $\Delta ABC \equiv \Delta CDA$ by the $\perp hs$ criterion because ΔABC and ΔCDA have the common side AC, and it is given that the hypotenuse of ΔABC equals the hypotenuse of ΔCDA. ◈

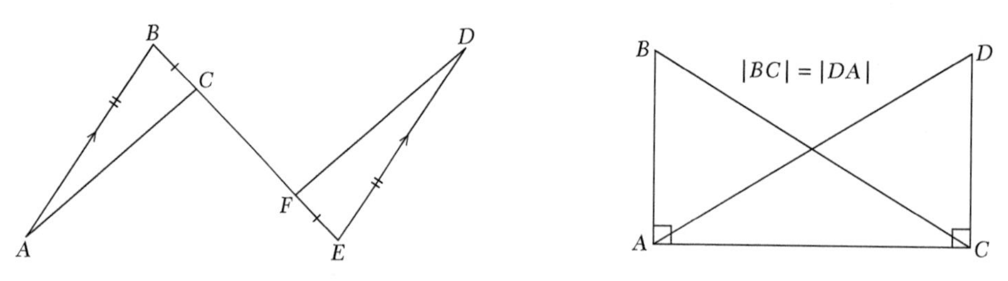

FIGURE 9.18. Examples of congruent triangles.

THEOREM 9.5.7.

(a) The angles opposite the equal sides of an isosceles triangle are equal.

(b) (Converse) If two angles of a triangle are equal, then the sides opposite them are equal.

The proof of theorem 9.5.7(a) requires the construction of congruent triangles; for example, if ΔABC is isosceles because $|AB| = |AC|$, then a line can be drawn from the vertex A to a point D on the base BC so that the angle at A is bisected. This creates two congruent triangles ΔBAD and ΔCAD by the $s\angle s$ criterion. Thus, we can conclude that in ΔABC the angle at B equals the angle at C. (Draw a diagram!) The proof of theorem 9.5.7(b), that is, the converse of theorem 9.5.7(a), is almost identical, and so is left as an exercise.

REMARK 9.5.1. *It is a corollary of theorem 9.5.7(a) that an equilateral triangle has all angles equal to* $60°$, *and it is a corollary of theorem 9.5.7(b) that, if all angles of a triangle equal* $60°$, *then the triangle is an equilateral triangle.*

EXAMPLE 9.5.3. We will demonstrate, by means of figure 9.19, that an angle can be trisected using a marked ruler (or straightedge). Suppose that the angle to be trisected is $B\hat{O}D$, and B and D are points on a circle centered at O. A ruler, on which the radius of the circle is marked, is aligned from B to a point A (which is collinear with D and O) in such a way that $|AC|$ is a radius of the circle and C is a point of the circle. If O and C are joined, then triangle ΔACO is isosceles and, therefore, $O\hat{C}B = 2O\hat{A}C$. Furthermore, ΔOCB is also isosceles (OB is another radius) and, therefore, $O\hat{C}B = O\hat{B}C$. Now, because $D\hat{O}B$ is an exterior angle to ΔOAB, we have

$$\angle D\hat{O}B = O\hat{A}C + O\hat{B}C = O\hat{A}C + O\hat{C}B = O\hat{A}C + 2O\hat{A}C = 3O\hat{A}C.$$

We thus conclude that a line drawn from O parallel to AB will trisect $B\hat{O}D$. ◈

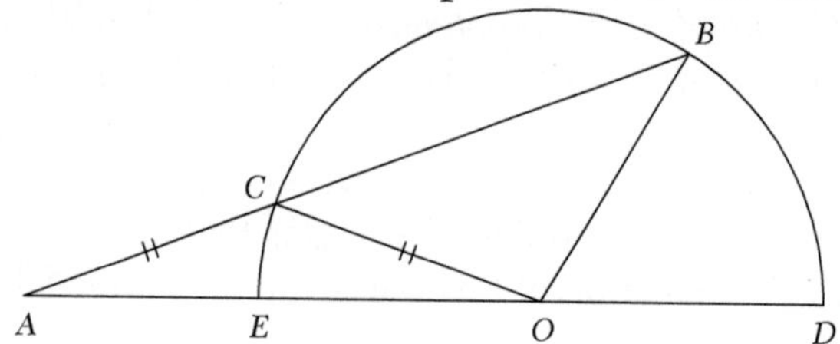

FIGURE 9.19. Trisecting an angle using a marked ruler.

In 1837, it was proved by the French Mathematician Pierre Wantzel that it is generally impossible to trisect and angle using only a compass and an unmarked straightedge. This put an end to the attempts by geometers for thousands of years to find a way to do it!

A fundamental theorem in Euclidean geometry is the Pythagorean Theorem (theorem 9.5.8). There are many different and surprising ways to prove it. In this chapter, we present four different proofs.

THEOREM 9.5.8.

(a) *The square of the length of the hypotenuse of a right triangle is equal to the sum of the squares of the lengths of the other two sides (the* Pythagorean Theorem*).*

(b) *(Converse) If the square of the length of one side of a triangle is equal to the sum of the squares of the lengths of the other two sides, then the angle opposite the first side is a right angle.*

Our first proof of theorem 9.5.8(a) is a visual proof involving the rearrangement of four identical copies of the same right triangle, as shown in figure 9.20.

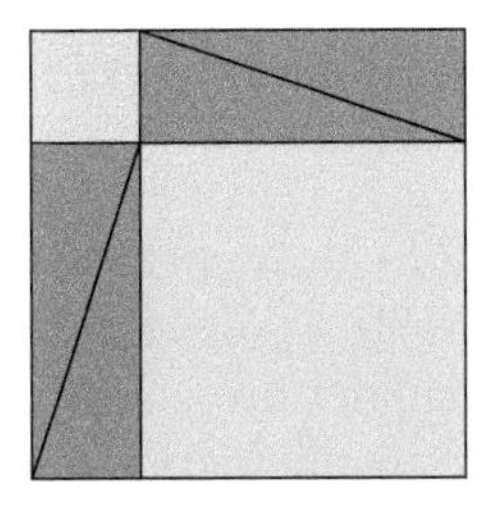

FIGURE 9.20. A proof of the Pythagorean Theorem.

Theorem 9.5.8(b), the converse of the Pythagorean Theorem, is the final proposition of book 1 of Euclid's Elements. Here is Euclid's proof: in ΔABC in figure 9.21, let the square of the side BC equals the sum of the squares of the sides BA and AC. It will now be demonstrated that $B\hat{A}C$ is a right angle. Draw AD from the point A at right angles to the line AC so that $|AD|=|BA|$ and join D to C. (Note that we cannot assume that B, A, and D are collinear.) Then, the sum of the squares on AD and AC equals the square on CD, by theorem 9.5.8(a). What's more, because the square on AD equals the square on AB, and AC is a common side, it follows from the hypothesis that the squares on CD and CB are the same. In this way, we conclude that $|CD|=|CB|$. The triangles ACD and ACB are now seen to be congruent by the *sss* condition and, therefore, $B\hat{A}C$ must be a right angle.

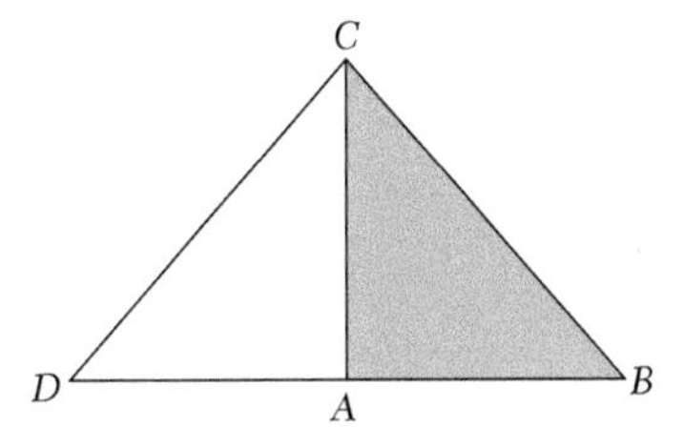

FIGURE 9.21. The converse of the Pythagorean theorem.

9.5.3 Parallelograms and Parallel Lines

THEOREM 9.5.9.

(a) *The opposite sides of a parallelogram have the same length.*

(b) *(Converse) If the opposite sides of a quadrilateral have the same length, it is a parallelogram.*

(c) *The opposite angles of a parallelogram are equal.*

(d) *(Converse) If the opposite angles of a quadrilateral are equal, then it is a parallelogram.*

(e) *The diagonals of a parallelogram bisect each other.*

(f) *(Converse) If the diagonals of a quadrilateral bisect each other, then it is a parallelogram.*

(g) *If both members of one pair of opposite sides of a quadrilateral are parallel and have the same length, it is a parallelogram.*

(h) *The diagonals of a rectangle have the same length.*

(i) *The diagonals of a rhombus bisect each other at right angles and bisect the angles of the rhombus.*

Theorem 9.5.9(a) and (c) can be proved by joining a pair of opposite vertices of a parallelogram with a diagonal line to create a pair of triangles. The two pairs of alternate angles formed in this way are equal, and so the triangles are congruent by the $\angle\angle s$ criterion. Thus, both members of a pair of opposite angles of the parallelogram are equal, and both members of a pair of opposite sides are equal. Similarly, the members of the other pairs of opposite angles and opposite sides are equal. The proof of theorem 9.5.9(e) is also a consequence of the formation of congruent triangles, when both diagonals are drawn. The proofs of the converse statements, that is, theorem 9.5.9(b), (d), (f), and (g), are left as exercises. Theorem 9.5.9(h) and (i) are two special cases that can be verified easily.

EXAMPLE 9.5.4. In figure 9.22, the line AF is a median of both ΔABC and ΔADH because the diagonals of the parallelogram $BDCH$ bisect each other at F. ◈

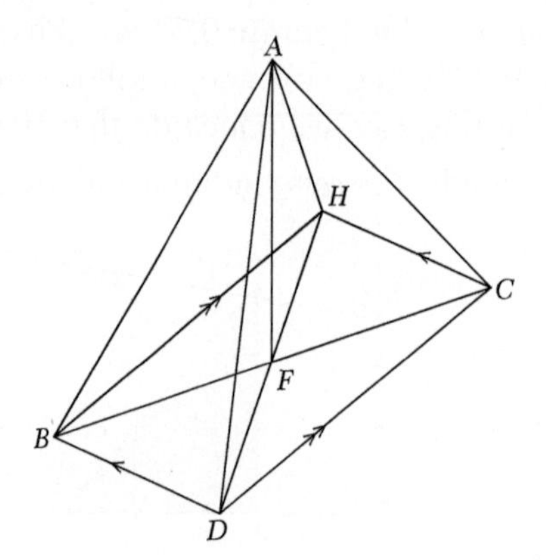

FIGURE 9.22. Diagram for Example 9.5.4.

THEOREM 9.5.10.

(a) *The area of a parallelogram is bisected by each diagonal.*

(b) *A parallelogram and a rectangle on the same base and between the same parallels have equal areas.*

(c) *The area of a triangle is equal to one-half the area of a parallelogram on the same base and between the same parallels.*

A diagonal of a parallelogram divides the parallelogram into two congruent triangles, and so each of these triangles is half the area of the parallelogram (this proves theorem 9.5.10(a)).

If a pair of parallel lines is drawn through a pair of opposite sides of a rectangle and, if one of these sides of the rectangle is fixed as the *base* of the rectangle, then we can distort the rectangle into a parallelogram by sliding the side opposite the base of the rectangle along the other parallel line. This is called a *shear deformation* of the rectangle. The *height* of any parallelogram obtained by a shear deformation of a rectangle, as described above, is the (shortest) distance between the parallel lines. Because the area of a parallelogram is defined as the length of the base times the height of the parallelogram, shearing a parallelogram does not change its area. This is exactly the statement of theorem 9.5.10(b).

Similarly, any side of a triangle can be taken as the base of the triangle and a line parallel to the base can be drawn through the vertex that is opposite the base so that the height of the triangle is the distance between the parallel lines. If the opposite vertex is allowed to slide along the parallel line, the result is a shear deformation of the triangle that keeps the area of the triangle unchanged (this proves theorem 9.5.10(c)).

EXAMPLE 9.5.5. In figure 9.23, the points A, B, C, and D are on a line parallel to a line containing the points E and F. The vertex A of ΔAEF can slide to any of the points B, C, or D, and the corresponding triangles, that is, ΔBEF, ΔCEF, and ΔDEF are shear deformations of ΔAEF that have the same area as ΔAEF. ◈

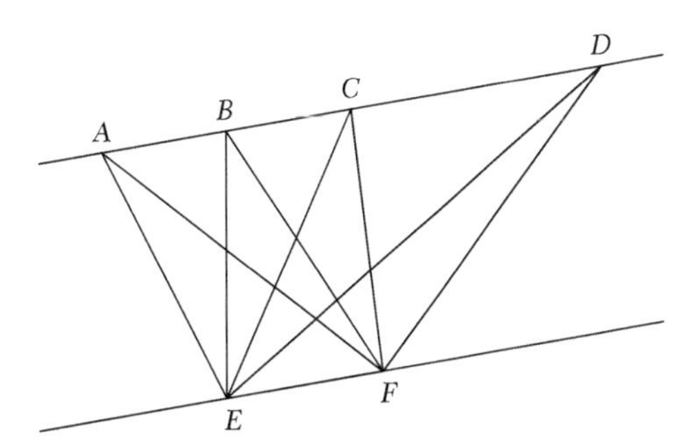

FIGURE 9.23. Shear deformation of a triangle.

The next example is another proof of the Pythagorean Theorem, which employs sheared triangles.

EXAMPLE 9.5.6. In figure 9.24, the right triangle ΔABC has the squares $ADEC$, $BCIH$, and $ABGH$ situated on its sides AC, BC, and AB, respectively. What needs to be shown is that the areas of the smaller two squares, that is, the first two squares, add up to the area of the largest square, that is, the third square. We start with a shear transformation of ΔACD to ΔABD by sliding the vertex C along the line EB (note that AD and BE are parallel lines). Now ΔADB is congruent to ΔACF (rotate ΔADB around the point A onto ΔACF) and ΔACF can be transformed by a shear to ΔAJF by sliding vertex C along the dotted line that has been inserted parallel to AF. The end result is that the dark-shaded areas are equal and, for the same reason, the light-shaded areas are equal. This proves the Pythagorean Theorem. ◈

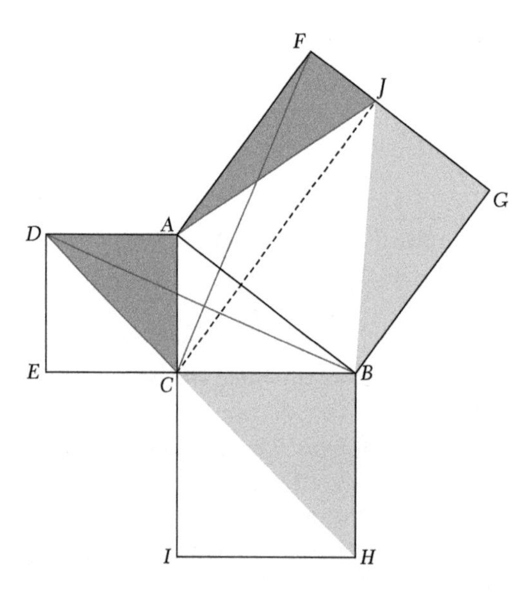

FIGURE 9.24. A proof of the Pythagorean Theorem.

Here is another theorem involving parallel lines and triangles.

THEOREM 9.5.11.

(a) *If three or more parallel lines cut off equal line segments on one transversal, then they cut off equal line segments on any other transversal.*

(b) *If a line drawn parallel to the base of a triangle bisects one of the sides of the triangle, then it bisects the third side of the triangle.*

(c) *The line segment joining the mid-points of two sides of a triangle is parallel to the third side and equal to half the third side (this is known as the* mid-point theorem).

We will use the properties of parallelograms to prove theorem 9.5.11(a). Figure 9.25 shows three parallel lines cut by transversals AC and DF. We need to prove that, if $|AB|=|BC|$, then $|DE|=|EF|$. The method of the proof is to create parallelograms $ABXD$ and $BCYE$ by constructing the lines DX and EY parallel to AC, as shown. We can conclude that the angles marked with a bullet are equal and the angles marked with asterisks are equal (corresponding angles), and the segments DX, AB, BC, and EY are equal (opposite sides of a parallelogram). It follows that $\Delta DXE \equiv \Delta EYF$ and so we conclude that $|DE|=|EF|$.

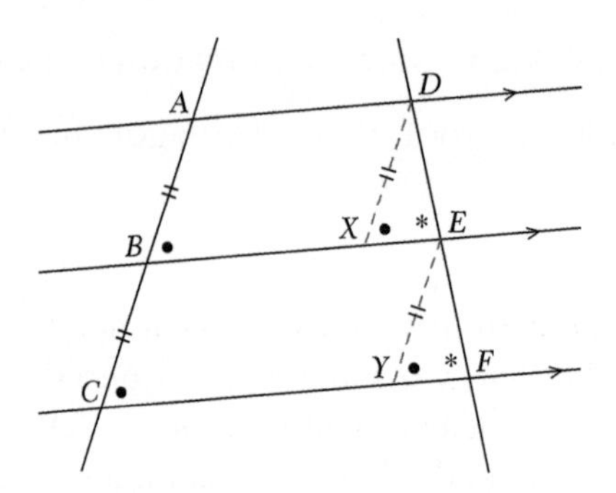

FIGURE 9.25. Proof of theorem 9.5.11(a).

Theorem 9.5.11(b) follows immediately from theorem 9.5.11(a) once a third parallel line is drawn through the vertex that is opposite the base of the triangle. The proof of theorem 9.5.11(c) is left as an exercise.

9.5.4 Concurrency, Proportionality, and Similarity

THEOREM 9.5.12.

(a) *The internal bisectors of the angles of a triangle are concurrent, and the point of concurrence is the in-center of the triangle.*

(b) *The perpendicular bisectors of the sides of a triangle are concurrent, and the point of concurrence is the circum-center of the triangle.*

(c) (i) *The medians of a triangle are concurrent, and the point of concurrence is the centroid of the triangle.*

(ii) *Furthermore, the centroid is one-third of the distance from the opposite side to the vertex along any median.*

(d) *The altitudes of a triangle are concurrent, and the point of concurrence is the ortho-center of the triangle.*

The proofs of theorem 9.5.12(a) and (b) are exercises at the end of this chapter.

Theorem 9.5.12(c) can be proved by means of a clever construction: figure 9.26 shows two medians DA and EC of a triangle ABC. A third line, passing through B and the intersection point P of the two medians meets AC at F. (We will prove that FB is a median.) EG and DH are constructed, as shown, so that they are each parallel to BF. Now in ΔAPB, EG is parallel to PB and bisects AB, and so, by theorem 9.5.11(b), EG also bisects AP. Similarly, DH bisects CP. We can now conclude, by the mid-point theorem (theorem 9.5.11(c)), that $GH \parallel AC$. It is also a consequence of the mid-point theorem that $ED \parallel AC$. Thus, $EGHD$ is a parallelogram (both pairs of opposite sides are parallel) and the diagonals EH and DG bisect each other (theorem 9.5.9(e)). This proves that $|AG| = |GP| = |PD|$ and $|CH| = |HP| = |PE|$. We have therefore established that any two medians of a triangle trisect one another. Consequently, because FB trisects DA (it passes through P), it is also a median; that is, the medians of a triangle are concurrent, proving part (i). Evidently we have also proved part (ii).

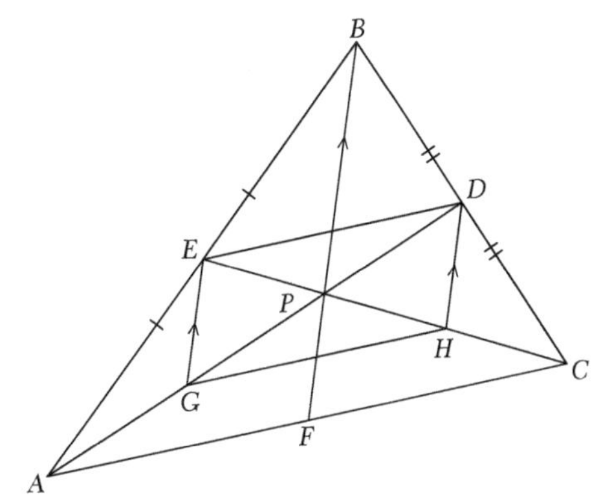

FIGURE 9.26. The proof of theorem 9.5.12(c).

The proof of theorem 9.5.12(d) is left as an exercise.

Theorem 9.5.12(a)–(d) is illustrated by means of figures 9.27 and 9.28. The in-center, circum-center, centroid, and ortho-center of a triangle ABC are located at the intersection of the dotted lines.

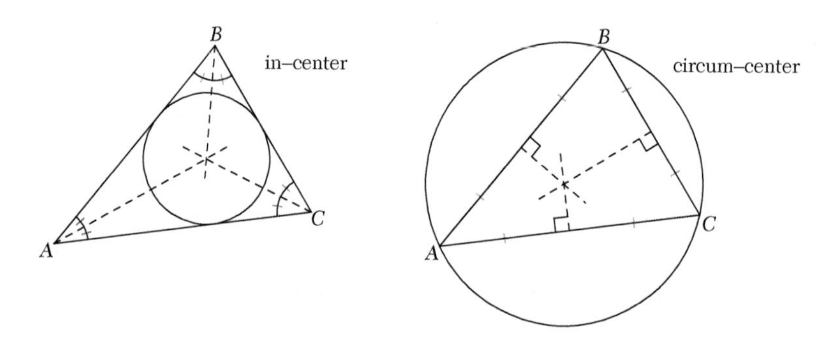

FIGURE 9.27. The in-center and circum-center of a triangle.

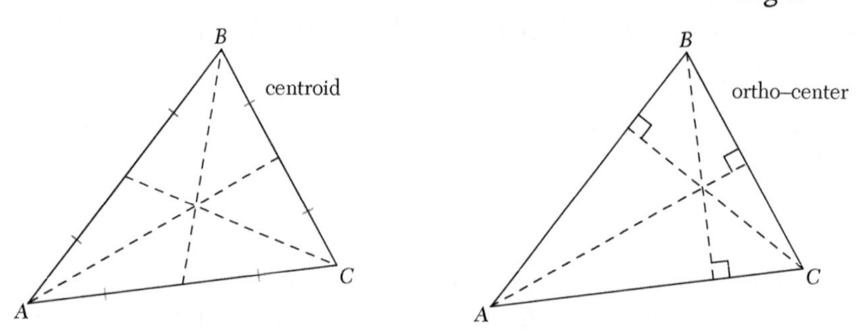

FIGURE 9.28. The centroid and ortho-center of a triangle.

THEOREM 9.5.13.

(a) *A straight line parallel to one side of a triangle divides the other two sides proportionally.*

(b) *(Converse) If a line cuts two sides of a triangle so as to divide them in the same ratio, then it is parallel to the third side.*

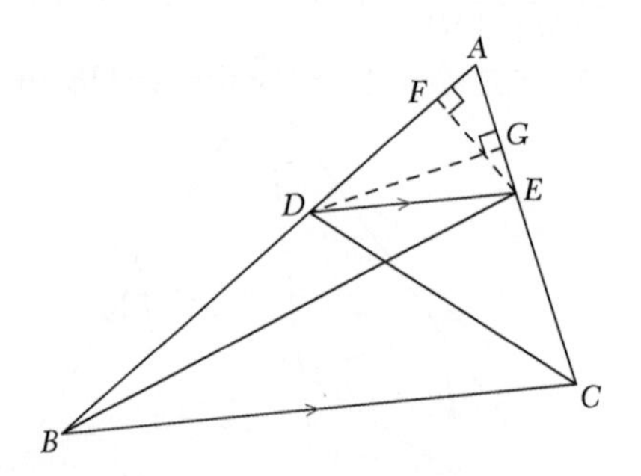

FIGURE 9.29. The proof of theorem 9.5.13(a).

Theorem 9.5.13(a) can be proved with the help of figure 9.29. A line DE is drawn parallel to the side BC of ΔABC and we want to prove that it divides the sides AB and AC of ΔABC proportionally; that is, we want to prove that $\frac{|AD|}{|DB|}=\frac{|AE|}{|EC|}$. The first observation we make is that ΔBDE and ΔCDE have the same area because they lie on the same base (DE) and between the same parallel lines (DE and BC). Also, in figure 9.29, the common altitude of triangles ADE and BDE is EF, and the common altitude of triangles ADE and CDE is DG. Therefore, we can write the following sequence of equations:

$$\frac{|AD|}{|DB|}=\frac{\frac{1}{2}|EF|\cdot|AD|}{\frac{1}{2}|EF|\cdot|DB|}=\frac{\text{Area }\Delta ADE}{\text{Area }\Delta BDE}=\frac{\text{Area }\Delta ADE}{\text{Area }\Delta CDE}=\frac{\frac{1}{2}|DG|\cdot|AE|}{\frac{1}{2}|DG|\cdot|EC|}=\frac{|AE|}{|EC|}.$$

EXAMPLE 9.5.7. An application of theorem 9.5.13(a) is the following: suppose an angle bisector is drawn from vertex A of a triangle ABC and meets the side BC at D. We will prove that $\frac{|AC|}{|AB|}=\frac{|CD|}{|BD|}$.

Refer to the first diagram in figure 9.30. If a line is drawn from the vertex B parallel to AD, meeting CA extended at E, as shown in the second diagram, then the angles marked at B and E are alternate and corresponding angles, respectively, to the equal angles marked at A. The triangle ABE is therefore isosceles, with $|AB|=|AE|$ and, by theorem 9.5.13(a), we have $\frac{|CD|}{|DB|}=\frac{|CA|}{|AE|}=\frac{|AC|}{|AB|}$, which is what we wanted to prove. ◈

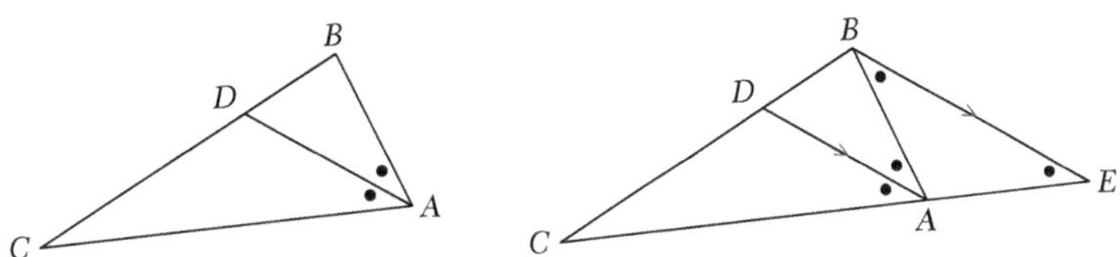

FIGURE 9.30. Diagram for Example 9.5.7.

The statement of theorem 9.5.13(a) can be expressed in a slightly different way: if we add 1 to both sides of the equation $\frac{|DB|}{|AD|}=\frac{|EC|}{|AE|}$ (refer to figure 9.29) and add the fractions on both sides, then $\frac{|DB|+|AD|}{|AD|}=\frac{|EC|+|AE|}{|AE|}$, which is equivalent to $\frac{|AB|}{|AD|}=\frac{|AC|}{|AE|}$. This states that the ratio of the length of a side of the large triangle ABC to the length of the side of the small triangle ADE is the same for two different pairs of sides. We now prove that the same equality holds for the third ratio of sides, namely the ratio $\frac{|BC|}{|DE|}$.

In figure 9.31, a line is drawn from E, parallel to the side AB, meeting the side BC at the point K. By theorem 9.5.13(a), $\frac{|BC|}{|BK|}=\frac{|AC|}{|AE|}=\frac{|AB|}{|AD|}$. Now, because $BDEK$ is a parallelogram, $|BK|=|DE|$, and so $\frac{|AB|}{|AD|}=\frac{|AC|}{|AE|}=\frac{|BC|}{|DE|}$. Note that triangles ΔABC and ΔADE are similar.

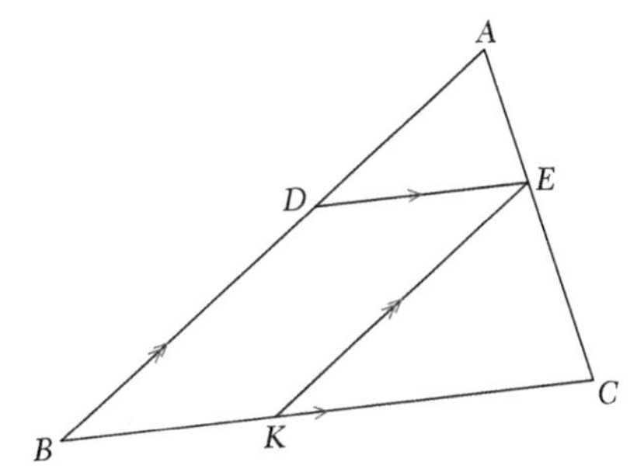

FIGURE 9.31. ΔABC is similar to ΔADE.

We have, therefore, proved the following theorem.

THEOREM 9.5.14.

(a) *If two triangles are similar, then their sides are in proportion.*

(b) *(Converse) If the sides of two triangles are in proportion, then they are similar triangles.*

Whenever we state that two triangles are similar, we label the vertices in the order in which corresponding angles are equal. For instance, we write $\Delta EFG \,|||\, \Delta PGR$ to mean that the angle at E equals the angle at P, the angle at F equals the angle at Q, and the angle at G equals the angle at R. Theorem 9.5.14(a) states that the corresponding sides EF and PQ, EG and PR, and FG and QR are in the same proportion, that is, we can write a set of equations as we did above: $\frac{|EF|}{|PQ|}=\frac{|EG|}{|PR|}=\frac{|FG|}{|QR|}$.

When we write these ratios, any pair of letters we choose from "*EFG*" is exactly matched with a pair of letters from "*PQR*." With this in mind, any pair of correct ratios can be written down automatically. For example, we can write $\frac{|EF|}{|FG|}=\frac{|PQ|}{|QR|}$.

We now present a third proof of the Pythagorean Theorem, which involves similar triangles.

EXAMPLE 9.5.8. In figure 9.32, a line is drawn through vertex C of a right triangle parallel to the side AB, and the line segments AD and BE are perpendicular to the parallel lines. Two pairs of alternate angles are marked equal. This construction creates two triangles, ADC and CEB, both similar to ΔBCA. It follows that $\frac{|DC|}{|CA|}=\frac{|CA|}{|AB|}$ and $\frac{|CE|}{|BC|}=\frac{|BC|}{|AB|}$. If we set $l=|CD|$, $m=|CE|$, $a=|AC|$, $b=|BC|$, and $c=|AB|$, then the preceding ratios can be expressed as $l=\frac{a^2}{c}$ and $m=\frac{b^2}{c}$, respectively. Because $c=l+m$, it follows that $c^2=a^2+b^2$. ◈

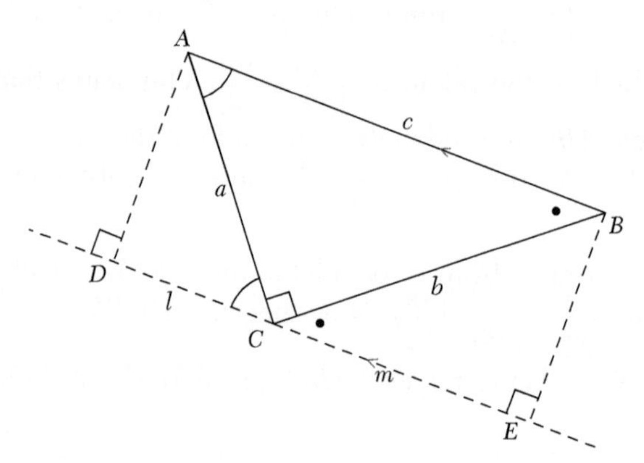

FIGURE 9.32. A proof of the Pythagorean Theorem.

Here is another property of right angles, involving similar triangles, that is useful to remember.

THEOREM 9.5.15. *The perpendicular drawn from the vertex of a right angle of a right triangle to the hypotenuse divides the triangle into two triangles that are similar to each other and similar to the original triangle.*

The proof of theorem 9.5.15 is example 9.4.3.

A proof of the final theorem of this section and its converse was published by the Italian Mathematian Giovanni Ceva in 1678. (We will not provide the proof here.) It is not known whether he knew about a proof dating back to the eleventh century, by an Arabic ruler of Spain.

THEOREM 9.5.16.

(a) *If* D, E, *and* F *are three points on the sides of a triangle* ABC, *such that* D *is on the side opposite* A, E *is on the side opposite* B, *and* F *is on the side opposite* C, *and such that* AD, BE, *and* CF *are concurrent, then* $\frac{|AF|}{|FB|}\cdot\frac{|BD|}{|DC|}\cdot\frac{|CE|}{|EA|}=1$ *(Ceva's Theorem).*

(b) *(Converse) If* D, E, *and* F *are three points on the sides of a triangle* ABC, *such that* D *is on the side opposite* A, E *is on the side opposite* B, *and* F *is on the side opposite* C, *and such that* $\frac{|AF|}{|FB|}\cdot\frac{|BD|}{|DC|}\cdot\frac{|CE|}{|EA|}=1$, *then* AD, BE, *and* CF *are concurrent.*

REMARK 9.5.2. Theorem 9.5.12(c) is a special case of theorem 9.5.16(b), because, if $|AF|=|FB|$ and $|BD|=|DC|$ (i.e., CF and AD are medians), then it follows that $|CE|=|EA|$ (i.e., BE is a median).

9.6 ELEMENTARY THEOREMS RELATING TO CIRCLES

Most of the elementary theorems relating to circles that we present here (theorems about subtended angles, theorems about cyclic quadrilaterals, and theorems about tangent lines) are proved in book 3 of the Elements. This section also contains a fourth proof of the Pythagorean Theorem. The theorems are restated in the Appendix for ease of reference.

9.6.1 Chords and Subtended Angles

THEOREM 9.6.1.

(a) *The line segment joining the center of a circle to the mid-point of a chord is perpendicular to the chord.*

(b) *(Converse) The perpendicular drawn from the center of a circle to a chord bisects the chord.*

(c) *(Corollary) The perpendicular bisector of a chord passes through the center of the circle.*

Theorem 9.6.1(a) states that a line through the center of a circle that bisects a chord is perpendicular to the chord. The reason is that, if the end points of the chord are joined to the center of the circle, then two congruent triangles are formed and the equal angles at the mid-point of the chord lie on a straight line and are therefore both equal to right angles. (Draw a diagram!)

Theorem 9.6.1(b), the converse theorem, can be proven in the same way, and theorem 9.6.1(c) can be proved by contradiction: if the perpendicular bisector of a chord does not pass through the center of the circle, then another line passing through the center of the circle is also a perpendicular bisector of the chord and, by theorem 9.6.1(a) and (b), it is obviously not possible for two such lines to exist.

REMARK 9.6.1. It is a consequence of theorem 9.6.1(c) that, if two circles intersect one another, then the line joining the centers of the circles is a perpendicular bisector of the common chord joining the points at which the circles intersect, as shown in figure 9.33.

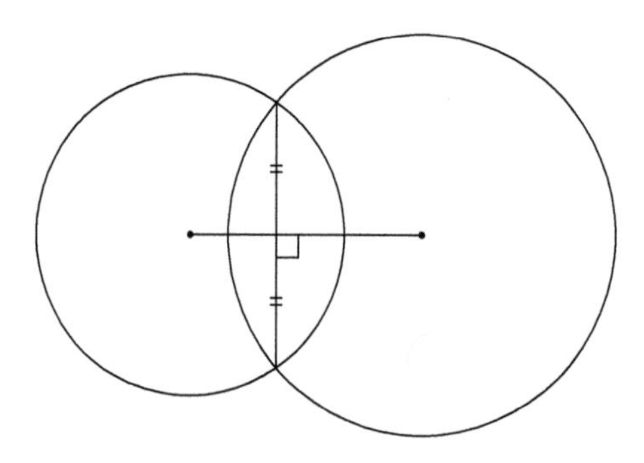

FIGURE 9.33. Circles with a common chord.

THEOREM 9.6.2. *The angle that an arc of a circle subtends at the center of the circle is twice the angle it subtends at any point of the circle.*

For the proof of theorem 9.6.2, refer to the two diagrams in figure 9.34. The arc AC subtends the angle $A\hat{O}C$ at the center of the circle and the angle $A\hat{B}C$ on the circumference of the circle. (By the arc AC, we mean the minor arc AC in the first diagram and the major arc AC in the second diagram.) We need to prove that $A\hat{O}C = 2A\hat{B}C$ in each case. The technique of the proof is to construct the dotted lines, as shown in each diagram, that create the angles marked s and t at B, and the angles marked α and β at O. Note that ΔAOB and ΔCOB are isosceles triangles, in each case. By theorem 9.5.5(a), $\alpha = s + s = 2s$ and $\beta = t + t = 2t$. Thus, $A\hat{O}C = \alpha + \beta = 2s + 2t = 2(s+t) = 2A\hat{B}C$, in each case.

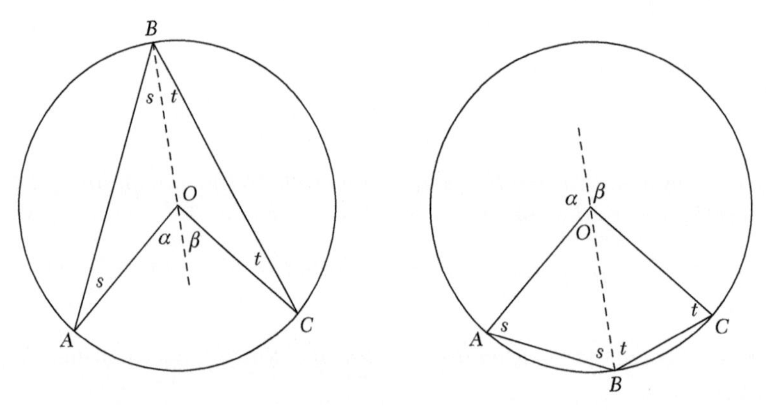

FIGURE 9.34. The Proof of theorem 9.6.2.

It is not always easy to see when theorem 9.6.2 can be applied, and so care should be taken to apply it correctly.

EXAMPLE 9.6.1. In figure 9.35, RS is a diameter of a circle centered at A. A second circle centered at B cuts at the points R and S on the circumference of the first circle. The line BA joining the centers of the circles is extended to a point C on the circumference of the second circle and the radii RB and SB of the second circle are drawn. R and S are joined to C. If $C\hat{R}S = \alpha$, what is $C\hat{B}S$? What is $C\hat{B}R$? Determine the magnitudes of the other angles in the diagram. ◈

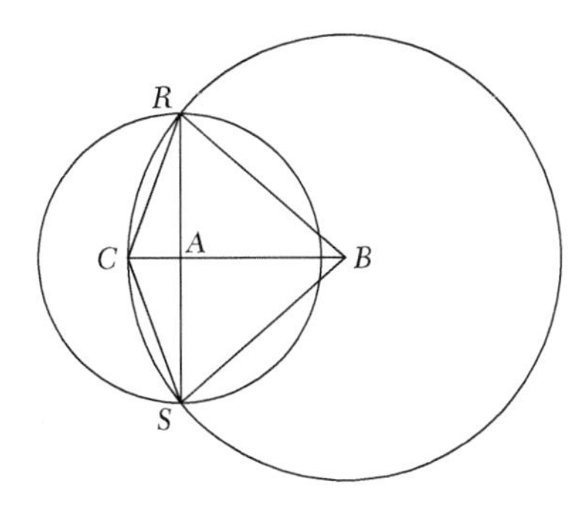

FIGURE 9.35. Diagram for example 9.5.8.

THEOREM 9.6.3.

(a) *The diameter of a circle subtends a right angle at the circumference.* (Thales' Theorem.)

(b) *(Converse) If the angle subtended by an arc of a circle at a point of the circle is a right angle, then the arc is a semicircle.*

(c) *(Converse) If the hypotenuse of a right triangle is taken as the diameter of a circle, then the circle passes through the vertex containing the right angle.*

Theorem 9.6.3(a) can be proved using theorem 9.6.2, with the help of figure 9.36, in which AC is a diameter subtending a point B at a point on the circumference. It is clear that α and β add up to $90°$ (because 2α and 2β add up to $180°$). This means that $A\hat{B}C$ is a right angle.

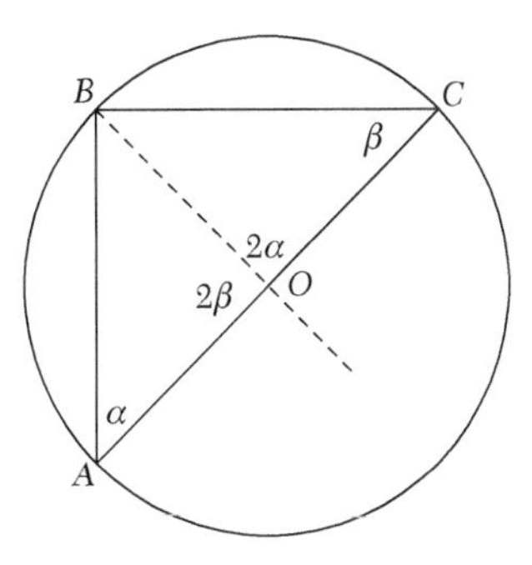

FIGURE 9.36. The proof of Thales' Theorem.

Theorem 9.6.3(b) can be proved by contradiction: if the arc is not a semicircle, it is possible to subtend an angle at the center of the circle equal to $180°$, by theorem 9.6.2, but this angle would not lie on a straight line.

Theorem 9.6.3(c) can also be proved by contradiction. If the vertex of the right triangle did not lie on the circumference of the circle with hypotenuse as diameter, then it would be possible to extend or shorten (depending on whether the vertex is inside or outside the circle, respectively) one side of the right triangle to the circumference of the circle in order to create another right triangle and then a contradiction would be obtained, by theorem 9.5.5(b). (Draw a diagram!)

EXAMPLE 9.6.2. Suppose a triangle ABC has a circumscribed circle centered at O (the circum-center of the triangle) and AD is a diameter of the circle, as shown in figure 9.37. If B and C are joined to D and altitudes BE and CF are drawn intersecting at H (the ortho-center of the triangle), then $BHCD$ is a parallelogram. This can be deduced from theorem 9.6.3(a) because the diameter AD subtends the right angle $A\hat{B}D$ and, because $B\hat{F}C$ is also a right angle, CF and BD are parallel (part (iii) of theorem 9.5.3(b)). Similarly, DC and BE are parallel. ◈

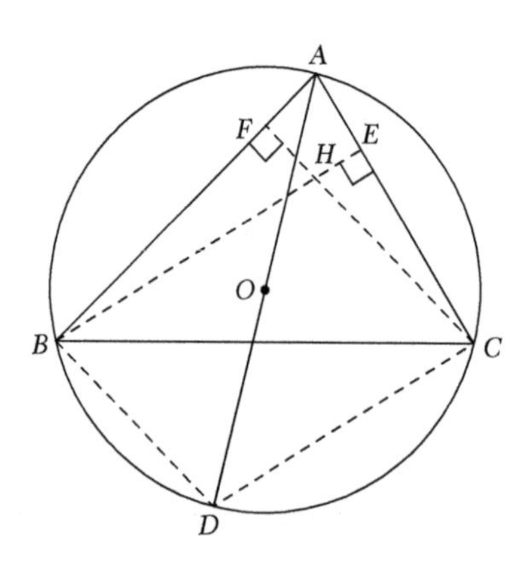

FIGURE 9.37. Diagram for example 9.6.2.

THEOREM 9.6.4.

(a) *Angles in the same segment of a circle are equal.*

(b) *(Converse) If a line segment joining two points subtends equal angles at two other points on the same side of the line segment, then these four points are concyclic.*

(c) *(Corollary) The angles subtended by arcs of equal length in a given circle are equal.*

(d) *(Corollary) The angles subtended by arcs of equal length in two different circles with equal radii are equal.*

The statement of theorem 9.6.4(a) is illustrated in figure 9.38. All the angles marked with an arc are subtended by the arc AB, and they all lie on the same side of the chord joining A to B (the dotted line); that is, they are all in the same segment. These angles are equal, by theorem 9.6.2, because they are all half the magnitude of the angle subtended by the arc AB at the center of the circle (not shown in the diagram). The proof of the converse statement and corollaries is left as an exercise.

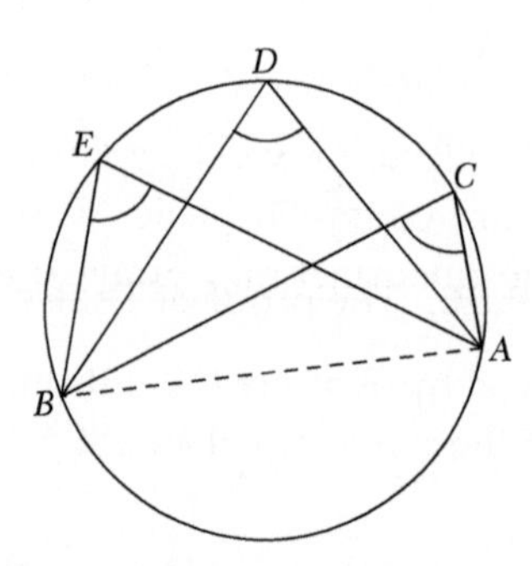

FIGURE 9.38. Angles in the same segment.

EXAMPLE 9.6.3. In figure 9.39, $ABCD$ is a cyclic quadrilateral with diagonals AC and BD. The two angles subtended by the arc joining A to B are equal, by theorem 9.6.4(a). There are three other pairs of equal angles at the vertices of the cyclic quadrilateral. (Make sure you can find them!) ◈

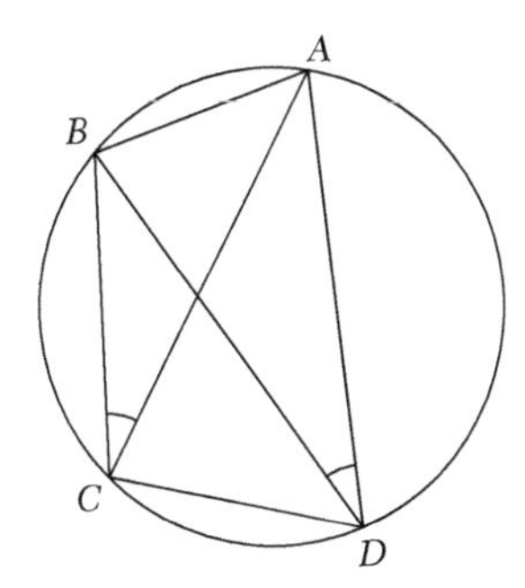

FIGURE 9.39. Angles in a cyclic quadrilateral.

EXAMPLE 9.6.4. In figure 9.40, a circle with center O passes through the points Q and R, and another circle passes through the points O, Q, and R. Thus, QR is a chord common to the two circles. A chord OS of the second circle intersects QR at P. The radii OQ and OR are drawn, and Q is joined to S. The angles at R and S are equal (angles in the same segment), and these are also equal to $O\hat{Q}R$, because ΔROQ is isosceles. Also, note that $\Delta OQS \,|||\, \Delta OPQ$. Hence $\frac{|OQ|}{|OS|} = \frac{|OP|}{|OQ|}$. Another way to write this equation is $|OQ|^2 = |OP| \cdot |OS|$. If the radius of the circle centered at O is r, then we have proved that $|OP| \cdot |OS| = r^2$. ◈

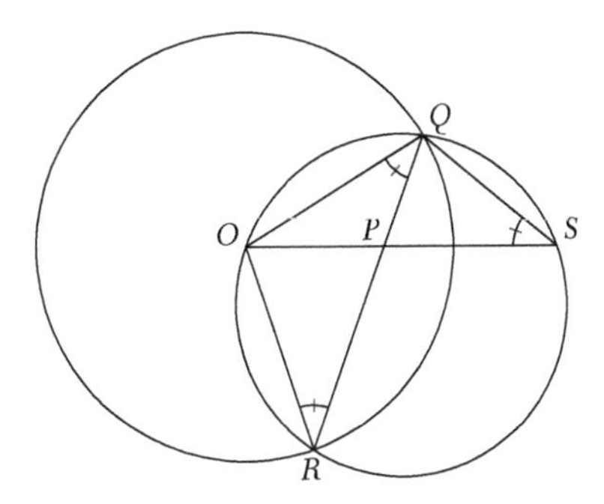

FIGURE 9.40. Diagram for example 9.6.4.

Theorem 9.6.4(b) (which can easily be proved by contradiction) is the first of several conditions that determine when four points are concyclic (which is the same as saying that the four points determine the vertices of a cyclic quadrilateral). Some other conditions are theorems 9.6.5(b), 9.6.6(b), and 9.6.7(b).

9.6.2 Cyclic Quadrilaterals

THEOREM 9.6.5.

(a) *The opposite angles of a cyclic quadrilateral are supplementary.*

(b) *(Converse) If one pair of opposite angles of a quadrilateral are supplementary, then the quadrilateral is a cyclic quadrilateral.*

The proof of theorem 9.6.5(a) is an application of theorem 9.6.2: refer to figure 9.41, in which the major arc QS subtends $\hat{O}_1$ (which is equal to $2Q\hat{R}S$) and the minor arc QS subtends $\hat{O}_2$ (which is equal to $2Q\hat{P}S$). Because $\hat{O}_1$ and $\hat{O}_2$ add up to 360°, it follows that $Q\hat{R}S + Q\hat{P}S = 180°$; that is, the opposite angles R and P of the cyclic quadrilateral $PQRS$ are supplementary. Similarly, the opposite angles Q and S are supplementary.

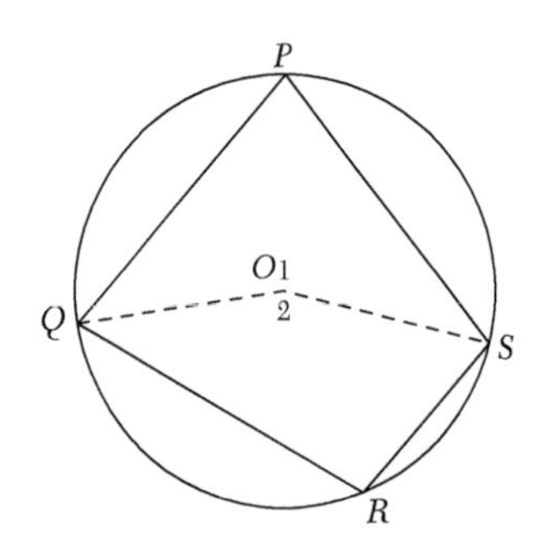

FIGURE 9.41. The proof of theorem 9.6.5(a).

THEOREM 9.6.6.

(a) *An exterior angle of a cyclic quadrilateral is equal to the interior opposite angle.*

(b) *(Converse) If an exterior angle of a quadrilateral is equal to the interior opposite angle, then the quadrilateral is cyclic.*

Theorem 9.6.6(a) follows immediately from theorem 9.6.5(a): figure 9.42 shows one side QP of a cyclic quadrilateral $PQRS$ extended in the direction of T to form an exterior angle $T\hat{P}S$. Because the angles $Q\hat{P}S$ and $T\hat{P}S$ are supplementary and the angles $Q\hat{P}S$ and $Q\hat{R}S$ are supplementary, it follows that $T\hat{P}S$ and $Q\hat{R}S$ are equal, as marked. Try to draw another seven possible exterior angles in the diagram and mark the interior angles to which they are equal.

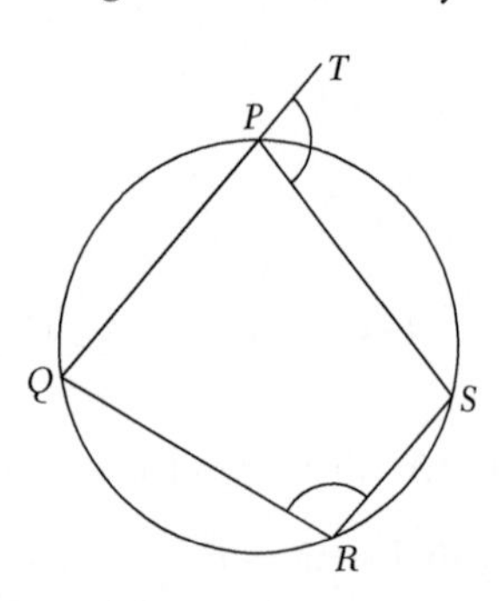

FIGURE 9.42. The exterior angle of a cyclic quadrilateral.

EXAMPLE 9.6.5. In a triangle ABC, let D, E, and F be any points on the sides opposite A, B, and C, respectively. Prove that the circle through the points A, E, and F, the points B, D, and F, and points C, D, and E, intersect at the same point.

Answer: Figure 9.43 shows the first two circles intersecting at a point M. By theorem 9.6.6(a), the cyclic quadrilateral $AFME$ has an exterior angle at F equal to the interior angle at E and, similarly, the cyclic quadrilateral $BDMF$ has an exterior angle at D equal to the interior angle at F. Therefore, the quadrilateral $CEMD$ has an exterior angle at E that is equal to the interior angle at D and so, by theorem 9.6.6(b), we can conclude that $CEMD$ is also a cyclic quadrilateral; that is, the circle through the points C, E, and D also passes through M. ◈

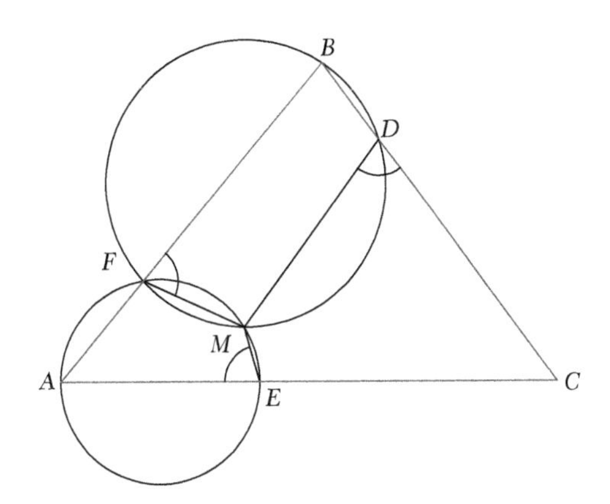

FIGURE 9.43. Diagram for example 9.6.5.

The next theorem is named after Claudius Ptolemaeus (Ptolemy), an Astronomer, Geometer, and Mathematician, who lived in the city of Alexandria (Egypt) in the second century AD.

THEOREM 9.6.7.

(a) *The sum of the products of the lengths of the two pairs of opposite sides of a cyclic quadrilateral equals the products of the lengths of its diagonals* (Ptolemy's Theorem).

(b) *(Converse) If the sum of the products of the lengths of the two pairs of opposite sides of a quadrilateral equals the products of its diagonals, then the quadrilateral is cyclic.*

Ptolemy's Theorem (theorem 9.6.7(a)) can be proved using similar triangles with the help of figure 9.44. We need to prove that $|PQ|\cdot|SR|+|QR|\cdot|PS|=|PR|\cdot|QS|$. A useful trick is to insert a line segment QA, so that $\angle A\hat{Q}R = S\hat{Q}P$, as marked in the diagram. By theorem 9.6.4(a), $\Delta SPQ \,|||\, \Delta RAQ$, and so $\frac{|QR|}{|AR|}=\frac{|QS|}{|PS|}$, or $|QR|\cdot|PS|=|AR|\cdot|QS|$. Similarly, by theorem 9.6.4(a), $\Delta PQA \,|||\, \Delta SQR$, and so $\frac{|PQ|}{|PA|}=\frac{|SQ|}{|SR|}$, or $|PQ|\cdot|SR|=|PA|\cdot|SQ|$. If we add these two pairs of equations, then

$$|QR|\cdot|PS|+|PQ|\cdot|SR|=(|AR|+|PA|)\cdot|QS|=|PR|\cdot|QS|.$$

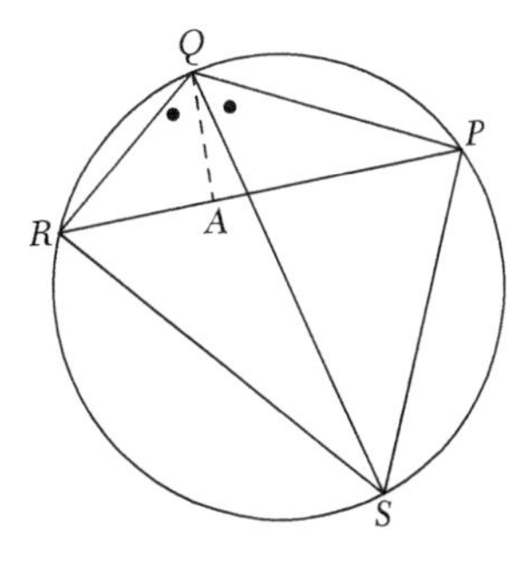

FIGURE 9.44. A proof of Ptolemy's Theorem.

REMARK 9.6.2. If Ptolemy's Theorem is applied to a regular pentagon (inscribed in circle), then it proves that the ratio of the length of a chord of the pentagon to the length of a side is the golden ratio, which we define in section 9.7.

9.6.3 Tangent Lines and Secant Lines

THEOREM 9.6.8.

(a) *A tangent to a circle is perpendicular to the radius at the point of contact.*

(b) *(Converse) A line drawn perpendicular to a radius at the point where the radius meets the circle is a tangent to the circle.*

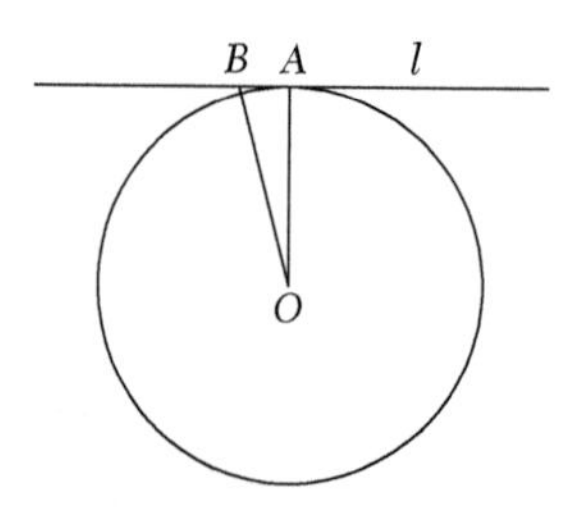

FIGURE 9.45. Proof of theorem 9.6.8(a).

Theorem 9.6.8(a) can be proved with the help of figure 9.45, in which a line l is tangent to a circle at the point A. If we suppose that the radius OA is not perpendicular to l, then, by exercise 9.3, OA is not the shortest distance from O to A and so there is some other point B on l such that $|OB|$ is the shortest distance from O to l (and OB is perpendicular to l). This results in a contradiction, because any point on l besides the point A is a point outside the circle (recall that a line that is tangent to a circle touches the circle at a single point only) and so the distance from O to that point is greater than the radius of the circle. Therefore, it must be the case that A is the point on l that is closest to O and OA must be perpendicular to l. The proof of theorem 9.6.8(b) is left as an exercise.

EXAMPLE 9.6.6. In figure 9.46, the centers of two touching circles are labeled O and P, and the touching point is labeled A. The common tangent line is constructed, and the centers O and P are joined to A. By theorem 9.6.8(a), the four angles at A are right angles, and so the line segments OA and PA join to form a straight line. In other words, we have proved that the line joining the centers of two touching circles passes through the touching point. ◈

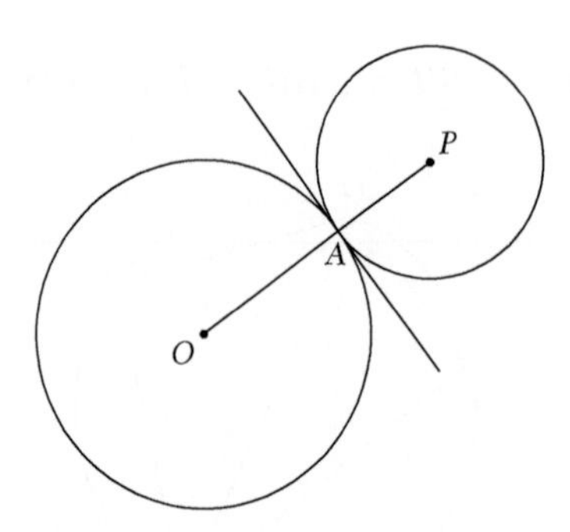

FIGURE 9.46. Touching circles.

EXAMPLE 9.6.7. Two intersecting circles are said to be orthogonal to each other if the tangent lines to the respective circles at the intersecting points are perpendicular to one another, as shown in figure 9.47. It follows from theorem 9.6.8(a) that the tangent line to either of the circles extends through the center of the other circle, as illustrated. Note that the quadrilateral $OAPB$ is a kite. It is also a cyclic quadrilateral, by theorem 9.6.5(b). ◈

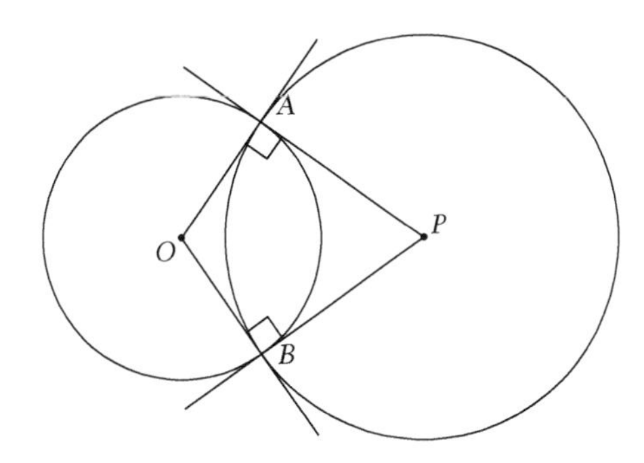

FIGURE 9.47. Circles intersecting orthogonally.

THEOREM 9.6.9.

(a) *The angle between a tangent to a circle and a chord drawn from the point of contact is equal to an angle in the alternate segment of the circle.*

(b) *(Converse) If an angle between a chord of a circle and a line through the end of that chord is equal to an angle in the alternate segment, then the line is a tangent to the circle.*

The statement of theorem 9.6.9(a) needs to be understood very well. The theorem applies to an angle formed between a tangent line and a chord at the point of contact of the tangent line with the circle, and the theorem states that this angle is equal to any angle subtended by the chord in the *alternate segment*; that is, the segment of the chord that is not on the same side of the chord as the specified angle between the chord and the tangent line. This is illustrated in the first diagram in figure 9.48. The angle between the tangent line PQ and the chord AB is $\angle P\hat{A}B$, and the angle in the alternate segment is $A\hat{C}B$.

The proof of this statement is demonstrated in the second diagram in figure 9.48, in which a diameter is drawn from A to D, and AD subtends a right angle at B (theorem 9.6.3(a)). Thus, $B\hat{A}D$ and $A\hat{D}B$ are complementary angles. By theorem 9.6.8(a), DA is perpendicular to PQ, and so $B\hat{A}D$ and $P\hat{A}B$ are also complementary angles. Therefore, $P\hat{A}B$ and $A\hat{D}B$ are equal, as marked. Now, by theorem 9.6.4(a), $A\hat{D}B$ and $A\hat{C}B$ are equal, as marked, and so we conclude that $P\hat{A}B - A\hat{C}B$.

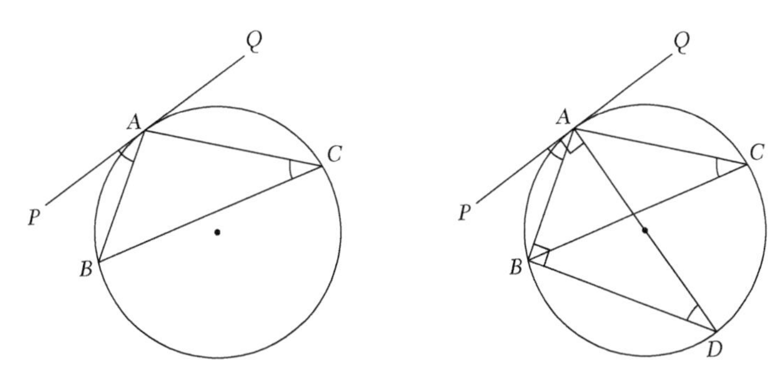

FIGURE 9.48. Proof of theorem 9.6.9(a).

THEOREM 9.6.10.

(a) *If a point P is outside a circle and two secant lines from P pass through the circle at A and D, and B and C, respectively, then* $|AP|\cdot|DP| = |BP|\cdot|CP|$ *(the* theorem of intersecting secants*).*

(b) *(b) (Corollary) The tangent to a circle from an external point is the* mean proportional (geometric mean) *of the lengths of the segments of any secant from the external point.*

(c) *If* A, B, C, *and* D *are distinct points on the circumference of a circle such that chords* AD *and* BC, *extended, intersect at a point P, then* $|AP|\cdot|DP| = |BP|\cdot|CP|$ *(the* theorem of intersecting chords*).*

The proofs of theorem 9.6.10(a) and (c) are exercises using similar triangles (write the proofs yourself!). The statement of theorem 9.6.10(b) is the limiting case of theorem 9.6.10(a) when the points B and C, for example, get closer and closer as the secant line PC containing these points cuts the circle more and more finely, that is, PC becomes a tangent line and B and C coincide. In this limit, the equation $|AP|\cdot|DP|=|BP|\cdot|CP|$ becomes the equation $|AP|\cdot|DP|=|CP|^2$, or $|CP|=\sqrt{|AP|\cdot|DP|}$ (that is, $|CP|$ is the *mean proportional* or *geometric mean* of $|AP|$ and $|DP|$).

Our fourth proof of the Pythagorean Theorem is an application of theorem 9.6.10(b).

EXAMPLE 9.6.8. Figure 9.49 shows a right triangle ABC in which the sides AC and BC are the diameters of two circles. If P is a point on AB such that CP is perpendicular to AB, then, by theorem 9.6.3(c), both circles pass through P. We denote $|BC|$, $|AC|$, $|BP|$, and $|AP|$ by a, b, x, and y, respectively, and we let $c=x+y$. By theorem 9.6.10(b), $a^2=x\cdot c$ and $b^2=y\cdot c$. If we add these equations, then $a^2+b^2=x\cdot c+y\cdot c=(x+y)\cdot c=c\cdot c=c^2$. ◈

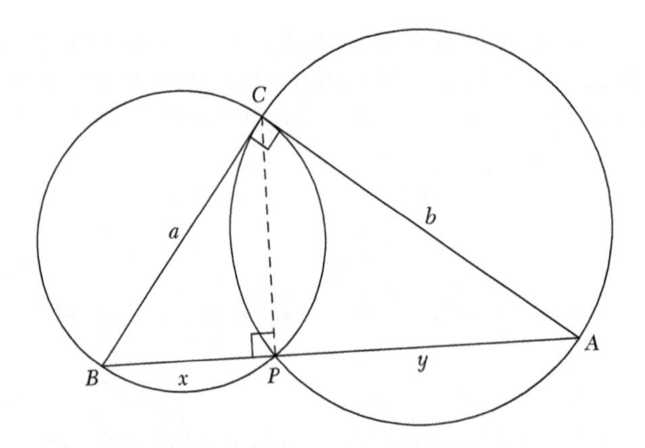

FIGURE 9.49. A proof of the Pythagorean Theorem.

9.7 EXAMPLES AND APPLICATIONS

There is a natural and meaningful interplay between geometry and algebra. We will demonstrate this by means of an example in which the Golden Ratio is constructed geometrically, and another example in which certain *classical means* are shown to be related geometrically.

In the third example of this section, we will build on examples 9.5.4 and 9.6.2 to prove the interesting fact that the circum-center, ortho-center, and centroid of any triangle are collinear (i.e., they always lie on a straight line).

The fourth example is a demonstration of a compass and straightedge construction of a tangent line to two given circles. The method and proof of this construction make use of many of the theorems of this chapter.

We end this section with another proof of Ptolemys's Theorem (theorem 9.6.7(a)).

EXAMPLE 9.7.1. The *Golden Ratio*, usually designated ϕ, is the positive number whose square is one more than itself, that is, $\phi^2 = \phi + 1$. The positive solution of this equation is $\phi = \frac{1+\sqrt{5}}{2}$. If a line segment AB is cut into two segments AC and CB, with AC being longer than CB, so that $\frac{|AC|}{|AB|} = \frac{|CB|}{|AC|}$, that is, the proportion of the longer segment to AB is the same as the proportion of the shorter segment to the longer segment, then it is called a golden cut of AB. If the length of CB is 1 unit, that is, $|CB| = 1$, then $|AC| = \phi$.

We now prove that a golden cut can be constructed using a 3-4-5 triangle (recall that this is a right triangle with side-lengths equal to 3, 4, and 5). In figure 9.50, $|BC| = 3$, $|AC| = 4$, and $|AB| = 5$ in ΔABC. The line bisecting the angle at B passes through a point O that lies on AC. The point D is the intersection of a perpendicular to AB through O. Thus, ΔODB is congruent to ΔOCB (by $\angle\angle s$), and so OD and OC are radii of a circle centered at O. This circle meets the bisector of B at the points labeled S and T. Furthermore, because $|OC| = |OD|$ and $\Delta ADO \,|||\, \Delta ACB$, we have $\frac{|AO|}{|OC|} = \frac{|AO|}{|OD|} = \frac{|AB|}{|BC|} = \frac{5}{3}$. However, $|AO| + |OC| = 4$ and this implies, together with the previous equation, that $|OC| = \frac{3}{2}$ and $|AO| = \frac{5}{2}$. Now, by theorem 9.6.10(a) (the theorem of intersecting secants), $|BT| \cdot |BS| = |BC|^2$. This in turn implies that $|BT|(3 + |BT|) = 9$. If we solve this equation, then the positive solution is $|BT| = \frac{6}{1+\sqrt{5}}$. It is now easy to verify that the circle intersects the line segment BS in the golden cut at the point T (i.e., $\frac{|ST|}{|BT|} = \phi$). ◈

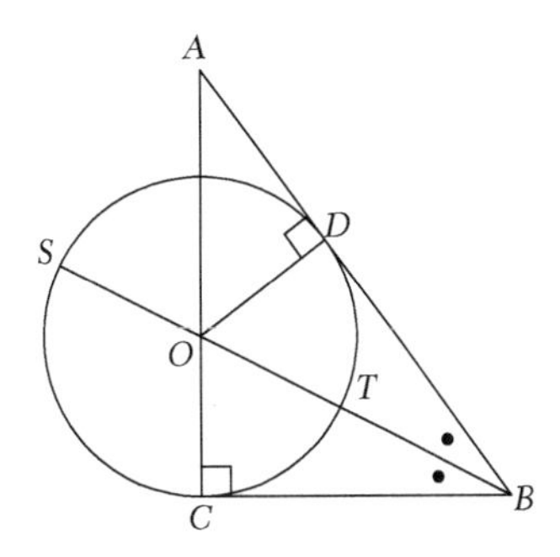

FIGURE 9.50. The Golden Cut.

EXAMPLE 9.7.2. If a and b form a pair of positive numbers, then the classical means are

- the arithmetic mean $A = \frac{1}{2}(a+b)$
- the geometric mean $G = \sqrt{ab}$
- the quadratic mean $Q = \frac{\sqrt{a^2+b^2}}{2}$.
- the harmonic mean $H = \frac{2}{\frac{1}{a}+\frac{1}{b}}$

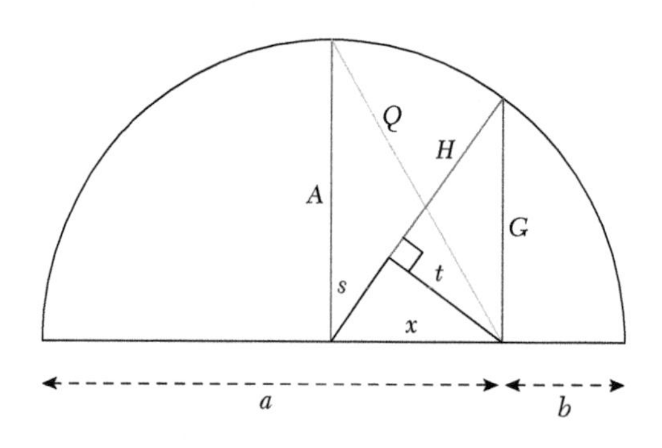

FIGURE 9.51. The classical means.

A, G, H and Q relate to each other geometrically as demonstrated in figure 9.51, where each mean is the length of a line segment. Because the diameter of the semicircle is $a+b$, the radius of the semicircle is the arithmetic mean A of a and b.

In the right triangle with $s+H$ as hypotenuse, we can state the Pythagorean Theorem as $G^2+\left(\frac{a+b}{2}-b\right)^2=\left(\frac{a+b}{2}\right)^2$. This simplifies to the equation $4G^2+(a-b)^2=(a+b)^2$, which further simplifies to $4G^2=4ab$, or $G=\sqrt{ab}$.

Next, it can be verified by means of the steps in table 9.4 that *H* is the harmonic mean of a and b.

The proof that *Q* is the quadratic mean of *a* and *b* is exercise 9.32.

TABLE 9.4. The harmonic mean

Statement	**Reason**
$H\cdot s=t^2$	Consequence of theorem 9.5.15
$H\cdot s+H^2=t^2+H^2=G^2$	Pythagorean Theorem
$H\cdot(s+H)=G^2$	
$H\cdot A=G^2$	$s+H$ and A are radii (i.e., $A=S+H$)
$H=\dfrac{G^2}{A}=\dfrac{ab}{\frac{1}{2}(a+b)}=\dfrac{2}{\frac{1}{a}+\frac{1}{b}}$	

◈

Example 9.7.3. We will prove that the circum-center (*O*), centroid (*P*), and ortho-center (*H*) of ΔABC in figure 9.52 are collinear. A part of the diagram is a reconstruction of figure 9.37. It was proved in example 9.5.4 that *BHCD* is a parallelogram, so, if we include the diagonal *HD* of this parallelogram, intersecting the diagonal *BC* at *F*, then, as in figure 9.22, the line segment *AF* is a median of ΔABC and a median of ΔADH. Furthermore, the line segment *OH* is another median of ΔADH. Because medians trisect one another (theorem 9.5.12(c)), the intersection point *P* of *OH* and *AF* is one-third of the distance from *F* to A. This means that *P* is also the centroid of ΔABC. Thus, the circum-center, centroid, and ortho-center of ΔABC are collinear. ◈

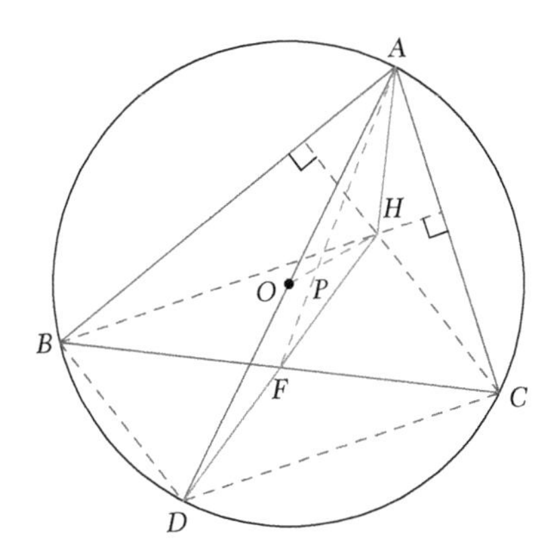

FIGURE 9.52. The circum-center, centroid, and ortho-center are collinear.

Example 9.7.4. Compass and straightedge constructions often lead to interesting and useful insights. In this example, we demonstrate that it is possible to use a compass and a straightedge to construct a common tangent line to two circles.

There are several cases that can be considered and we will consider only the case demonstrated in figure 9.53, in which the two circles (with centers B and C) are separated from each other (i.e., do not intersect each other), and the common tangent line is the line segment AD. A compass and a straightedge can be used to construct the perpendicular bisector of the line segment EF. The point labeled O, which is the mid-point of segment BC can also be obtained using a compass and a straightedge. The circle (not shown) centered at O with diameter BC intersects the perpendicular bisector of EF at G. A circle with center at G and radius GE passes through the points A, F, and D, as shown. We prove now that the line AD is a common tangent line to the two given circles:

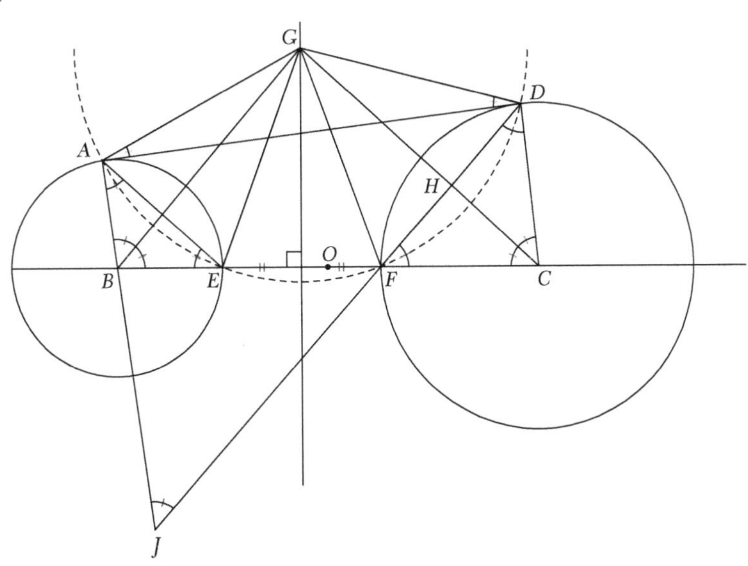

FIGURE 9.53. Constructing a common tangent to two circles.

First, draw lines AJ and DJ passing through B and F, respectively, and also construct the other line segments shown in figure 9.53. The angles marked at B with tick marks are equal because triangles GAB and GEB are congruent (sss). Similarly, the angles marked at C with tick marks are equal because triangles CFG and CDG are congruent (sss). As a consequence, triangles CFH and CDH are also congruent, and so the angles at H are right angles. Another right angle is $B\hat{G}C$ (by Thales' Theorem and the construction of G).

This implies that $GB \parallel DJ$, which means that the two pairs of complementary angles at B and F and B and J are equal. The angles marked at D and F are also equal, and this proves that $AJ \parallel DC$ (the alternate angles at J and D are equal). Consequently, the pair of conterior angles $B\hat{A}D$ and $C\hat{D}A$ are supplementary. Furthermore, because ΔGEF is isosceles, $G\hat{E}B = G\hat{F}C$, and so, by the congruence of triangles identified above, $G\hat{A}B = G\hat{D}C$. We conclude that the angles marked at A and D are also equal (because GAD *is* an isosceles triangle) and, therefore, $B\hat{A}D = C\hat{D}A$. This means that $B\hat{A}D$ and $C\hat{D}A$ are both right angles. Our conclusion that AD is a common tangent line to the two given circles now follows from theorem 9.6.8(b). ◈

A problem that can be solved using geometric methods could also be solvable using trigonometric and/or algebraic methods. A good example is the following alternative proof of Ptolemy's Theorem (theorem 9.6.7(a)):

Proof: A cyclic quadrilateral is shown in figure 9.54. We need to prove that $xy = ac + bd$. To this end, we begin with the cosine rule (formula (4.20)) to obtain two expressions for x^2:

$$x^2 = a^2 + b^2 - 2ab\cos(\theta) \quad \text{and} \quad x^2 = c^2 + d^2 - 2cd\cos(\pi - \theta)$$

We can solve for $\cos\theta$ in each of these equations (using the identity $\cos(\pi - \theta) = -\cos(\theta)$), and equate the resulting expressions:

$$\frac{x^2 - a^2 - b^2}{-ab} = \frac{x^2 - c^2 - d^2}{cd}$$

Now we can solve for x^2 and, by means of a clever regrouping of terms, arrive at an expression that is close to what we are looking for:

$$\begin{aligned} x^2 &= \frac{a^2cd + b^2cd + abc^2 + abd^2}{ab + cd} \\ &= \frac{ac(ad + bc) + bd(bc + ad)}{ab + cd} \\ &= \frac{(ac + bd)(ad + bc)}{ab + cd} \end{aligned} \tag{9.1}$$

By analogous reasoning (refer to the second diagram)

$$y^2 = \frac{(ac + bd)(ab + cd)}{ad + bc} \tag{9.2}$$

When formula (9.1) is multiplied by formula (9.2) the factors in the denominators cancel with factors in the numerators, and the result is

$$x^2y^2 = (ac + bd)^2.$$

The proof is complete when we take the square root on each side. □

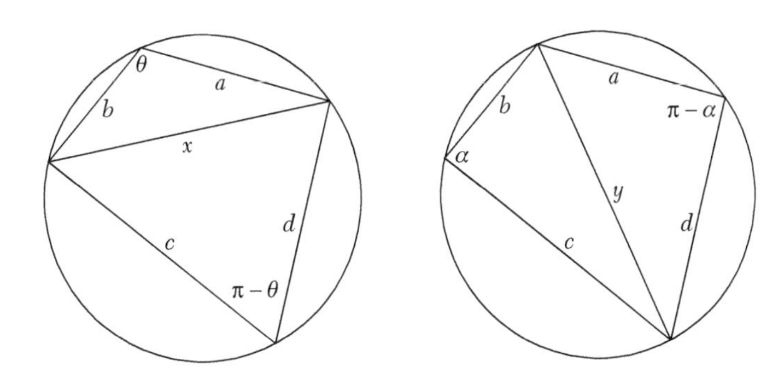

FIGURE 9.54. A proof of Ptolemy's theorem.

EXERCISES

9.1. Prove that, in any triangle, the angle opposite the greater side is greater than the other two angles. (Proposition 18, book 1, of The Elements.)
(Hint: Construct an isosceles triangle with a vertex on the greater side and use theorem 9.5.5(a).)

9.2. Prove that, in any triangle, the side opposite the greater angle is longer than the other two sides.
(Proposition 19, book 1, of The Elements.)

9.3. Prove that the line segment from a point to a line that is perpendicular to the line is the shortest line segment from that point to the line.
(Hint: This follows from exercise 9.1 and theorem 9.5.5(c).)

9.4. Prove that in any triangle the sum of any two sides is greater than the remaining one. (Proposition 20, book 1, of the elements). (Hint: Use the property stated in exercise 9.2 to prove that $|AB|+|AC|>|BC|$ in figure 9.55.)

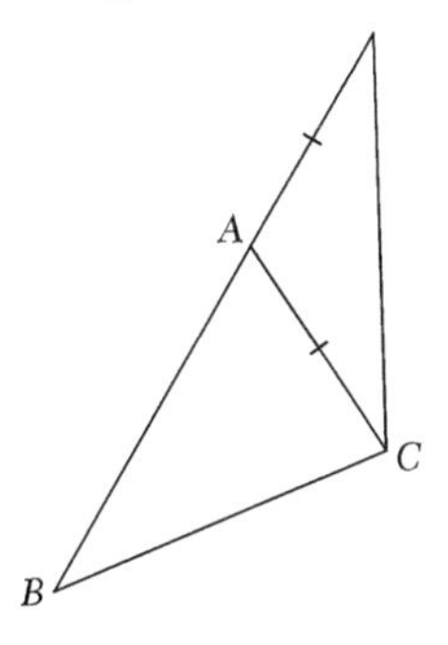

FIGURE 9.55. Diagram for exercise 9.4.

9.5. A circle with radius r is drawn tangent to each of a pair of intersecting lines, as shown in figure 9.56.

(a) Verify that this circle is centered on the dotted line that bisects the angle between the intersecting lines.
(Hint: Use theorem 9.6.8(a).)

(b) A second, larger circle (not shown), with radius R, is also tangent to each of the intersecting lines, and its center is a distance t from the smaller circle. Prove that $R < t + r$.

(Hint: draw appropriate triangles and use the properties stated in exercises 9.2 and 9.4.)

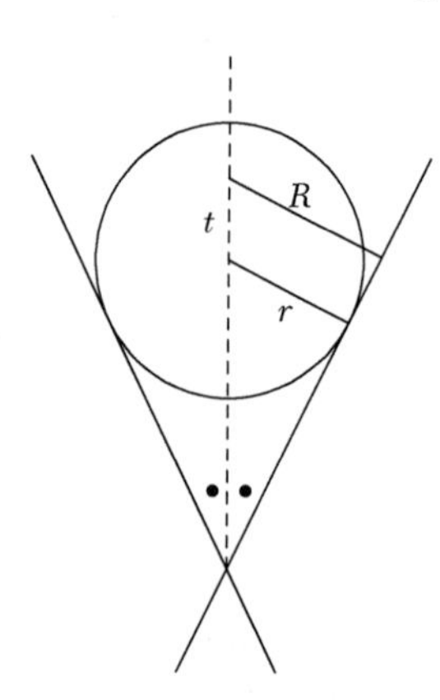

FIGURE 9.56. Diagram for exercise 9.5.

9.6. Demonstrate that the perpendicular bisector of a given line segment can be constructed using a compass and a straightedge.

9.7. Demonstrate that the angular bisector of a given angle can be constructed using a compass and a straightedge.

9.8. Prove theorem 9.5.12(a).

9.9. Prove first that any two angle-bisectors both pass through the in-center of the triangle; then prove that the three angle-bisectors are concurrent.

9.10. Prove theorem 9.5.12(b).

9.11. Figure 9.57 shows a triangle RUV with a point Q on the side RU such that $|QV| = |UV|$. A line PQ is drawn parallel to VU from a point P on RV, and extended to a point T so that $|QT| = |QV|$. The point V is joined to T, and a line SQ is drawn parallel to VT from another point S on RV. Prove that SQ is perpendicular to RU.

(Hint: Find congruent triangles.)

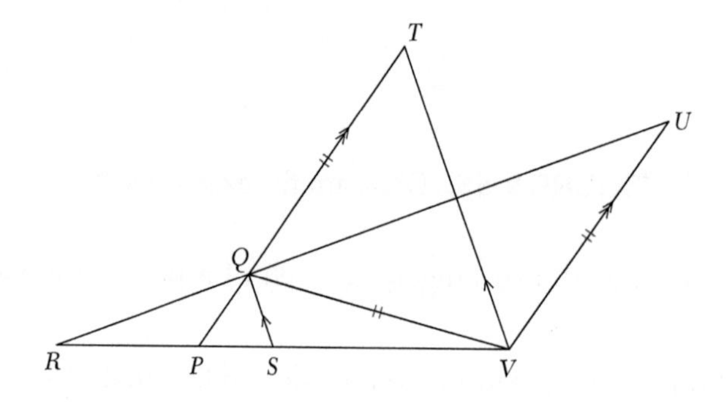

FIGURE 9.57. Diagram for exercise 9.11.

9.12. Prove theorem 9.5.7(b).

9.13. Figure 9.58 shows an equilateral triangle ABC. The arcs AB, BC, and CA are arcs of circles centered at C, A, and B, respectively. The line segments PQ, QR, and RP are tangent to the arcs AB, BC, and CA, respectively, and parallel to the sides AB, BC, and CA of triangle ABC, respectively, so that an equilateral triangle PQR contains the equilateral triangle ABC. If the length of each side of triangle ABC is 1, prove that the length of each side of triangle PQR is $2(\sqrt{3} - 1)$.

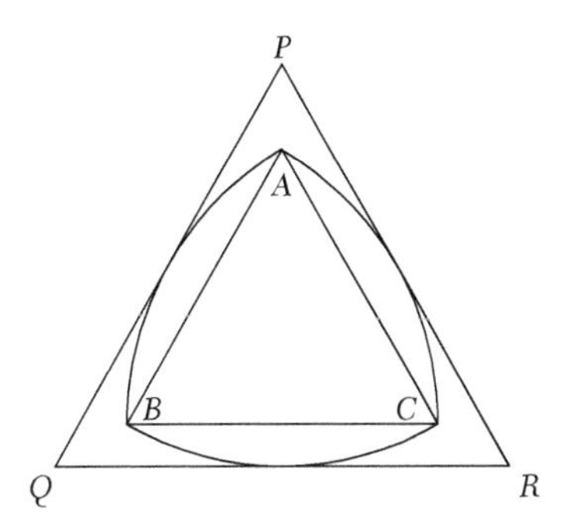

FIGURE 9.58. Diagram for exercise 9.13.

9.14. Prove theorem 9.5.9(b), (d), (f), and (g).

9.15. In a right triangle, with right angle at C, the mid-point P of side AB is joined to C. Prove that $|AP|=|CP|$.

(Hint: Use theorem 9.5.9(e) and (h).)

9.16. Prove theorem 9.5.11(c).

(Hint: The proof requires the construction of a parallelogram. Use theorem 9.5.9(f) and (g).)

9.17. Redraw the diagrams showing the in-center, circum-center, centroid, and ortho-center for an obtuse instead of acute triangle.

9.18. Prove theorem 9.5.13(b).

9.19. Prove theorem 9.5.14(b).

9.20. Demonstrate how you would use a compass and straightedge to draw the circle passing through any three (noncollinear) points in the plane.

9.21. Prove theorems 9.6.4(b), 9.6.5(b), and 9.6.6(b).

9.22. In figure 9.59, the pairs of opposite sides AD and BC, and AB and DC, of a cyclic quadrilateral $ABCD$ are extended to meet at P and Q, respectively. The lines bisecting the angles at P and Q intersect the cyclic quadrilateral at points L and J and K and M, respectively. Prove that $JKLM$ is a rhombus.

(Hint: mark equal angles and look for congruent triangles.)

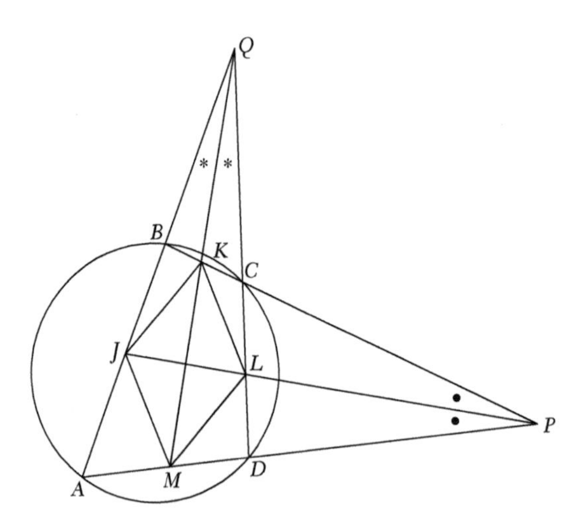

FIGURE 9.59. Diagram for exercise 9.22.

9.23. A trapezoid is called an *isosceles trapezoid* if a pair of base angles (i.e., the angles on one of the parallel sides) is equal. Prove that a trapezoid is a cyclic quadrilateral if and only if it is an isosceles trapezoid.

(Hint: This is a consequence of theorem 9.6.5(a) and (b).)

9.24. Prove theorems 9.6.8(b) and 9.6.9(b).

9.25. When two touching semicircles are centered on the diameter of a larger semicircle so that they are also touching the larger semicircle, then the plane region between the semicircles, as shaded in figure 9.60, is called an *arbelos*. In figure 9.60, the diameter of the larger semicircle is AC and the diameters of the two smaller semicircles are AD and CD. The two smaller semicircles touch each other at D and a line from D to a point B of the larger semicircle is a tangent to the two smaller semicircles at D. The circle with diameter BD is labeled S. Prove the following:

(a) The area enclosed by the circle S is equal to the area of the arbelos.

(b) If E and F are the intersection points of S with the two smaller semicircles, then the line segments AB and CB pass through E and F, respectively, as shown in the diagram.

(c) The line segment through the points E and F is a common tangent line to the two smaller semicircles.

(Hint: (a) By calculation, the area of the arbelos is πrs and the area enclosed by S is also πrs. (b) Join D to E and D to F and use Thales' Theorem. (c) Look for parallel lines (theorem 9.6.9(a) and (b) are helpful).)

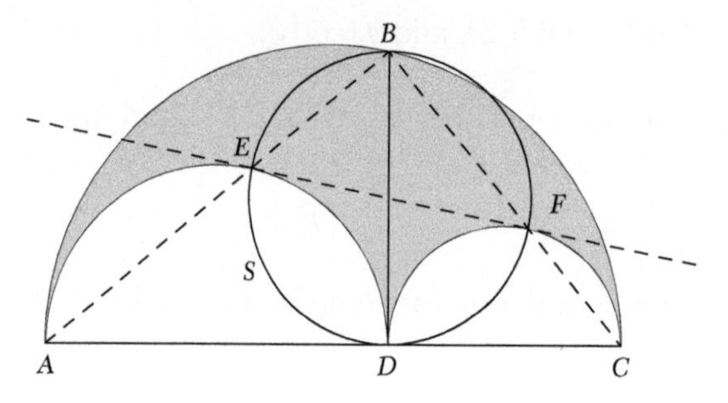

FIGURE 9.60. An arbelos.

9.26. In figure 9.61, the common tangent lines to two intersecting circles pass through the points S and T and U and V, respectively. The points S and U are joined by a line, and the points T and V are joined by a line. Prove that these two lines are parallel.

(Hint: it will help to extend the tangent lines to a point where they meet and, from there, construct one other line.)

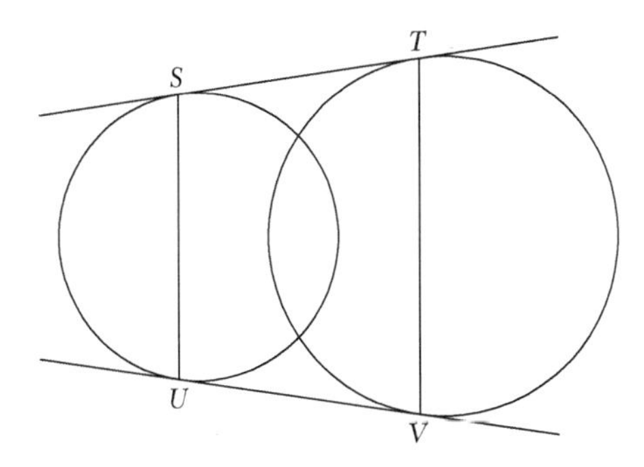

FIGURE 9.61. Diagram for exercise 9.26

9.27. Describe how you would construct a common tangent line to two intersecting circles (as shown in figure 9.61).

(Hint: this can be done by means of a method that is exactly analogous to the method that was used to construct the common tangent line for two separated (i.e., nonintersecting) circles in section 9.7.)

9.28. Figure 9.62 shows a circle with radius r and center O. A point P is a point inside the circle and OP is perpendicular to a chord RT. OP is produced to meet a tangent line from T at Q. Prove that $|OP| \cdot |OQ| = |OT|^2 = r^2$. (This means that Q is the *inverse* of P with respect to the circle.)

(Hint: Find similar triangles in the diagram.)

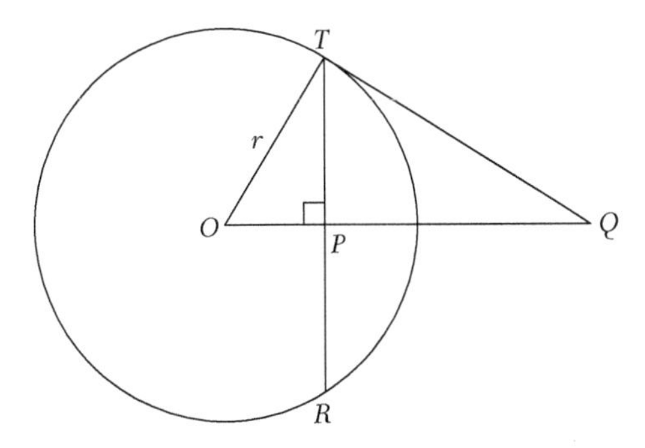

FIGURE 9.62. Diagram for exercise 9.28.

9.29. This exercise demonstrates a second construction of the inverse point Q of a point P inside a circle. Figure 9.63 shows a circle with center O and diameter ST perpendicular to OP. TP is produced to a point R on the circle, and the chord SR is produced to a point Q on the line extending through P from O. Prove that $|OP| \cdot |OQ| = |OT|^2$.

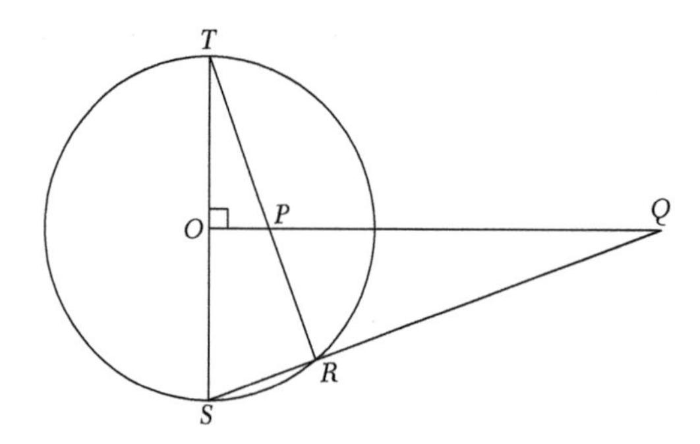

FIGURE 9.63. Diagram for exercise 9.29.

9.30. This exercise demonstrates a third construction of the inverse point Q of a point P inside a circle. Figure 9.64 shows a circle with center O and radius OT perpendicular to OP. A circle with diameter OT meets TP at R, and in this smaller circle, a chord SR is drawn parallel to OT. TS produced meets OP produced at Q. Prove that $|OP| \cdot |OQ| = |OT|^2$.
(Hint: Construct a common tangent line to the circle at T.)

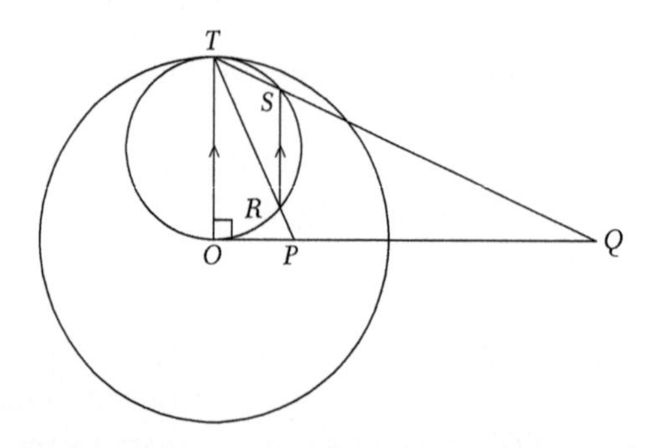

FIGURE 9.64. Diagram for exercise 9.30.

9.31. If Q, A, G, and H are the quadratic, arithmetic, geometric, and harmonic means of two positive numbers a and b, respectively, prove that $Q > A > G > H$. Make use of figure 9.51. Also prove that Q in figure 9.51 is the quadratic mean of a and b.

9.32. Prove theorem 9.6.10(a).
(Hint: Join A to B and C to D. Use theorem 9.6.6(a) to find similar triangles.)

9.33. Prove theorem 9.6.10(c).

9.34. If P, Q, R, and S are four distinct points such that the line segments QP and RS both extend to a point O, and $OP \cdot OQ = OS \cdot OR$, prove that the points P, Q, R, and S are concyclic.
(Hint: What happens if the circle through P and Q and S does not pass through R?)

9.35. Two circles are orthogonal to each other if and only if a pair of radii of the two circles, drawn to either of the intersection points of the two circles, are orthogonal to each other. (This means each of these radii will be a tangent to the other circle.) Let O be a circle, P a point inside O, and l a secant line passing through P. Use a compass and ruler to construct a circle A orthogonal to O such that l is tangent line to A at P.

(Hint: Construct the point S (shown in figure 9.40) as explained in example 9.6.4. Theorem 9.6.10(b) will be helpful.)

9.36. Prove theorem 9.6.10(b).

9.37. In figure 9.65, the distance between O and G is 10 units and the distance between G and E is 2 units. Solve for x, that is, find the distance between E and C.

(Hint: Apply Ceva's Theorem to triangles REO and RCO in order to express x in terms of the magnitudes $|RP|, |PE|, |RT|$, and $|TC|$, and then apply the Sine Rule (formula (4.19) to ΔOTC, ΔOPE, and ΔPRT.)

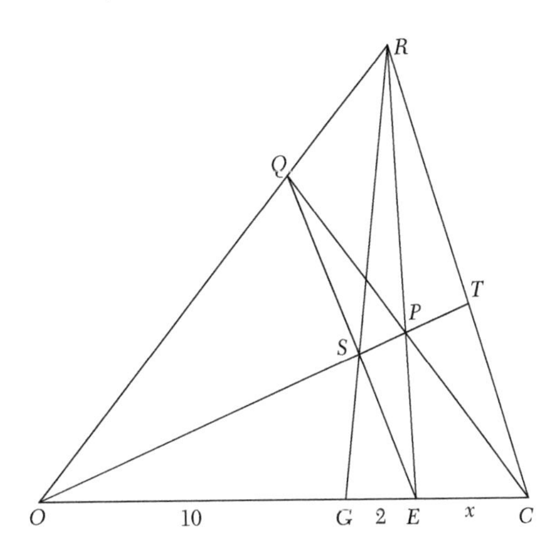

FIGURE 9.65. Diagram for exercise 9.37.

9.38. In figure 9.66, PR is the diameter of a circle with center O, Q is another point on the circle such that $Q\hat{O}P = \alpha$ is an acute angle, and QS is a chord perpendicular to PR, with T on PR.

Prove that $\tan\left(\frac{\alpha}{2}\right) = \frac{\sin(\alpha)}{1+\cos(\alpha)}$.

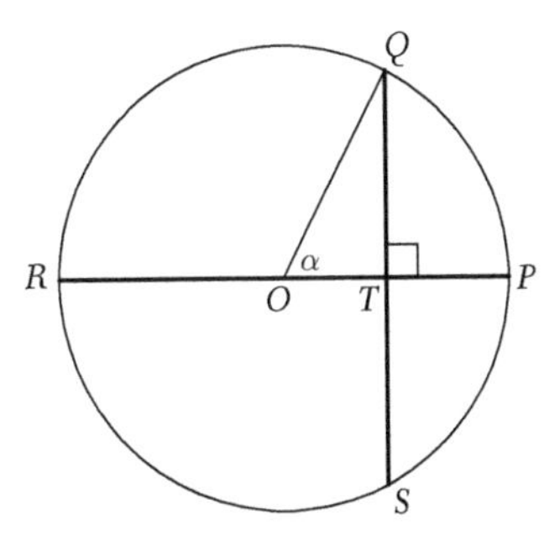

FIGURE 9.66. Diagram for exercise 9.38.

9.39. In figure 9.67, prove that $|AC| = 2x\sqrt{1+15\cos^2\theta}$.

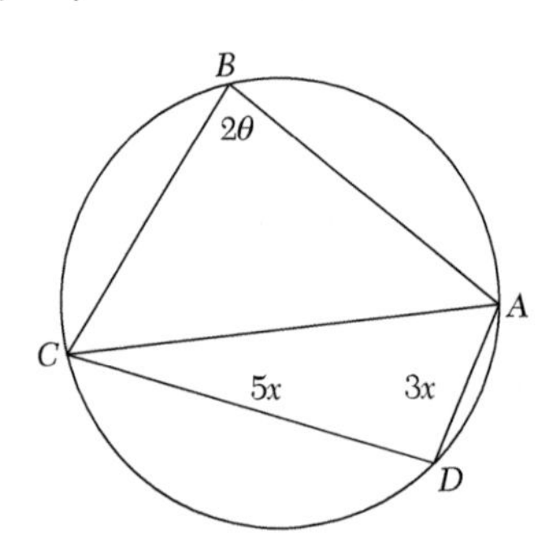

FIGURE 9.67. Diagram for exercise 9.39.

9.40. In figure 9.68, O is the center of the circle, $|QR| = 2$ and ΔPRS has an angle of magnitude x at R and an angle of magnitude $2x$ at S. Prove that $r = \csc(3x)$, where r is the radius of the circle.

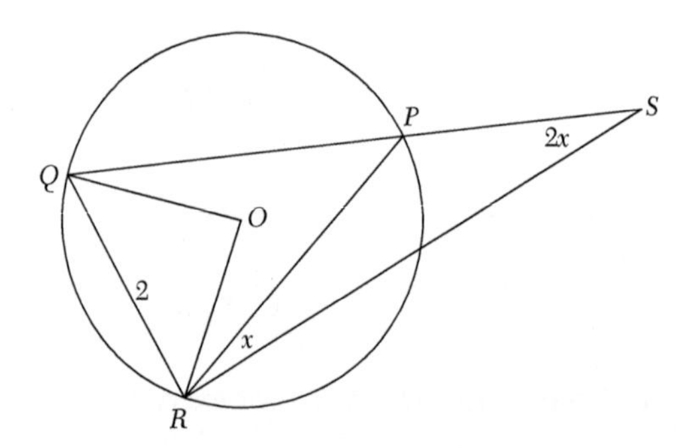

FIGURE 9.68. Diagram for exercise 9.40.

CHAPTER APPENDIX: GEOMETRY THEOREMS

Theorem 9.5.1

(a) If two straight lines intersect, then the sum of any pair of adjacent angles is 180°.

(b) (*Converse*) If the sum of any pair of adjacent angles is 180°, then their non-common sides lie on the same line.

Theorem 9.5.2

If two lines intersect, then the vertically opposite angles are equal.

Theorem 9.5.3

(a) angles are equal to one another.
(ii) If a transversal intersects two parallel lines, then pairs of alternate angles are equal to one another.
(iii) If a transversal intersects two parallel lines, then pairs of co-interior angles are supplementary.

(b) **(i)** If a transversal intersects two parallel lines, then pairs of corresponding
(i) (Converse) If two lines are intersected by a transversal such that a pair of corresponding angles are equal, then the two lines are parallel.
(ii) (*Converse*) If two lines are intersected by a transversal such that a pair of alternate angles are equal, then the two lines are parallel.
(iii) (*Converse*) If two lines are intersected by a transversal such that a pair of co-interior angles are supplementary, then the two lines are parallel.

Theorem 9.5.4 Lines that are parallel to the same line are parallel to each other.

Theorem 9.5.5
(a) The exterior angle of a triangle is greater than either of the interior opposite angles.
(b) The exterior angle of a triangle is equal to the sum of the interior opposite angles (ext. $\angle$ of Δ).
(c) The sum of the angles of a triangle is $180°$.

Theorem 9.5.6
(a) If two sides and the included angle of one triangle are respectively equal to two sides and the included angle of another triangle, then the two triangles are congruent ($s\angle s$).
(b) If two angles and a side of one triangle are respectively equal to two angles and the corresponding side of another triangle, then the triangles are congruent ($\angle\angle s$).
(c) If three sides of one triangle are equal to three sides of another triangle, then the triangles are congruent (sss).
(d) If, in two right-angled triangles, the hypotenuse and one side of the one are respectively equal to the hypotenuse and one side of the other, then the triangles are congruent ($\perp hs$).

Theorem 9.5.7
(a) The angles opposite the equal sides of an isosceles triangle are equal.
(b) (*Converse*) If two angles of a triangle are equal, then the sides opposite them are equal.

Theorem 9.5.8
(a) The square of the hypotenuse of a right-angled triangle is equal to the sum of the squares of the other two sides. (Pythagorean Theorem.)
(b) (*Converse*) If the square of one side of a triangle is equal to the sum of the squares of the other two sides, then the angle opposite the first side is a right angle.

Theorem 9.5.9

(a) The opposite sides of a parallelogram have the same length.

(b) (*Converse*) If the opposite sides of a quadrilateral have the same length, it is a parallelogram.

(c) The opposite angles of a parallelogram are equal.

(d) (*Converse*) If the opposite angles of a quadrilateral are equal, then it is a parallelogram.

(e) The diagonals of a parallelogram bisect each other.

(f) (*Converse*) If the diagonals of a quadrilateral bisect each other, then it is a parallelogram.

(g) If one pair of opposite sides of a quadrilateral are both parallel and have the same length, it is a parallelogram.

(h) The diagonals of a rectangle have the same length.

(i) The diagonals of a rhombus bisect each other at right angles and bisect the angles of the rhombus.

Theorem 9.5.10

(a) The area of a parallelogram is bisected by each diagonal.

(b) A parallelogram and a rectangle on the same base and between the same parallels have equal areas.

(c) The area of a triangle is equal to one-half the area of a parallelogram on the same base and between the same parallels.

Theorem 9.5.11

(a) If three or more parallel lines cut off equal line segments on one transversal, then they cut off equal line segments on any transversal.

(b) If a line drawn parallel to the base of a triangle bisects one of the sides of the triangle, then it bisects the third side of the triangle.

(c) The line segment joining the midpoints of two sides of a triangle is parallel to the third side, and equal to half the third side (the midpoint theorem).

Theorem 9.5.12

(a) The internal bisectors of the angles of a triangle are concurrent, and the point of concurrence is the *in-center* of the triangle.

(b) The perpendicular bisectors of the sides of a triangle are concurrent, and the point of concurrence is the *circum-center* of the triangle.

(c) The medians of a triangle are concurrent and the point of concurrence, the *centroid* of the triangle, is one third of the distance from the opposite side to the vertex along any median.

(d) The altitudes of a triangle are concurrent, and the point of concurrence is the *ortho-center* of the triangle.

Theorem 9.5.13

(a) A straight line parallel to one side of a triangle divides the other two sides proportionally.

(b) (*Converse*) If a line cuts two sides of a triangle so as to divide them in the same ratio, then that line is parallel to the third side.

Theorem 9.5.14

(a) If two triangles are similar, then their sides are in proportion.

(b) (*Converse*) If the sides of two triangles are in proportion, then they are similar triangles.

Theorem 9.5.15

The perpendicular drawn from the vertex of a right angle of a right-angled triangle to the hypotenuse, divides the triangle into two triangles that are similar to each other and to the original triangle.

Theorem 9.5.16

(a) If D, E, and F are three points on the sides of a triangle ABC, such that D is on the side opposite A, E is on the side opposite B, and F is on the side opposite C, and such that AD, BE, and CF are concurrent, then $\frac{|AF|}{|FB|}\cdot\frac{|BD|}{|DC|}\cdot\frac{|CE|}{|EA|}=1$. (Ceva's Theorem.)

(b) (*Converse*) If D, E, and F are three points on the sides of a triangle ABC, such that D is on the side opposite A, E is on the side opposite B, and F is on the side opposite C, and such that $\frac{|AF|}{|FB|}\cdot\frac{|BD|}{|DC|}\cdot\frac{|CE|}{|EA|}=1$, then AD, BE, and CF are concurrent.

Theorem 9.6.1

(a) The line segment joining the center of a circle to the mid-point of a chord is perpendicular to the chord.

(b) (*Converse*) The perpendicular drawn from the center of a circle to a chord bisects the chord.

(c) (*Corollary*) The perpendicular bisector of a chord passes through the center of the circle.

Theorem 9.6.2

The angle that an arc of a circle subtends at the center of the circle is twice the angle it subtends at any point of the circle.

Theorem 9.6.3

(a) The diameter of a circle subtends a right angle at the circumference. (Thales' Theorem.)

(b) (*Converse*) If the angle subtended by a chord at a point of the circle is a right angle, then the chord is a diameter.

(c) (*Converse*) If the hypotenuse of a right-angled triangle is taken as the diameter of a circle, then the circle passes through the vertex containing the right angle.

Theorem 9.6.4

(a) Angles in the same segment of a circle are equal.

(b) (*Converse*) If a line segment joining two points subtends equal angles at two other points on the same side of the line segment, then these four points are concyclic.

(c) (*Corollary*) The angles subtended by arcs of equal length in a given circle, are equal.

(d) (*Corollary*) The angles subtended by arcs of equal length in two different circles with equal radii, are equal.

Theorem 9.6.5

(a) The opposite angles of a cyclic quadrilateral are supplementary.

(b) (*Converse*) If one pair of opposite angles of a quadrilateral are supplementary, then the quadrilateral is a cyclic quadrilateral.

Theorem 9.6.6

(a) An exterior angle of a cyclic quadrilateral is equal to the interior opposite angle.

(b) (*Converse*) If an exterior angle of a quadrilateral is equal to the interior opposite angle, then the quadrilateral is cyclic.

Theorem 9.6.7

(a) The sum of the products of the lengths of the two pairs of opposite sides of a cyclic quadrilateral equals the products of the lengths of its diagonals. (Ptolemy's Theorem.)

(b) (*Converse*) If the sum of the products of the lengths of the two pairs of opposite sides of a quadrilateral equals the products of its diagonals, then the quadrilateral is cyclic.

Theorem 9.6.8

(a) A tangent to a circle is perpendicular to the radius at the point of contact.

(b) (*Converse*) A line drawn perpendicular to a radius at the point where the radius meets the circle, is a tangent to the circle (line $\perp$ radius).

Theorem 9.6.9

(a) The angle between a tangent to a circle and a chord drawn from the point of contact is equal to an angle in the alternate segment of the circle.

(b) (*Converse*) If an angle between a chord of a circle and a line through the end of that chord is equal to an angle in the alternate segment, that line is a tangent to the circle.

Theorem 9.6.10

(a) If a point P is outside a circle and two secant lines from P pass through the circle at A and D, and B and C, respectively, then $|AP|\cdot|DP|=|BP|\cdot|CP|$. (The theorem of intersecting secants.)

(b) (*Corollary*) The tangent to a circle from an external point is the mean proportional (geometric mean) of the lengths of the segments of any secant from the external point.

(c) If A, B, C, and D are distinct points on the circumference of a circle such that chords AD and BC, extended, intersect at a point P, then $|AP|\cdot|DP|=|BP|\cdot|CP|$. (The theorem of intersecting chords.)

CHAPTER 10

SPHERICAL TRIGONOMETRY

10.1 INTRODUCTION

Spherical trigonometry is used for computing angles and distances in terrestrial navigation and celestial astronomy. For the purposes of terrestrial navigation, we regard the Earth as a *sphere* (a perfectly round ball), although, correctly speaking, the shape of the Earth is approximately a *geoid*, not a sphere.

An early pioneer of spherical trigonometry was Mohammed ibn Mûsâ al-Khowârizmî in the ninth century AD. The sine rule that is basic to spherical trigonometry was discovered by another Arabic Astronomer and Mathematician Abu al-Wafa al-Buzjani (tenth century AD). The development of spherical trigonometry continued in the Islamic Iberian Peninsula (Spain) in the eleventh century and in Iran in the thirteenth century.

The objective of this chapter is to derive formulas for solving triangles on the sphere, as demonstrated in sections 10.4 to 10.6. By traingles on the sphere, we mean triangles formed by arcs of *great circles*, that is, a path between two points on a sphere that takes the shortest distance.

All the terminology that is needed for doing geometry and trigonometry on the sphere is introduced in section 10.2. The properties of vectors in space that are introduced in section 10.3 lay the groundwork for proving the sine and cosine rules in section 10.4 for triangles on the sphere.

In this chapter, we will investigate some relationships between objects (e.g., spheres, planes, and lines) in three-dimensional space. It is possible to give a formal and mathematical description of three-dimensional space using a three-dimensional coordinate system, that is, an extension of the Euclidean plane by means of a third axis; however, for our purposes, it will be satisfactory to regard the objects as "objects in space."

While this is primarily a chapter on spherical trigonometry, there is also a development of *spherical geometry*. Spherical geometry is an example of a *non-Euclidean* geometry. Some of the differences between Euclidean geometry and spherical non-Euclidean geometry will be commented on.

The notation used in chapter 9 relating to angles, lines, and triangles will be used here too.

10.2 PLANES AND SPHERES

10.2.1 Planes in Space

In section 10.2.2 the angles between curves on a sphere are defined in terms of angles between planes intersecting the sphere. Therefore, it is appropriate to begin with some remarks about planes in space.

A basic fact is that two planes that are not parallel to each other intersect in a straight line, which is contained in each of the planes. If three *distinct* (i.e., different) planes pass through a unique point in space (called a *vertex*), then the three lines created by the (pairwise) intersecting planes form the boundaries of three triangular faces of a *pyramid*, as shown in figure 10.1. The base of the pyramid can be taken to be any other triangle formed by a fourth plane that intersects the three planes and does not pass through the vertex V. Each of the three triangular faces has an angle at V called a *face angle* (of the pyramid). Our first theorem is a statement regarding the relative sizes of the three face angles of a pyramid.

THEOREM 10.2.1. *The sum of any two face angles of a pyramid is greater than the third face angle.*

PROOF. We only need to prove that the sum of the two smaller face angles is greater than the largest face angle. In the first diagram in figure 10.1, we suppose that $A\hat{V}C$ is greater than both of $A\hat{V}B$ and $B\hat{V}C$. This means that a line can be drawn in the face AVC from V toward the base of the pyramid to a point D so that $A\hat{V}D = A\hat{V}B$ and $|VD| = |VB|$. In the second diagram in figure 10.1, the base of the pyramid is adjusted so that the line AC passes through D.

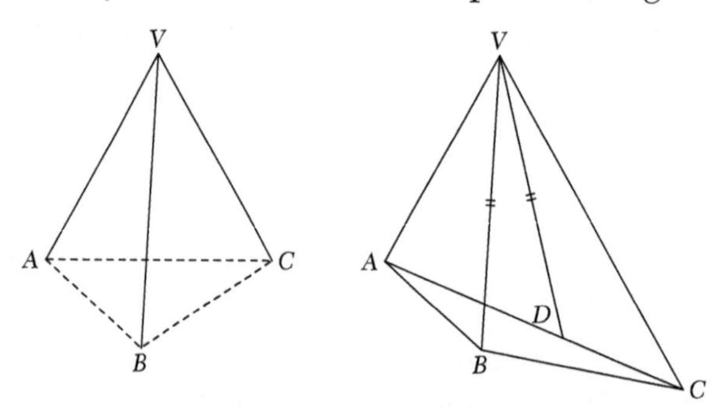

FIGURE 10.1. Three planes intersecting in a vertex.

In the triangle ABC forming the base of the pyramid,

$$|AB| + |BC| > |AC| \quad \text{and} \quad |AC| = |AD| + |DC| \tag{10.1}$$

but we see that $\Delta AVD \equiv \Delta AVB$ (*sas*), and so $|AB| = |AD|$. By formula (10.1), this allows us to conclude that $|BC| > |DC|$, and it follows from this that $B\hat{V}C > D\hat{V}C$ (why?). Now we can add $A\hat{V}B$ to both sides of the last inequality to obtain $B\hat{V}C + A\hat{V}B > D\hat{V}C + A\hat{V}B$. Therefore, $B\hat{V}C + A\hat{V}B > D\hat{V}C + A\hat{V}D$ (because $A\hat{V}B = A\hat{V}D$). This is equivalent to $B\hat{V}C + A\hat{V}B > A\hat{V}C$, which is what we had to prove. □

10.2.2 Spheres

DEFINITION 10.2.1. *A sphere is the collection of all points (or surface) in space that are a specified distance from a given point (called the center of the sphere).*

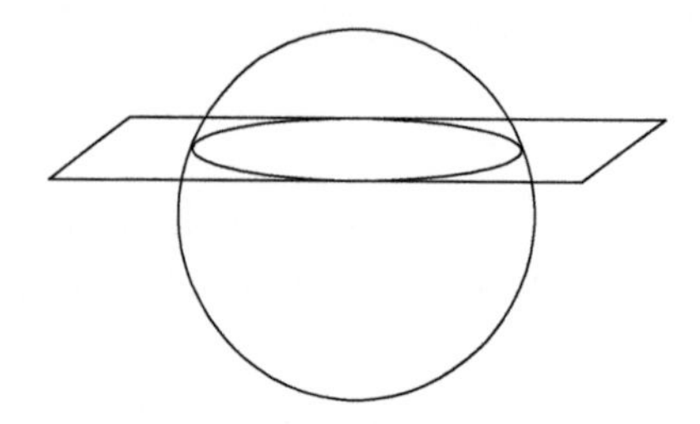

FIGURE 10.2. A plane intersecting a sphere.

If a plane intersects a sphere, the intersection is always a circle in the sphere (as shown in figure 10.2). Conversely, any circle in the sphere is contained in a plane intersecting the sphere.

DEFINITION 10.2.2. *If a plane that intersects the sphere passes through the center of the sphere, the intersection is called a great circle; if not, it is called a small circle.*

EXAMPLE 10.2.1. The equator and meridians on the Earth (regarded as a sphere) are great circles. ◈

DEFINITION 10.2.3. *A hemisphere is either of two halves of a sphere obtained when any plane passes through the center of the sphere.*

DEFINITION 10.2.4. *The poles of a great circle are the two ends of the diameter of the sphere that is perpendicular to the plane containing the great circle (see figure 10.3).*

Any two points that are the end points of a diameter of the sphere can be referred to as *antipodal points*. Thus, the poles of a great circle are antipodal points.

EXAMPLE 10.2.2. The North Pole and the South Pole are poles of the equator (the line connecting the North Pole to the South Pole through the center of the Earth is perpendicular to the plane of the equator). ◈

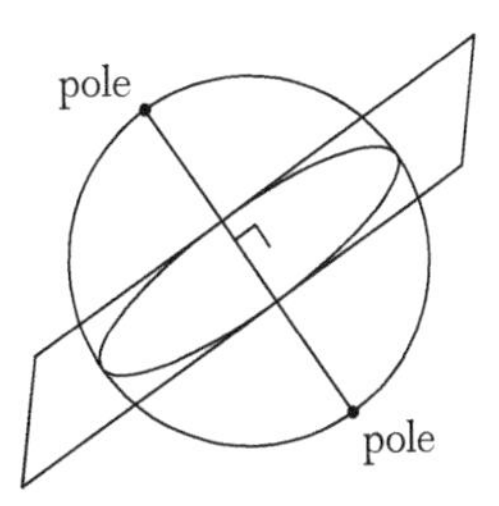

FIGURE 10.3. The poles of a great circle.

REMARK 10.2.1. Because the relationship of the equator to the poles of the Earth is familiar to us, it will be useful in the continuation of this chapter to refer to the great circle in a plane perpendicular to the diameter of the sphere joining two specified antipodal points, as the "equator," and to refer to the arcs of great circles joining them, as "meridians".

An important geometrical fact about the sphere is that there is a *unique* great circle passing through any two given points on the sphere and that the shortest distance between any specified pair of points is the length of the shorter arc of the great circle passing through the two points. For this reason, we think of great circles on the sphere as the analogues of straight lines in the plane.

DEFINITION 10.2.5. *A tangent plane to a sphere is a plane that touches the sphere at only one point, and a tangent line to a sphere is a line (contained in a tangent plane) that touches the sphere at only one point (see figure 10.4).*

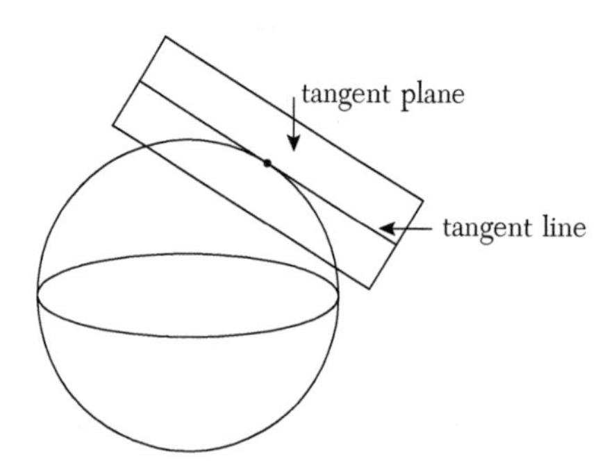

FIGURE 10.4. A tangent plane and a tangent line.

Note that for any specified point p on a sphere, there is a *unique* tangent plane denoted as T_p that is tangent to the sphere and passes through p. All tangent lines to the sphere passing through p are contained in T_p.

DEFINITION 10.2.6. *A tangent line to a circle on the sphere is a line that is tangent to the sphere and also contained in the plane that contains the circle (see figure 10.5).*

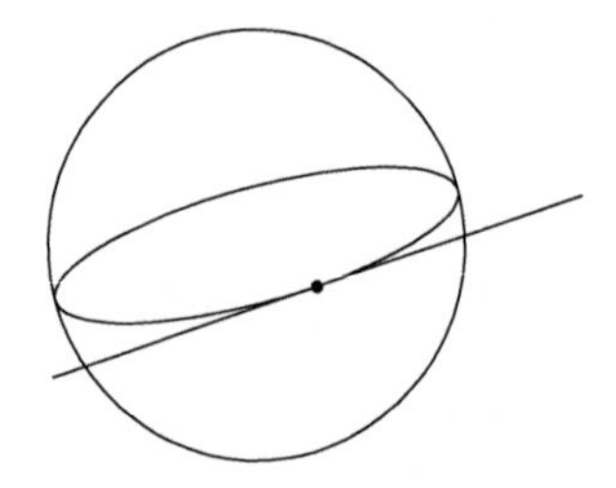

FIGURE 10.5. A tangent line to a circle on the sphere.

If two distinct circles on a sphere intersect (at one or two points), then the angle between the circles at an intersection point is defined to be the smallest angle formed between the two tangent lines to the circles at the point of intersection. If there are two intersection points, then the angle will be the same at each point, and so we can talk about the *angle between the circles*. This is shown in figure 10.6, in which the angle, labeled θ, is measured in the tangent plane that contains the tangent lines. If two circles meet at one point then the circles are touching and they have a common tangent line at the touching point, which means the angle between them is zero.

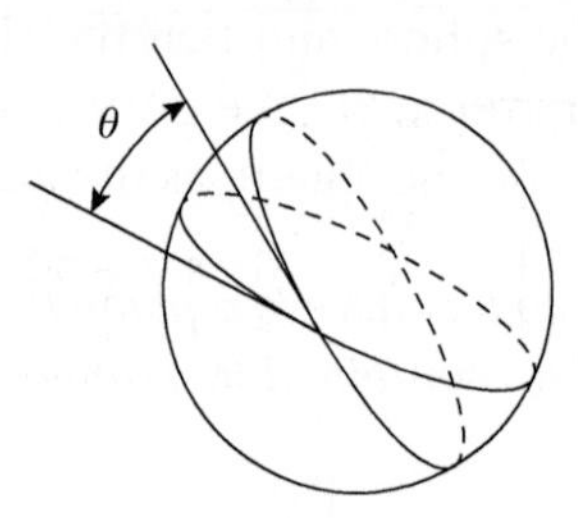

FIGURE 10.6. An angle between two tangent lines.

The following fact is not hard to verify geometrically (think about the definition of a great circle):

REMARK 10.2.2. Any two distinct great circles intersect at two antipodal points, called vertices, on the sphere.

Therefore, we have the following definition:

DEFINITION 10.2.7. *The angle between any two distinct great circles (i.e., the angle at each vertex) is referred to as a* spherical angle.

REMARK 10.2.3. Any two distinct great circles on a sphere divide the sphere into four regions, separated by four meridians, and each region is called a *lune*.

REMARK 10.2.4. In order to simplify the statements of the basic formulas in spherical trigonometry, we will assume, for the remainder of this chapter, that the sphere is a unit sphere, which means that all points are a unit distance from the center of the sphere. A consequence of this is that all great circles on the sphere are unit circles (i.e., circles with unit radius) that are centered at the center of the sphere.

REMARK 10.2.5. Because the radian measure of an angle is defined as the length of the arc of the unit circle corresponding to the angle (see definition 4.2.1), we can identify the length of an arc of a great circle with the angle subtended by the arc at the center of the sphere. This identification can be made in radians or in degrees.

In figure 10.7, the length of an arc of a great circle is 30° ($\pi/6$ radians).

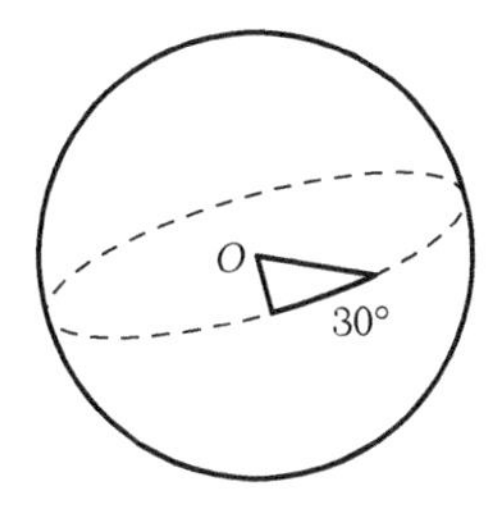

FIGURE 10.7. Measuring an angle.

We have the following useful interpretation of the radian measure of a spherical angle.

REMARK 10.2.6. The radian measure of the spherical angle formed by any two distinct great circles is the radian measure of the angle subtended at the center of the sphere by the smaller arc of the equator crossing the great circles. (That is, the equator with respect to the vertices, as explained in remark 10.2.1).

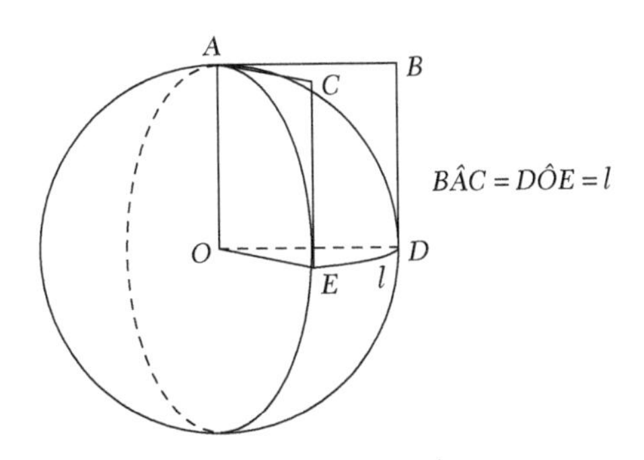

FIGURE 10.8. A spherical angle.

It is helpful to imagine the spine of an opened book aligned along the diameter joining the vertices and the opened pages of the book aligned with the planes containing the great circles, then the spherical angle is the angle between the opened pages in a plane that is perpendicular to the spine of the book.

Examine figure 10.8, in which the radian measure of the angle $D\hat{O}E$ (the radian measure of the spherical angle) is precisely the length of the arc of the equator from E to D (labeled l).

10.2.3 Spherical Triangles

DEFINITION 10.2.8. *A spherical triangle is a triangle on the sphere bounded by three arcs of great circles.*

In figure 10.8, the arcs AE, AD, and ED are the circular arc boundaries of the spherical triangle AED with vertices at A, E, and D.

We need to think a little bit about the geometry of spherical triangles, because the construction of spherical triangles is more complicated than the construction of triangles in the plane. If three distinct lines in the plane are not concurrent (i.e., do not all pass through the same point) and no two lines are parallel to each other, then they determine a single triangle. On the sphere, however, any three distinct great circles that do not all pass through the same pair of vertices determine a number of spherical triangles. For example, figure 10.9 shows three distinct great circles on a sphere. Can you count the number of spherical triangles?

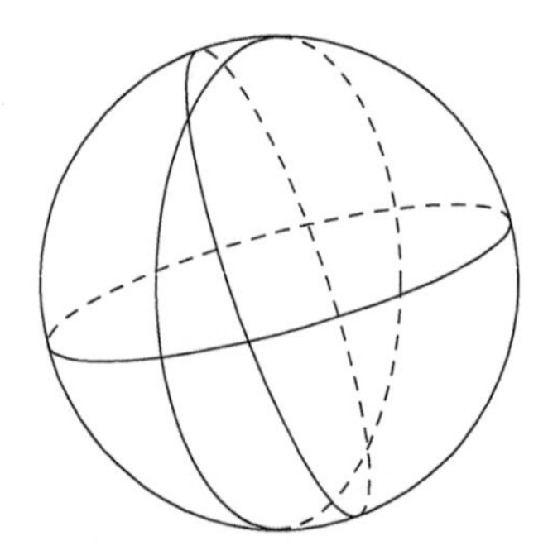

FIGURE 10.9. Spherical triangles on a sphere.

We can simplify this complicated situation by designating as *proper spherical triangles*, those spherical triangles that can fit on a hemisphere (that is, they do not wrap too much around the sphere). It is clear that there are eight proper spherical triangles in figure 10.9 because any two distinct great circles divide a sphere into four lunes, which are each smaller than a hemisphere, and the third great circle divides each lune into two proper spherical triangles.

Henceforth, we will assume that all spherical triangles are, in fact, proper spherical triangles, but we should not forget that we are restricting the definition of a spherical triangle.

The theorems below state some important facts about spherical triangles. The angle at each vertex of a spherical triangle is the spherical angle (as in definition 10.2.7). We typically label the vertices of a spherical triangle as A, B, and C, and the sides opposite these vertices as a, b, and c. We also refer to the angle at a vertex and the length of a side by means of their label. For instance, theorem 10.2.2 states that for any spherical triangle $a+b<c$, $a+c<b$, and $b+c<a$.

THEOREM 10.2.2. *The length of any side of a spherical triangle is less than the sum of the lengths of the other two sides.*

PROOF. The three distinct planes containing the three sides of the spherical triangle form a vertex at the center of the sphere. The length of each side is exactly a face angle at the center of the sphere, and so the result follows from theorem 10.2.1. □

THEOREM 10.2.3.

(a) *The length of any side of a spherical triangle is less than* 180° (π).

(b) *The sum of the lengths of the sides of a spherical triangle is less than* 360° (2π).

PROOF. Both of these properties follow from the fact that a spherical triangle is contained within a hemisphere. In particular, (a) states that any side of a spherical triangle is less than half of a great circle (this is obvious) and (b) states that the perimeter of a proper spherical triangle is less than the perimeter of a hemisphere.

In more detail, to see why (b) is true, consider figure 10.10, in which the two sides a and b of a spherical triangle ABC extend to two meridians intersecting at the pair of vertices C and C^*, and the spherical triangle ABC is contained in the lune between these meridians. The third side c of the spherical triangle ABC divides the lune into two spherical triangles. The sides of the second spherical triangle are a^*, b^*, and c. By theorem 10.2.2, $c < a^* + b^*$, therefore $a+b+c < a+b+a^* + b^*$, which can be restated as $a+b+c < (a+a^*)+(b+b^*)$. Because each of $a+a^*$ and $b+b^*$ is a meridian with length π, we can conclude that $a+b+c < 2\pi$. □

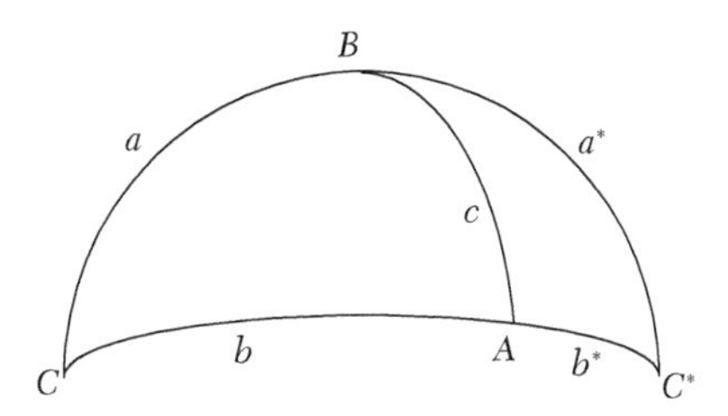

FIGURE 10.10. A lune.

Here is one property that planar triangles and spherical triangles have in common. (We do not provide the proof.)

THEOREM 10.2.4. *Let the lengths of the sides and the spherical angles of a spherical triangle be labeled as* a,b,c *and* A,B,C, *respectively. Then, the order of magnitudes of the sides is the same as the order of magnitudes of the angles, that is, if* $a < b < c$, *then* $A < B < C$ *(and conversely).*

We now introduce the notion of a *polar spherical triangle*, which is very useful for proving theorems about spherical triangles.

DEFINITION 10.2.9. *If we regard the three vertices of a given spherical triangle* ABC *as the poles for three great circles, then arcs of these great circles form a second, unique triangle, called the polar spherical triangle* A′B′C′ *of the first, in which the arc* A′B′ *is contained in the equator of* C, *the arc* A′C′ *is contained in the equator of* B, *and the arc* B′C′ *is contained in the equator of* A; *furthermore,* A *and* A′ *are in the same hemisphere determined by the great circle passing through* B *and* C, B *and* B′ *are in the same hemisphere determined by the great circle passing through* A *and* C, *and* C *and* C′ *are in the same hemisphere determined by the great circle passing through* A *and* B.

Figure 10.11 is an illustration of a spherical triangle ABC and its polar spherical triangle $A'B'C'$.

Here is a basic theorem relating to polar spherical triangles:

THEOREM 10.2.5. *If* A′B′C′ *is the polar spherical triangle of a spherical triangle* ABC, *then* ABC *is the polar spherical triangle of the spherical triangle* A′B′C′. □

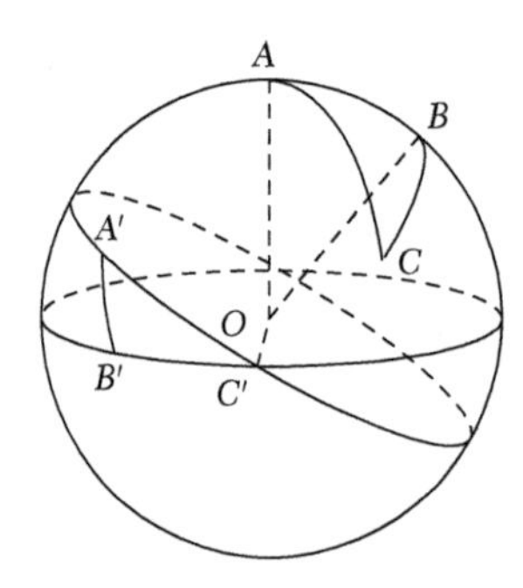

FIGURE 10.11. Mutually polar spherical triangles.

PROOF. In figure 10.11, O is the center of the sphere. The line AO is perpendicular to the plane through O containing the arc $B'C'$ and, therefore, perpendicular to the line $C'O$. Similarly, the line BO is perpendicular to the plane through O containing the arc $A'C'$ and, therefore, also perpendicular to the line $C'O$. This means that $C'O$ is perpendicular to the plane through O containing the arc AB, from which we conclude that C' is a pole for the arc AB. Similarly, B' and A' are poles for the arcs AC and BC, respectively. Therefore, ABC is the polar spherical triangle of spherical triangle $A'B'C'$. □

We can describe ABC and $A'B'C'$ as *mutually polar spherical triangles*. The statement of the next theorem relating to mutually polar spherical triangles might be confusing at first, but the proof of the theorem will make it clear.

THEOREM 10.2.6. *In two mutually polar spherical triangles, an angle of one is the supplement of the side opposite the corresponding angle of the other.*

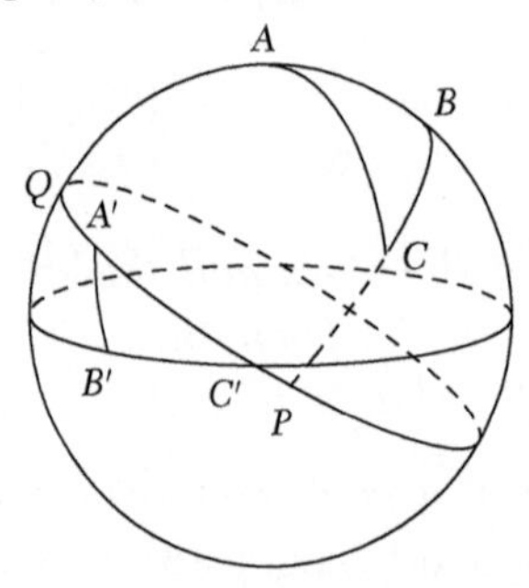

FIGURE 10.12. Mutually polar spherical triangles.

PROOF. Figure 10.12 is a modification of figure 10.11, in which the arc BC has been extended to meet the equator of B at the point P. Similarly, the arc BA meets the equator of B at the point Q. By the definition of a spherical angle, the magnitude of the angle B is the same as the length of the (shorter) arc from P to Q. Now because BP is contained in the equator for A', and BQ is contained in the equator for C', we can conclude that $A'P = 90°$ and $C'Q = 90°$. Therefore,

$$B = PQ = PA' + A'Q = 90° + (90° - A'C') = 180° - A'C'.$$

Similarly, $A = 180° - B'C'$ and $C = 180° - A'B'$. This proves the theorem. □

It is worthwhile to make a list of all the relationships contained in the statement of theorem 10.2.6. If we denote by a', b', c' the arcs of spherical triangle $A'B'C'$ opposite the vertices A', B', C', respectively, then

$$A = 180° - a' \qquad A' = 180° - a \qquad a = 180° - A' \qquad a' = 180° - A$$

$$B = 180° - b' \qquad B' = 180° - b \qquad b = 180° - B' \qquad b' = 180° - B$$

$$C = 180° - c' \qquad C' = 180° - c \qquad c = 180° - C' \qquad c' = 180° - C$$

We are ready now to prove a theorem that exhibits another way in which the geometry of spherical triangles is drastically different from the geometry of planar triangles.

THEOREM 10.2.7.

(a) *Every angle in a spherical triangle is less than* 180° (π).

(b) *The sum of the angles of a spherical triangle is less than* 540° (3π).

(c) *The sum of the angles of a spherical triangle is greater than* 180° (π).

PROOF. (a) We have already proved that any side of a spherical triangle is less than 180°. Therefore, the length of a side subtracted from 180° is also less than 180°. If we look at the first column of equations above, this means that A, B, and C are each less than 180°. (b) By adding A, B, and C in the first column of equations above, we find that $A + B + C = 540° - (a' + b' + c')$, and so the right-hand side of this equation is less than 540°. (c) Furthermore, by theorem 10.2.3(b) above, $a' + b' + c' < 360°$. Therefore,

$$A + B + C = 540° - (a' + b' + c') > 540° - 360° = 180°. \qquad \square$$

The amount by which the sum of angles of a spherical triangle exceeds 180° is known as the *spherical excess* of the spherical triangle. It is usually denoted as E, that is

$$E = (A + B + C) - 180°.$$

The following theorem states something surprising about the spherical excess.

THEOREM 10.2.8. *On the unit sphere, the spherical excess of a spherical triangle is equal to the area of the spherical triangle* (Girard's Theorem).

PROOF. Figure 10.13 shows a spherical triangle with vertices at A, B, and C formed by two great circles meeting at C and a third great circle (the boundary of the hemisphere) passing through A and B. The hemisphere is divided into four regions and their areas are labeled u, v, w, and x.

We will make use of the following fact: if the angle between two great circles is denoted by α, then the area of each of the smaller two lunes between the great circles is equal to $\frac{\alpha}{2\pi} \cdot 4\pi = 2\alpha$. (Recall that the surface area of the unit sphere is 4π.) Applied to figure 10.13, this means that

$$u + x = 2\hat{A}, \quad v + x = 2\hat{B}, \quad w + x = 2\hat{C},$$

where $\hat{A}$, $\hat{B}$, and $\hat{C}$ are the angles at the vertices of the spherical triangle. If we add the three equations above, we obtain

$$u + v + w + 3x = 2(\hat{A} + \hat{B} + \hat{C}).$$

However, we know that $u + v + w + x = 2\pi$ (the area of a hemisphere). Therefore,

$$2x = 2(\hat{A}+\hat{B}+\hat{C}) - 2\pi$$

or

$$x = (\hat{A}+\hat{B}+\hat{C}) - \pi$$

which is what we had to prove. □

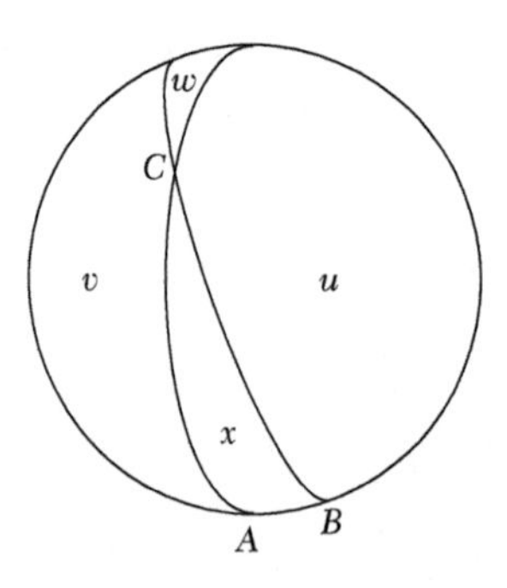

FIGURE 10.13. A proof of Girard's Theorem.

Theorem 10.2.8 is named after Albert Girard, a French-born mathematician who lived from 1595 to 1632.

10.3 VECTORS IN SPACE

Vectors in the Cartesian plane were introduced in chapter 2 and studied further in chapter 4. Here, we look at vectors in space and define an operation called the cross product of two vectors. The properties of the cross product that we discover in this section will make it possible for us to prove some important formulas relating to spherical triangles in section 10.4.

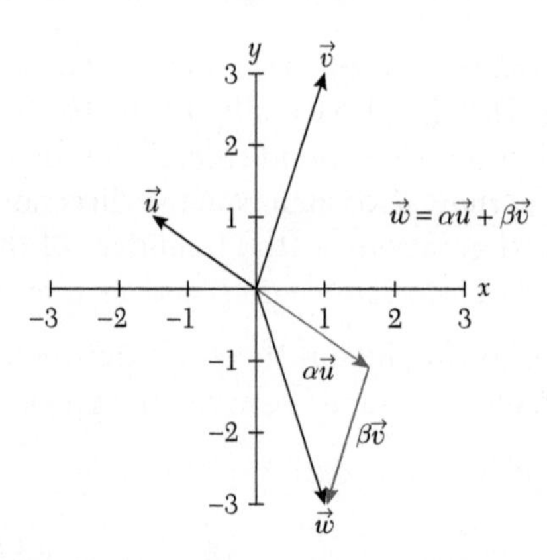

FIGURE 10.14. Two vectors spanning the Cartesian plane.

Before proceeding, it is important to know that any two nonparallel vectors in the Cartesian plane *span* the Cartesian plane. This means that any other vector in the Cartesian plane can be expressed as a sum of appropriate scalar multiples of each of the given vectors.

To illustrate, suppose that the two (nonparallel) vectors are $\vec{u} = \langle -1.5, 1 \rangle$ and $\vec{v} = \langle 1, 3 \rangle$, and $\vec{w} = \langle 1, -3 \rangle$ is some other arbitrary vector. It is easy to demonstrate, as shown in figure 10.14, that if $\vec{u}$ is stretched in the opposite direction (i.e., multiplied by a suitable negative scalar α) and $\vec{v}$ is shortened in the opposite direction (i.e., multiplied by another suitable negative scalar β), then they add up to the vector $\vec{w}$. In so doing, the vectors $\alpha\vec{u}$ and $\beta\vec{v}$ *span a parallelogram* with $\vec{w}$ as a diagonal vector and we say that the vector $\vec{w}$ is expressed as a *linear combination* of vectors $\vec{u}$ and $\vec{v}$.

In the same way that two nonparallel vectors in the Cartesian plane span the Cartesian plane, any two nonparallel vectors in space span a plane in space, as shown in figure 10.15.

The angle between two vectors in space can be determined in the plane spanned by the vectors. By convention, when we refer to the angle between two vectors, we always mean the smaller of the two angles; thus, the angle between two vectors can be 0° (when the vectors point in the same direction), an acute angle (as shown in the first diagram in figure 10.15), 90° (when the vectors are orthogonal), an obtuse angle (as shown in the second diagram in figure 10.15), or 180° (when the vectors point in opposite directions).

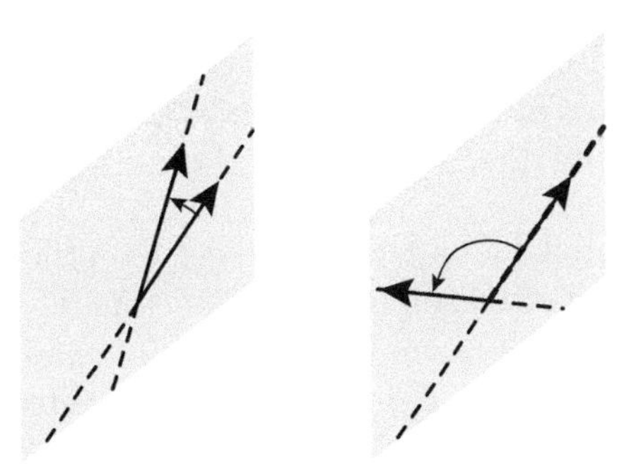

FIGURE 10.15. The angle between two vectors.

10.3.1 The Cross Product of Two Vectors

Recall that the dot product applied to two vectors produces a scalar (a real number). The *cross product* (notation: $\vec{u}\times\vec{v}$) of two vectors $\vec{u}$ and $\vec{v}$, however, is a new *vector* so we have to specify its direction and magnitude.

First, $\vec{u}\times\vec{v}$ is perpendicular (orthogonal) to both of the given vectors $\vec{u}$ and $\vec{v}$ (which means it is also perpendicular to the plane spanned by $\vec{u}$ and $\vec{v}$), so there are two possible directions for $\vec{u}\times\vec{v}$, each the opposite of the other (as shown in figure 10.16). The appropriate direction is chosen according to a convention called the *right-hand rule*: if the fingers of the right-hand curl through the (smaller) angle from $\vec{u}$ to $\vec{v}$, then $\vec{u}\times\vec{v}$ points in the direction of the thumb.

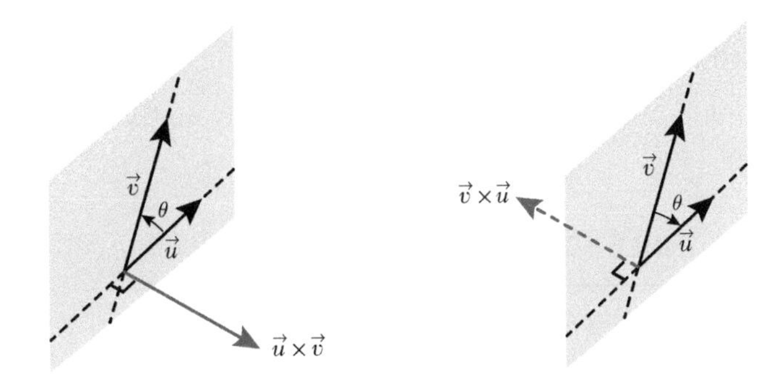

FIGURE 10.16. The right-hand rule.

Second, the magnitude of $\vec{u}\times\vec{v}$ is calculated by means of the formula

$$\boxed{|\vec{u}\times\vec{v}| = |\vec{u}||\vec{v}|\sin(\theta)} \tag{10.2}$$

where θ is the angle between vectors $\vec{u}$ and $\vec{v}$ (note that $0 \leq \theta \leq \pi$). (This resembles the formula for the dot product given in formula (4.22).) The factor $|\vec{v}|\sin(\theta)$ is the magnitude of the *perpendicular projection of* $\vec{v}$ *onto* $\vec{u}$, which is the length labeled h in figure 10.17.

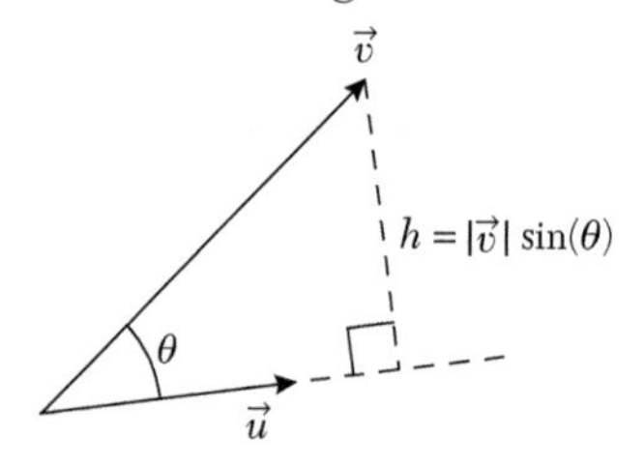

FIGURE 10.17. The perpendicular projection of a vector.

REMARK 10.3.1. Unlike the dot product, the order of the vectors in the cross product does matter; in fact $\vec{u}\times\vec{v}=-\vec{v}\times\vec{u}$ (if the fingers of the right-hand curl from $\vec{v}$ to $\vec{u}$ instead of from $\vec{u}$ to $\vec{v}$, then the thumb points in the opposite direction). This is shown in figure 10.16.

REMARK 10.3.2. Vectors are parallel if and only if their cross product is the zero vector. (In particular, the cross product of a vector with itself is the zero vector.) The reason for this is that two vectors are parallel if and only if the angle between them is zero or π, and in either case the sine ratio of this angle is zero, and so, by formula (10.2), the magnitude of the cross product is zero.

REMARK 10.3.3. It is important to remember that a dot product of two vectors is always a scalar (a real number), whereas the cross product is always a vector (which can be the zero vector).

If the magnitude of a vector is equal to 1, that is, it has a length of one unit, then it is called a *unit vector*.

REMARK 10.3.4. If two unit vectors are perpendicular to each other, then their cross product is again a unit vector. This follows from formula (10.2): if the angle θ between the unit vectors $\vec{u}$ and $\vec{v}$ is a right angle, that is, $\theta=\frac{\pi}{2}$, then

$$\boxed{|\vec{u}\times\vec{v}| = |\vec{u}||\vec{v}|\sin\left(\frac{\pi}{2}\right) = \sin\left(\frac{\pi}{2}\right) = 1}. \tag{10.3}$$

REMARK 10.3.5 If $\vec{v}$ is perpendicular to $\vec{u}$, then the direction of $\vec{u}\times\vec{v}$ can be found by rotating $\vec{v}$ an angle 90° around an axis containing the vector $\vec{u}$, according to the right hand rule.

In table 10.1 are some properties of the cross product. Note that resemblance to properties of the dot product in table 2.3.

TABLE 10.1. Properties of the cross product

(I)	$\vec{u}\times\vec{v}=-(\vec{v}\times\vec{u})$	
(II)	$(c\vec{u})\times\vec{v}=c(\vec{u}\times\vec{v})=\vec{u}\times(c\vec{v})$	A scalar distributes through a cross product
(III)	$\vec{w}\times(\vec{u}+\vec{v})=(\vec{w}\times\vec{u})+(\vec{w}\times\vec{v})$	The cross product distributes through addition
(IV)	$(\vec{u}+\vec{v})\times\vec{w}=(\vec{u}\times\vec{w})+(\vec{v}\times\vec{w})$	The cross product distributes through addition

The first property has already been explained. The proof of the second property is left as an exercise. The third and fourth properties express the distributive law for the cross product. They are difficult to prove using geometric methods, so their proofs will not be given here.

Recall that the sum of two vectors $\vec{u}$ and $\vec{v}$ can be expressed geometrically as the completion of a parallelogram. Because $|\vec{u}\times\vec{v}|=|\vec{u}||\vec{v}|\sin(\theta)=|\vec{u}|(|\vec{v}|\sin(\theta))$, and the factor in parenthesis is the magnitude of the perpendicular projection of $\vec{v}$ onto $\vec{u}$, we ascertain the following.

REMARK 10.3.6. The magnitude of the cross product of two vectors is the area of the parallelogram spanned by the vectors.

10.3.2 Parallelepipeds and Cross Product Identities

In the same way that two nonparallel vectors span a parallelogram in a plane, three vectors span a *parallelepiped* (pronounced "parallelpiped") in space so long as the three vectors are not coplanar (that is, they are not all contained in the same plane). A parallelepiped has six faces, each of which is a parallelogram spanned by two of the vectors.

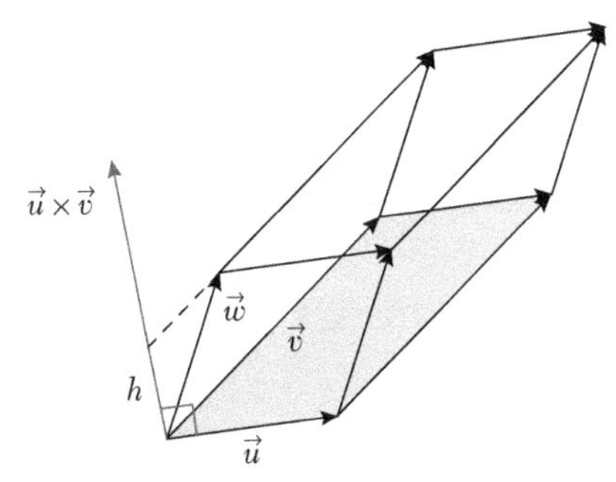

FIGURE 10.18. A parallelepiped.

In figure 10.18, a parallelepiped is spanned by vectors $\vec{u}$, $\vec{v}$, and $\vec{w}$. The parallelogram spanned by $\vec{u}$ and $\vec{v}$ can be taken as the base of the parallelepiped (shaded in figure 10.18), and so $\vec{u}\times\vec{v}$ is perpendicular to the base. The magnitude of the projection of $\vec{w}$ onto $\vec{u}\times\vec{v}$ (denoted h in the diagram) is the *height* of the parallelepiped. Therefore, the *volume* of the parallelepiped, which is defined as the *area of the base* times the *height*, can be expressed as $h|\vec{u}\times\vec{v}|$. According to the geometric interpretation of the dot product (see section 4.14.2), an expression for h is $\frac{|(\vec{u}\times\vec{v})\cdot w|}{|\vec{u}\times\vec{v}|}$. Consequently, we can state the following:

REMARK 10.3.7. The volume of a parallelepiped spanned by three non-coplanar vectors $\vec{u}$, $\vec{v}$, and $\vec{w}$, is $|(\vec{u}\times\vec{v})\cdot w|$.

In fact, either of the permutations $|(\vec{u}\times\vec{w})\cdot v|$ and $|(\vec{v}\times\vec{w})\cdot u|$ is also the volume of the parallelepiped (in each case a different parallelogram is taken as the base of the parallelepiped). Keep in mind that a volume is a positive number so we have taken the *absolute value* of the dot product in each case. By means of a more careful analysis (which we will not do here), it is possible to prove the *scalar triple product identity*:

$$\boxed{\vec{u}\cdot(\vec{v}\times w)=(\vec{u}\times\vec{v})\cdot w} \tag{10.4}$$

Because a cross product of vectors is also a vector, it is possible to form cross products of cross products, for example, $\vec{u}\times(\vec{u}\times\vec{v})$, $(\vec{u}\times\vec{v})\times u$, $(\vec{u}\times\vec{v})\times(\vec{u}\times\vec{w})$, and so on. We might need to simplify an expression like $\vec{u}\times(\vec{u}\times\vec{v})$, so it is helpful to think about what this means geometrically. We know that $\vec{u}\times\vec{v}$ is a vector that is perpendicular to the plane spanned by vectors $\vec{u}$ and $\vec{v}$. But this means that any vector that is perpendicular to $\vec{u}\times\vec{v}$, particularly the cross product of $\vec{u}$ with $\vec{u}\times\vec{v}$, would be contained in this plane.

Therefore, as explained in the beginning of this section, we can write $\vec{u}\times(\vec{u}\times\vec{v})$ as a linear combination of the vectors $\vec{u}$ and $\vec{v}$, that is, $\vec{u}\times(\vec{u}\times\vec{v})=c\vec{u}+d\vec{v}$ for some scalars c and d.

Mathematicians like to encapsulate their results as *lemmas*. These are usually stepping stones to more important results and are usually technical in nature. We will find the values of c and d above in the special case that $\vec{u}$ and $\vec{v}$ are unit vectors (vectors with unit length) and express the result in the form of a lemma:

LEMMA 10.3.1. *If* $\vec{u}$ *and* $\vec{v}$ *are unit vectors, that is,* $|\vec{u}|=|\vec{v}|=1$, *then*

$$\boxed{\vec{u}\times(\vec{u}\times\vec{v})=(\vec{u}\cdot\vec{v})\vec{u}-\vec{v}}. \tag{10.5}$$

PROOF. As explained above, we can write

$$\boxed{\vec{u}\times(\vec{u}\times\vec{v})=c\vec{u}+d\vec{v}} \tag{10.6}$$

where c and d are unknown scalars. We will find c and d as follows: because $\vec{u}\times(\vec{u}\times\vec{v})$ is perpendicular to $\vec{u}$, we have $\vec{u}\cdot(\vec{u}\times(\vec{u}\times\vec{v}))=0$ and then, by taking the dot product with $\vec{u}$ on both sides of formula (10.6):

$$\begin{aligned} 0&=\vec{u}\cdot(c\vec{u}+d\vec{v})\\ &=c(\vec{u}\cdot\vec{u})+d(\vec{u}\cdot\vec{v})\\ &=c+d(\vec{u}\cdot\vec{v}). \end{aligned}$$

Therefore $c=-d(\vec{u}\cdot\vec{v})$, and so we can write the following sequence of equations:

$$\begin{aligned} \vec{u}\times(\vec{u}\times\vec{v})&=-d(\vec{u}\cdot\vec{v})\vec{u}+d\vec{v}\\ \vec{v}\cdot(\vec{u}\times(\vec{u}\times\vec{v}))&=\vec{v}\cdot(-d(\vec{u}\cdot\vec{v})\vec{u}+d\vec{v})\\ (\vec{v}\times\vec{u})\cdot(\vec{u}\times\vec{v})&=-d(\vec{u}\cdot\vec{v})(\vec{v}\cdot\vec{u})+d(\vec{v}\cdot\vec{v})\\ -|\vec{u}\times\vec{v}|^2&=-d(\vec{u}\cdot\vec{v})^2+d|\vec{v}|^2=-d(\vec{u}\cdot\vec{v})^2+d. \end{aligned}$$

Note that the scalar triple product identity (formula (10.4)) was used to obtain the third equation. Now, we can solve for d:

$$d=\frac{|\vec{u}\times\vec{v}|^2}{(\vec{u}\cdot\vec{v})^2-1}.$$

If θ is the angle between $\vec{u}$ and $\vec{v}$, then this simplifies to

$$d=\frac{\left(|\vec{u}|\,|\vec{v}|\sin\theta\right)^2}{\left(|\vec{u}|\,|\vec{v}|\cos\theta\right)^2-1}=\frac{\sin^2\theta}{\cos^2\theta-1}=-1$$

and this completes the proof of the lemma. □

Here is another formula that can be used to find the volume of a parallelepiped in the case that the spanning vectors are unit vectors:

LEMMA 10.3.2. *If* $\vec{u}$, $\vec{v}$, *and* $\vec{w}$ *are unit vectors, then* $|(\vec{u}\times\vec{v})\times(\vec{u}\times\vec{w})|$ *is the volume of the parallelepiped spanned by* $\vec{u}$, $\vec{v}$, *and* $\vec{w}$. *Furthermore*

$$|(\vec{u}\times\vec{v})\times(\vec{u}\times\vec{w})|=|(\vec{v}\times\vec{u})\times(\vec{v}\times\vec{w})|=|(\vec{w}\times\vec{u})\times(\vec{w}\times\vec{v})|.$$

Proof. The vector $\vec{u}$ is perpendicular to both of the vectors $\vec{u}\times\vec{v}$ and $\vec{u}\times\vec{w}$. Stated differently, $\vec{u}$ is perpendicular to a plane spanned by $\vec{u}\times\vec{v}$ and $\vec{u}\times\vec{w}$. According to the definition of a cross product, this means that $(\vec{u}\times\vec{v})\times(\vec{u}\times\vec{w})$ is parallel to $\vec{u}$, that is

$$\boxed{(\vec{u}\times\vec{v})\times(\vec{u}\times\vec{w})=\alpha\vec{u}} \tag{10.7}$$

for some scalar value α. If we take the dot product with $\vec{u}$ on both sides of this equation, then

$$\vec{u}\cdot((\vec{u}\times\vec{v})\times(\vec{u}\times\vec{w}))=\alpha\vec{u}\cdot\vec{u}=\alpha|\vec{u}|^2=\alpha.$$

Now read the following set of equations carefully. (The scalar triple product identity is used to go from the first to the second line, and lemma 10.3.1 is used to go from the second to the third line. The remaining steps use properties of the dot and cross products.)

$$\begin{aligned}|\alpha| &= |\vec{u}\cdot((\vec{u}\times\vec{v})\times(\vec{u}\times\vec{w}))|\\ &= |(\vec{u}\times(\vec{u}\times\vec{v}))\cdot(\vec{u}\times\vec{w})|\\ &= |((\vec{u}\cdot\vec{v})\vec{u}-\vec{v})\cdot(\vec{u}\times\vec{w})|\\ &= |(\vec{u}\cdot\vec{v})\vec{u}\cdot(\vec{u}\times\vec{w})-\vec{v}\cdot(\vec{u}\times\vec{w})|\\ &= |0-\vec{v}\cdot(\vec{u}\times\vec{w})|\\ &= |\vec{v}\cdot(\vec{u}\times\vec{w})|.\end{aligned}$$

According to formula (10.7), this proves that

$$|(\vec{u}\times\vec{v})\times(\vec{u}\times\vec{w})|=|\alpha\vec{u}|=|\alpha|=|\vec{v}\cdot(\vec{u}\times\vec{w})|$$

which is the volume of the parallelepiped spanned by $\vec{u}$, $\vec{v}$, and $\vec{w}$. Similarly, the volume of the parallellpiped is equal to the other two cross product formulas. □

We will use lemma 10.3.2 to prove the basic identities relating the sides and angles of a spherical triangle in section 10.4. Before proceeding to do so we need to make some comments about normal vectors and the angle between two planes.

10.3.2 The Angle Between Two Planes

Any plane in space has two *unit normal vectors* (pointing in opposite directions) associated with it. A unit normal vector to a plane is a unit vector that is perpendicular to the plane, and it can be obtained by taking the cross product of any two nonparallel unit vectors contained in the plane and dividing the vector so obtained by its own length (or the negative of its own length depending on the choice of direction of the unit normal vector).

A useful property of unit normal vectors is that we can compute the *angle between two planes* as the angle between the two appropriate unit normal vectors. The direction of the unit normal vectors needs to be chosen depending whether the smaller angle or its supplement is chosen as the angle between the planes, if the planes are not perpendicular to each other. (Convince yourself that this is true by drawing a cross-sectional diagram of two intersecting planes.)

In figure 10.19, two planes intersect in a line L and a vector $\vec{u}$ is contained in L. The angle between the planes is the angle between the vectors $\vec{v}_0$ and $\vec{w}_0$, which are on the same side of L in each of the planes, and both are perpendicular to $\vec{u}$. According to remark 10.3.5, the angle between $\vec{u}\times\vec{v}_0$ and $\vec{u}\times\vec{w}_0$ will be the same as the angle between $\vec{v}_0$ are $\vec{w}_0$.

Now, if $\vec{v}$ is any vector contained in the same plane as $\vec{v}_0$ and points to the same side of L as $\vec{v}_0$ and, if $\vec{w}$ is any vector contained in the same plane as $\vec{w}_0$ and points to the same side of L as $\vec{w}_0$, then, according to the right-hand rule, $\vec{u}\times\vec{v}$ will point in the same direction as $\vec{u}\times\vec{v}_0$, and $\vec{u}\times\vec{w}$ will point in the same direction as $\vec{u}\times\vec{w}_0$. This tells us that the angle between the normal vectors $\vec{u}\times\vec{v}$

and $\vec{u}\times\vec{w}$ will be the same as the angle between the normal vectors $\vec{u}\times\vec{v}_0$ and $\vec{u}\times\vec{w}_0$, which is the angle between the planes.

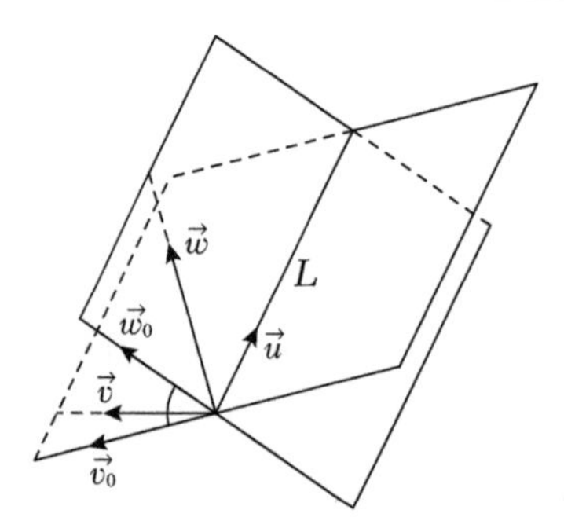

FIGURE 10.19. The direction of a normal vector.

Our conclusion from this discussion can be summarized as follows:

REMARK 10.3.8. If a vector is common to two intersecting, noncoinciding, planes and the cross products of this vector are taken with two vectors (in each of the planes) that point to the same side of the intersection line of the planes as the vectors that determine the angle between the planes, then the angle between the normal vectors obtained in this way is the same as the angle between the planes.

10.4 SOLVING SPHERICAL TRIANGLES

As in planar trigonometry, there are laws for solving spherical triangles. The first of these laws is called the *sine rule* (for spherical triangles). We will state it and then use the methods of vector geometry from section 10.3 to prove it. After that, we will state and prove the *cosine rule for sides* and state the *cosine rule for angles*.

THEOREM 10.4.1. *The sine rule: For a spherical triangle with spherical angles* A, B, *and* C *and arc lengths* a, b, *and* c,

$$\boxed{\frac{\sin A}{\sin a}=\frac{\sin B}{\sin b}=\frac{\sin C}{\sin c}}. \tag{10.8}$$

PROOF. Shown in figure 10.20 is a spherical triangle with vertices labeled A, B, C and the corresponding opposite sides labeled a, b, c, respectively. Also shown are unit vectors from the origin O of the unit sphere to the vertices A, B, and C, which are labeled $\vec{u}$, $\vec{v}$, and $\vec{w}$, respectively, that is, $\vec{u}=\overrightarrow{OA}$, $\vec{v}=\overrightarrow{OB}$, and $\vec{w}=\overrightarrow{OC}$, respectively.

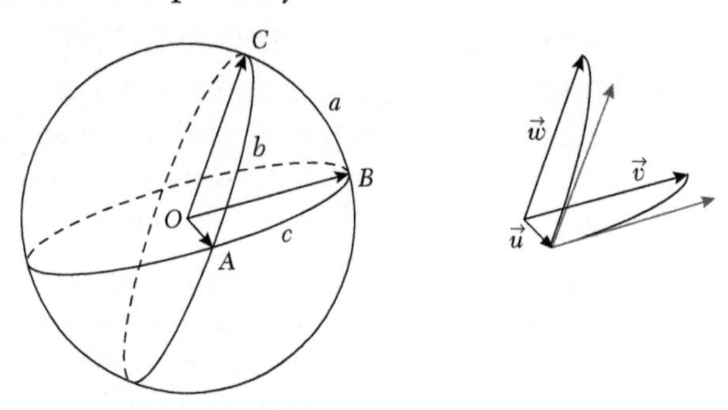

FIGURE 10.20. The proof of the sine rule.

The spherical angle A (the angle between the tangent vectors at A in figure 10.20) is the angle between the plane containing the vectors $\vec{u}$ and $\vec{v}$ and the arc c, and the plane containing the vectors $\vec{u}$ and $\vec{w}$ and the arc b. Because $\vec{u}\times\vec{v}$ and $\vec{u}\times\vec{w}$ are a pair of normal vectors to these planes, the angle between them is, by remark 10.3.8, equal to the spherical angle A (the vectors $\vec{v}$ and $\vec{w}$ point to the same side of $\vec{u}$ as the tangent vectors, which are perpendicular to $\vec{u}$) and so, by the definition of the cross product,

$$|(\vec{u}\times\vec{v})\times(\vec{u}\times\vec{w})| = |\vec{u}\times\vec{v}\,||\,\vec{u}\times\vec{w}\,|\sin(A).$$

Solving for $\sin(A)$ gives

$$\sin(A) = \frac{|(\vec{u}\times\vec{v})\times(\vec{u}\times\vec{w})|}{|\vec{u}\times\vec{v}\,||\,\vec{u}\times\vec{w}\,|}.$$

Another inference we can make from figure 10.20 is that the angle a is the angle between vectors $\vec{v}$ and $\vec{w}$. Thus

$$\sin(a) = \frac{|\vec{v}\times\vec{w}|}{|\vec{v}\,||\,\vec{w}\,|} = |\vec{v}\times\vec{w}\,|.$$

By dividing the last two equations, we obtain

$$\frac{\sin(A)}{\sin(a)} = \frac{|(\vec{u}\times\vec{v})\times(\vec{u}\times\vec{w})|}{|\vec{u}\times\vec{v}\,||\,\vec{u}\times\vec{w}\,||\,\vec{v}\times\vec{w}\,|}.$$

The corresponding formulas for the spherical angles at B and C are

$$\frac{\sin(B)}{\sin(b)} = \frac{|(\vec{v}\times\vec{u})\times(\vec{v}\times\vec{w})|}{|\vec{u}\times\vec{v}\,||\,\vec{u}\times\vec{w}\,||\,\vec{v}\times\vec{w}\,|} \quad\text{and}\quad \frac{\sin(C)}{\sin(c)} = \frac{|(\vec{w}\times\vec{u})\times(\vec{w}\times\vec{v})|}{|\vec{u}\times\vec{v}\,||\,\vec{u}\times\vec{w}\,||\,\vec{v}\times\vec{w}\,|}.$$

The denominator on the right-hand side is the same in each case and, by lemma 10.3.2, the numerators are the same in each case. This proves the sine rule for spherical triangles. □

THEOREM 10.4.2. *The cosine rule for sides:*

$$\boxed{\begin{aligned}\cos a &= \cos b\cos c + \sin b\sin c\cos A\\ \cos b &= \cos a\cos c + \sin a\sin c\cos B\\ \cos c &= \cos a\cos b + \sin a\sin b\cos C\end{aligned}} \tag{10.9}$$

PROOF. Refer again to figure 10.20. As given in the proof of the sine rule for spherical triangles, the angle between vectors $\vec{u}\times\vec{v}$ and $\vec{u}\times\vec{w}$ is the angle A. Therefore, according to the definition of the dot product,

$$\cos(A) = \frac{(\vec{u}\times\vec{v})\cdot(\vec{u}\times\vec{w})}{|\vec{u}\times\vec{v}\,||\,\vec{u}\times\vec{w}\,|}.$$

If we apply the scalar triple product identity (formula (10.4)) to the numerator of this fraction, then

$$\cos(A) = \frac{((\vec{u}\times\vec{v})\times\vec{u})\cdot\vec{w}}{|\vec{u}\times\vec{v}\,||\,\vec{u}\times\vec{w}\,|} = -\frac{(\vec{u}\times(\vec{u}\times\vec{v}))\cdot\vec{w}}{|\vec{u}\times\vec{v}\,||\,\vec{u}\times\vec{w}\,|}.$$

By lemma 10.3.1, this can be expressed as

$$\cos(A) = -\frac{((\vec{u}\cdot\vec{v})\vec{u}-\vec{v})\cdot\vec{w}}{|\vec{u}\times\vec{v}\,||\,\vec{u}\times\vec{w}\,|}.$$

This can be taken one step further by distributing the dot product with $\vec{w}$. This yields

$$\cos(A)=\frac{(\vec{v}\cdot\vec{w})-(\vec{u}\cdot\vec{v})(\vec{u}\cdot\vec{w})}{|\vec{u}\times\vec{v}||\vec{u}\times\vec{w}|}=\frac{\cos(a)-\cos(b)\cos(c)}{\sin(c)\sin(b)},$$

where the second equality follows from the fact that a is the angle between $\vec{v}$ and $\vec{w}$, b is the angle between $\vec{u}$ and $\vec{w}$, and c is the angle between $\vec{u}$ and $\vec{v}$. The terms in the equation above can be rearranged to give the first statement of the cosine rule for sides. The second and third statements of the cosine rule for sides can be proved similarly. □

The cosine rule for angles, stated next, can be proved by substituting the appropriate relationships from theorem 10.2.6 into the cosine rule for sides (theorem 10.4.2). This is left as an exercise.

THEOREM 10.4.3. *The cosine rule for angles:*

$$\boxed{\begin{aligned}\cos A&=-\cos B\cos C+\sin B\sin C\cos a\\ \cos B&=-\cos A\cos C+\sin A\sin C\cos b\\ \cos C&=-\cos A\cos B+\sin A\sin B\cos c\end{aligned}} \tag{10.10}$$

EXAMPLE 10.4.1. Find the side b in a spherical triangle if $a=76°$, $c=58°$, and $B=117°$.

Answer: Use the cosine rule for sides to solve for b:

$$\begin{aligned}\cos b&=\cos 76°\cos 58°+\sin 76°\sin 58°\cos 117°\\ \cos b&\approx -0.24537\\ b&\approx 104.20°\end{aligned}$$

◈

10.5 SOLVING RIGHT SPHERICAL TRIANGLES (I)

If one of the angles of a spherical triangle is a right angle, then it is called a *right spherical triangle*.

If the position of the right angle and two other elements of a right spherical triangle are given; for example, two sides or an angle and a side, then "solving the triangle," as in planar trigonometry, means to find the other three parts. In spherical trigonometry, the solution need not be unique (there can be more than one triangle). In planar trigonometry, the solution is always unique (only one possible triangle).

The formulas for solving a right spherical triangle will be derived from definitions of the planar trigonometric ratios by means of the creation of ordinary right triangles inside the sphere, having the same angles as the right spherical triangle. Precisely, how to do this will be explained next. Recall that the notation "$\perp$" is used to mean that one line segment is perpendicular to another line segment.

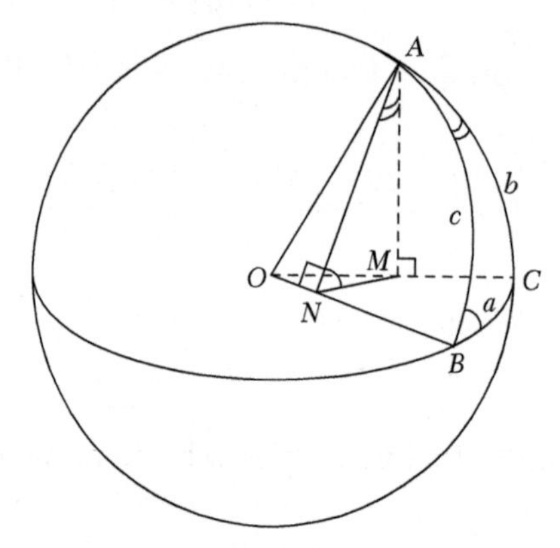

FIGURE 10.21. Right triangles with common vertex at *A*.

Figure 10.21 shows a unit sphere and a spherical triangle *ACB* with a right angle at *C*. Furthermore, the plane *AOC* is perpendicular to the plane *BOC*, $AN \perp OB$, and $AM \perp OC$. The points *M* and *N* are joined to create the triangle *MON*. It is also true that the triangle *MNO* has a right angle at *N*. This can be verified by an application of the converse of the Pythagorean Theorem (theorem 9.5.8(b)):

$$|ON|^2 + |MN|^2 = (|OA|^2 - |AN|^2) + (|AN|^2 - |AM|^2)$$
$$= |OA|^2 - |AM|^2 = |OM|^2 .$$

We can now conclude that *MN* and *AN* are parallel to the tangent vectors that form the spherical angle at B and, therefore, the spherical angle at *B* is equal to $A\hat{N}M$.

In triangle *AOM*, the angle at *O* is equal to *b*; therefore,

$$\sin b = \frac{|AM|}{|OA|} = \frac{|AM|}{1} = |AM| \tag{10.11}$$

$$\cos b = \frac{|OM|}{|OA|} = \frac{|OM|}{1} = |OM|. \tag{10.12}$$

In triangle *AON*, the angle at *O* is equal to *c*; therefore

$$\sin c = \frac{|AN|}{|OA|} = \frac{|AN|}{1} = |AN| \tag{10.13}$$

$$\cos c = \frac{|ON|}{|OA|} = \frac{|ON|}{1} = |ON|. \tag{10.14}$$

In triangle *MON*, the angle at *O* is equal to *a*; therefore (using formula (10.12)),

$$\sin a = \frac{|MN|}{|OM|} = \frac{|MN|}{\cos b} \tag{10.15}$$

and so

$$|MN| = \sin a \cos b. \tag{10.16}$$

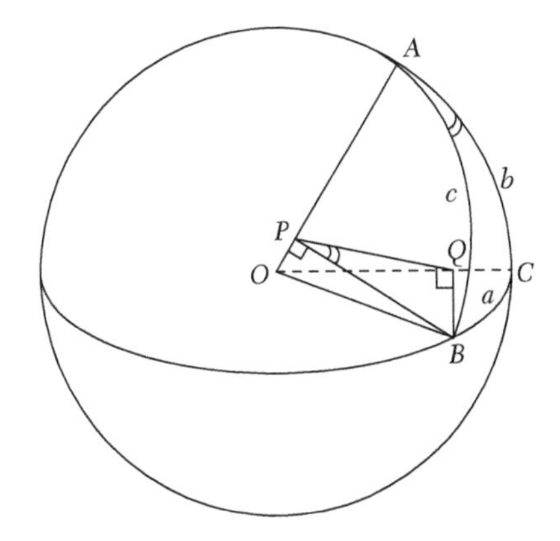

FIGURE 10.22. Right triangles with common vertex at *B*.

In figure 10.22, $BP \perp OA$, $BQ \perp OC$, and the points P and Q are joined to create triangle BPQ. Triangle OPQ has a right angle at P. (This can also be proved by applying the converse of the Pythagorean Theorem.)

We can conclude now that the spherical angle at A is equal to $B\hat{P}Q$ (definition of a spherical angle). We can also write down trigonometric ratios of the angles a, b, and c:

In triangle BOQ, the angle at O is equal to a; therefore,

$$\sin a = \frac{|BQ|}{|OB|} = \frac{|BQ|}{1} = |BQ| \tag{10.17}$$

$$\cos a = \frac{|OQ|}{|OB|} = \frac{|OQ|}{1} = |OQ|. \tag{10.18}$$

In triangle BOP, the angle at O is equal to c; therefore,

$$\sin c = \frac{|BP|}{|OB|} = \frac{|BP|}{1} = |BP| \tag{10.19}$$

$$\cos c = \frac{|OP|}{|OB|} = \frac{|OP|}{1} = |OP|. \tag{10.20}$$

In triangle POQ, the angle at O is equal to b; therefore (using formula (10.18)),

$$\sin b = \frac{|PQ|}{|OQ|} = \frac{|PQ|}{\cos a} \tag{10.21}$$

and so

$$|PQ| = \sin b \cos a. \tag{10.22}$$

We now derive ten identities for right spherical triangles. These are all the identities needed for solving right spherical triangles.

From figure 10.21, we derive the following identities (using formulas (10.11) and (10.13) in the first case and formulas (10.11) and (10.16) in the second case).

$$\sin B = \frac{|AM|}{|AN|} = \frac{\sin b}{\sin c}, \qquad \tan B = \frac{|AM|}{|MN|} = \frac{\sin b}{\sin a \cos b} = \frac{\tan b}{\sin a}.$$

From these two equations, we obtain

$$\boxed{\sin b = \sin c \sin B} \tag{10.23}$$

and

$$\boxed{\sin a = \tan b \cot B}. \tag{10.24}$$

Using triangle BPQ in figure 10.22, we derive the following identities (using formulas (10.17) and (10.19) in the first case and formulas (10.17) and (10.22) in the second case).

$$\sin A = \frac{|BQ|}{|BP|} = \frac{\sin a}{\sin c}, \qquad \tan A = \frac{|BQ|}{|PQ|} = \frac{\sin a}{\sin b \cos a} = \frac{\tan a}{\sin b}$$

From these two equations, we obtain

$$\boxed{\sin a = \sin c \sin A} \tag{10.25}$$

and

$$\boxed{\sin b = \tan a \cot A}. \tag{10.26}$$

From the third equation of the cosine rule for sides (theorem 10.4.2) applied to the spherical triangle ABC (with $C = 90°$), we derive the identity

$$\boxed{\cos c = \cos a \cos b}. \tag{10.27}$$

By replacing the left-hand side of formula (10.26) with the right-hand side of formula (10.23) and then eliminating the factor $\sin c$ according to formula (10.25), we obtain

$$\boxed{\cos A = \sin B \cos a}. \tag{10.28}$$

By replacing the left-hand side of formula (10.24) with the right-hand side of formula (10.25) and then eliminating the factor $\sin c$ according to formula (10.23), we obtain

$$\boxed{\cos B = \sin A \cos b}. \tag{10.29}$$

By using formula (10.23) to replace the factor $\sin B$, and formula (10.27) to replace the factor $\cos a$ on the right-hand side of formula (10.28), we obtain

$$\boxed{\cos A = \tan b \cot c}. \tag{10.30}$$

By using formula (10.25) to replace the factor $\sin A$, and formula (10.27) to replace the factor $\cos b$ on the right-hand side of formula (10.29), we obtain

$$\boxed{\cos B = \tan a \cot c} \tag{10.31}$$

and finally, by using formula (10.28) to replace the factor $\cos a$, and formula (10.29) to replace the factor $\cos b$ on the right-hand side of formula (10.27), we obtain

$$\boxed{\cos c = \cot A \cot B}. \tag{10.32}$$

This situation of having 10 different equations, that is, formulas (10.23)–(10.32), for solving right spherical triangles is different from the situation in planar trigonometry where essentially three equations are enough to solve any right triangle. (Can you write down three such equations? You may assume that an equation such as $\cos\theta = \frac{a}{c}$ makes it possible to solve for θ if a and c are known.)

10.6 SOLVING RIGHT SPHERICAL TRIANGLES (II)

10.6.1 Rules for Quadrants

Some useful facts regarding right spherical triangles can be gleaned from formulas (10.23) to (10.32). They are expressed in the form of three rules below.

We will refer to an angle that is not a right angle or a straight angle as an *oblique angle*.

LEMMA 10.6.1. *Rule 1 for quadrants: In a right spherical triangle, an oblique angle and the side opposite are in the same quadrant (i.e., either the first or the second quadrant).*

PROOF. Consider formula (10.29) from section 10.5, which states: $\cos B = \sin A \cos b$. Because $\sin A > 0$ (any angle in a spherical triangle is less than 180°), $\cos B$ and $\cos b$ have the same sign, so B and b are in the same quadrant (the first quadrant if $\cos B$ and $\cos b$ are both positive, and the second quadrant if $\cos B$ and $\cos b$ are both negative). □

LEMMA 10.6.2. *Rule 2 for quadrants: When the hypotenuse (c) of a right spherical triangle is less than 90°, then the other two arcs are in the same quadrant; if the hypotenuse is greater than 90°, then the other two arcs are in different quadrants.*

PROOF. Consider equation formula (10.27) from section 10.5, which states that $\cos c = \cos a \cos b$. If $0° < c < 90°$ (i.e., $\cos c$ is positive), then either (i) $\cos a$ or $\cos b$ are both positive and, consequently, a and b are both in the first quadrant, or (ii) $\cos a$ or $\cos b$ are both negative and, consequently, a and b are both in the second quadrant. On the other hand, if $90° < c < 180°$ (that is, $\cos c$ is negative), then either (i) $\cos a$ is positive and $\cos b$ is negative or (ii) $\cos a$ is negative and $\cos b$ is positive—in either case, we see that a and b are in different quadrants. □

LEMMA 10.6.3. *Rule 3 for quadrants: When two given parts of a right spherical triangle are a side and its opposite angle, and they are not equal, then there are always two solutions (i.e., two triangles). If they are equal, then there is only one solution (i.e., one triangle).*

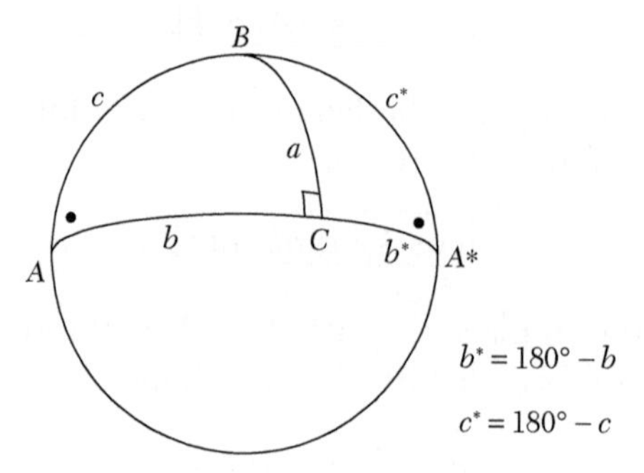

FIGURE 10.23. A lune intersected by a great circle.

PROOF. Consider the lune AA^* (in figure 10.23) that is intersected by the arc BC of a great circle (not necessarily the equator) perpendicular to one side of the lune (i.e., with a right angle at C). In the resulting spherical triangles ABC and A^*BC, the angles A and A^*, being vertices of the lune, are equal. The arc a is opposite A in the triangle ABC and opposite A^* in the triangle A^*BC. We observe that unless BC divides the lune exactly in half (in which case $a = A = A^*$ and the triangles ABC and A^*BC are identical), the two spherical triangles ABC and A^*BC are different triangles with the side a in common and equal angles at A and A^*.

Thus, when an angle and the side opposite are given, and they are not equal, then there are always two possible right spherical triangles, together forming a lune. □

EXAMPLE 10.6.1. Given $a = 46°$ and $A = 59°$, solve the right spherical triangle.

Answer: By rule 3 for quadrants (lemma 10.6.3), there are two solutions (two triangles). We apply formulas (10.25), (10.28), and (10.26), that is

$$\sin c = \sin a \csc A \qquad \sin B = \sec a \cos A \qquad \sin b = \tan a \cot A$$

and then calculate the first solution: $c = 57.056°$, $B = 47.853°$, and $b = 38.478°$. (Here a and b are in the same quadrant by rule 2 for quadrants (lemma 10.6.2) and b and B are in the same quadrant by rule 1 for quadrants (lemma 10.6.1).) The second solution is $c = 122.94$, $B = 132.15°$, and $b = 141.523$. (Here, a and b are in different quadrants, by rule 2 for quadrants (lemma 10.6.2), and b and B are in the same quadrant, by rule 1 for quadrants (lemma 10.6.1).) As an exercise, check these answers using formula (10.23), that is, $\sin b = \sin c \sin B$. ◈

EXAMPLE 10.6.2. Given $c = 109°$ and $B = 27°$, solve the right spherical triangle.

Answer: We use rule 2 for quadrants (lemma 10.6.2) and formulas (10.23), (10.31), and (10.32), that is

$$\sin b = \sin c \sin B \qquad \tan a = \tan c \cos B \qquad \cot A = \cos c \tan B$$

and then calculate the solution: $b = 25.420°$, $a = 111.13°$, and $A = 99.419°$. (Here, b and B are in the same quadrant by rule 1 (lemma 10.6.1), a and b are in different quadrants, by rule 2 (lemma 10.6.2), and a and A are in the same quadrant, by rule 1 (lemma 10.6.1).) This answer can be checked using formula (10.26), that is, $\sin b = \tan a \cot A$. ◈

10.6.2 Napier's Rules

Napier's Rules describe a method for easily finding the equations needed to solve a right spherical triangle by means of a simple diagram, so that it is not necessary to search through the list of ten equations derived in section 10.5 to find the correct equations.

We make use of the following notation: Let $\bar{A}$ denote $90°–A$ (i.e., $\bar{A}$ is the complement of A), $\bar{B}$ denote $90°–B$ (i.e., $\bar{B}$ is the complement of B), and $\bar{c}$ denote $90°–c$ (i.e., $\bar{c}$ is the complement of c). (Recall from definition 9.3.5 that two angles are *complementary* if they add up to $90°$.) We write the parts $a, \bar{B}, \bar{c}, \bar{A}, b$, in a "pie" diagram (*Napier's pentagon*) as shown in figure 10.24: Two or three consecutive parts in Napier's pentagon are called *adjacent*; for example, $\bar{A}, b, a$ and $\bar{B}, \bar{c}, \bar{A}$ are adjacent, whereas $\bar{B}, \bar{c}, b$ are not adjacent.

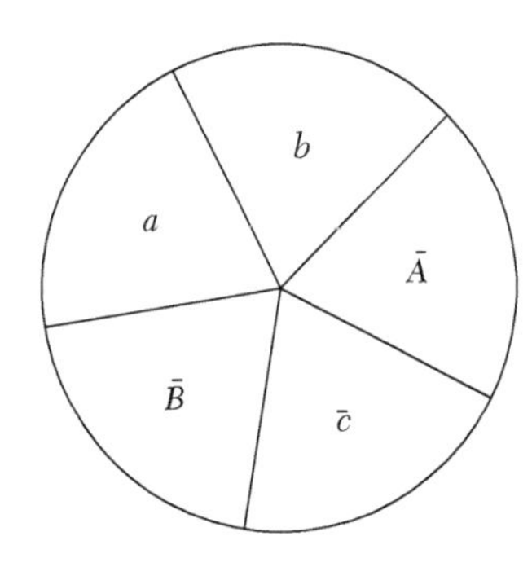

FIGURE 10.24. Napier's pentagon.

There are two rules (Napier's Rules) for writing down equations from Napier's pentagon:

LEMMA 10.6.4. *Napier's Rules (I): If three parts of Napier's pentagon are not adjacent, then the sine of the separate part is the product of the cosines of the opposite parts.*

EXAMPLE 10.6.3. $\sin \bar{c} = \cos a \cos b$ ◈

LEMMA 10.6.5. *Napier's Rules (II): If three parts of Napier's pentagon are adjacent, then the sine of the middle part is the product of the tangents of the two adjacent parts.*

EXAMPLE 10.6.4. $\sin\overline{B} = \tan a \tan\overline{c}$ ◈

An equation for determining the missing part of any right spherical triangle (given any two parts) can now be determined using Napier's Rules.

EXAMPLE 10.6.5. Using the data in example 10.6.2, that is, $c = 109°$ and $B = 27°$, find b, a, and A using Napier's Rules.

Answer: To determine b, note that $\overline{B}$ and $\overline{c}$ are opposite b in Napier's pentagon. Therefore, we apply rule 1 (lemma 10.6.4):

$$\begin{aligned}\sin b &= \cos\overline{B}\cos\overline{c}\\ &= \cos(90° - B)\cos(90° - c)\\ &= \sin B \sin c\\ &= \sin 27° \sin 109°\end{aligned}$$

and so $b = 25.420°$, as in example 10.6.2. To determine a, we note that $a, \overline{B}, \overline{c}$ are adjacent and $\overline{B}$ is the middle part; therefore, by rule 2 (lemma 10.6.5),

$$\sin\overline{B} = \tan a \tan\overline{c}$$

which is equivalent to

$$\begin{aligned}\tan a &= \cot\overline{c}\sin\overline{B}\\ &= \cot(90° - c)\sin(90° - B)\\ &= \tan c \cos B\\ &= \tan 109° \cos 27°\end{aligned}$$

and so $a = 111.13°$, as in example 10.6.2.

To continue, use Napier's Rules to obtain $\cot A = \cos c \tan B$, in order to solve for A. ◈

EXERCISES

10.1. If a spherical triangle has vertices A, B, and C, use the statements of theorems 10.2.2, 10.2.6, and 10.2.7 to prove that $-\pi < A + B - C < \pi$. (Similarly, we have $-\pi < B + C - A < \pi$ and $-\pi < A + C - B < \pi$.)

10.2. Use the statement and method of proof of lemma 10.3.1 to prove the following more general identity for vector products, called the *vector triple product*.

$$\boxed{\vec{u} \times (\vec{v} \times \vec{w}) = (\vec{u} \cdot \vec{w})\vec{v} - (\vec{u} \cdot \vec{v})\vec{w}}$$

(Hint: Let $\vec{x} = \vec{u} \times (\vec{v} \times \vec{w})$ and set $\vec{x} = c\vec{v} + d\vec{w}$. Calculate $\vec{u} \cdot \vec{x}$ and $(\vec{u} \times \vec{x}) \cdot \vec{v}$ in order to solve for c and d. (Verify that lemma 10.3.1 still applies if $|\vec{v}| \neq 1$.))

10.3. Prove the cosine rule for angles (theorem 10.4.3).

10.4. The purpose of this exercise is to prove the following half angle identity for spherical triangles:

$$\boxed{\begin{aligned}&\tan\frac{c}{2} = K\cos(S - C)\\ &\text{where } K^2 = \frac{-\cos S}{\cos(S - A)\cos(S - B)\cos(S - C)} \quad \text{and } S = \frac{A + B + C}{2}\end{aligned}} \qquad (33)$$

To prove this identity, start from the identity $\tan^2\frac{C}{2}=\frac{1-\cos c}{1+\cos c}$, which can be derived from the half angle identities for sine and cosine (formulas (4.17) and (4.18)) and then use a reformulation of the cosine rule for angles (theorem 10.4.3) to replace cos c by $\frac{\cos C+\cos A\cos B}{\sin A\sin B}$. After simplifying the expression, make use of the addition identities and identities for sums and differences of cosine ratios in planar trigonometry. Notice that the factors $\cos(S-A)$, $\cos(S-B)$, and $\cos(S-C)$ are each positive because of the property proved in exercise 10.1, and $-\cos(S)$ is positive by theorem 10.2.7(b) and (c).

10.5. Write out neatly the derivations of formulas (10.28)–(10.32).

10.6. Show that if a spherical triangle has two right angles, then the sides opposite the right angles are each 90° and the third side equals the third angle.
(Hint: Assume $B=C=90°$ and use formulas (10.23), (10.28) and (10.32).)

10.7. If a spherical triangle is identical to its polar triangle, what can you say about the spherical triangle?

10.8. In a spherical triangle ABC,

(a) given $B=65°$, $b=47°$, $C=79°$, find c using the sine rule.

(b) given $a=118°$, $c=68°$, $B=65°$, find b using the cosine rule for sides.

10.9. Derive the following form of the cosine rule for angles.

$$\begin{aligned}\cos B&=\frac{-\cos A\cos C\pm\cos a\sin C\sqrt{1-\sin^2 a\sin^2 C-\cos^2 A}}{1-\sin^2 a\sin^2 C}\\ \cos A&=\frac{-\cos C\cos B\pm\cos c\sin B\sqrt{1-\sin^2 c\sin^2 B-\cos^2 C}}{1-\sin^2 c\sin^2 B}\\ \cos C&=\frac{-\cos B\cos A\pm\cos b\sin A\sqrt{1-\sin^2 b\sin^2 A-\cos^2 B}}{1-\sin^2 b\sin^2 A}\end{aligned}\tag{10.34}$$

(Hint: To derive the first formula in formula (10.34), replace "$\sin(B)$" with "$\sqrt{1-\cos^2 B}$" in the first equation in formula (10.10) and then use algebraic methods to solve for $\cos(B)$.)

10.10. Derive the following form of the cosine rule for sides:
(Hint: rewrite the formulas from the previous exercise in terms of the corresponding polar triangle.)

$$\begin{aligned}\cos b&=\frac{\cos a\cos c\pm\cos A\sin c\sqrt{1-\sin^2 A\sin^2 c-\cos^2 a}}{1-\sin^2 A\sin^2 c}\\ \cos a&=\frac{\cos c\cos b\pm\cos C\sin b\sqrt{1-\sin^2 C\sin^2 b-\cos^2 c}}{1-\sin^2 C\sin^2 b}\\ \cos c&=\frac{\cos b\cos a\pm\cos B\sin a\sqrt{1-\sin^2 B\sin^2 a-\cos^2 b}}{1-\sin^2 B\sin^2 a}\end{aligned}$$

10.11. Use the sine rule, the cosine rules and the reformulations of the cosine rules in exercises 10.9 and 10.10 to to solve the following spherical triangles with the information given. The solutions should satisfy the statements of theorems 10.2.2, 10.2.3, 10.2.7, and especially theorem 10.2.4. In some cases there can be *two* solutions (two triangles) and in some cases there can be *no* solution. The number of solutions is indicated in parentheses because we have not presented enough theory in this chapter to determine from the given information how many solutions there should be.

(a) $a=135°$, $b=45°$, $c=120°$ (one solution)

(b) $A=145°$, $B=120°$, $C=150°$ (one solution)

(c) $c=84°$, $A=B=82°$ (one solution)

(d) $a=76°$, $A=B=50°$ (one solution)

(e) $a=36°$, $c=84°$, $B=22°$ (one solution)

(f) $b=49°$, $B=141°$, $C=42°$ (no solution)

(g) $a=81°$, $b=68°$, $B=56°$ (two solutions)

(h) $b=132°$, $B=128°$, $C=55°$ (two solutions)

10.12. Given $C=90°$, $A=35°$, $c=114°$, make use of rule 1 for quadrants (lemma 10.6.1) to find a, and then make use of rule 2 for quadrants (lemma 10.6.2) to find b.

10.13. Solve (if possible) each of the following right spherical triangles (with $C=90°$) with the given information by selecting from the ten identities derived in section 10.5.

(a) $a=49°$, $b=61°$

(b) $A=62°$, $c=71°$

(c) $A=125°$, $B=108°$

(d) $b=139°$, $c=112°$

(e) $a=73°$, $c=112°$

(f) $A=63°$, $B=20°$

(g) $a=90°$, $c=90°$

10.14. Use Napier's rules (lemmas 10.6.4 and 10.6.5) to solve the triangles in exercise 10.13.

APPENDIX A: ANSWERS TO SELECTED EXERCISES

CHAPTER 1 SELECTED ANSWERS

2. (a) $\left\{-\frac{4}{7},\ \pi,\ \frac{22}{7},\ -\frac{7}{4}, 3.1\right\}$

5. (a) $(b,c]$
(b)
(c) $[b,d)$
(d)
(e) $\{b\}$

8. (a) 133
(b)
(c) $2x^2+11x-6$
(d)
(e) 2

11. (a) $\frac{11}{12}$
(b)
(c) $\frac{a+3}{17(a+4)}$
(d)
(e) $\frac{118}{119}$
(f)
(g) $\frac{43}{11}$

14. (a) $\frac{13}{21}$
(b)
(c) $\frac{263}{143}$
(d)
(e) $\frac{53}{14}$
(f)
(g) $\frac{4+81y^2}{18y}$
(h)
(i) $\frac{20}{17}$

16. (a) 0.07
(b)
(c) 0.077
(d)
(e) 0.125
(f)
(g) 0.625
(h)
(i) $0.\dot{1}$
(j)
(k) $0.\dot{7}$
(l)
(m) $0.\dot{7}$
(n)
(o) $3.\overline{142857}$
(p)
(q) $2.\overline{71811}$

22. (a) prime
(b)
(c) composite 23^2
(d)
(e) composite $13 \cdot 14$
(f)
(g) composite $3^2 \cdot 3803 \cdot 3607$

25. (a) $\frac{18}{35}$
(b)
(c) $\frac{13}{300}$
(d)
(e) $\frac{295}{1717}$
(f)
(g) $\frac{83641}{41820}$

28. (a) $x\sqrt[3]{x^2}$
(b)
(c) $|xy|\sqrt{y}$
(d)
(e) $-2(xy)^3$
(f)
(g) $|y|\sqrt[14]{77x^3y}$

CHAPTER 2 SELECTED ANSWERS

6. (a) $\frac{4}{15}$
(b)
(c) $\frac{15}{4}$

10. (a) $y=-\frac{1}{4}x+\frac{1}{2}$
(b)
(c) $y=-6x-1$

13. $y=\frac{31}{33}x+\frac{716}{165}$

16. (a) 1
(b)
(c) $\sqrt{2}$

19. (a) $y=-6x+\frac{19}{2}$

22. (a) $(x-2)^2+y^2=5$
(b)
(c) $\left(x-\frac{\sqrt{3}}{3}\right)^2+y^2=\frac{4}{3}$

30. (a) $\langle 2,2\rangle$
(b)
(c) $\langle 4,-1\rangle$

CHAPTER 3 SELECTED ANSWERS

1. (a) 10
(b)
(c) no solution
(d)
(e) $-\frac{5}{4}$

4. (a) $(x+1)^2-1$
(b)
(c) $2(x+54)^2-\frac{33}{8}$

7. (a) $-6x^3+3x^2-10x+5$
(b)
(c) $1-16x^4$
(d)
(e) $1-4x^2$
(f)
(g) $1-8x+24x^2-32x^3+16x^4$
(h)
(i) $64-192x+240x^2-160x^3+60x^4-12x^5+x^6$

11. -18

19. (a) $2(x+2)(7x-1)$
(b)
(c) $-6(11x+5)(2x-1)$

21. (a) $-4(x-1)^2$
(b)
(c) $(6u-19v)(6u+19v)$
(d)
(e) $3(9t-2)(9t+2)$

23. (a) -1, 2
(b)
(c) $\pm\frac{2}{9}$

29. **(a)** $2x^2+4xy$
(b)
(c) $x=2, y=3$

33. **(a)** $\begin{bmatrix} 17 & 2 \\ 2 & 8 \end{bmatrix}$
(b)
(c) $\begin{bmatrix} 1 & 0 \\ 0 & 1 \end{bmatrix}$
(d)
(e) $\begin{bmatrix} 198 & -72 \\ 18 & 216 \end{bmatrix}$
(f)
(g) $\begin{bmatrix} 0 & 0 \\ 2 & 3 \end{bmatrix}$

39. **(a)** $4x^2-4x+10$
(b)
(c) $x^3+3x^2+52x+50$

42. $(x+2i)(x-2i)(x+i)(x-i)$

46. $-2\pm\sqrt{3}$ and $2\pm\sqrt{6}$

CHAPTER 4 SELECTED ANSWERS

1. **(a)** 315°
(b)
(c) 282.857°
(d)
(e) 57.296°

3. **(a)** second quadrant

6. **(a)** 2
(b)
(c) $\frac{\sqrt{2}}{2}$
(d)
(e) $\sqrt{3}$
(f)
(g) $-\frac{2}{\sqrt{3}}$

19. **(a)** $\frac{\sqrt{2}+\sqrt{6}}{4}$
(b)
(c) $\frac{\sqrt{6}-\sqrt{2}}{4}$
(d)
(e) $\frac{1}{4}\sqrt{8-2\sqrt{6}+2\sqrt{2}}$

22. 20.415 meters

27. 4.226 meters

37. $|\vec{F}|=11.072$ N, $\theta=44.26°$

CHAPTER 5 SELECTED ANSWERS

8. **(a)** $\{x\in R|x\neq-3\}$
(b)
(c) $\{t\in R|t\neq-7, t\neq\frac{-1}{3}\}$
(d)
(e) $\{u\in R|u\geq0, u\neq4\}$
(f)
(g) $\{r\in R|r\leq3 \text{ or } r\geq11\}$

17. **(a)** 49
(b)
(c) 6
(d)
(e) 2

35. (a) $f \circ g(x) = \frac{1}{\sqrt{x-1}} + 1$, the domain is $\{x \in R | x > 1\}$

(b) $g \circ f(x) = \sqrt{\frac{1}{x} + 1}$, the domain is $\{x \in R | x > 0\}$

(c) $g \circ g(x) = \sqrt{\sqrt{x-1} - 1}$, the domain is $\{x \in R | x \geq 2\}$

51. (a) $5^3 = 125$
(b)
(c) $16^{\frac{1}{4}} = 2$
(d)
(e) $\log_{\frac{1}{64}} 4 = -\frac{1}{3}$

54. (a) 8
(b)
(c) -1
(d)
(e) -1
(f)
(g) -3

56. (a) $\frac{3}{2}$
(b)
(c) $\left(\frac{3}{2+\log_2 3}\right)^2$

64. (a) x
(b)
(c) $\sqrt{1-4x^2}$
(d)
(e) $\frac{y}{\sqrt{1+y^2}}$
(f)
(g) $\frac{\sqrt{3}y}{2}\sqrt{4-3y^2}$

CHAPTER 6 SELECTED ANSWERS

1. (a) $\frac{2x^3+3x^2+x}{1-2x}$
(b)
(c) $\frac{r^2-3r-18}{r^2-3r-10}$

4. (a) $(1-x^{\frac{1}{3}})(1+x^{\frac{1}{3}})$
(b)
(c) $2(3\sqrt{x}+7)(\sqrt{x}+1)$
(d)
(e) $(1-x^{\frac{1}{2}})(1+x^{\frac{1}{3}}+x^{\frac{2}{3}})$

7. (a) $\frac{x-9}{x+4}$
(b)
(c) $\frac{3t-15}{15t+20}$

13. $x = 6$

16. (a) -1
(b)
(c) $\pm 1, \pm\frac{\sqrt{21}}{3}$
(d)
(e) $-3, 5, 1\pm\sqrt{2}$
(f)
(g) $v \leq 1,\ v \geq 3$

19. (a) 9
(b)
(c) 7
(d)
(e) 1, 4
(f)
(g) $-1, 2$
(h)
(i) 2
(j)
(k) $\sqrt{3}$

22. (a) $n\pi,\ \frac{\pi}{6}+2n\pi,\ \frac{5\pi}{6}+2n\pi$

(b)

(c) $\frac{n\pi}{2},\ n\pi$

(d)

(e) $\frac{2\pi}{3}+2n\pi,\ \frac{4\pi}{3}+2n\pi$

23. (a) $\frac{\pi}{12}+2n\pi,\ \frac{5\pi}{12}+2n\pi$

(b)

(c) $0.581+2n\pi,\ 2.561+2n\pi$

27. (a) $1-\frac{4}{x+4}$

(b)

(c) $1-\frac{3}{x+6}$

(d)

(e) $1-\frac{7}{2x+7}$

(f)

(g) $\frac{1}{2}+\frac{1}{2(2x+1)}$

30. (a) $\frac{1}{14(w-4)}-\frac{1}{10(w-2)}+\frac{1}{35(w+3)}$

(b)

(c) $\frac{3}{8(w-3)}+\frac{3}{8(w+1)}-\frac{3}{4(w+3)}$

(d)

(e) $\frac{28}{9(2w+7)}-\frac{8}{5(w+4)}+\frac{2}{45(w-1)}$

(f)

(g) $\frac{1}{24(3w+4)}+\frac{7}{24(3w-5)}$

33. (a) $\left(\frac{9}{5},\infty\right)$

(b)

(c) $(-\infty,-4)\cup(-4,-3)\cup(-1,\infty)$

(d)

(e) $\left(-\infty,-\frac{1}{2}\right]\cup\left[\frac{1}{3},2\right]$

(f)

(g) $(-\infty,-\sqrt{6})\cup(\sqrt{6},\infty)$

36. (a) $[1,\infty)\cup\{0\}$

(b)

(c) $\left[-\frac{2}{7},2\right]$

42. (a) $x^2+y^2=4$

(b) $-\frac{1}{3}(x+2)(x-6)$

(c) $\left\{y<-\frac{1}{3}(x+2)(x-6),\ x>\sqrt{4-y^2},\ y>-\frac{1}{2}|x-1|\right\}$

(d) $[0, 7]$

(e) $[-3, 5]$

CHAPTER 7 SELECTED ANSWERS

4. (a) 0

(b) 1

(c) 1

(d) 0

10. (a) 1

(b)

(c) 0

(d)

(e) 0

(f)

(g) 0

13. (a) i) $-\frac{1}{2}$ ii) $\frac{3}{2}$

16. $f(x)=\begin{cases} \frac{1}{2} & \text{if } x<0 \\ 3 & \text{if } x>0 \end{cases}$ $\lim_{x\to 0} f(x)$ does not exist

20. $f(x)=\begin{cases} -\frac{1+x^2}{1+x} & \text{if } x<0 \\ -\frac{1+x^2}{1-x} & \text{if } x>0 \end{cases}$ $\lim_{x\to 0} f(x)=-1$

25. (a) $-\frac{1}{7}$
(b)
(c) $\frac{3}{16}$
(d)
(e) 108
(f)
(g) 0
(h)
(i) –2

35. (a) $\frac{5}{8}$
(b)
(c) 1
(d)
(e) $\frac{1}{49}$
(f)
(g) 2

CHAPTER 8 SELECTED ANSWERS

2. (a) 2
(b)
(c) 6
(d)
(e) $\frac{\sqrt{2}}{4}$

5. (a) $f'(1)$, for $f'(x)=\sqrt{x}$

13. (a) $-2x^{-1}$
(b)
(c) $33x^2+\frac{18}{x^2}$
(d)
(e) $-\frac{1}{x^2}-\frac{2}{x^3}-\frac{3}{x^4}$

16. (a) $-\frac{2}{(1+x)^2}$
(b)
(c) $\frac{1}{(1+x)^2}$
(d)
(e) $-\frac{2}{(1+x)^2}$

20. (a) $a-\frac{c}{x^2}$
(b)
(c) $\frac{x(ax+2b)}{(ax+b)^2}$

24. (a) $\left(1,-\frac{1}{2}\right)$

28. –2.67, –0.58 and 3.25

33. **(a)** $\left(\frac{\pi}{4}+n\pi, \sqrt{2}\right)$, for $n \in \mathbb{Z}$

(b)

(c) $\left(\frac{2\pi}{3}+2n\pi, \frac{\sqrt{3}}{3}\right)$ and $\left(\frac{4\pi}{3}+2n\pi, -\frac{\sqrt{3}}{3}\right)$, for $n \in \mathbb{Z}$

(d)

(e) $\left(\frac{\pi}{4}+n\pi, 2\right)$ and $\left(-\frac{\pi}{4}+n\pi, -2\right)$ for $n \in \mathbb{Z}$

37. 300 by 600

43. $\frac{30\pi}{\pi+4} \approx 13.2$ inches bent into a circle, and $\frac{120}{\pi+4} \approx 16.8$ inches bent into a square

48. **(a)** $10x$

(b)

(c) $-\pi\sin(\pi x)$

(d)

(e) $2(x+\pi)$

51. **(a)** 2304

(b)

(c) $-\pi$

(d)

(e) $\sqrt{8}$

(f)

(g) 0

(h)

(i) $-\frac{1}{9}$

(j)

(k) 0

53. **(a)** $\frac{1}{4\sqrt{x}\sqrt{1+\sqrt{x}}}$

(b)

(c) $-\frac{1}{2\sqrt{1+x^2}\sqrt{1-\sqrt{1+x^2}}}$

(d)

(e) $-\frac{\cos\left(\frac{1}{x}\right)}{x^2}$

(f)

(g) $\frac{-\cos\left(\sec\left(\frac{1}{x}\right)\right)\sec\left(\frac{1}{x}\right)\tan\left(\frac{1}{x}\right)}{x^2}$

57. **(a)** $\langle -\sin(t), \cos(t)\rangle$

(b)

(c) $\langle -2\pi\sin(2\pi t), 3\pi\cos(3\pi t)\rangle$

(d)

(e) $\langle 2, 2t\rangle$

(f)

(g) $\langle 2, 3\rangle$

(h)

(i) $\left\langle -\frac{1}{(1+t)^2}, 2\right\rangle$

60. **(a)** $2u\exp(u^2)$

(b)

(c) $(\sin(u)+\cos(u))e^{u\sin(u)}$

(d)

(e) $-\frac{1}{u^2}e^{\frac{1}{u}} - \frac{2}{u^3}e^{\frac{1}{u^2}}$

63. **(a)** $\ln(6)6^w$

(b)

(c) $3\ln(2)w^2 2^{w^3} + 2\ln(3)w3^{w^2}$

(d)

(e) $\ln(2)2^{\sin(w)}\cos(w) - \ln(3)3^{\cos(w)}\sin(w)$

(f)

(g) $\frac{2^{2+2w}}{(1+2^{2w})^2}\ln(2)$

66. (a) $\frac{1}{1+x}$
(b)
(c) $\frac{2}{1-u^2}$
(d)
(e) $\frac{1+e^y}{y+e^y}$
(f)
(g) $\frac{1}{2x\ln(2)}$
(h)
(i) $\frac{-2u}{1-u^2}$
(j)
(k) $\frac{5^y-5^{-y}}{5^y+5^{-y}}$

CHAPTER 10 SELECTED ANSWERS

11. (a) $A = 125.26°,\ B = 54.74°,\ C = 90°$
(b)
(c) $a = 153.09°,\ b = 43.10°,\ c = 156.77°$
(d)
(e) $C = 85.23°,\ a = b = 81.21°$
(f)
(g) $A = 62.02°,\ c = 123.45°,\ C = 131.75°$ or $A = 117.98°,\ c = 24.92°,\ C = 22.13°$ (check that $\frac{\sin C}{\sin c} = \frac{\sin B}{\sin b} = 0.8942$ in each case)

13. (a) $c = 71.45°,\ B = 67.30°,\ A = 52.75°$
(b) $a = 56.60°,\ b = 53.74°,\ B = 58.52°$

INDEX

H

I

L

M

N

P

U

V

Z